PULP AND PAPER MANUFACTURE

THIRD EDITION

VOLUME 4

SULFITE SCIENCE & TECHNOLOGY

Edited by

Dr. Otto V. Ingruber, *Consultant*
Retired Senior Research Associate
CIP Research Ltd., Hawkesbury, Ontario, Canada

Dr. Michael J. Kocurek, *Professor and Chairman*
Department of Paper Science & Engineering
University of Wisconsin at Stevens Point, U.S.A.

and

Al Wong, *P.E., Specialist Technology Management*
Arbokem, Inc., Montreal, Quebec, Canada

Prepared under the direction of and published by

THE JOINT TEXTBOOK COMMITTEE OF THE PAPER INDUSTRY

TAPPI	CPPA
Technology Park/Atlanta	Sun Life Bldg., 23rd Floor
P.O. Box 105113	1155 Metcalfe Street
Atlanta, Georgia U.S.A. 30348-5113	Montreal, Que. Canada H3B 2X9

PULP AND PAPER MANUFACTURE

A series of volumes detailing the principles and practice of pulping and papermaking
for mill staff and university students
by contributors from the manufacturing, academic and supply sectors of the industry

THIRD EDITION

Volume 1. Properties of Fibrous Raw Materials and their Preparation for Pulping
Volume 4. Sulfite Science & Technology
Volumes on mechanical, kraft, bleaching, stock treatment, papermaking,
boardmaking and services, in preparation
under the direction of

THE JOINT TEXTBOOK COMMITTEE OF THE PAPER INDUSTRY 1985

Lyle J. Gordon, Chairman	Robert B. Barker	Peter M. Nobbs
R.A. Joss, Secretary/Treasurer	F. Keith Hall	William R. Saxton
William L. Cullison	Michael J. Kocurek	John Tasman

PUBLISHING HISTORY OF THE SERIES

Initiated in 1918 at a joint meeting of the Education committees
of the Technical Section CPPA and of TAPPI.
Published under the Joint Textbook Committee in the following editions:

1921 The Manufacture of Pulp and Paper	1st ed 5v. Ed. J.N. Stephenson	McGraw-Hill
1925 The Manufacture of Pulp and Paper	2nd ed 5v. Ed. J.N. Stephenson	McGraw-Hill
1930 La fabrication des pâtes et papiers	1st ed 5v. Trans. L. Vidal	McGraw-Hill
1938 The Manufacture of Pulp and Paper	3rd ed 5v. Ed. J.N. Stephenson	McGraw-Hill
1950 Pulp and Paper Manufacture	1st ed 4v. Ed. J.N. Stephenson	McGraw-Hill
1969 Pulp and Paper Manufacture	2nd ed 3v. Eds. R.G. Macdonald, J.N. Franklin	McGraw-Hill

PULP AND PAPER MANUFACTURE VOL. 4
SULFITE SCIENCE & TECHNOLOGY
With bibliographic references and index.

Canadian Cataloguing in Publication Data

Main entry under title.
  Pulp and paper manufacture

Contents: v. 4. Sulfite science and technology/edited by M.J. Kocurek, O.V.
  Ingruber and A. Wong.
ISBN 0-919893-04-X (set). – ISBN 0-919893-22-8 (v. 4).

1. Paper making and trade. 2. Wood-pulp. I. Kocurek, M.J. II. Stevens, Frederick.
III. Joint Textbook Committee of the Paper Industry of the United States and
Canada.

TS1105.P84 1983   676   C83-090155-8

CONTENTS

M.J. KOCUREK

Dr. Michael Kocurek is a Professor and Chairman of the Paper Science & Engineering Department at the University of Wisconsin — Stevens Point. He has been a central figure in the education of the industry's engineers and scientists for 15 years. Dr. Kocurek received his B.S., M.S. and Ph.D. in Paper Science and Engineering from the SUNY College of Forestry at Syracuse University. He has served as a consultant/instructor to over 30 different pulp and paper companies. He has been and still is very active in TAPPI activities, particularly in the area of continuing education and professional development. He is a member of the Joint Textbook Committee and is presently on the Board of Directors of TAPPI. He was elected a TAPPI Fellow in 1985. Dr. Kocurek is the author of numerous papers.

PREFACE

A. WONG

Al Wong has a B.A.Sc. in chemical engineering from the University of British Columbia and an M.Eng. in chemical engineering from McGill University. His professional experience includes working as a kraft mill process engineer at Crown Zellerbach Canada Ltd., following which he worked at the Pulp and Paper Research Institute of Canada as the Head of the Water Research Section. He recently formed his own consulting company, Arbokem Inc. which is involved in the technology redevelopment of older sulfite mills and construction of new small-scale sulfite mills. Mr. Wong has authored over 85 technical and scientific publications and has been very active in the Sulfite Committees of both TAPPI and the Technical Section, Canadian Pulp and Paper Association.

In 1982, the Tappi Sulfite and Semichemical Committee undertook to coordinate the preparation of sulfite pulping manuscripts for a modern pulping textbook, to be published by the Joint Textbook Committee of the Paper Industry.

The technical domain of sulfite has changed and expanded considerably since the publication of the Second Edition of Pulp and Paper Manufacture in 1969. A single "Sulfite Chapter" was considered inadequate. Consequently, the Sulfite and Semichemical Committee decided that all new manuscripts should be prepared to cover a) conventional sulfite, b) high yield sulfite, c) neutral sulfite semichemical and d) dissolving sulfite pulping technologies separately. A group of distinguished Industry experts was assembled for the task. A second group of mill personnel was organized to review the finished manuscripts before publication. The principal aim was to develop a textbook which describes the present state of the art, for the use of both novices and experts.

During the 1970's, many Industry pundits were eagerly watching the rapid demise of the sulfite pulping industry. There were many closures of dilapidated sulfite mills in North America and Scandinavia. The unsolvable problem of pollution control of smaller sulfite mills was the main driving force. Since that period, there have been very few replacement sulfite mills constructed. The Industry trend was towards the construction and operation of very large kraft pulp mills. Unlike the kraft process, the prevailing sulfite processes were limited to the pulping of only certain wood species.

Sulfite has survived. It has emerged from the pit into a phase of new growth and technological developments. Its practice is more widespread than ever. Refer to Appendix I for a comprehensive list of sulfite mills worldwide, their processes, capacities, and paper products. Most Industry statistics underestimate its extent. Unfortunately, sulfite pulp is still associated, in some Industry circles, with inferior pulp produced by inefficient and polluting means. This representation is largely inaccurate. Modern sulfite mills can be very efficient, low-pollution and profitable operations. The pulp can be very competitive technically.

With over 100 years of operating experience, sulfite pulping is still unmatched in versatility. The basic process is practiced today for the production of pulp ranging from about 40% to nearly 95% yield, and with cooking pH spanning from less than 1 to over 14. Various wood and non-wood raw materials are used in sulfite pulping throughout the world. Its unusual process features even permit the profitable operation of small production units.

With recent improvements in technology, sulfite pulp can now be custom-made, from broader sources of cellulosic raw materials, to a wider range of specifications of product yield, strength, softness, bleachability and printability.

In the realm of very high yield pulps, the basic process can produce a pulp with desired properties of both chemical and mechanical pulps, for the production of wood-containing papers and paperboards. Chemimechanical pulping (CMP) and its many variations are, in fact, versions of high yield sulfite pulping: a judicious combination of chemical and mechanical treatments of cellulosic materials. Virtually all CMP and derivatives in vogue use a sulfite chemical to treat the cellulosic raw materials. Interestingly, this concept has already been patented and practiced, in one form or another, in sulfite mills in the 1880's. To date, no chemical substitute of economic importance has been discovered. So what is new? There have been substantial improvements in the ways and means of fiberizing sulfite-treated cellulosic materials. It is now possible to achieve full development of essential fibre properties to meet specific and new end use needs.

Sulfite as a full-chemical pulping process can also be optimized to the end use requirements. The pulping process can be varied to manufacture a wide range of pulp grades which include newsprint-reinforcement, packaging, wood-free printing, tissue/fluff, and dissolving. In contrast, the standard kraft process is essentially a fixed process producing a pulp which is "over-qualified" for many end use needs. In the kraft option, the opportunity to improve mill profitability in face of changing market competition is severely restricted. With the advent of sulfite pulping in the alkaline pH ranges, sulfite pulp is offering a credible challenge to the classical kraft pulp. It provides comparable physical strengths at significantly higher pulp yields.

Although the sulfite process holds many new promises, it has also many technical problems to overcome. Two notable examples are i) alkaline sulfite pulping (even with the addition of anthraquinone) requires much longer cooking time than standard kraft pulping, and ii) chemical recovery for very high yield sulfite-mechanical processes is not resolved. We hope that this textbook will create new interests and initiatives for continued advancements of sulfite pulping science and technology.

HISTORICAL DEVELOPMENT

O.V. INGRUBER

Dr. Ingruber has over 30 years experience in process conception, development, implementation and control. He received his Ph.D. in pulp and paper chemistry from the Technical University of Graz, Austria. His professional experience includes 8 years at the Pulp and Paper Research Institute of Canada, and 23 years at Canadian International Paper Research. He has authored or co-authored 22 papers, 9 patents, and 3 books on sulfite and kraft pulping, equipment, automation, recovery, by-product utilization, process measurement and control in systems at elevated temperatures and pressures. In 1983 he retired as a Senior Research Associate from CIP, and is now an independent consultant.

Industrial pulping, that is the large-scale production of cellulosic fiber for paper making in pressurized equipment at high temperature, was introduced in the second half of the ninteenth century, in the period from 1853 to 1884, occurring in the middle of the industrial revolution in Europe and North America.

Alkaline pulping of plant material which had been practiced since antiquity in Asia and had reached Europe via the Arabs after 1000 A.D. was the first (1853) to be industrialized in the form of the soda pulping process, i.e. using caustic soda as chemical agent and straw as the raw material.

The next process, acid sulfite pulping (first patented by Tilghman in 1866/67) deserves special consideration since it presents a true invention without prior art. Again sulfur dioxide and its aqueous solutions had been known and used since antiquity, not for pulping but for their properties of disinfectant, food preservative, and bleach. Industrial production of calcium bisulfite for sugar manufacture from limestone and sulfur dioxide gas had begun before 1850. The idea of using sulfur dioxide or metal sulfites as pulping chemicals came to Tilghman in Pennsylvania, and later to Ekman in Sweden indirectly from their application as bleaching agents: in the first case, by way of a tap made of softwood which appeared somewhat bleached, softened and fibrous, and used on a barrel in Paris where fats were treated with sulfur dioxide-water. In Ekman's case (1872, Sweden), it was simply a case of trying bisulfite, which he studied as bleaching chemical, also as cooking chemical. Kellner's invention of acid sulfite pulping occurred at about the same time in a soda mill in Austria, in an even more sudden and unpredictable fashion: his technician performing soda pulping experiments in the laboratory took by mistake a bottle from the shelf containing sodium bisulfite and used it for liquor preparation rather than the caustic soda prescribed for the cook.

Of these three originators of the acid bisulfite pulping process, Tilghman used calcium bisulfite from the beginning. Ekman declared magnesium bisulfite to be the best, thus foreshadowing the present trend in the Swedish sulfite industry. Led by his technician, Kellner started off with sodium base, thus foreshadowing the important role this strong base would assume in later sulfite pulping developments.

Ammonium, the fourth of the sulfite pulping bases of commercial importance, was neglected during this initial period. Processes appearing after 1900 used neutral ammonium sulfite. The first suggestion of ammonium bisulfite pulping was patented in the U.S. (Sammet and Merill, 1912) using gaseous sulfur dioxide, ammonia, and steam, but the first clear proposal of the acid ammonium bisulfite process is found in patents granted to Marusawa in 1917 in Japan and the U.S.

The third major development which took place in this period concerns the kraft (sulfate) process (patented 1884 by Dahl in Germany), where sodium sulfide was added to the caustic soda cook. The result of adding this chemical was better bleachability and less cellulose degradation (higher fiber strength than obtained in the severe caustic soda environment). Adding the sulfide also caused the offensive smell by which the sulfate or kraft cook has been known and plagued since.

The early period of development of the three leading processes was marked by substantial fluctuations of their relative importance. By 1907 calcium base bisulfite, with its inexpensive base and its relative ease of operation without chemical recovery, had seriously challenged the initially leading soda process. The latter had gained temporarily with the new impulse given by the kraft (sulfate) which, in turn, abated again with improvements in soda pulp bleaching and lower cost of caustic soda.

The first real sulfite mill had been built by Ekman in 1874 using rotary digesters and indirect heat by means of a steam jacket to produce magnesium base sulfite pulp. In 1882 Wheelwright brought the Ekman process to America and built the first sulfite mill on the continent at East Providence, Rhode Island. The Mitscherlich process, using calcium base similar to Tilghman, and indirect heating, was developed in Germany about 1880 and installed soon after in the Fletcher mill at Alpena, Michigan. A calcium base sulfite process in upright digesters with direct steam heating was effected in Austria in 1878 by Ritter-Kellner. It was introduced into Canada by Russell and Riordon, and the first sulfite mill built in 1885 at Merriton, Ontario.

After the introduction of these sulfite processes into America, progress was rapid and in a few years large quantities of the new pulp were available for the uses of the paper maker. By 1920, sulfite pulp production had risen to 1 586 000 tons in the United States and to 678 000 tons in Canada.

In 1925 the total production of chemical wood fiber in the U.S. of four million tons shows a distribution of 20% soda, 20% sulfate and 60% unbleached and bleached sulfite. While soda pulping regressed further, sulfite production capacity was overtaken by kraft capacity from 1936 on (1960 in Canada). The world ratio of

sulfite to kraft (sulfate) tonnage was 1:2 in 1967 and sulfite has levelled off at close to 20% of the 100 million tons of world pulp production. Appendix I contains a list of North American and foreign sulfite mills, their processes and capacities (1).

While the kraft process remained virtually unchanged, modifications of the sulfite process, away from the original acid bisulfite process and into higher pH levels, began to appear after 1930. The first such process was high yield neutral sulfite pulping, mainly of hardwoods, developed for packaging material. After 1950, sulfite technology moved into the pH range between acid and neutral with the development of bisulfite high yield pulps mainly for the news sector. These advances which put new life into sulfite technology as a whole were possible only because the traditional adherence to inexpensive and "disposable" calcium base was broken under the pressure of environmental concerns, increasing shortage of spruce wood and paper strength demands. The semi-soluble magnesium base and the soluble sodium and ammonium bases were finally introduced on a larger scale, resulting in new processes with recovery of chemicals which became known under the names of Magnefite, Weyerhaeuser FB and HO, Stora, Sivola, Rauma and others. At the same time, Canada assumed leadership in the newly opened field of high yield bisulfite pulping of northern softwoods for newsprint, to be expanded in the seventies to ultra high yield sulfite pulping with yields from wood in the 90% range.

Finally, in the late sixties technological investigation began to open up the remaining area of sulfite pulping in the alkaline field. This led first to the development of the alkaline sulfite process which for the first time challenged the position of kraft pulping in the field of highest paper strength, offering at the same time easy bleaching and operation at low inoffensive odor. This was followed in 1978 by a sulfite process at lesser alkalinity but with anthraquinone as pulping catalyst. Pulps produced by this most recent extension of sulfite pulping are better delignified, brighter, less degraded and stronger than kraft pulps; industrial implementation is now in progress.

Thus the whole field of sulfite pulping has been explored and opened up technologically within the last 100 years, with the result that the potential and capabilities of the four common bases at various pH levels are known and employed. With the introduction of full chemical recovery the use of calcium, the original base is being reserved for specialty pulp mills using advanced processes for the manufacture of by-products of high market value, like lignin, alcohol, a yeast from the spent cooking liquor. With the final extension into the alkaline range, based on sodium and the most recent addition of highly effective anthraquinone accelerator, the sulfite process now provides all possible pulp grades required for paper manufacture, as well as complete facilities for the recycle of process heat and chemicals.

Relative to kraft, sulfite pulp production has been standing still since 1945. At the same time, technological advances in sulfite pulping have been numerous and outstanding, notably in Canada, Scandinavia, and, with the advent of anthraquinone accelerator, also in Japan, Australia and South Africa.

Never before have there been so many choices of direction in which the application of sulfite pulping can proceed, arising from the inherent flexibility of the process comprising two bivalent and two monovalent bases and the complete range of acid, neutral and alkaline solutions.

The need to follow, evaluate and use this technology calls for a more thorough knowledge of its fundamental organic and physical chemistry, as well as more knowledge of existing systems in operation, than has been available in textbooks so far.

REFERENCES
1. TAPPI Sulfite and Semichemical Pulping Committee, 1984.

PART ONE

SULFITE SCIENCE

I

SULFITE PULPING COOKING LIQUOR AND THE FOUR BASES

O.V. INGRUBER

Retired Senior Research Associate from CIP, now an independent consultant

MANUSCRIPT REVIEWED BY
David Johnston, Manager Pulping
Nova Scotia Forest Industries
Port Hawkesbury, Nova Scotia

A. SULFUR DIOXIDE-WATER, THE ELUSIVE ACID SYSTEM

Unlike the kraft process, which is based on sodium at a pH of the fresh cooking liquor of about 13.5, the sulfite process is characterized by covering the whole range of pH from below one for sulfur dioxide solutions in water, to above 13 for sodium sulfite solutions with free sodium hydroxide added. Within this range lies the versatility of the process of producing pulps of different yield and properties covering all commercial requirements.

"Acidum volatile", as solutions of sulfur dioxide in water were aptly called in previous centuries, abstracts their two distinctive properties: the tendency to gas-off and their elusive acidity. While the tendency of acid sulfite solutions to give off pungent SO_2 gas is understood quantitatively and is responsible for high system pressures at elevated temperature, the second property and its relation to the first was explained only more recently.

The required information on the ionic changes and equilibria in sulfite solutions was provided by introducing a technique of direct pH measurement in the system under conditions of actual pulping operating conditions. By their nature, these labile equilibria adjust instantaneously to changes of temperature, pressure and concentration (due to reactions or decomposition), and cannot be measured by methods of chemical analysis or photometry useful only at, or close to, room temperature. It was established by direct pH measurement that temperature and pH, that is the concentration of hydrogen and hydroxyl ions, are the principal kinetic factors in sulfite pulping, and are linked to the concentration of both sulfur dioxide and base. With the help of this tool it was possible to remove much of the confusion existing in the field of sulfite solutions, due to not being able to separate out the effect of the pH factor.

The key to understanding acid sulfite solutions was found in the fact that the compound H_2SO_3, "sulfurous acid", does not exist [1] and that therefore the SO_2-H_2O system lacks the undissociated stable acid which is typical for common acid systems and the ionic dissociation constants derived for them.

That "H_2SO_3" does not exist has been suspected for a long time since no trace of it could be found in UV, IR and Raman spectra of solutions of sulfur dioxide and the sulfites, even

(See Appendix 3 for Conversion Tables.)

in frozen systems [49]. Its non-existence was proven by pH measurements in heated sulfite systems under pressure [2] and confirmed by IR absorption spectra and electrical conductivity measurements [50].

Rather, the acidity of aqueous SO_2, is due to hydrogen ion $(H+)$[1] and bisulfite ion (HSO_3^-) formed from hydrated SO_2 $(SO_2 \cdot H_2O, SO_2$ aq). The mechanism may be direct by association with hydroxyl ion:

$$SO_2 + OH^- = SO_2 \cdot OH^-$$

or indirect via an activated SO_2 hydrate[1]:

$$SO_2 \cdot HOH = SO_2 \cdot \underset{\dot{H}}{OH} = SO_2 \cdot OH^- + H^+$$

Sulfite ion, SO_3^{--}, forms from HSO_3^- ion by secondary dissociation: $HSO_3^- = H^+ + SO_3^{--}$, which is favored increasingly as the pH is raised to values above 4 [2].

In the absence of a distinct acid compound, the ionic equilibria, and therefore the pH values and concentrations of the sulfite ions are extremely labile and respond immediately to changes of temperature, pressure, pH, concentration of chemical; and the strength of the cation present, hydrogen ion being the weakest.

In view of the peculiar nature of the sulfite system, it is best to avoid the terms "sulfurous acid" or "H_2SO_3" altogether and to use aqueous SO_2 or SO_2 aq, or SO_2 hydrate instead, to avoid misconceptions of practical consequence. For the same reason the classical term, "acid dissociation constant", should be replaced by acid dissociation (or ionization) ratios (R) for any particular set of conditions of the system.

The acid dissociation ratios in the SO_2-H_2O system at room temperature are:

$$\frac{(H^+) \cdot (HSO_3^-)}{(SO_2 \text{ aq})} = R_1 = 1.3 \cdot 10^{-2} \quad \text{and}$$

$$\frac{(H^+) \cdot (SO_3^{--})}{(HSO_3^-)} = R_2 = 6.3 \cdot 10^{-8}$$

Note that those values have been quoted as the acid dissociations constants of "sulfurous acid" in the literature.

---

[1] Note that hydrogen ions, H^+, are always associated with water molecules to form hydronium ions, H_3O.

The technological consequence of ionization based on a loosely associated gas hydrate rather than a distinct acid in solution is the great thermodynamic instability of the concentration equilibria of the ions, particularly the bisulfite ion, HSO_3^-.

$$\overrightarrow{\text{TEMPERATURE}}$$

$$HSO_3^- \rightleftarrows SO_2 + OH^-; H + HSO_3^-$$
$$\rightleftarrows H_2O + SO_2$$

$$SO_3^= + H_2O \rightleftarrows HSO_3^- + OH^-;$$
$$\rightleftarrows H_2O + SO_2$$

$$\overleftarrow{\text{PRESSURE}}$$

Thus acid sulfite solutions are de-acidified by rising temperatures shown in Fig. 1 [3] and they are acidified by rising system pressure. For the same reason, the pressure of confined acid bisulfite solutions rises steeply on heating. For example, in the SO_2-H_2O system containing 7% dissolved SO_2; gas pressure increases from zero psig at room temperature to 1.38 MPa (200 psig) at 150°C; as compared with 0.37 MPa (54 psig) for water alone at 150°C.

The presence of stable metal cation in the system stabilizes the bisulfite or sulfite ion and reduces the vapor pressure of a solution containing a given amount of SO_2. The stabilizing effect of ammonium cation on the sulfite anion is less than that of metal cations because of its own thermodynamic instability which will be discussed in the section on ammonium base.

The temperatures used in sulfite pulping range from 130°C to 185°C and it is clear from the above properties of the system that the actual concentrations of sulfite ions at cooking temperature will be substantially lower than those measured at room temperature by analysis. From dissociation data obtained in heated

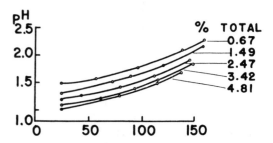

Fig. 1. Effect of temperature rise on the pH of aqueous SO_2 solutions.

sulfite solutions, it can be extrapolated that an ammonium sulfite system will lose its content of disssociated sulfite ions and will react neutral somewhere between 200 and 300°C and consist of the three components NH_3, SO_2 and H_2O (H , OH^-) only. Because sodium is a much stronger base, it will stabilize the sulfite system more than another base. This system will lose its sulfite ions probably between 300 and 400°C and consist of SO_2 gas and a weak NaOH solution of about pH 8 at that temperature.

It is important here to consider that the deactivation of the sulfite ions by higher cooking temperature does not remove the chemicals from the system. Any consumption of sulfite ions in the chemical reaction with wood will upset the equilibrium and draw more ions from the undissociated reservoir.

B. ANALYTICAL CONSIDERATIONS

Traditionally, sulfite cooking solutions are characterized analytically by their content of total SO_2, free SO_2 and combined SO_2. Total SO_2 is determined conveniently and precisely by iodometric titration, using starch or potentiometric end points. Commonly, the Palmrose iodate method is used for titration of acid solutions below pH 2.5, and all sulfite solutions can be analyzed by this method if they are acidified to this level before the test. However, it is important not to lose volatile free SO_2 from the sample by shaking or delaying the test as long as the concentration of SO_2 in solution is high. More elaborate methods are available if high precision is wanted or if cooking liquors with very high SO_2 content are involved. In such cases it is recommended collecting the samples in pipettes with stopcocks, adding it to dilute NaOH solution in order to bind all free SO_2, and titrating without acidification with iodine solution. The latter method is used also for sulfite liquors from above pH 4 to alkaline sulfite. In this direct titration method frequent standardization of the iodine solution is required since it loses volatile iodine. Alternatively, an excess of iodine solution may be added to the sulfite solution sample and the consumption determined by back titration with standard thiosulphate.

See Chapter III,B,2 for a discussion of analysis techniques and procedures.

The term combined SO_2 carries the historical ambiguity of its meaning: an analytical result obtained in calcium base acid bisulfite solution expressed in terms of insoluble calcium sulfite.

Free SO_2 is found in sulfite solutions only

below about pH 4, where the contents of base and bisulfite in the liquor are equivalent. It is commonly determined in acid bisulfite liquors in a neutralizing titration with NaOH after the bisulfite and SO_2 content has been oxidized to bisulfate and sulfuric acid by iodine.

In order to express the concentration of base and of base-bound bisulfite and sulfite ion alike, the term "combined SO_2" is used in the sulfite industry although it is a misnomer as is sometimes the case with technical terms. The term originates from the results obtained with the Palmrose method when applied to calcium base acid bisulfite cooking liquor by subtracting the value of free SO_2 from total SO_2. It expresses, in fact, the equivalent base concentration over the whole pH range: percent combined SO_2 · 0.3125 = normality of base. Its relation, however, to the amount of SO_2 bound in bisulfite and/or sulfite ion depends on the pH level. Up to the bisulfite inflection at $\cong$ pH 4 the relation is % combined SO_2 · 0.3125 = molarity of SO_2 (in bisulfite ion) whereas at and above the sulfite inflection at $\cong$ pH 10 (for sodium) the relation becomes % combined SO_2 · 0.15625 = molarity of SO_2 (in sulfite ion). Intermediate values will be found between the bisulfite and the sulfite level depending on the ratio of the HSO_3^- and $SO_3^=$ concentrations.

The above relations show that the common Palmrose liquor test for acid bisulfite pulping (TAPPI Standard T604 pm-79, CPPA Standard J-1) measures only half of the base actually present and augments the value of free SO_2 by the other half of the base equivalent. This is why the terms "true free SO_2" and "true combined SO_2" are used when the actual free SO_2 and base content are required for calculations: % Free SO_2 – % combined SO_3 = % true free SO_2; % combined SO_2 · 2 = % true combined SO_2. The above is summarized in terms of titrant volume as follows:

Total SO_2 (iodine)	Free SO_2 (NaOH)	Combined SO_2 ([$SO_3^=$])
a	b	a - b
	True Free SO_2 2b - a	True Combined SO_2 ([HSO_3^-]) 2(a - b)

In bisulfite mill operations the term "square acid" is used, denoting a bisulfite solution containing 2% combined SO_2 and 2% free SO_2 as determined by the Palmrose titration method. Hence the actual composition is 4% total SO_2, 4% true combined SO_2 and zero free SO_2.

The above methods and relations were developed in support of calcium base acid bisulfite pulping, the dominant process until the mid 1950's. To be useful over the complete sulfite range, it has been suggested to introduce the term "bound SO_2" for characterizing a sulfite pulping liquor regarding the molecular concentration of sulfonating anion over the whole pH range, as determined by iodine titration. Below the bisulfite inflection, "bound SO_2" is twice the value of "combined SO_2" as determined by the Palmrose methods and gives the concentration of bisulfite ion. Between the bisulfite and sulfite inflections, "bound SO_2" represents the molecular concentration of any mixture of bisulfite and sulfite ion in solution; above the sulfite inflection it gives the concentration of sulfite ion. The fact that the monovalent bisulfite ion changes to bivalent sulfite ion above pH 4 makes it clear that twice the base concentration is required on the sulfite level than on the bisulfite level in order to maintain a required concentration of sulfonating anions. As will be explained, the sulfonic acid group which forms in the reaction of lignin with sulfite solution is monovalent and carries 1 mol SO_2, independent of whether sulfonation was achieved by bisulfite or sulfite ion. The term "bound SO_2" is therefore related directly to the potential of the liquor to achieve the desired effect in a cook.

C. NOMENCLATURE AND DEFINITIONS OF TERMS. A SULFITE SPECTRUM

From the analytical discussions in Section B it would appear that the classical definitions for sulfite liquor strength can be confusing, not only for students and new engineers but also for teachers and professionals. As long as the sulfite process was limited to acid bisulfite pulping, these definitions were conventionally accepted and useful for cooking control. With the expansion of sulfite pulping to bisulfite, neutral sulfite and alkaline sulfite, they became useless because they have no general validity over the total range of cooking liquors.

What SO_2 species are present if there is excess "free SO_2" above that required to form bisulfite with the base? Conversely, what SO_2 species are present if the pH is above 4 and there is insufficient SO_2 present to form all bisulfite?

To assist in the understanding of the structure of the total sulfite system Tables 1A and 1B were drawn up. Table 1A provides a quick reference to the approximate pH and chemical composition of the four major sulfite processes. Table B gives complete information, in a simplified form, on the varying composition ratios of the SO_2 species throughout the sulfite pH spectrum, and the related terms. It gives also the approximate position of the characteristic limiting pH for each base where either precipitation occurs (Ca, Mg) or the maximum alkaline strength is reached (NH_4^+).

Both tables are considered for inclusion in the revised version of Tappi Standard 1201 "Definition of terms in the sulfite pulping process." Refer also to Chapter VII,A,1.

D. THE FOUR BASES
1. Calcium

Calcium base was dominant in sulfite pulping

Table 1A. Typical sulfite cook specifications

Chemical	Acid	Bisulfite	Neutral	Alkaline
Total SO_2 charge, % on o.d. wood	20	20	20	20
Free SO_2, % of total SO_2 (true free SO_2)	80	0	0	0
Bound SO_2, % of total SO_2 (true combined SO_2)	20	100	100	100
Liquor to wood ratio (o.d. wood)	4:1	4:1	4:1	4:1
Buffer: Na_2CO_3, Na_2O equivalents ratio	- - -	- - -	0.20	- - -
Alkali: NaOH, Na_2O equivalents ratio	- - -	- - -	- - -	0.10-0.50
Additive(s) - - AQ - % on o.d. wood	- - -	- - -	- - -	0.02-0.20
Starting pH	1-2	3-5	7-9	10-13
Palmrose Liquor Analysis (g SO_2/100 mL liq.)				
Total SO_2	5	5	5	5
Free SO_2	4	2.5	—	—
Combined SO_2	1	2.5	—	—
Actual SO_2 Concentration				
SO_2 as excess 'H_2SO_3'	3	0	0	0
SO_2 as $MHSO_3$	2	5	2.5	0
SO_2 as M_2SO_3	0	0	2.5	5

from its inception in the 1870's until the mid 1950's. The enduring preference for this base was due to its low cost (from limestone, $CaCO_3$) and the absence of stringent environmental quality regulations. In fact, calcium base suffers from severe chemical limitations which impede further process development.

The above examples are intended as a simplification. A more extensive number of chemical species are actually present.

(1) Calcium is soluble only below pH 2.3 and only in its bisulfite form,

$$Ca + 2HSO_3 \underset{P}{\overset{T}{\rightleftarrows}} Ca(HSO_3)_2 \underset{P}{\overset{T}{\rightleftarrows}} CaSO_3 + SO_2 \cdot H_2O$$

[this is analogous to the dissolution of limestone as calcium bicarbonate, $Ca(HCO_3)_2$, in the presence of atmospheric CO_2], occurring in nature.

(2) When calcium bisulfite solutions are heated, deactivation of the bisulfite takes place, as illustrated in the above reaction, and the equilibrium moves to the right precipitating calcium sulfite. This compound, moreover, exhibits characteristically reversed solubility – it becomes even less soluble with rising temperature.

Thus high charges of free SO_2 and low cooking temperatures must be maintained which confine the process to a narrow range of two units at the bottom of the pH scale and impair improve-

ments in production rate and flexibility in pulp quality.

(3) The third major deficiency of calcium base chemistry is the formation of calcium sulfate in

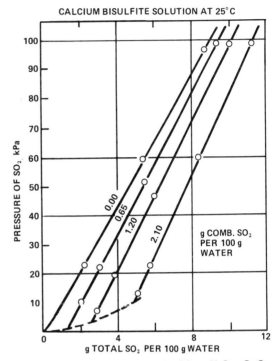

Fig. 2. Solubility relations in the $SO_2 - H_2O - CaO$ system [51].

Table 1B. Spectrum of sulfite cooking liquors
M = monovalent cation (Na^+, NH_4^+, ½ Ca^{++}, ½ Mg^{++})

Fresh Cooking Liquors Approximate pH (*)	1	2	4	6	7	10	11.5	14
Maximum pH (base)		Ca		Mg		NH_4^+,Na —————— Na		
Free SO_2(**) (True free (SO_2))	100	50	0 —————————————————— 0					
Bound SO_2(**), HSO_3^- (True combined SO_2)	0	50	100	50		0 ————————— 0		
Bound SO_2(**), SO_3^{--}	0 ——————————— 0				50	100	85	50
Cardinal forms of sulfite(***)	"H_2SO_3" (A)	(A)+(B)	$MHSO_3$ (B)	[(B) +(C)]		M_2SO_3 (C)	[+Na_2CO_3][+NaOH] Alkaline Sulfite (with or	
Types of cook	Acid Bisulfite		Bisulfite "Square Acid"	Neutral Sulfite			without additives; AQ)	

Notes: (*) Non-linear scale, approximate positions only. Actual pH will vary with base used, temperature, etc.
      (**) As percent of total SO_2.
      (***) Cardinal form of sulfite chemicals. Between cardinal points, the mixtures shown in brackets [ ] are found.
          Beyond point (C), only sodium monosulfite (Na_2SO_3) mixed with alkali will be found.

the recovery process. Back conversion of calcium sulfate to cooking chemical is not practical since a temperature of over 1150°C is required for achieving its complete decomposition to calcium oxide and sulfur dioxide.

Solubility relations in calcium base and bisulfite solutions at room temperature [4], covering the practical range of SO_2 and base concentration, are shown in Fig. 2. In Fig. 3 the demarcation is drawn for liquors of increasing temperature where the base is completely dissolved as bisulfite and where insoluble $CaSO_3$ separates [5]. A very large excess of free SO_2 needs to be maintained above 120°C for preventing the bisulfite ion concentration to be

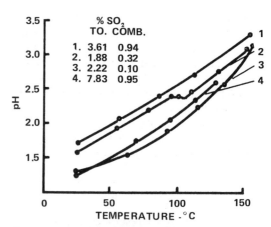

Fig. 4. Effect of temperature rise on the pH of acid calcium bisulfite solution [3].

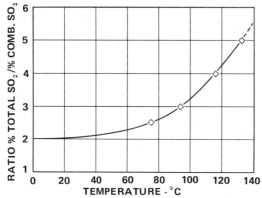

Fig. 3. SO_2 ratio of saturated $Ca(HSO_3)_2$ solution as a function of temperature [5].

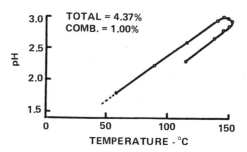

Fig. 5. pH during $CaSO_3$ precipitation [3].

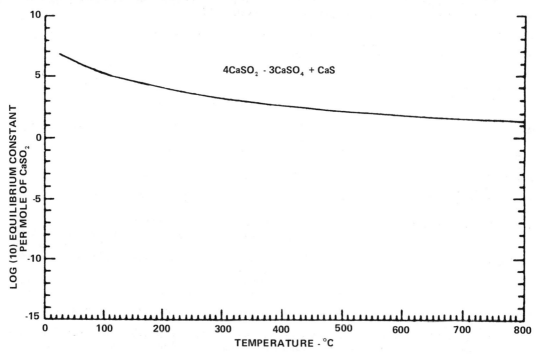

Fig. 6. Calcium sulfite disproportionation [6].

lowered to the point where the tendency to convert to sulfite ion and to precipitate with the calcium ion prevails. Large system pressure is required for maintaining this excess of SO_2, especially in the early stages of the cook where little of the bisulfite has yet been consumed.

The situation upon heating is illustrated in Figs. 4 and 5 [3]. The acidity of the liquor decreases by 1.5 pH as the temperature rises to 150°C, denoting a drop in hydrogen ion concentration of 30 times. The kink at 100°C in the line for a liquor at 1.88% total SO_2 and 0.32% combined SO_2 in Fig. 3 indicates precipitation of some $CaSO_3$ due to insufficient SO_2 content of the system.

Precipitation, which is of practical interest in calcium base sulfite pulping regarding scale formation, "liming up", of digester circulation lines and heat exchangers, is demonstrated again by direct pH measurement in Fig. 5. In a calcium base liquor containing 4.37% total SO_2 and 1.00% combined SO_2, the pH began to fall off at 145°C indicating a marked change of the existing ionic equilibrium. At 152°C heating was stopped and the solution allowed to cool. After 10 min at 150°C the pH settled down on another straight line below the heating curve. The drop in pH at 145°C was 0.23 unit corresponding to a 70% increase in hydrogen ion concentration and higher system pressure, indicating more free SO_2. On opening the autoclave $CaSO_3$ precipitate was found which is known to redissolve very slowly.

The practical limit of calcium base concentration is 1.3% combined SO_2 or 40 g $Ca(HSO_3)_2$ per liter.

Apart from solutions, the thermochemistry of decomposition of calcium sulfite and sulfate at the temperatures prevailing in recovery furnaces is of interest. It has been known since the last century that metal sulfites inclusive of $CaSO_3$ and Na_2SO_3 disproportionate into sulfate and sulfide, the latter being oxidized at least partly to sulfate. At the same time decomposition of the sulfites into oxide and sulfur dioxide takes place to a greater or lesser degree. The reactions for calcium are [6]:

Disproportionation (Fig. 6)
$$4CaSO_3 \rightarrow 3CaSO_4 + CaS$$

Decomposition (Fig. 7)
$$CaSO_3 \rightarrow CaO + SO_2$$

Calcium sulfite anhydride has a tendency to disproportionate over the whole temperature range and calcium sulfate formed is stable in the furnace atmosphere up to 1150°C. The decomposition temperature of $CaSO_3$, on the other hand, is about 1000°C while CaO is the stable phase above 1150°C. These are the thermodynamic reasons that the combustion product from the burning of spent calcium base sulfite liquor (fly ash) consists of about 50% calcium sulfate and 50% calcium oxide and that the regeneration

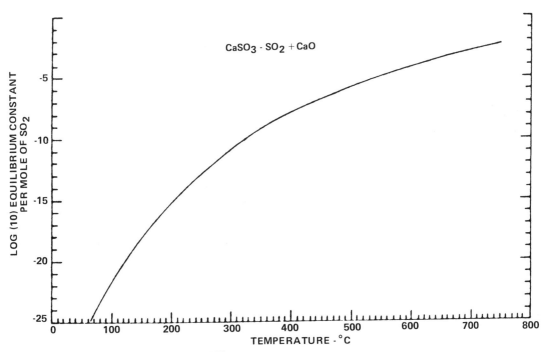

Fig. 7. Calcium sulfite decomposition [6].

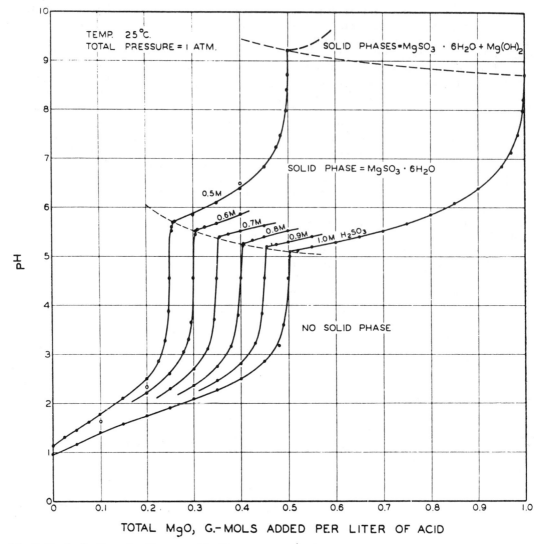

Fig. 8. Neutralization of sulfurous acid with magnesium hydroxide [8].

of the cooking liquor from it is both difficult and uneconomical regarding energy, and equipment cost and maintenance.

For these reasons calcium base sulfite pulping contributed in 1980 less than 20% of sulfite capacity and is now reserved for pulp mills with advanced specialty pulp and by-product manufacture (refer to Chap. VIII) while the share of the semisoluble magnesium base and the soluble ammonium and sodium bases rose in the last 40 years from almost nil to over 80% of the world sulfite capacity of 20 million tons per year.

2. Magnesia

Magnesium base sulfite pulping was introduced by Ekman in Sweden in 1872 but it was not until the 1960's that this base gained

acceptance in large-scale commercial operations, and only in combination with the Tomlinson-Wilcoxson recovery process first described in 1940. The reason for this is obviously found in the fact that the cost of MgO is at least ten times that of CaO and therefore full chemical recovery is mandatory for economical reasons. In 1951, the first calcium base mill was converted to magnesium base with recovery [7].

Apart from cost, magnesium base has significant advantages in sulfite pulping due its chemical nature. Although next to calcium in the bivalent group of elements and therefore similar in some of its properties, the differences are enough to greatly improve its usefulness in the field of pulping technology.

Compared to calcium bisulfite where the limit

of solubility is at pH 2.3, magnesium bisulfite solutions are stable up to pH 5.6 and can therefore be used for acid-bisulfite, bisulfite and bisulfite-sulfite pulping operations. Magnesium sulfite is much more soluble than calcium sulfite and has normal, that is increasing, solubility in water with increasing temperature. Solubility data express this difference (grams per 100 mL solution, °C):

MgO	0.00062, 20°	0.0086, 30°
$MgSO_3$	0.65, cold water	3.3, 66°
$Mg(HSO_3)_2$	4.7, 25°	5.8, 66°
CaO	0.13, 10°	0.07, 80°C
$CaCO_3$	0.0043, 18°	0.0011, 100°C
$Ca(HSO_3)_2$	4.1, 25°C[1, see p.11]	

Magnesia base cooking liquor consists of the primary components $Mg(OH)_2$ (from MgO), SO_2 and H_2O. Magnesium hydroxide obtains when magnesia (MgO) is slaked with water. It is sparsely soluble and its suspension in water is known as milk of magnesia; the pH is 10.4 and the dissociation constant $1.95 \cdot 10^{-11}$, compared to $1.73 \cdot 10^{-3}$ for $Ca(OH)_2$, a much stronger base. Cooking liquors of various concentrations and pH values are prepared by treating milk of magnesia with the required quantities of SO_2 gas

[8] as shown in Fig. 8. The following reactions describe the solutions used in the magnesia base pulping process for preparing pulps of different properties:

$H_2O + SO_2 \rightarrow SO_2$ aq., free acid solution

$Mg(OH)_2 + SO_2$ aq. (excess) $\rightarrow Mg(HSO_3)_2$ + SO_2 aq., acid bisulfite

$Mg(OH)_2 + 2SO_2 \rightarrow Mg(HSO_3)_2$, bisulfite, Magnefite

$Mg(HSO_3)_2$ (excess) $+ Mg(OH)_2 \rightarrow$ $Mg(HSO_3)_2 + MgSO_3$, "neutral Magnefite"

$Mg(HSO_3)_2 + Mg(OH)_2 \rightarrow$ $2 MgSO_3 + 2H_2O$, magnesium sulfite solution

$MgSO_3 + SO_2 + H_2O \rightarrow$ $Mg(HSO_3)_2$, bisulfite

When all $Mg(OH)_2$ has been neutralized by SO_2 aq, the substance found in the system as suspension or precipitate at pH values between 9.5 and 5.6 is magnesium sulfite, $MgSO_3$. Below pH 5.6 all $MgSO_3$ present is in the soluble form and at pH 4 the system consists entirely of magnesium bisulfite dissolved in water. Acid magnesium bisulfite solutions obtain when further SO_2 is added to the bisulfite solution. In this area, at pH levels below 2, magnesium base sulfite solutions meet with the original calcium base sulfite solutions in properties and cooking action. The specific gravities of magnesium base sulfite solutions over a practical range of compositions is given [9] in Fig. 9.

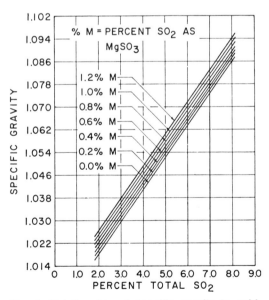

Fig. 9. Relationship of specific gravity to acid composition at room temperature (20 to 25°C) [9].

[1] This is equivalent to 1.3% combined SO_2, the practical limit of base concentration in calcium bisulfite cooking acid considering precipitation and scale formation during the cook.

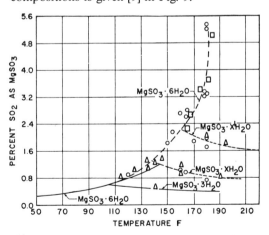

Fig. 10. Solubility of magnesium sulfite in water. — Literature data; - - - Babcock & Wilcox research data; O By adding MgO to $Mg(HSO_3)_2$ solution; ☐ By adding SO_2 to MgO slurry; Δ By adding $MgSO_3 \cdot 6H_2O$ to water [9].

As far as the solubility of MgSO₃ is concerned, the system is rather complicated (see Fig. 10). Firstly, three crystalline forms, MgSO₃·6H₂O, MgSO₃·3H₂O, MgSO₃·2H₂O are involved between room temperature and 80°C and secondly, the amount of MgSO₃ which can be kept in aqueous solution depends on the solubility of MgSO₃ in Mg(HSO₃)₂ solution as well as in water [9]. This second situation concerns pH levels between 4 and 5.6 where bisulfite and sulfite coexist. For practical purposes the solubility of MgSO₃ in water is taken to be about 0.65% at room temperature, 1.7% at 55°C and 5.2% at 80°C. This greatly increased

solubility at higher temperature can be utilized for preparing stronger cooking liquors in mill operations. Alternatively cooks at pH values higher than the saturation point can be performed by preparing liquors with excess MgSO₃ or even some Mg(OH)₂ in suspension [10]. Because of the low molecular weight of Mg compounds such slurries can be handled industrially by providing sufficient turbulence in tanks and pipelines. As the temperature rises and more and more chemical is consumed in the reaction with the wood such slurry liquors clear

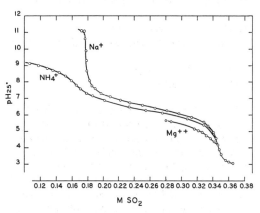

Fig. 11. pH vs molar SO₂ content for magnesium bisulfite solution [11].

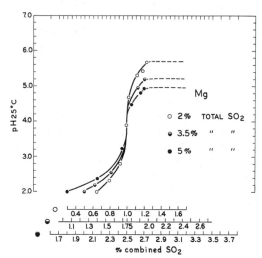

Fig. 12. Relation between combined SO₂ and pH for magnesium base [11].

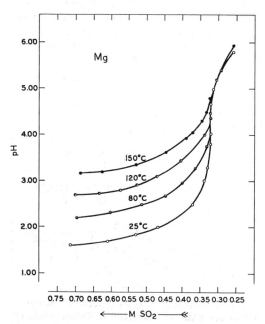

Fig. 13. Titration of magnesium bisulfite solution with SO₂ at 4 temperatures; 0.315 N Mg (1.01% combined SO₂) [11].

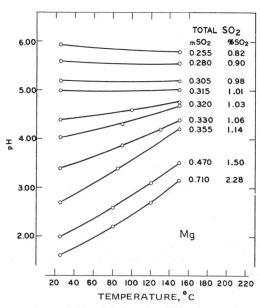

Fig. 14. Temperature-pH relations in the soluble portion of the Mg(OH)₂ - SO₂ system; 0.315 N Mg (1.01% combined SO₂) [11].

up during the cook. It has also been demonstrated that the solubility of MgSO₃ extends to higher pH levels when wood is present in the system. This effect is due to soluble sugars and pectic substances from the plant material.

The position and pH range of the magnesia base sulfite system at room temperature is shown relative to the systems of the two soluble bases, NH₄OH and NaOH, in Fig. 11. The data were obtained by titrating solutions containing 0.344 N base which is equivalent to 1.10% combined SO₂ (Palmrose test) with SO₂ gas through the sulfite break (soluble bases only) and the bisulfite break for all three bases. Both the advantage of magnesia base over calcium base (maximum pH 2.3) and its limitation compared with the soluble bases is clearly seen [11].

A closer look at the bisulfite inflection exposes two anomalies typically found in sulfite systems. At the point of inflection, the base and SO₂ concentrations are not exactly equivalent but contain an excess of 0.02% SO₂. This then causes the smell of the bisulfite salts and their solutions. The second anomaly is the distinctly lower position of the point of inflection for magnesium ion (3.80) than for sodium or ammonium ion (4.20). Both anomalies are due to the unstable nature of the acid, SO₂ aq., on which the sulfite system is based and the difference in strength of the basic nature of the cation.

Figure 12 provides complete information on composition and pH of magnesium base sulfite liquors at room temperature over the range of 2-5% total SO₂ and 0.3-3.7% combined SO₂ (0.9-6.0% MgSO₃). The difference between the base concentration data below the inflection point and the data below the solubility limits give the amount of MgSO₃ in solution. These data will be useful for calculations of chemical requirements and for control of liquor making in the mill [11]. Actual examples of these calculations are given in Chapter VII.

The effect of heating on the pH value of magnesium base sulfite solutions is demonstrated for a liquor containing 1.01% combined SO₂ (0.315N Mg) in Figs. 13 and 14 from room temperature to 150°C [11]. For reasons explained earlier, the pH of the hot solution rises steeply by 0.012 pH/°C from the lowest pH levels where large amounts of SO₂ aq. are present, but only at a rate of 0.06 pH/°C at the bisulfite level at pH 4. Zero slope occurs between pH 5 and 6 where the system performs as a high temperature buffer solution where the heat induced charges of the ionic equilibria are such that the hydrogen ion concentration (activity) remains rather constant.

Apart from its usefulness up to pH 5.6, the other advantage of the magnesium base sulfite system over the calcium base system lies in its

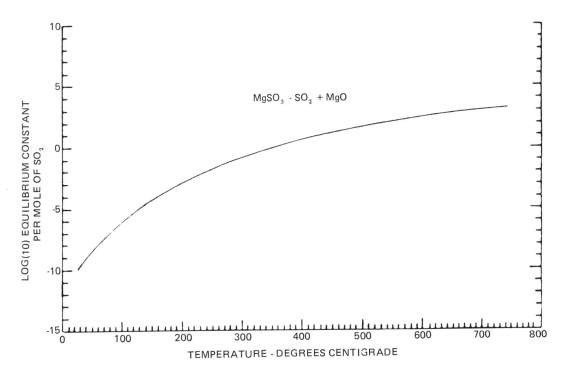

Fig. 15. Magnesium sulfite decomposition [6].

thermochemical behavior [6]. In contrast to the calcium system, and most of the other sulfites, the thermal decomposition products of $MgSO_3$ contain no, or only very little, sulfide, MgS. The principal reaction, $MgSO_3 = MgO + SO_2$ (Fig. 15) occurs at 360°C, compared with 1000°C for $CaSO_3$. Side reactions of the $MgSO_3$ decomposition form some sulfate, thiosulfate, and elemental sulfur below 500°C operating temperature; and sulfate, elemental sulfur and magnesium oxide, MgO, above 500°C.

The disproportionation reaction of $MgSO_3$, $4MgSO_3 \rightarrow 3MgSO_4 + MgS$ (Fig. 16) occurs under similar conditions than that for $CaSO_3$ (Fig. 6). The apparent peculiarity of the magnesium sulfite is therefore a consequence of its relatively low decomposition temperature, the negligible rate of reaction below 600°C, and the stability of the thiosulfate up to 500°C. Figure 17 shows a three-dimensional diagram of the $Mg\text{-}SO_2\text{-}O_2$ system over the temperature range of 500°C to 1100°C, constructed from available data.

3. Ammonium base

By 1925 the use of ammonium sulfite, $(NH_4)_2SO_3$, had attracted the attention of several investigators. In one of the early processes [Braun, 1902] wood chips were cooked in neutral $(NH_4)_2SO_3$ solution to "a bright, completely defibered pulp of good yield which is readily bleached. The NH_3 was recovered from the spent liquor by distillation with lime".

The above illustrates that complete pulping processes were already conceived at an early date, but could not compete with the inexpensive lime process. Today ammonium base pulp capacity is about 10% of total sulfite capacity.

Ammonia is an asphyxiating gas under atmospheric conditions. It is supplied in liquified form in steel containers at 1700 kPa for industrial applications. The flammable limits in air are 15-79%, the ignition temperature 650°C. The gas dissolves readily in water as ammonium hydrate:

NH_3 solubility in water, wt %:

0°C	42.8
20°C	33.1
40°C	23.4
60°C	14.1

Density of aqueous ammonia at 15°C:

NH_3, wt %	g/L
8	0.970
16	0.947
32	0.889
50	0.832
75	0.733
100	0.618

The analogy between the aqueous sulfur dioxide to form the unstable sulfite anion and the aqueous ammonia system to form the unstable

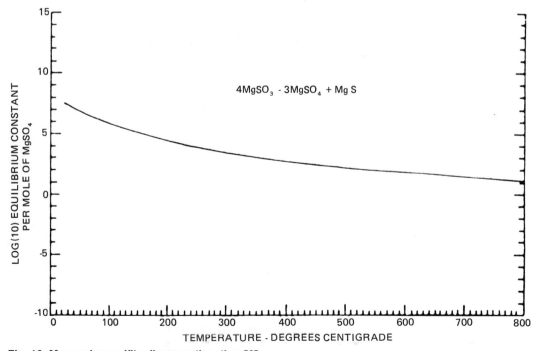

Fig. 16. Magnesium sulfite disproportionation [6].

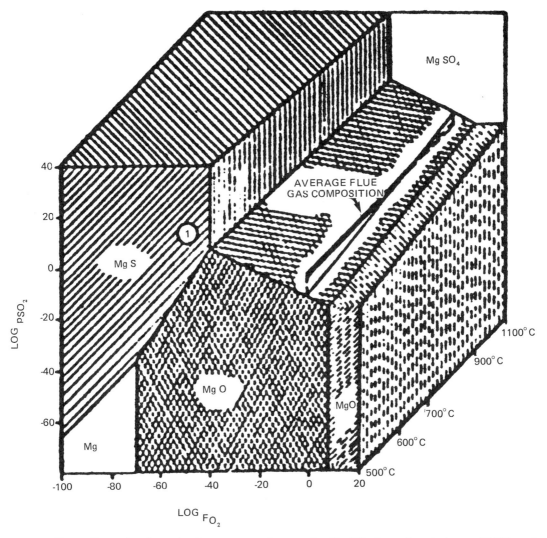

Fig. 17. Three-dimensional predominance area diagram for Mg-SO₂-O₂ system between 500°C and 1100°C [6].

ammonium cation, NH_4^+, has been suspected for some time and was eventually proven by IR absorption, pH and electrical conductivity measurements [50]. Thus ammonium hydroxide, the formal base of the system does not exist as such and the base ions are formed directly:

$$NH_3 \cdot H_2O(NH_3\ aq) \rightleftarrows NH_4^+ + OH^-$$

Aqueous ammonium is a weak base with anionization ratio of 1.80×10^{-5} and a pH of 11.6 of the 1N solution at room temperature. Thus the concentration of base ion is only two hundredths of that in 1N NaOH. The solution readily absorbs SO_2 gas forming bisulfites and sulfites:

$$NH_3\ aq + SO_2\ aq = NH_4^+ + HSO_3^-\ \text{in water}$$

$$2NH_3\ aq + SO_2\ aq = 2NH_4^+ + SO_3^{--}\ \text{in water}$$

At 0°C, 100 mL of water dissolve as much as 267 g of NH_4HSO_3 or 128 g of $(NH_4)_2SO_3$ suggesting that the ions are highly hydrated. The pH values of the chemical equivalence points for the NH_3-SO_2 neutralization curve at 25°C are shown in Fig. 11 for solutions containing 1.10% combined SO_2 (0.344 N NH_3): pH 4.22 for the bisulfite and pH 7.85 for the sulfite. It is seen immediately that ammonia and sodium give identical bisulfite inflections but that the alkalinity attainable with ammonium hydroxide is far lower than that attainable with sodium hydroxide. These curves also show clearly the nature of the bi-gaseous ammonium-sulfite system where the ammonium cation is strengthened in more acid environment and the sulfite anion is

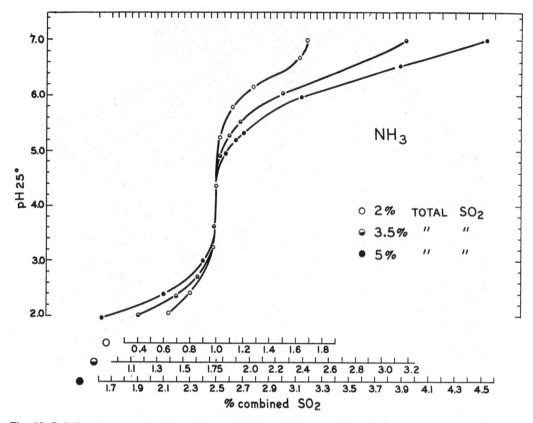

Fig. 18. Relation between combined SO₂ and pH for ammonium base [11].

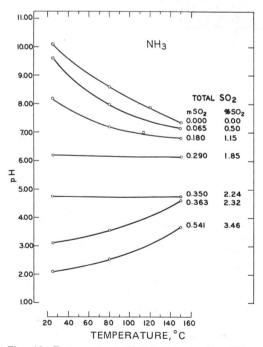

Fig. 19. Temperature-pH relations in the NH₃-SO₂-H₂O system; 0.344 N NH₃ (1.10% combined SO₂) [11].

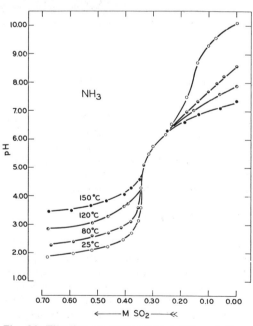

Fig. 20. Titration of ammonium sulfite solutions with SO₂ at 4 temperatures; 0.344 N NH₃ (1.10% combined SO₂) [11].

strengthened in more alkaline environment. Obviously the ammonium cation is becoming very weak above pH 7 and consequently also the sulfite anion. The technological usefulness of the two monovalent bases, ammonium and sodium, is directly related to these dissimilarities in ionic strength with increasing pH level.

The exact relations between SO_2 content, base content and pH at 25°C in the ammonium sulfite system are given in Fig. 18. The data are corrected for volume changes due to titrant addition and sampling. It is seen that the pH of the sulfite inflection shifts to higher levels as the concentration of sulfite is increased, an anomaly typical for the NH_3-SO_2 system.

The great thermal sensitivity of ammonium base sulfite solutions was explained before and is illustrated in Figs. 19 and 20 by the changes of pH occurring between room temperature and 150°C.

The differences between this system and the sodium-base system at higher temperatures are striking. At 80°C sulfite ions have already disappeared and at 150°C even the bisulfite ion appears rather weak. The temperature-pH slopes are similar to sodium only below pH 3 (cold). Similar to the Mg-SO_2 system, the pH is not influenced by temperature between 5 and 6. The maximum pH obtainable in heated ammonium sulfite solutions will be about 8 at 120°C and 7.5 at 160°C, this to be taken with reference to the neutral pH of water, 6.0 and 5.85 at these temperatures. In agreement with these findings, system pressures are high; increasing the digester pressure accelerates the rate of pulping, and, for example, relief gases from ammonium bisulfite solutions of pH 5-6 heated to 160°C smell of free ammonia. Thus on heating, the acid or basic associations of SO_2-H_2O and NH_3-H_2O fail and these failures are enhanced by a simultaneous decline in acid or base potency favouring these associations. It can be estimated that an ammonia-sulfite solution will react neutral (=pH 5.6) above 200°C and consist of SO_2 and NH_3 gas dissolved in water. As in the case of SO_2-H_2O, the equilibrium will be strongly influenced by pressure:

$$NH_4^+ \underset{P}{\overset{T}{\rightleftarrows}} NH_3 \text{ aq} + H^+.$$

Strongly alkaline ammonium base systems are conceivable but may not be economical because of the extremely high pressures required.

With regard to recovery of chemicals and heat, the nature of ammonia demands radically different methods than developed for the three metal bases. Two properties are utilized: its ability to burn in air with generation of much heat,

$$4NH_3 + 3O_2 \rightarrow 2N_2 + 6H_2O; \ 1.84 \text{ MJ/kg } NH_3$$

and the ease of liberating it from its compounds and solutions. In most of today's ammonium base sulfite mills the first of the two properties is exploited by concentrating the spent liquor from the pulping process in evaporators and burning it in steam boilers. The ammonia left in the spent liquor as well as the ammonia which combined with the wood lignin during the cook are both burnt together with the organic material adding to the amount of process steam generated. During the burning process a large portion of the sulfur dioxide is liberated also and can be reclaimed for preparation of fresh liquor. Since no ash is formed, as it is in the case of metal base liquors, the furnace operation is clean (cf. Chap. IX,C,4 and Chap. X,D,3).

In ammonium recovery operations submicroscopic particles of ammonium salts present in gaseous emissions can cause a bluish haze in the atmosphere under certain weather conditions. This environmental problem can be eliminated by means of demisting equipment such as described in Chap. X,D,6.

Chemical processes developed for recovering ammonia by liberating it from the spent liquor take two forms:
1. Utilizing the tendency of NH_3 to gas-off, the liquor can be treated with a strong alkali, such as from the kraft cooking process, or it can be treated with a weak but stable alkali such as milk of magnesia and the ammonia stripped off by heat and turbulence (steam). Due to the volatile nature of ammonia in its compounds the latter succeeds, although ammonium base is ten thousand times stronger than magnesium base.
2. Ammonia can be extracted from spent acid bisulfite liquor by passing it through a column filled with cation exchange resin. The ammonium ions are adsorbed by the resin and hydrogen ions released. The base is then reclaimed from the resin by passing acid SO_2 water through the column.

Both methods have been developed to the mill level and operated for some time in isolated cases, but have not found continued use so far.

4. Sodium base

Sodium is the sixth most abundant element in the hydrosphere and lithosphere of the earth, after calcium and before potassium and magnesium. It is commercially available in large deposits and from sea water in the form of its salts: NaCl (sodium chloride, table salt), Na_2SO_4 (sodium sulfate, salt cake), Na_2CO_3 (sodium carbonate, soda). Of these, salt cake and soda are important in pulping technology. The other major sodium chemical for use in pulping is NaOH (sodium hydroxide, caustic, or caustic soda) which is obtained from NaCl by electrolytical processes or from Na_2CO_3 by treatment with lime. The electrolytical processes which produce NaOH from salt yield at the same time the chlorine compounds used for the bleaching of pulps in the manufacture of fine papers.

Calcium base used to be about 10 times less expensive than sodium base but is now only 4 times less expensive. Sodium base has several clear chemical advantages which offset the economic disadvantage when used under the appropriate technological conditions.

Firstly, the sodium base sulfites ($NaHSO_3$ and Na_2SO_3) are easily soluble over the whole pH range:

Sodium bisulfite   very soluble
Sodium sulfite   12.54%, 0°C; 28.3%, 80°C

Secondly, due to the basic strength and stability of the sodium cation, Na^+, the stability of the acid bisulfite and sulfite anions is greatly increased. The useful range of liquor pH at room temperature extends from 1 to 4 with free SO_2, from 4 to 10 for the sulfite alone, to 11.6 with carbonate and to 14 with the addition of free NaOH.

Thirdly, spent liquors are amenable for burning, recovery of the inorganic chemicals, and of by-products.

Sodium was the first base used in industrial pulping processes in the form of hydroxide and it was the first sulfite base used in Tilghman's and Kellner's laboratories in the last century. A combination of hydroxide and sulfite, alkaline sulfite, has been used for pulping of straw and other annual plant material since the turn of the century [52]. All significant aspects of sodium base sulfite pulping were identified by extensive experimentation already in the last century, and the advantages of this base established regarding versatility, effectiveness in the cook, and bleachability of the pulps. Extensive industrial application did not materialize, however, until several decades later, the major obstacle being the development of practical spent liquor recovery processes for sodium sulfites, whether to be used alone, or in acid or alkaline solutions.

Even before the general acceptance of a

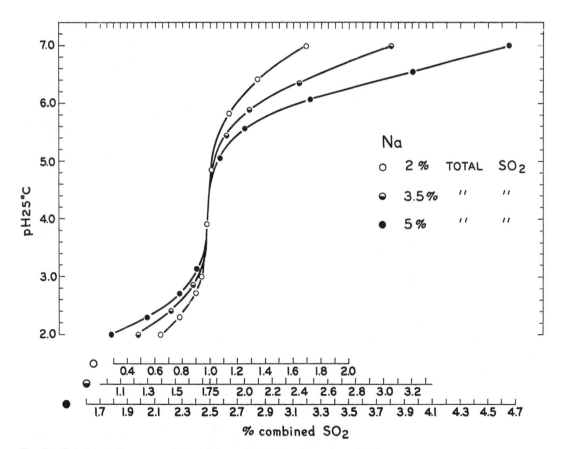

Fig. 21. Relation between combined SO_2 and pH for sodium base [11].

satisfactory recovery process, sodium sulfite with the addition of sodium bicarbonate and sodium carbonate was introduced in the 1930's as the NSSC pulping process (for neutral sulfite semichemical) for the manufacture of high yield corrugated boxboard from hardwoods, and high yield sodium bisulfite pulping of northern softwood in the 1950's for newsprint manufacture (together with magnesium bisulfite), mainly as replacement for calcium base liquors. Only in very recent years has the rising dissatisfaction with the conventional kraft process and the successful introduction of sodium base sulfite recovery prepared the ground for considering a major expansion of sodium base sulfite pulping in the alkaline range.

In order to reconcile sometimes conflicting concentration and pH data found in the literature of sodium base sulfite solutions at room temperature, the bisulfite and sulfite

equivalence inflections were measured carefully at 25°C. The results in Fig. 11 bring out the anomalies between sodium, ammonium and magnesium ion at a concentration of 0.344 N base ($= 1.1\%$ combined SO_2). Figure 21 shows the effect of base: sulfur dioxide ratio on solution pH at different concentrations (% combined $SO_2 \cdot 0.3125 = N$ base). It was mentioned in the earlier section on magnesia base that the bisulfite solutions contain a small amount of free SO_2 at the equivalent point and smell of SO_2.

Another anomaly of importance in liquor preparation and process control is the lowering of the pH value of sodium bisulfite solution with increasing chemical concentration as in Fig. 22. In weak solutions at about 0.1M base equivalent, the pH is 4.55, the value calculated for $NaHSO_3$: $1/2\,pK_1 + 1/2\,pK_2 = pH$. In cooking liquors of 1-4% combined SO_2 the bisulfite equivalent pH is between 4.30 and 4.05; it settles at a minimum of 3.70 above 20% combined SO_2. A consequence of this unexplained effect [12] is the reduction of the free SO_2 required for preparing acid bisulfite cooking liquor of increasing concentration at a given pH value.

Due to the ionic strength of sodium base, its sulfite salts and their solutions in water are least affected by increasing temperature when com-

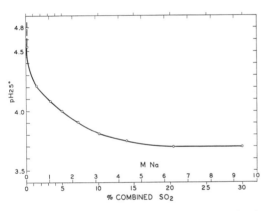

Fig. 22. Effect of concentration on the pH of sodium bisulfite solutions [11].

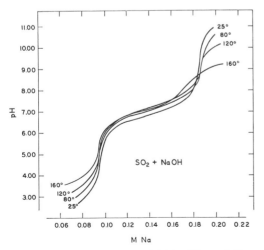

Fig. 23. Titration of an acid bisulfite solution (0.097 MSO₂) with 4 N NaOH at four temperatures [11].

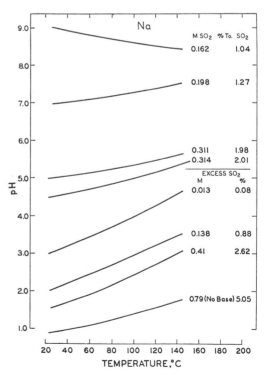

Fig. 24. Temperature-pH relations in the NaOH-SO₂ system; 0.312 N NaOH (1.0% combined SO₂) [11].

pared with the other three commercial bases. Figures 23 and 24 illustrate that both the bisulfite and the sulfite ions are quite stable in this system; the first beyond 160°C and the second beyond 120°C with obvious deterioration below 160°C. At about pH 8 in Fig. 24 the pH slope becomes zero and a sodium sulfite solution of this composition functions as a high temperature buffer for calibration purposes. The same illustration shows that the pH of a solution of SO_2 gas in water rises less under the influence of temperature than that of an acid bisulfite solution below pH 4 at room temperature. The above makes it quite clear that in a sodium sulfite cook the major anionic species will be bisulfite, HSO_3^-, and that at pH levels above 8 much of the sulfite ion, $SO_3^=$, will revert to bisulfite ion at 160°C and higher. It can be extrapolated that even the sodium sulfite ionic system will not be stable above 350°C and will consist mainly of water, SO_2 and partly dissociated NaOH.

In view of the expansion of sulfite pulping into the range of medium and high alkalinity, beyond its intrinsic pH of 10, it is of importance to include the strong base sodium hydroxide into this discussion. The use of sodium bicarbonate and carbonate was mentioned in connection with NSSC pulping. An aqueous solution of sodium bicarbonate has a pH of 8.4, a sodium carbonate solution a pH of 11.6. Both are buffering substances, bicarbonate between pH 5 and 11 and carbonate between pH 8.4 and 11.6.

In order to raise the pH of a sodium sulfite solution above pH 11, free caustic soda must be present for supplying the additional amount of hydroxyl ions required: $NaOH \rightarrow Na^+ + OH^-$. Aqueous solutions of NaOH at a molar concentration of about one, as used in pulping, are highly dissociated at room temperature but their ionic equilibrium is highly influenced by the increasing dissociation of water on heating,

$$2H_2O \underset{\rightleftarrows}{T} H_3O^+ + OH^-$$

The result of this, as seen, is a substantial decrease of the solution pH with rising temperature of any alkaline pulping solution containing sodium hydroxide. Figure 25 illustrates the drop in pH of a NaOH, a Na_2SO_3 and an alkaline sulfite solution, all at the same Na_2O concentration of 60 g/L, due to raising the temperature from 25 to 85°C [13]. A solution of sodium sulfite has no buffering capacity above its own pH level and the addition of strong alkali raises the pH of the mixture to its own level. In actual cooks natural wood acids and acids formed in the process of lignin solubilization will add to the effect of decreasing the alkalinity of the cooking liquor (see following section).

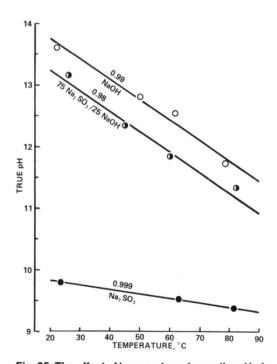

Fig. 25. The effect of temperature rise on the pH of solutions of alkaline sulfite pulping chemicals at 60 g/L Na_2O [11].

Table 2. Changes of ionic concentration of solutions of NaOH on heating

		Temperature pW	22°C 14.00		120°C 11.90		200°C 11.39	
			pH	pOH	pH	pOH	pH	pOH
0.01	NaOH		12.00	2.00	9.90	2.00	9.39	2.00
0.0001	NaOH		10.00	4.00	7.904*	3.996*	7.398*	3.992*
0.000001	NaOH		8.00	6.00	6.14*	5.76*	5.81*	5.58*

*Values near the neutral point are calculated for the electroneutrality conditions $[OH^-] = [H^+] + [Na^+]$:

$$[H^+] = \frac{-[Na^+] + [Na^+]^2 + 4 K_w}{2}$$

All sulfite pulping chemicals inclusive of carbonate and caustic soda are regenerated individually by the sodium base sulfite recovery system. This system requires the addition of causticizing equipment only if more free NaOH is used in the cook than is available from the chemical make-up of the process.

For make-up purposes, sodium carbonate is available on the market as powder or crystals; sodium hydroxide as pellets, flakes; or 50% solution in water, the most convenient form for continuous operation.

The thermochemical behavior of sodium base pulping liquors has been investigated primarily in connection with kraft (sulfide) process technology [14] and the results are also useful for understanding the thermal relations of the sulfite system. For accurate calculations of the equilibria, however, it is necessary to consider that sulfide is a reduced sulfur compound, whereas sulfite is partly oxidized, carrying three oxygen atoms per molecule.

Similar to Ca and unlike Mg, the sodium system tends to form sulfide and sulfate rather than oxide. The complicated mixture of chemicals in the spent cooking liquor break down in the burning or pyrolysis process almost instantaneously into a few simple inorganic substances: Na_2SO_4, Na_2S, Na_2CO_3, Na_2O and NaOH as solids or liquids depending on the temperature, and O_2, CO, CO_2, H_2, H_2O, S_2, SO_2, SO_3, Na, Na_2 in the hot gas phase. Other sulfur compounds will be present, but in amounts which are not significant in the present discussion. Nitrogen and inert gases introduced with the combustion air will act only as diluents for the gas phase.

The results from equilibrium calculations [14] can be presented in the form of logarithmic partial pressure diagrams [53]. Figure 26 at 727°C is pertinent for liquor decomposition by pyrolysis to obtain H_2S gas and solid sodium carbonate, and Fig. 27 at 1127°C for liquor burning to obtain Na_2S + Na_2CO_3 as smelt. The phase diagram of Na_2S-Na_2CO_3 [15] is given in Fig. 28. The system has an eutectic melting point of 762°C at 0.40 mol fraction Na_2S (i.e. 40% smelt sulfidity). There is evidence for limited solid solubility, particularly on the Na_2CO_3 side of the system. When 5 mol % of the Na_2CO_3 were replaced by Na_2SO_4 nearly the same melting point and eutectic temperature were obtained.

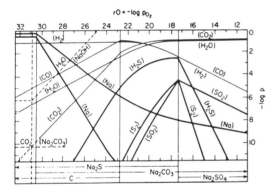

Fig. 26. Condensed phases and partial pressures of gases at 727°C. NaOH and Na₂O cannot exist at this or higher temperatures [14].

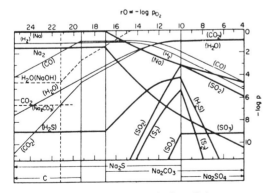

Fig. 27. Condensed phases and partial pressures of gases at 1127°C. The partial pressure of metallic sodium is becoming more and more important as the temperature increases, and will lead to high fly ash losses [14].

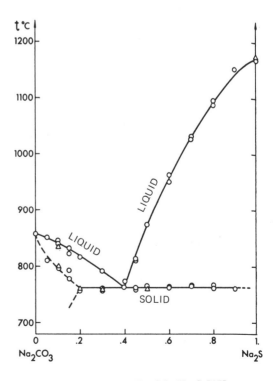

Fig. 28. Phase diagram Na₂CO₃-Na₂S [15].

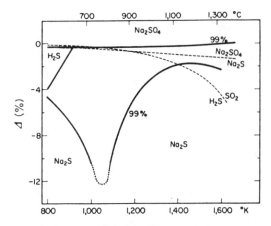

Fig. 29. Compounds obtained at various temperatures and oxygen deficiencies [14].

When comparing the two diagrams, it is seen that the conditions in Fig. 26 are more suitable for liberating a major portion of sulfur in the form of H_2S with Na_2CO_3 as the solid phase. By maintaining the oxygen concentration in the furnace at a partial pressure corresponding to $-\log O_2 = 17$-18 the formation of H_2S is optimized and the formation of both Na_2SO_4 and Na_2S is minimized.

Alternatively, as seen in Fig. 27, Na_2S and Na_2CO_3 smelt required by the common "green liquor" systems can be recovered by operating at the higher temperature. A diagram of percent oxygen deficiency vs. operating temperature given in Fig. 29 shows clearly that the most favorable operating conditions for optimum Na_2S recovery will be near $1100°C$ and 2% less O_2 than is necessary for complete combustion. The solid lines mark the regions where 99% of the sulfur present will be in the form of Na_2SO_4, Na_2S or H_2S. The dashed lines indicate the equilibrium conditions between Na_2SO_4 and Na_2S and between SO_2 and H_2S.

As in all thermochemical systems, other compounds might be formed during cooling of the phases after reaction. This can be prevented by quick separation of gases and smelt or ash and rapid cooling.

5. DECOMPOSITION AND OXIDATION OF SULFITE SOLUTIONS[1]

The tendency of acid bisulfite solutions to decompose spontaneously has been known for over 200 years. Heating of the solution greatly

---

(1) This discussion uses sodium base bisulfite solutions as an example for which ample data are available. Magnesium base solutions are as prone to auto-decompositions, but ammonium base solutions much less so.

increases the rate of decomposition. The products of complete auto-decomposition are sulfuric acid, bisulfate and sulfate, and elemental sulfur:

$$3SO_2 + 2H_2O \rightarrow 2H_2SO_4 + S;$$
$$3NaHSO_3 \rightarrow NaHSO_4 + Na_2SO_4 + S + H_2O$$

The pH values of aqueous solutions of $NaHSO_4$ are:

$$0.1\ M = pH_{25}1.75;\ 1.0\ M = pH_{25}1.10$$

The decomposition mechanism involved is much more complicated than indicated by the above net equations and involves thiosulfate, $Na_2S_2O_3$, and polythionate, $Na_2S_2O_6$, as intermediate chemical compounds as well as chemicals formed from wood during the cook, such as sugars, formic acid and terpenes as accelerators [16]. The chief agent in this autocatalytic decomposition is thiosulfate which forms spontaneously, and is always present to some degree in acid bisulfite and bisulfite solutions. Its control is essential in bisulfite liquor recovery.

The other fundamental agent influencing the decomposition reaction is the hydrogen ion concentration (acidity) of the liquor. Pure bisulfite solutions at pH 4 remain stable for many hours at 140°C. In acid bisulfite solutions below pH 4 the decomposition tendency increases greatly with the concentration of free SO_2. The strongly acid decomposition products further increase the hydrogen ion concentration (decrease in pH) until the latter counteracts the formation of thiosulfate

$$5Na_2S_2O_3 + 6H^+ + 3SO_4^{--} \rightarrow 2Na_2S_5O_6 + 3Na_2SO_4 + 3H_2O$$

and also displaces SO_2 from the solution

$$NaHSO_3 + 2H^+ + SO_4^{--} \rightleftarrows SO_2 + NaHSO_4 + H_2O$$

thus decreasing the bisulfite concentration.

Although decomposition comes to a standstill, the liquor now lacks HSO_3 cooking chemical. Sulfur precipitates in the digester and the acidity is high enough to result in a situation which is appropriately called "burnt cook". The wood chips turn dark and assume an acrid smell and the cellulose of the fibers degrades rapidly, leading to a complete loss of the cook.

The SO_2 stability of a bisulfite cooking liquor is normally greater than that of an acid bisulfite liquor, however the bisulfite cook is more sensitive to decomposition catalysts, and its rate of decomposition is greater because of its higher bisulfite concentration and operating temperature. Cobalt, molybdenum and selenium are accelerators of the decomposition reaction. Selenium, in particular, forms selenosulfate whose catalyzing effect is considerably greater than that of thiosulfate. Poor circulation conditions can be responsible for liquor breakdown because liquor from overcooked parts of the digester charge contains more thiosulfate than from undercooked parts. For the same reason liquor decomposition is of much greater concern in extended cooks to bleachable pulp than in short high yield cooks for news furnish. Similarly, stagnant liquor in some recesses of the digester can produce enough thiosulfate to cause a loss of the cook.

Hardwood cooks produce more thiosulfate than comparable softwood cooks. Ammonium bisulfite cooks form considerably less thiosulfate than cooks with the other three bases and consequently consume less sulfur.

With so many factors involved, the level at which thiosulfate in the liquor must be controlled will vary from case to case.

In order to avoid costly production upsets it is sound practice to control thiosulfate content in commercial bisulfite cooks below the dangerous level, particularly if chemical recovery is practiced.

An example of a simple analytical test used for determining thiosulfate in fresh and used bisulfite mill liquors is given here:

Procedure for Thiosulfate Determination

Pipette 10.00 mL bisulfite cooking liquor into a 250 mL flask, add 50 mL distilled water and 10 mL 10% formaldehyde. Add 10.00 mL 0.1 N iodine solution standardized against 0.1 N thiosulfate. Titrate with 0.1 N thiosulfate solution until most of the iodine color has faded, then add soluble starch indicator. Continue the titration dropwise until the dark blue color disappears.

Calculation
Thiosulfate in bisulfite cooking liquor, g $Na_2S_2O_3$ per liter $= (mL\ 0.1\ N\ iodine - mL\ 0.1\ N\ thiosulfate) \cdot 1.5811$

When the pH of sulfite solutions is raised above the bisulfite level the tendency for liquor decomposition soon disappears. This is for the most part due to the quickly diminishing hydrogen ion concentration (by a factor of 10 for every pH unit) and also the increasing replacement of bisulfite ion by sulfite ion. Also, sulfonation of lignin in neutral and alkaline sulfite liquors takes place in the presence of thiosulfate. Therefore neutral sulfite and alkaline sulfite cooks remain unaffected even by large amounts of thiosulfate in the liquor whose control ceases to be critical and is concerned only with economical reasons for avoiding unnecessary amounts of chemical ballast.

Acid bisulfite and bisulfite cooking liquors are protected against oxidation by the partial pressure of free SO_2 at the surface of the solution. The protective effect depends on the amount of free SO_2 in the liquor and therefore also on the pH of the solution. It is strong for acid bisulfite solutions and becomes very weak from the bisulfite level of pH 4 upward, as the free SO_2 diminishes towards zero. Most of the sulfate found by analysis in acid bisulfite and bisulfite solutions is due to impurity of the sulfite purchased, or due to recycling from a recovery process.

In sulfite solutions above the bisulfite level, the rate of oxidation by air increases rapidly with the pH. Pure Na_2SO_3 solution is oxidized easily by air. Stirring or turbulence of such solutions increase the surface area of solution and cause very rapid oxidation of sulfite to sulfate. The addition of bases, such as Na_2CO_3 or NaOH increases the stability of a sulfite solution. Divalent iron ion, Fe^{++}, has a strong catalytic effect on the oxidation of Na_2SO_3 and this effect is reduced or eliminated by the addition of NaOH which precipitates ferrous hydroxide, $Fe(OH)_2$.

If neutral sulfite or alkaline sulfite liquors are stored at all, it is necessary to provide protection from contact with air. This can be done by filling vessels with tight fitting covers to the top, by replacing the air in a covered vessel with nitrogen, or by covering the solution with a film of paraffin oil or similar medium.

REFERENCES
References for Chapters I and II are found at the end of Chapter II.

II

THE SULFITE COOK

O.V. INGRUBER

Retired Senior Research Associate from CIP, now an independent consultant

MANUSCRIPT REVIEWED BY
David Johnston, Manager Pulping
Nova Scotia Forest Industries
Port Hawkesbury, Nova Scotia

The purpose of the pulping of wood is to liberate fibers from the wood structure with the least amount of damage. This can be greatly facilitated with the help of chemicals for softening, reducing or eliminating the lignin binder which holds the fiber structure together.

A. IMPREGNATION STAGE

1. Effect of wood structure

In the first volume of this series, the complex porous structure of wood has been illustrated and described. It consists predominantly of capillary-like cells or fibers with closed ends embedded in lignin and arranged longitudinally in the wood. The voids of these cells, the lumen, are interconnected by pit membranes through which limited diffusion of liquids and gases can take place. Hardwoods, in addition, contain vessel cells which are larger in diameter than ordinary wood cells or fibers and, when open, permit more rapid penetration of liquids or gases into the interior of the wood than in softwood. Vessel cells are often blocked by tyloses, thin-walled bag-like growths in the lumen. However, these obstructions interefere less with the cooking of hardwoods than anticipated.

Since the effectiveness and efficiency of a wood cook depends on the degree of interaction of wood surface and pulping chemicals, it is of utmost importance to achieve complete filling of the capillary structure and the voids of the wood with cooking liquor as early as possible in the cook.

The true density of the solid wood substance itself is 1.3 g/cm^3 and varies little with species, type of growth or chemical composition. The porosity of wood, on the other hand, and hence the apparent density of dry wood vary greatly. The commercial pulp woods from North American softwood and hardwood species cover the density range of 0.3-0.6; corresponding to 20%-40% by volume of wood substance, and 60%-80% by volume of void space in the lumina of fibers, vessels and ducts. For the typical case of spruce wood at 0.4 specific gravity, the pore volume will be $1/0.4 - 1/1.5 = 1.83$; this wood can take up about 180% of its own dry weight in water and will then have a moisture content of 65%.

Compared with the long cooking schedules used in the first half century of sulfite pulping, the second half has seen a strong tendency for reducing digester cooking time in order to increase production. When cooking times were long, penetration was not considered to be a serious problem; in fact, it was stated in 1932 that it was nearly impossible to bring a commercial digester of any size to 110°C before penetration was complete. With more rapid temperature rise called for by present-day operations, such schedules would produce non-uniform pulp with centers of the chips under-cooked or burnt and high reject rates on pulp screens. As a consequence, the mechanisms by which wood chips are filled with cooking liquor and take up cooking chemicals have been studied by several investigators.

2. Mass flow and diffusion mechanisms

Any impregnation of wood chips with liquid takes place by two processes:
1) mass flow through capillaries of greatly variable diameters, and
2) diffusion from these capillary ducts into the cell wall structure.

The force of capillary rise from the cut ends of chips produces the first step of impregnation, and continuation of capillary flow into the interior of the wood chip is the most important requirement. Diffusion from the lumen into the fiber wall is a relatively rapid process consider-ing the minute distances involved.

When a liquid flows through a capillary, a stationary boundary layer forms adjacent to the wall. In the immediate vicinity of this boundary layer, laminar flow is observed followed by turbulent flow towards the center. The move-ment of material from the capillary wall (wood) into the boundary layer and from there into the layer of laminar flow (cooking solution) follows first the path of diffusion and thereafter the path of convection [4].

In experiments with water, 14% of the uptake by wood was accounted for by diffusion, and 86% by capillary rise. All measures which facilitate the capillary penetration by liquids are therefore favorable for the cooking process and its results.

Liquid movement through membranes takes place through sub-microscopic openings or cracks. A capillary sieve effect can occur allowing smaller molecules to pass while holding back larger molecules. The state of charge of the membranes is determined by the kind of ionized chemicals. The greater the permeability of the membrane, the smaller the electromotive effect. In the sub-microscopic range, a variable struc-ture exists besides the permanent structure which becomes apparent only in the swollen state. While liquid movement can occur through this variable sub-microscopic structure subse-quent to, or simultaneous with the movement through the permanent capillary structure, gases can penetrate wood only through the permanent capillary structure.

Initially, acid sulfite liquors penetrate into the longitudinal wood structure 50-100 times faster than in the transverse direction. This movement comes quickly to a standstill, however, if the wood capillaries are only partly filled with liquid (at >25% wood moisture, the fiber saturation point) and contain liquid-air interfaces (Fig. 30). Hydrostatic pressurization of the system reduces the size of the air bubbles somewhat and brings more wood substance in contact with the liquor, but the effect is small since extremely high system pressures would be required to overcome the resistance of the surface tension forces of multiple liquid-air interfaces in the minute capillaries.

In the special case of acid bisulfite cooks with excess free SO_2 (pH <4), the gas can advance

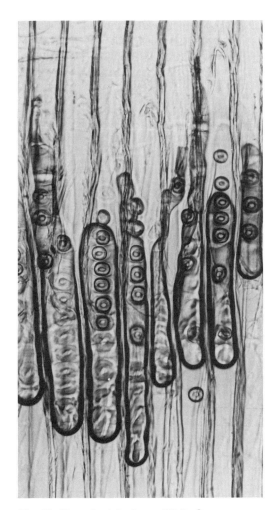

Fig. 30. Air-water interfaces. White Spruce, never dried. X480. (CIPRL file).

faster and further into the wood structure than the liquor and, in the absence of base, causes acidic conditions which lead to lignin condensation with rising temperature. The condensation reaction blocks lignin sites preventing its sulfonation and dissolution and produces "dark centers" in the cooked chips.

Several studies of the rate of diffusion of acid bisulfite and neutral sulfite pulping chemicals into blocks of wood saturated with water have been reported. It was found in all cases that the rate of diffusion was faster in the longitudinal than in the radial or tangential direction, but the rate difference was much smaller than that of liquor flow. Rate ratios of 3-10:1 have been quoted. Pit membranes account for about 90% of the resistance to transverse diffusion. Chip length is therefore important in the case of acid-to-weakly alkaline sulfite cooks but should not be decreased below 15 mm in order to avoid undue fiber damage.

Liquid movement within the cell wall occurs by way of diffusion and is therefore independent of the hydrostatic pressure applied. In the case of a soluble gas such as free SO_2 or ammonia, on the other hand, hydrostatic pressure will affect the chemical concentration of these gases in the penetrating liquor.

A major transition in the radial and tangential rate of diffusion occurs in alkaline solutions above 12.5 pH, as shown in Fig. 31. While acid, neutral and moderately alkaline solutions do not change the effective cross-sectional area of the capillary system, alkali greatly swells the wood fibers and produces a major transition in the permeable structure of the cell wall and pit membranes. Hence in strongly alkaline sulfite liquor diffusion will occur with almost equal rate in the three directions. In such cases chip

thickness becomes more important than chip length regarding uniformity of the cook.

Another significant difference between the effects of contacting wood with acid or alkaline solution is the rapid sorption of alkali by the wood components, both carbohydrate and lignin. At alkaline liquor concentrations used in pulping, wood absorbs about 5% of its weight in NaOH. Consequently, as the liquor penetrates into a chip the alkali concentration decreases and a gradient develops between the outside and the center of the chip where insufficient alkali may be available for a complete cook. This sorption effect is a major reason for using thin chips in alkaline pulping.

To increase the alkali concentration and liquor-to-wood ratio for minimizing the concentration gradient has not been found practical in commercial batch operations. In continuous digesters more fresh cooking liquor can be brought in contact with the chips by increasing the circulation rate in the impregnation zone. To increase the temperature of impregnation raises the rate of diffusion but also the rate of the reaction of the alkali with the wood. Too high a temperature may cause an even steeper concentration gradient.

Estimates of diffusion constants at 20°C of some sulfite liquor components are given below:

Component	Diffusion Constant cm^2/day
Sulfur Dioxide	2.4
Calcium bisulfite	0.88
Magnesium bisulfite	0.83
Ammonium bisulfite	1.66

3. Practical methods and experimental techniques

From the results of investigations done in the fifties mainly in Canada and Sweden, practical methods were developed for preparing the wood chips and to assure proper impregnation at the start of the cook. Suitable methods proposed for removing air from the voids of the wood were *steaming, evacuation,* and *pressure impregnation.* At the same time the importance of *chip size* was stressed and also the *duration and temperature* of the impregnation stage in the case of cooking chemicals which react with wood components [42].

The several methods and combinations developed and in some cases brought to commercial trials or short time use including presteaming, evacuation, pressure impregnation, "Va-purge"

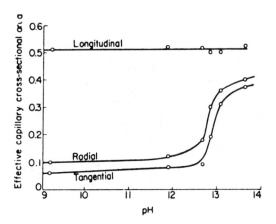

Fig. 31. Effect of pH on the effective capillary cross-sectional area of aspen [46].

(steam purges) and "Vilamo" (hydraulic pressure pulses) were finally put to a critical test to assess the effectiveness of impregnation [43]. The two criteria used were direct weight increase of individual chips, and the level of screening rejects after a standard cook. It was concluded that pre-evacuation was ineffective with wet wood and that the more complicated processes had no advantage over the most commonly used impregnation technique used in sulfite mills: atmosphere steaming following by pressure impregnation (Fig. 32).

It must be added that presteaming followed by impregnation with cooking liquor is now practiced widely for the purpose of optimum effectiveness and uniformity of cooking, and includes all pulping processes from acid to alkaline and all yield levels. Proper impregnation is critical in shortened acid bisulfite cooks; in liquid phase cooks with rapid temperature rise including continuous cooking; and in vapor (steam) phase cooks where the total amount of chemical required by the cook must be charged into the wood by impregnation prior to the cook. In alkaline cooking, be it conventional kraft or alkaline sulfite, the resistance to radial and tangential diffusion of cooking liquor into the wood is much smaller than in acid cooks and therefore the degree of air purging is of lesser importance.

The conditions for optimum chip impregnation derived from these investigations are:
(1) Steaming at atmospheric or slightly elevated pressure (100-105°C) until the temperature of the chip charge causes evaporation of the water in the chip in order to displace the air by steam. Steaming above 120°C leads to lignin condensation and cellulose degradation by hydrolysis.
(2) Submerging the presteamed chips in cooking liquor below 100°C in order to condense the water vapor in the chips and to fill the evacuated volume with liquor (optimum rate at 80-90°C, but in acid bisulfite cooks SO_2 tends to penetrate chips ahead of base).
(3) (Optional for optimum results) Pressurization of the completely filled vessel to 350-700 kPa either by a cooking liquor pump developing sufficient pressure or by starting the heating cycle of the cook for expanding the liquor volume. In either case excess pressure build up is controlled by automatic top relief valves.

In batch cooking the digester is used as impregnation vessel whereas in continuous cooking two-vessel systems are used.

B. TEMPERATURE RISE STAGE

Far from being standardized or unimportant, the period of raising the temperature of the digester from completed impregnation to cooking at maximum temperature is an important variable affecting chemical and heat economy, delignification, uniformity of product, and pulp yield. Both its length and slope vary greatly between acid bisulfite cooking (pH 1.2-1.5) and bisulfite (pH 3-5) or sulfite cooking at higher pH levels.

Slow heating to relatively low cooking temperatures of 130-145°C is necessary in the acid bisulfite processes in order to assure complete filling of the wood with liquor carrying base ion.

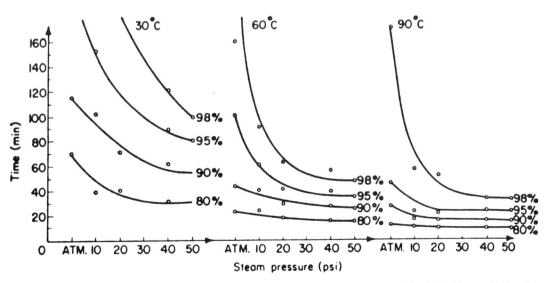

Fig. 32. Effect of steaming pressure (steaming for 10 min) on the time needed for 80-98% penetration by water of 30-90°C and 2 kp/cm² hydraulic pressure of Spruce [46].

Volatile free SO_2 gas from the surface of the liquor advances faster into the wood and, in the absence of base, can create strongly acid conditions locally, which cause at elevated temperatures, high residual lignin content due to condensation, darkening, and weakening of the resulting pulp fibers through hydrolytic degradation of cellulose.

The higher pH level used in bisulfite cooks, and the greater stability of the bisulfite ion stoichiometrically balanced with base ion, reduce the effect of cellulose degradation in cooking. Provided that impregnation with cooking liquor is adequate, bisulfite cooks can be raised more rapidly to higher cooking temperature than permitted in acid bisulfite cooks. Using temperatures of 160-165°C is not by choice but by necessity since the rate of delignification in the acid range of sulfite cooking decreases steeply with increasing pH.

Experimental evidence from diffusion studies suggests that bisulfite ($Na^+ + HSO_3^-$) enters a wet chip almost exclusively through the cut ends of the chip. Therefore chips should be as short as consistent with fiber quality. Steaming alone, however, may already double the permeability of wood in the tangential and radial directions. In hot liquor the wood structure is then opened up more. After light cooking, as during a low rise to 110°C, rates of diffusion across the wood grain (longitudinal fiber structure) may be increased by a factor of 5. Accordingly, and because a chip of conventional shape exposes more side than end grain, a considerable portion of the bisulfite enters through the sides during temperature rise.

Another factor worth mentioning is that the rate of diffusion of a solute in aqueous solution does not increase as steeply with increasing temperature as do rates of most chemical reactions. The diffusion rate doubles by a rise of about 30°C whereas the delignification rate in acid bisulfite pulping doubles every 10°C by rule of thumb, and almost triples in the strongly alkaline soda cook.

In so-called "neutral" sulfite cooks (pH 7-9) and in alkaline sulfite cooks practical limitations to the rate of temperature rise do not exist. Long chips are of little concern because, as shown, liquor penetration rates in the three dimensions become almost equal with rising pH, particularly when sodium hydroxide with its strong swelling is present.

In fact, liquid phase cooks of any of the known chemical pulping processes with the exception of acid bisulfite pulping may be raised to maximum temperature at any practicable rate after impregnation. The limiting factor in all these cooks will be the generating capacity of the steam plant.

The major application of bisulfite pulping and neutral sulfite pulping is in the field of high yield and ultra high yield pulps, from softwood for newspaper furnish; and from hardwoods for corrugated medium board of multiwall containers. A high degree of sulfonation of the lignin in the chip in short cooks is necessary for achieving both required pulp quality and economy of operation. In many cases, part or the whole cooking stage is performed in the vapor (steam) phase after withdrawing the impregnating liquor. When applicable, vapor phase cooking saves heat, time and chemicals. For such processes perfect impregnation of the wood material with cooking chemicals before and during temperature rise is therefore essential.

1. Side relief

An important feature of acid bisulfite, bisulfite and "neutral" sulfite pulping in batch digesters is the practice of withdrawing a large portion of the cooking liquor charged initially at some predetermined time during temperature rise when impregnation is known to be accomplished. This liquor is returned to storage and re-used in the preparation of fresh cooking liquor utilizing residual chemical. Significant savings in heat energy and chemicals result from this practice and no undesirable effects are caused in subsequent cooks. In mills without chemical recovery systems the effluent load is greatly reduced and, in mills with recovery of chemicals, a more concentrated spent liquor enters the evaporation plant.

The practice of "hesitation" during temperature rise of acid bisulfite and bisulfite batch cooks in order to allow more time for impregnation has been discontinued in most mills with the use of more efficient presteaming and impregnation techniques which permit shorter digester cycles.

C. OVERVIEW OF PULPING REACTIONS OF WOOD

Chemical and material changes of the wood substance begin as soon as contact with cooking liquor is made during impregnation. The rate of these changes is very small at ambient temperatures; it increases during temperature rise but becomes of practical significance for the outcome of the cook only above the boiling point.

Studies of this subject reported in the existing literature have been aimed almost exclusively at the two major pulping processes, sulfite and kraft, where sulfite stands for acid bisulfite. With the extension into the alkaline range with the help of sodium hydroxide, sulfite pulping now covers the whole range of the pH scale. Knowledge of the interaction of wood with acid bisulfite liquor is incomplete. A discussion of kraft reactions to explain the alkaline sulfite field is not correct, because kraft pulping uses sodium sulfide which produces chemical and rate effects which are quite different from those of sulfite or sodium hydroxyde (soda) cooking liquors. However, until detailed information on the reactions of alkaline sulfite liquors become available, some effects may be viewed as being similar to those with soda liquors or even kraft liquors, but always keeping in mind that there is no sulfonation taking place in soda cooks, and that the lignin reaction mechanism in kraft pulping is entirely different. A schematic diagram of the dimensions and composition of the softwood cell walls and middle lamella is given in Fig. 33. Changes due to pulping have been studied under the microscope using polarized or UV light, staining methods, and microtome sections.

The middle lamella consists of 70-82% of lignin and is estimated to be 15-25% more porous than the cell wall. A detailed discussion of lignin is found in the following section on lignin reactions and in Volume 1. Lignin incrustation is also found throughout the cell wall, more concentrated near the middle lamella and decreasing towards the lumen. About ½-⅔ of softwood lignin is located in the middle lamella and the adjacent part of the cell wall. The middle lamella contains also a few percent hemicellulose which appears to form a network in the amorphous mass of lignin.

Of the carbohydrates, cellulose is concentrated in the inner half of the secondary wall (S2) and hemicellulose in the outer half.

The first purpose in chemical pulping (45-50% yield range) is to remove the lignin material of the middle lamella, and lignin from within the fiber wall. In high yield chemimechanical pulping, the goal is to preserve as much of the lignin in the middle lamella and the fiber wall as is consistent with adequate sulfonation of the lignin. In the special case of dissolving pulps, the hemicelluloses are also removed and the remaining cellulose softened with the least amount of damage to the fiber.

A surprising finding of earlier studies of acid bisulfite pulping of wood was that the lignin content of the middle lamella decreased more rapidly than that of the fiber wall. It seemed that the structural changes started and proceeded from the middle lamella, although the cooking liquor enters the structure through the fiber lumen. This was resolved by UV microscopy of sections from wood cooked to different degrees of delignification [22]. In the early stages of acid bisulfite or alkaline cooks, the lignin is removed mainly from the cell wall, but later in the cook the middle lamella dissolves rapidly while residual lignin is left in the cell wall.

Neutral sulfite cooks are different in that the relative rate of lignin removal from the secondary wall and middle lamella is about equal from the start.

The innermost tertiary wall dissolves almost completely in acid bisulfite cooking. Its remnants, however, can cause impurity problems in

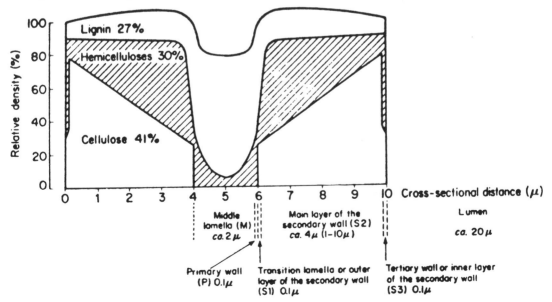

Fig. 33. Thickness and composition of the cell walls and the middle lamella of softwoods [46].

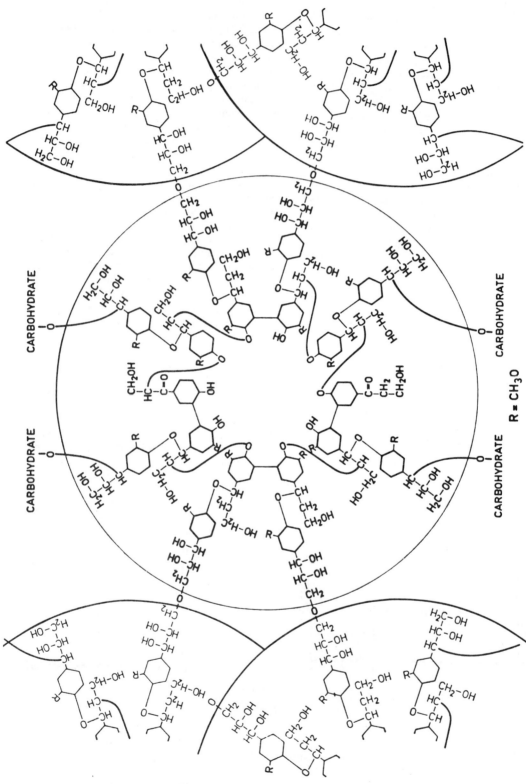

Fig. 34. Formulation for the repeating unit of spruce lignin (by Forss, Fremor and Stenlund in 1966). Note that actual structure is three-dimensional [44].

dissolving pulp manufacture. Well-preserved tertiary walls are found in alkaline pulps, since xylan is more resistant to alkali than to acid. Alkaline sulfite dissolving pulps require acid prehydrolysis to low pH prior to the cook in order to meet purity specifications by removing pentosan-type hemicelluloses.

When most of the lignin is removed in the cook, acid bisulfite fibers lose their rigidity and collapse to flat ribbons, whereas alkaline fibers are round with the cell wall swollen towards the center with only a very narrow lumen space remaining.

In both acid and alkaline processes more carbohydrate material is dissolved from the outer parts of the fiber wall than from parts near the lumen. Differences between the processes exist, however, as to the degree of polymerization and chemical composition of the carbohydrates. In the acid bisulfite cook, the main carbohydrate reaction is hydrolysis of the glucosidic chains, particularly of the hemicellulose, causing depolymerization and dissolution. Consequently, the chain length (DP) distribution of cellulosic material is uneven and the low-molecular material is concentrated in the outer parts of the cell wall. This partly explains the much greater tendency to swelling and hydration of acid bisulfite pulps compared with alkaline pulps where the chain length distribution is more even throughout the fiber wall. In neutral sulfite pulping at pH 7-9, hydrolysis of carbohydrate material is much reduced. The resulting pulps are high in hemicellulose content, promoting bonding and strength as required for stiff corrugated board. The nature and effect of the carbohydrate reactions in neutral sulfite pulping has not been investigated.

Alkaline peeling of cellulosic material is known from the traditional alkaline pulping processes and is bound to take place also in sulfite pulping with strongly alkaline liquor. The peeling reaction removes successively end units of cellulosic chains reducing both chain length and yield and is unavoidable in regular cooks since it requires a lower temperature than alkaline delignification. Thus, because in the acid bisulfite cook delignification is more rapid than in the alkaline cook, lower temperatures can be used which, in turn, reduce hydrolytic attack on carbohydrates. The resulting yields from softwood are 3-5% higher than from alkaline cooks.

D. LIGNIN REACTIONS

When wood substance is impregnated with a solution of sulfite cooking chemicals, a complex reaction mechanism is set in motion which causes chemical and physical changes leading eventually to the liberation of the cellulosic fibers from the lignin structure of the tree.

In the field of sulfite pulping, the modification of lignin by sulfonation is the most important reaction. It has two major aspects: to soften lignin and to render it more hydrophilic, as required by high yield chemimechanical pulps or to remove most of the lignin from the fibers, as required for high strength and bleachable pulps at lower yields from wood.

An important consideration at this point is that high yield pulps, where lignin and carbohydrate components of the wood are intentionally preserved and modified, and low yield pulps where lignin in particular and some carbohydrate fractions are removed for the purpose of pulp purification, yield fibers of the same number but of different weight from the same weight of dry wood. Since paper is made on a weight basis, this means that papers from high yield pulps will have fewer fibers.

Wood is characterized by the presence of lignin in the fiber structure. Its formation, lignification, occurs together with the main growth of wood fiber and may be a growth process in itself producing one large three-dimensional network molecule extending throughout the wood structure.

Information on the chemical structure of the lignin molecule has been accumulated for many years. Recently, supplemented by data from advanced analytical techniques, model compound studies and computer evaluations, the existing information was used [17] to construct the probable model of softwood shown in Fig. 34. The main building blocks can be seen in Fig. 35 to be phenyl propane units carrying one methoxyl group, $-OCH_3$; they are called guaiacyl units. Guaiacyl units build also one half of hardwood lignin, while the other half consists of syringyl units carrying two methoxyl groups,

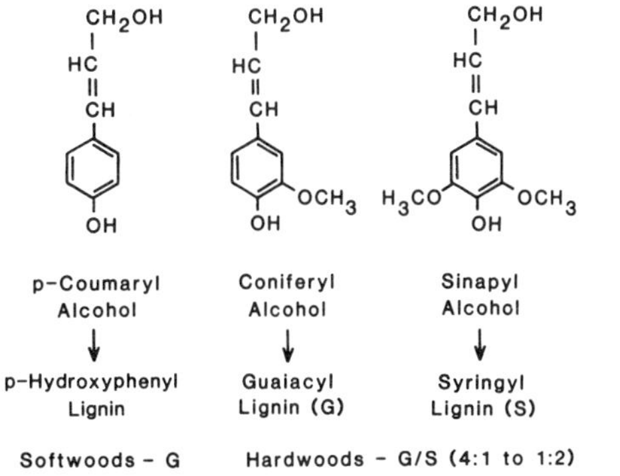

Softwoods – G   Hardwoods – G/S (4:1 to 1:2)

Fig. 35. Chemical precursors probably involved in lignin biosynthesis. [45].

as shown. These units are linked together in many different ways, predominantly by oxygen (ether) bridges connecting the alpha or beta carbons of the side chain of one unit with the phenyl ring of the other.

A description of the reactions of lignin in the various sulfite processes from acid to alkaline has been published recently [18]. It has been found that, in spruce and birch, one half of the structures combining two phenyl propane units are of the beta-phenolic ether type shown in Fig. 36, and must therefore play a major role in the process of sulfonation and fragmentation of the lignin macro molecule. Using this example, a simplified model of the reactions in a neutral sulfite cook is given in Fig. 36.

At any pH level used in sulfite pulping the alpha-carbon position in the phenyl propane unit is the most active site for reaction with the negatively charged ions used in conventional pulping processes, such as hydroxide, bisulfide, bisulfite and sulfite, as well as aqueous sulfur dioxide. In native lignin, only a few sites are ready for reaction with these pulping reagents but reactive alpha-sites are caused by transformation of the phenolic structure into quinone

methide structures. The beta-carbon position is less active and frequently engaged in beta-phenolic ether linkages, but can be sulfonated in neutral and weakly alkaline media after cleavage of the oxygen link (sulfitolysis), induced and accelerated by sulfonation in the adjacent alpha-carbon position. Thus, in this case, two kinds of phenyl propane sulfonic acids result, one sulfonated in both the alpha and beta carbon positions, the other only in the alpha position.

Under acid bisulfite conditions the alpha-carbon site assumes the nature of a positively charged ion and immediately reacts with the negative bisulfite ion in solution, resulting in extensive sulfonation. Sulfonic acid groups are also introduced into the side chains of oxidized lignin structures while degradation is restricted to the cleavage of oxygen (ether) linkages. On the other hand, condensation reactions between phenyl propane structures also occur in the acid environment.

Condensation results in the formation of carbon-carbon linkages between lignin structures. These linkages are resistant to acid hydrolysis, inhibiting the progress of lignin depolymerization to smaller, soluble com-

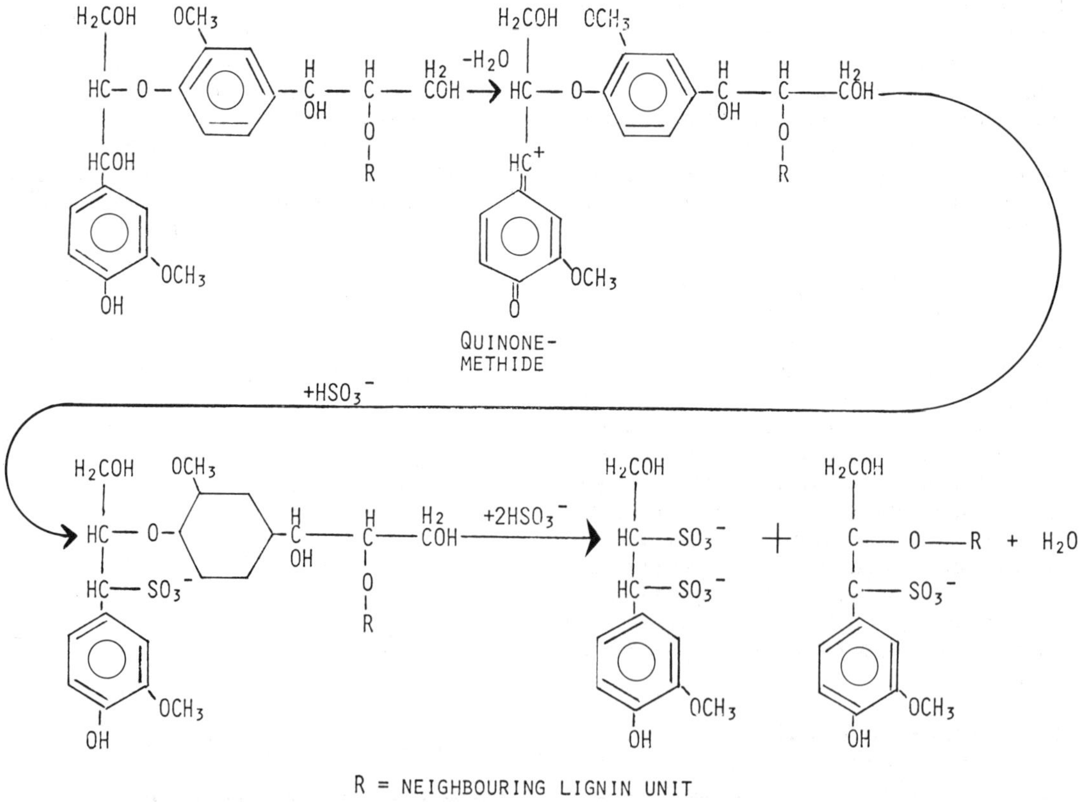

R = NEIGHBOURING LIGNIN UNIT

Fig. 36. Model of lignin sulfonation and degradation in "neutral" sulfite pulping.

pounds. It is also for reasons of formation of condensation products with lignin that resinous woods, such as pines, with a high content of non-lignin phenolic substances resist pulping by the acid bisulfite process.

In bisulfite pulping, lignin sulfonation and dissolution has not been studied. The process uses sodium or magnesium base cooking liquors and serves predominantly for manufacturing high yield chemimechanical pulps from softwood for use as the long-fiber component of publication papers. The emphasis is therefore on sufficient sulfonation, as measured by sulfur content of the pulp, for the purpose of making the lignin binder soft and hydrophilic rather than to dissolve the lignin, and thereby to facilitate the liberation of the fibers and to increase their bonding strength. It is known that at pH 4 the phenyl propane-beta-phenolic ether structure is easily sulfonated at the alpha-carbon. Dissolution of the lignin is, however, much slower than in acid bisulfite pulping, but more rapid than in "neutral" sulfite pulping.

Interest in the low-odor alkaline sulfite pulping process in the pH range 10-13.5 has been increasing considerably in recent years, since it is capable of producing very strong chemical pulps which are more easily bleached than kraft pulps. Results of definitive studies of the chemical mechanisms involved have not come forth at this time. From the point of view of our knowledge of the chemistry of the sulfite process and of the alkaline processes, kraft and soda, the new sulfite process appears to be attractive. (A concise summary of the delignification mechanics in common alkaline systems is found in Casey, 3d Edition, Vol. 1, p. 69. [47].) Combining the potential of two strong pulping reagents, sulfite ions and hydroxyl ions, lignin bonds should be effectively attacked and secondary condensation reactions of lignin structures should be reduced by sulfonation and hydroxylation of intermediate structures isolated during the cook. Thus alkaline sulfite pulping may offer a combination of the phenol-dissolving and depolymerizing action of soda (NaOH) pulping with the creation of hydrophilic properties of the fiber typical of sulfite pulping.

In its natural state, lignin is insoluble in water and hydrophobic (non-wettable). When reacted with sulfite chemicals it is converted to lignosulfonate with radically different properties. With the introduction of sulfonic acid groups into the phenyl propane structure the formerly neutral and ionically inactive compound becomes the second strongest organic acid known with a dissociation constant of $3 \cdot 10^{-1}$, 15 times stronger than a solution of SO_2 in water of

K $2 \cdot 10^{-2}$ at room temperature. Sulfonation takes place over the whole pH range, but decreases rapidly towards the strongly alkaline range where the activity of the bisulfite ion has stopped and the lesser activity of the sulfite ion is overtaken by the rapid rise of the activity of the hydroxyl ion as delignification agent.

The strong and stable lignosulfonic acid anion $L\text{-}SO_3^-$ is thus present at many sites throughout the lignin macromolecule due to sulfite pulping treatment and transforms the native lignin into a large multi-branched macro-anion, a schematic model of which is shown in Fig. 37 [19]. (Note that, in hardwood bisulfite cooks, up to 30% of the sulfur content of the lignin is not in the form of lignosulfonic acid.)

Viewed from another angle, lignin becomes a strong anionic ion exchange complex with greatly increased hydrophilic (wettable, swellable) properties. This is probably the reason that unbleached sulfite pulps develop good strength with relatively little mechanical energy input.

In high yield pulping, more or less of this sulfonated complex remains in the chip with the fibers until fiberization of the structure by mechanical means. No matter whether in the chip or fragmented with the fibers, it carries an equivalent amount of positive base ion (Na^+, Mg^{2+}, Ca^{2+}, NH_4^+) to satisfy the electrostatic principle of electroneutrality. When the water is removed from such a structure the positive base ions join the negative macro ion and form lignosulfonate salt.

When cooking pulps of greater purity the same applies as above, except that the original micromolecule is fragmented by chemical action and a large portion of the fragments is reduced to

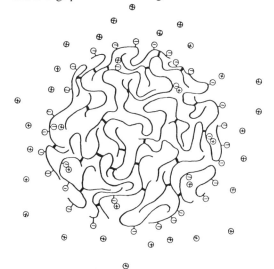

Fig. 37. Idealized spherical model of the spruce lignin sulfonate macromolecule [19].

molecular sizes which are soluble in water. Electron microscopy of sodium lignosulfonates revealed flattened globules and also larger spherical aggregates formed of the globules [20]. This property has implications in spent liquor recovery and in the preparation and uses of lignin by-products. Lignin solubility covers a wide range of polymolecularity and includes both solubility and apparent solubility as colloids (particles visible under the micro-scope).

In a contiuous flow digester experiment, for example [21] different time fractions of sulfite lignin were collected from the same wood. A variety of tests applicable to lignosulfonate, such as methoxyl content, UV absorption, sulfur content, refractive index change and neutraliza-tion equivalents of sulfonic acid groups showed no significant differences. Still, the average molecular weight increased from 10 000 (about 50 lignin units) to 140 000 (700 units) in acid bisulfite liquor and from 10 000 to 77 000 (about 400 units) in bisulfite pulping liquor [21].

Although the pattern may not be as clear in actual cooks, this shows firstly that fairly large fragments of the sulfonated lignin macro mole-cule can go into solution; and secondly, that in the process of sulfonation and degradation the larger fragments are released only after smaller fragments from readily accessible locations were removed from the structure.

It has been shown that in acid bisulfite pulping the sulfur content, as a measure of the number of sulfonic acid groups, must rise to at least 3.5% for dissolution to occur. This indicates that about one sulfonic acid group for every four lignin building units is required to produce sufficient fragmentation for solution to occur.

Previous concepts that either sulfonation or acid hydrolysis should determine the rate of lignin dissolution are not convincing any more since the mechanisms involved were derived exclusively from acid bisulfite pulping studies, and change radically over the available range of pH; and also, because cleavage of chains between lignin units can occur due to the introduction of a sulfonic acid group (sulfitoly-sis) rather than hydrolysis (action of hydrogen ions).

Complete delignification in acid or alkaline cooks would result in very low yield and weak fibers. Therefore cooking is always completed with a few percent lignin remaining in the fibers.

It is necessary at this point to insert a note on the use of anthraquinone in alkaline sulfite pulping. The use of this agent produces an extraordinary increase of the rate of delignifica-tion combined with a major reduction of carbohydrate degradation, in other words a pulp of unusually high yield for a given lignin content, or vice versa. Moreover, an anomalous narrow area of sulfite concentration has been discovered recently, at 80-85% Na_2SO_3 as Na_2O and 15-20% Na_2CO_3 and/or NaOH, both as Na_2O, where pulp yield of softwood cooks is at a maximum and both lignin content and carbohydrate degradation are at their minima. This stands in stark contrast to the relationship found in conventional pulping processes and the results of studies to expose the underlying reaction mech-anisms will be followed with great interest (refer to Section G of the Chapter).

E. CARBOHYDRATE REACTIONS

About 80% of the wood fiber produced in the world serve the market for pure chemical cellulose, for bleached writing or printing papers, and for high strength unbleached pack-aging papers. These products are valued for their content of pure cellulose in man-made fiber manufacture or for their content of cellulose and hemicellulose in fine or strong paper manufac-ture. The remaining 20% serve the market of publication papers requiring only limited quality and durability. Here the valued properties are high yield from wood including both carbo-hydrates and lignin, and sufficiently high unbleached brightness.

Cellulose is a polymer consisting of long chains of ring-shaped glucose units (containing 6 carbon atoms, hexose) linked together by oxygen atoms (glucosidic bonds), as in Fig. 38A. The average number of glucose units in a chain of native spruce cellulose is estimated to be 2400, or as high as 15 000.

Hemicelluloses are also polymers made of chains of ring-shaped sugar units, but the chain length is far shorter than in cellulose (e.g. 200 units) and the units consist of different sugars containing 5 (pentose) or 6 (hexose) carbon atoms (Fig. 38B). Hemicelluloses are highly important components of the fiber, imparting desirable properties to the paper or, in the case of chemical cellulose, determine the degree of pulp purification achieved.

The major hemicellulose component of soft-wood species is galacto glucomannan shown schematically in Fig. 38B. All sugars contain 6 carbon atoms, and acetyl groups found in softwoods are attached to these chains. Also present are xylan polymers formed of xylose units containing 5 carbon atoms (a pentose) and glucuronic acid and arabinose units attached to the xylan chains, as shown.

Hardwoods contain predominately xylan-type hemicellulose but without arabinose and with the xylose units carrying glucuronic acid units and acetyl groups (also shown in Fig. 38B).

When the carbohydrates of wood are exposed to the action of hot cooking liquor hydrolytic reactions take place, splitting main chains or reducing their length, and eliminating side units. The mode and severity of the actual reactions, however, and their results on the carbohydrate composition and quality of the pulp vary greatly along the pH scale. At the neutral pH level where both the concentration and activity of the hydrogen ions (H_3O^+) and hydroxyl ions (OH^-)

are minute (10^{-7} mol/L) hydrolytic action is at a minimum. The hydrolytic activity increases with the concentration of either the hydrogen ions towards more acid levels, or of the hydroxyl ions towards more alkaline levels, that is by a factor of 10 for every logarithmic pH unit.

In acid bisulfite pulping, the main carbohydrate reaction is hydrolysis of the bonds between glycosidic units. The long polymeric chains of sugar units forming cellulosic or hemicellulose material are shortened under the influence of low pH (high hydrogen ion activity) and temperature. Especially in the case of hemicelluloses which are less highly polymerized in the native

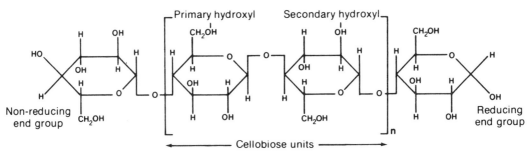

Fig. 38A. Cellulose structure.

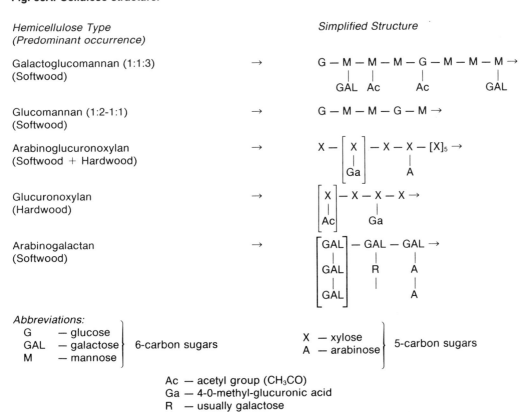

Fig. 38B. The types and simplified structures of the major hemicelluloses in wood [48].

state than cellulose, this leads to the dissolution of large portions of the material in the form of sugars. The long cellulose chains in spruce wood containing on the average 2400 glucose units (DP) are also cut by hydrolytic attack but not to the extent (for example DP 1400) that causes appreciable dissolution, except in extended cooks aimed at producing highly purified pulp for use as chemical cellulose.

In acid bisulfite cooks, galacto glucomannan is more resistant against hydrolytic action than xylan and its resistance can be further increased by a neutral or alkaline treatment stage prior to acid digestion (two- or three-stage sulfite pulping process) and the pulp yield thus increased. Acetyl side groups are split off the galacto glucomannan chains in these pretreatments at neutral or alkaline pH and the degree of stabilization of the hemicellulose can be monitored by the amount of acetyl removed. With or without stabilization, a substantial portion of the galacto glucomannans is dissolved in both acid bisulfite and bisulfite cooks.

Xylans of both softwood and hardwood are subject to major hydrolytic degradation in acid bisulfite cooks and consequently xylose is the predominant sugar found in the spent liquors. The softwood xylose chains lose the arabinose units early in the cook and xylose lineages are hydrolyzed also, reducing the chain length (DP). Compared with the arabinose side units, both glucuronic acid units and acetyl groups are more resistant against elimination from the hemicellulose chain.

With hydrolytic activity at a minimum at neutral pH, the adjacent regions on the pH scale are found to be most useful for the preparation of pulps with high cellulose and hemicellulose content. Retention of xylan, in particular, is important for these neutral sulfite pulps in order to impart high yield, stiffness and rigidity to the papers and boards made from them. Hardwoods contain about twice as much xylan and about one quarter less lignin than softwood and are therefore used almost exclusively for this process.

At the time of writing, specific information on carbohydrate reactions in highly alkaline sulfite cooks has not been reported. Information available from kraft and soda pulping with high charges of sodium hydroxide may be pertinent to some extent. The results from kraft studies suggest that the effects are due to alkalinity only, and are influenced insignificantly by sulfide content. A brief summary of general alkaline effects is given here pending detailed information from alkaline sulfite studies.

The reactions in alkaline pulping are entirely different from those in the acid sulfite range where hydrolytic splitting of polymeric sugar chains and of attached side groups by hdyrogen ion activity is the exclusive type of reaction. In alkaline medium the so-called "peeling" effect predominates. By "peeling" is meant the successive separation of single sugar units from the reducing end of cellulose and hemicellulose chains in hot alkaline environment. Once started, the "peeling" process will remove, on the average, 50-60 sugar units from the chain until it is terminated by the "stopping" reaction forming an alkali-stable carboxylic acid end group. The sugar units "peeled" off the carbohydrate chains do not enter the alkaline liquor as sugars but as isosaccharinic acid. The second important reaction is alkaline hydrolysis through the activity of hydroxyl ions at temperatures of 160-180°C. This shortens the chains of the cellulose and of the remaining hemicellulose in the pulp fibers and can have some effect on the strength properties of the paper. Alkaline hydrolytic degradation is random but much more severe in extended soda (NaOH) cooks than acid hydrolytic degradation in sulfite cooks which tends to attack locations in the structure of the wood chip previously damaged by mechanical action.

Compared with acid bisulfite pulping, the stability of the two major hemicellulose components in softwood, xylan and glucomannan is reversed in alkaline pulping. In a kraft cook, for example, about three quarters of the original glucomannan is dissolved by the alkali "peeling" reaction whereas only about half of the xylan is dissolved directly in its native polymeric state ($\cong 200$ units). Later in the cook, when the temperature rises above 150°C, the dissolved xylan becomes degraded to an average chain length of about 40 units. Hardwoods contain on the average 2½ times more xylan than softwoods, and a correspondingly large portion remains in the fibers after an alkaline cook.

This natural resistance of xylans to dissolution in alkaline cooks increases the yield of paper pulps but makes it more difficult and costly to produce pure dissolving pulps when compared with the acid bisulfite process (refer to Chapter VIII, D, 2).

Acetyl groups in both softwood and hardwood are removed readily from the hemicellulose chains by reaction with alkali. Hardwoods contain more acetyl than softwoods and consumes therefore more alkali for their neutralization.

Table 3 and Fig. 39 summarize the carbohydrate compositions of softwood pulps. Hardwood pulps would be expected to contain more xylan.

F. SULFITE PULP QUALITY

Sulfite pulping comprises a family of processes which together are capable of producing pulp of any description regarding composition and properties for papermaking and chemical end uses. The parameters of sulfite pulping which link these processes together and permit this product flexibility from a given ligno-cellulosic material are the pH of the cooking liquor and the pulp yield, with a modifying effect added by the choice of type and concentration of the four bases. The choice and combination of ligno-cellulosic raw material, whether mainly long-fiber softwood, short-fiber hardwood, annuals, or mixtures thereof, further adds to the flexibility of product.

Before entering a discussion of details it will be useful to consider the results of the survey of the field using spruce wood and sodium base [13,23]. There are essentially four criteria for the quality and suitability of a pulp for commercial use: strength, optical properties, bleachability, and purity. Strength relates to papermaking, printing, and packaging materials; optical properties to brightness of unbleached pulps at different yields, and opacity to papers for printing; bleachability to the cost of market pulps; and purity to refined cellulose for chemical conversion.

Average variations of major pulp components throughout the sulfite field are shown in Fig. 40 [23]. Pulp viscosity is included as an indicator of

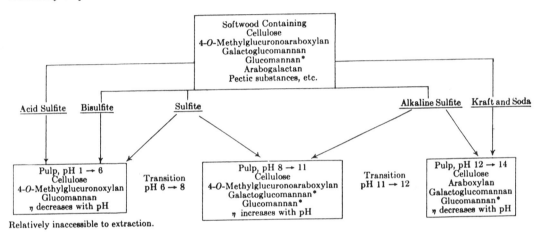

Fig. 39. The carbohydrate composition of single-stage full chemical pulps of spruce [41].

Table 3. A comparison of the carbohydrate contents of the residues obtained from western hemlock [40].

	Acid sulphite			Bisulphite			Conventional kraft		
	% in pulp	% Based on orig. wood	Loss, % of orig. component	% in pulp	% Based on orig. wood	Loss, % of orig. component	% in pulp	% Based on orig. wood	Loss, % of orig. component
Galactose	0[a]	0[a]	100	0[a]	0[a]	100[a]	0.6	0.3	90
Mannose	8.1	3.8	71	9.8	4.5	66	9.3	3.9	70
Arabinose	0	0	100	0	0	100	0.5	0.2	71
Xylose	2.2	1.0	68	3.2	1.5	52	5.7	2.4	23
Rhamnose[b]	0	0	100	0	0	100	0	0	100
4-O-Methyl-glucuronic acid[b]	P	P	L	P	P	L	0	0	100
Galacturonic acid[b]	0	0	100	0	0	100	0	0	100
Cellulose	87	41	5	83	38	12	81	34	21
Total carbohydrate	100	47	28	100	46	29	100	42	35
pH of spent liquors	1-1.5			2.5-4.0			12-13		
Temp., °C	130-150			150-170			160-175		

[a] Trace amounts noted.
[b] Qualitative estimation by paper partition chromatography: P = present, no quantitative estimate of loss known. L = extent of loss not known.

the weight average degree of carbohydrate polymerization. In order to avoid errors of the viscosity test over the wide yield range covered, it was necessary to extract the lignin from the pulp by a chlorite method which causes minimum pulp degradation [33].

The average carbohydrate responses have been developed with proper adjustments with the aid of the computer. A multiple covariance program was used, which corrected temperature, bound SO_2, and yield to the following overall average; 152.2°C, 5.6% bound SO_2, 56.7% yield (solid line). Since no one temperature was feasible over the whole pH range, and since levels

of bound SO_2 and yield were generally higher in cooks above pH 4 than below, two additional averages were developed with a split at the pH 4 level. The "high pH average" was 163.3°C, 6.8% bound SO_2, 58.9% yield, and the "low pH average" was 138.8°C, 4.2% bound SO_2, 54.1% yield. In the case of carbohydrate yield, three lignin, rather than yield, averages were used: 7.8% overall average, 9.5% "high pH" average, 5.7% "low pH" average. In this way a more realistic average response was obtained, particularly at the low pH levels. Generally, the effect of pH level on the various pulp properties is much stronger than the effects of temperature, bound

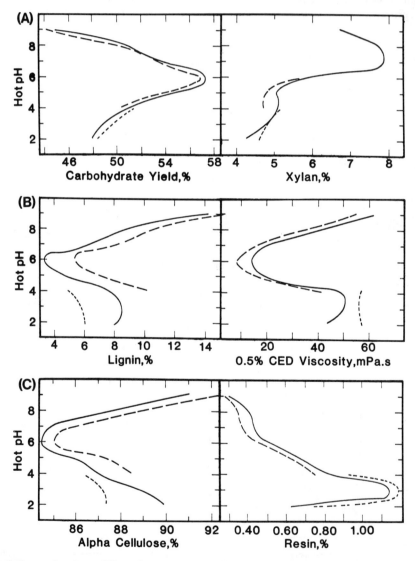

Fig. 40. Variations of major sulfite pulp components throughout the pH range [23].
Solid line 152°C, 5.6% bound SO_2, 56.7% yield
"high pH" 163°C, 6.8% bound SO_2, 58.9% yield
"low pH" 139°C, 4.2% bound SO_2, 54.1% yield.

SO$_2$, and yield, and the general shape of the response curve is preserved in the adjustments. Larger deviations are found only at the very low pH levels, where the actual cooking conditions were far removed from the grand average.

Average responses of optical and mechanical properties have also been developed and are shown in Figs. 41 and 42 [13]. The brightness of sulfite pulps in the acid-to-neutral range and from low to high yields from spruce, balsam fir and suitable hardwoods such as poplar has always been very satisfactory and has been the mainstay of the process during the great expansion of the kraft process after 1945.

This first and full-range evaluation of mechanical sulfite pulp properties brought two significant findings. The pulp strength rises steeply from the minimum of pH 6 (for spruce) as the cooking pH is raised into the alkaline range, and pulps having kraft-like strength can be produced by cooking in highly alkaline sulfite liquor. It

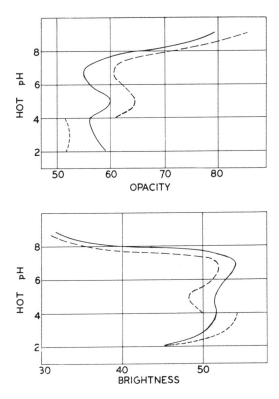

Fig. 41. Average responses of pulp opacity and brightness to cooking pH [13].
solid line 150°C, 4.7% bound SO$_2$, 53.5% yield
"high pH" 163°C, 5.4% bound SO$_2$, 53.8% yield
"low pH" 136°C, 3.9% bound SO$_2$, 53.5% yield.

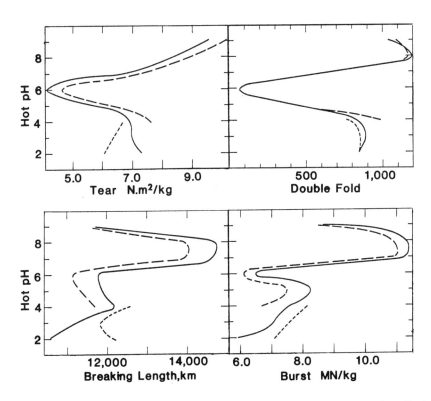

Fig. 42. Average variations of mechanical strength properties of sulfite pulp, with the pH of the cook [13].

had become evident that pulps having kraft strength can be produced by the sulfite process.

Regarding major features of the response curves it is seen that, with spruce, the pH 6 level plays a dominant role with a sharp maximum of carbohydrate yield and equally sharp minimum of lignin, viscosity, alpha cellulose (lignin free) and tear resistance as well as double fold. The carbohydrate maximum is supported by maxima of tests measuring mannan content. This leaves no doubt that both mannan retention and delignification are highly favored, and that a major effect of cellulose degradation prevails at this pH level. Alpha cellulose decreases from 90% in highly acid bisulfite cooks to a minimum of 84% at pH 6 and rises to 92% or higher in strongly alkaline sulfite cooks due to the purifying effect of the sodium hydroxide which is somewhat enhanced over that in kraft cooks.

The xylan content, on the other hand, is high in alkaline pulps and low in acid bisulfite pulps. Provided the wood quality is compatible, it is therefore easier to obtain pure cellulose by acid bisulfite pulping than by alkaline pulping where prehydrolysis with acidified water at high temperature is required to remove the bulk of the xylan (refer to Chap. VIII, D, 2). Prehydrolyzed alkaline sulfite dissolving pulp has been found to have a higher alpha-cellulose and lower xylan content than corresponding kraft pulp.

It has often been suggested in isolated cases that xylan content was connected with paper strength. The shape of the complete xylan curve makes it quite clear that, second to pH, xylan is directly responsible for the development of the two tensile strength properties, burst and breaking length and rather unrelated to tearing resistance. It is also seen that the maxima for xylan content and the tensile properties occur in the approximate pH range covered during a NSSC cook. Double fold measures the number of bends the fiber compound can take before breaking under a constant tension. It shows a very sharp minimum again at pH 6, the level of the weakest spruce pulp in the sulfite system.

Resin in pulp is generally undesirable, either regarding purity or regarding problems with "pitch" deposits in pulp processing lines. In unbleached acid bisulfite pulps, resin retention is seen to be rather high with a pronounced maximum below the bisulfite level and a drop in highly acid liquors. With increasing pH the resin level in pulp tends towards very low levels in strongly alkaline liquors. Thus for producing dissolving pulp, the acid bisulfite process has the advantage of xylan removal whereas the resin needs to be removed in a separate treatment with hot caustic soda and surfactant. On the alkaline side, increasing amounts of resin dissolve in the cooking liquor and are recovered as valuable by-products. Since a vary small amount of resin is beneficial in the rayon making process, controlled additions of resin are made to alkaline dissolving pulps after bleaching.

Unbleached brightness and opacity of sodium base sulfite pulps (Fig. 41) depend first on the pH level of the cook and are modified by cooking temperature. The shapes of the response curves are mirror images, depicting clearly the inverse relationship over the whole pH range. The opacity curve has minima at pH 4 and pH 7 and a maximum at pH 5 with the brightness curve showing the exact opposite. Higher cooking temperature increases opacity and lowers unbleached brightness. Lower temperature, as used in acid bisulfite cooks, has the opposite effect on the two optical properties. With increasing alkalinity (OH ion concentration) of the cook, the brightness drops and the opacity rises steeply. Unbleached alkaline sulfite pulps are almost as dark as kraft pulps, but bleach much more easily than kraft pulps.

Ammonium base acid bisulfite pulps are up to five points lower in brightness than pulps from the metal bases, with a grayish appearance which prevents their use in the unbleached state in publication and printing papers. They bleach, however, as easily as pulps from the other bases, and are manufactured in that form for specialty paper and high quality dissolving grade use.

Regarding the adaptation of the above pulp quality patterns throughout the sulfite field to R&D predictions and actual mill applications, it is necessary to remember that the clear and coherent results were obtained by pH level control throughout the cook. In common wood pulping this would not be immediately practical because of extra expenditures for instrumentation, control apparatus and maintenance. Thus ordinary cooks, whether batch or continuous, start with the pH of the fresh cooking liquor and follow different pH profiles to different pH end points under the influence of wood contact, heating, consumption and removal of cooking chemicals, formation of new compounds, and relief of gases. Thus, the digester pH of a cook traverses a band of true pH levels, and the resulting pulp properties will be the composite result of reactions occurring at different rates at different times from the start of the cook. In strongly alkaline and NSSC cooks the band of decreasing pH may be as wide as three pH units; in acid bisulfite cooks rise and fall may cover two pH units; and in bisulfite cooks less than one unit.

(a) Multiple stage sulfite pulping

Sulfite pulping in two or three (pH) stages has been developed industrially as a means of enhancing desirable pulp properties by controlled changes of the pH level during a cook. The pH bands used in these processes with sodium or magnesium base cover pH 1.5-4 and 4-7 in two-stage operations and pH 1.5-7.5 in three stage sodium base operation. Two-stage schedules which start at pH 4 greatly improve uniformity of lignin sulfonation, particularly in pines which are difficult to pulp in the strongly acid range, and increase pulp yield by glucomannan retention. A second stage at low pH can then follow by adding liquid SO_2 to the digester. This sequence produces pulps with high hemicellulose content which hydrate easily during beating. If, on the other hand, the pH in the second stage is raised to 6 or higher by addition of caustic soda or carbonate, the resulting pulps are low in hemicellulose and high in alpha cellulose content, hydrate slowly and develop high tear resistance or bursting strength (cf. Fig. 42). Such pulps are suitable for both strong papers or dissolving pulp manufacture (see Chap. IX, D, 1a and Chap. X, A, 3).

A further refinement was introduced technically in the form of a three-stage sodium base process where the first stage at pH 6 is followed by an acid stage at pH 1.5 and finally by a weakly alkaline stage. The second stage can be varied in intensity or left out. This sulfite process is extremely flexible with regard to Kappa No., yield, carbohydrate composition and end-use quality, from a variety of paper pulps to high grade dissolving pulp (see Chap. IX, D, 1b).

G. ALKALINE SULFITE PULPING WITH ANTHRAQUINONE

The use of anthraquinone as accelerator of delignification and protector of carbohydrates in soda and kraft pulping was proposed in 1972/73. Its usefulness in sulfite pulping was appreciated in a Japanese patent issued in 1976, and in its modified version in a Canadian patent issued in 1980 [35]. Broad claims made in the latter two patents assume applicability of anthraquinone and related compounds over the whole range of sulfite pulping, from acid to alkaline. While it has been found to be a most effective pulping aid in alkaline sulfite environment, there is no evidence from experiments or from the known chemistry of anthraquinone reactions to support this claim in the acid sulfite range. (Fig. 43).

Initial expectations that the use of AQ would raise the status of traditional soda pulping to that of kraft pulping and provide the desired low odor alternative have not yet been fulfilled. Some forty kraft mills, about half of which are situated in Japan, are now using AQ on a regular basis, primarily utilizing the acceleration effect and obtaining some yield benefits also.

Anthraquinone is a pale yellow powder of low toxicity which can be produced from anthracene present in coal tar or by synthesis from naphthalene which is also a component of coal tar, and more abundant than anthracene.

Anthraquinone has been an important industrial chemical since the 19th century as an intermediate for the manufacture of vat dyes. At present, worldwide production capacity is 40 000 t/y of which varying surplus quantities are available for the novel application in pulping. Implementation of plans for additional capacity will depend on increased demand and, therefore, only limited quantities will be available for some time. A mill producing 1000 t of pulp at 50% yield per day and applying 0.1% anthraquinone on an oven dry wood basis would consume 700 t/y of the substance.

The structure of anthraquinone is shown below; the molecular weight is 208. It is insoluble in water or pulping solutions but becomes soluble in alkaline solution when in contact with reducing substances such as sugars and carbohydrate material released from the wood into the liquor. As a result of recent investigations made at PAPRICAN and elsewhere the following model has been proposed to explain the anthraquinone effect in alkaline pulping:

In the reaction with reducing substances, the reduced compound, anthrahydroquinone, is formed which dissolves as the anion with dark red color.

| Anthra-quinone AQ | Anthrahy-droquinone AHQ | AHQ-ion (soluble) |

Anthrahydroquinone, in turn, reduces lignin fragments dissolved in the liquor in the cooking liquor and is oxidized back to anthraquinone, thus closing the catalytic cycle of its dual function as pulping acid:

reduction of dissolved lignin which prevents recondensation

AQ

AHQ

oxidation and stabilization of cellulosic material

The above model applies to strongly alkaline solutions such as used in kraft and soda pulping which have been studied since 1977 and in strongly alkaline sulfite pulping where the pH is on a similarly high level. A modification of sulfite pulping with anthraquinone using moderately alkaline liquor has been disclosed in 1979 [36] and further investigated and developed in Finland [37] and Canada [38]. Because of its advantages, it has found industrial acceptance rather quickly [39]. So far no reaction models for this modification have been proposed.

Only small amounts of the catalytic substance, of the order of 0.05-0.20% calculated on oven dry wood, are required for producing a major increase in pulping rate together with a protective effect on the cellulosic fibers.

For the following discussion of alkaline sulfite-anthraquinone pulping it will be useful to consider that with the addition of anthraquinone both the cooking kinetics and the composition and properties of the resulting pulps are profoundly altered. It is not possible therefore to extend the coherent treatment of the sulfite field presented in this text so far to these new processes. Because of the high paper strength they produce, comparisons are generally made with the common kraft process with and without anthraquinone, but they are distinctly different also from this process in chemistry, reaction mechanism, lignin to carbohydrate ratio, and degradation of cellulose and hemicelluloses.

Note that it is common practice in North America to express the concentration of chemicals in alkaline pulping on the common basis of sodium oxide, Na_2O. It is also necessary to adopt a consistent formula for expressing the chemical composition of alkaline sulfite liquors for the purpose of describing and characterizing the new processes and their modifications. The following formula has been found practical. The sodium base chemicals used are considered first as % total Na_2O and second as % active Na_2O applied to oven dry wood. For example, 22% total Na_2O and 18% active Na_2O. Their individual ratio is then expressed as % of total Na_2O in the following order: sodium sulfite, sodium carbonate and sodium hydroxide, for example "80-

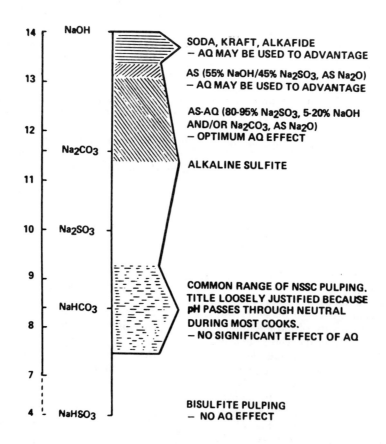

Fig. 43. Starting pH at room temperature of alkaline pulping processes [56].

10-10". It is important to consider both total and active chemical, since in the strongly alkaline sulfite and the kraft process only sodium hydroxide and the sulfur compounds (Na_2SO_3 or Na_2S) are active in the cook. Carbonate is passive in the kraft process, its presence being due to incomplete conversion in the chemical recovery cycle. In the moderately alkaline sulfite process, on the other hand, carbonate is also active as alkalinizing buffer and only total chemical charge needs to be quoted.

In Table 4 [38] chemical concentrations and pH will help to distinguish the two known sulfite anthraquinone processes from each other and from the common kraft process.

The strongly alkaline sulfite liquor has 14% sodium carbonate (as Na_2O) added to simulate recovery conditions. The alkali charge needs to be held on a similar level as the sulfite charge (both as Na_2O) for maintaining adequate delignification and kraft-like strength properties of the pulp. At the beginning of the cook the pH is the same as of the kraft liquor but one unit lower at the end. Regular kraft liquor uses 30%

more caustic soda and 40% less sulfur chemical than the strongly alkaline sulfite liquor.

The moderately alkaline version was discovered about eight years ago rather unexpectedly [36]. It presents a special case in the field of sulfite pulping in that maxima of cooking rate and pulp quality are obtained only within the narrow range of 75 to 90% sodium sulfite and 10 to 25% alkali in the form of sodium carbonate and/or sodium hydroxide and with anthraquinone accelerator. As seen in Table 4 the pH (cold) of this system with 10% sodium hydroxide is slightly lower than for the highly alkaline process at the start and drops to about 9.5 at the end of the cook. The starting pH depends on the ratio of carbonate to hydroxide used.

A comparison is made in Table 5 of spruce-type pulps at 50% yield prepared with strongly alkaline sulfite liquors without and with anthraquinone, and regular kraft liquor. It is seen that the addition of anthraquinone to the alkaline sulfite cook abruptly raises the rate of delignification while protecting cellulosic components against degradation. The resulting pulp beats

Table 4. Chemical ratios and pH level in AS and kraft liquors for bleachable pulp grades.

Chemicals as Na_2O, % of Charge	Na_2SO_3	Na_2CO_3	NaOH	pH of Cook start	end
Strongly Alkaline Sulfite 22% Total Alkali on Wood	43	14	43	13.6	11.1
Moderately Alkaline Sulfite 20-22% Total Alkali on Wood	80	10	10	13.2	9.5
	Na_2S				
Regular Kraft 18% Active Alkali (22% Total Alkali) on Wood	26	18	56	13.6	12.2

Table 5. The effect of 0.1% AQ on strongly alkaline sulfite cooks of northern softwood to 50% yield, and comparison with kraft.

	AS	AS-AQ	Δ% AS/AS-AQ AS=100	KRAFT
Active chemical (% Na_2O)	22	22	0	18
Ratio hydroxide/sulfite, Na_2O	1.12	1.12	—	—
Total chemical (% Na_2O)	22 (47-0-53)	25.6 (45-14-41)	+14	22
Anthraquinone, % on o.d. wood	—	0.1	—	—
Maximum temperature, °C	172	167	−3	166
Time at temperature, min	122	72	−41	55
Total Yield, %	50	50	0	50
Kappa No.	72	36	−50	50
Viscosity, mPa·s	29	34	+17	42
Beating time (PFI), 300 mL CSF	9.4	4.8	−49	6.4
Tear index, N·m²/kg	9.1	10.2	+12	10.5
Burst index, MN/kg	11.8	11.2	−5	10.3
Breaking length, km	13.2	14.7	+11	13.4
Brightness, ISO unbleached	23	28	+26	25

much more easily and reaches higher levels of mechanical strength. The efficient kraft cook still retains some advantage of temperature and time in delignification rate and in viscosity but is now deficient in tensile strength properties. The brightness of these pulps cooked in highly alkaline medium is similar.

A similar comparison in Table 6 of liner pulp cooks of Southern pine at equal time at temperature and similar pulp yield shows again the great improvement in cooking rate and viscosity due to adding 0.1% anthraquinone to the strongly alkaline sulfite cook. The kraft cook still retains the advantage of lower cooking temperature and has, with this wood, slightly higher tear but lower viscosity and tensile strength properties.

Recently, liner board pulp of 70% yield from northeastern softwood was produced in a mill-scale batch digester. The quality was similar to that of a 54% yield kraft pulp [55].

Unlike the strongly alkaline version which produces useful cooking results also without anthraquinone, the moderately alkaline version with its low charge of alkali and a much lower pH profile during the cook could be expected to require anthraquinone additive for producing usable pulp. In fact, the effect of adding the same amount of anthraquinone to this system is twice as great than with the strongly alkaline system. In other words, in the presence of anthraquinone the known action of alkalinity to accelerate cooking is replaced by a similar action of sulfite. The mechanism involved has not been explained as yet.

These effects are illustrated in Fig. 44 where the differences in delignification rate with and without anthraquinone are shown in comparison with a regular kraft cook [38]. When comparing cooks of two hours duration at 175°C, it is seen that without anthraquinone the rate in the moderately alkaline sulfite medium is very slow but twice as fast in the strongly alkaline sulfite medium. Adding 0.1% anthraquinone, calculated on dry wood, produces the unexpected effect of reducing the Kappa number of the moderately alkaline cook from 130 to 40 but that of the strongly alkaline cook from 65 to 20, or only by half as much.

This delignification maximum at 75-90% sulfite (Na_2O) is accompanied by equally unexpected maxima of pulp yield and viscosity shown in Fig. 45 [38].

It is also seen in Fig. 44 that the rates of a strongly alkaline sulfite cook with anthraquinone at 175°C and of a regular kraft cook at 166°C are similar.

The present status of moderately alkaline sulfite-anthraquinone pulping can be appreciated and discussed on the basis of the data in Tables 6 and 7 developed for northern softwood.

With the same charge of cooking chemicals, kraft pulping retains its superiority as a most efficient delignification medium, requiring lower temperature and less time to reach a given level of pulp yield. The superiority of the moderately alkaline sulfite-anthraquinone process, on the other hand, lies for bleached pulp grades (Table 7) in carbohydrate protection affording high pulp yield and viscosity, very easy bleaching to a higher brightness ceiling with less reversion,

Table 6. The effect of 0.1% AQ on AS cooks of southern pine liner pulp and comparison with kraft pulp.

	AS	AS-AQ	Δ% AS/AS-AQ AS=100	KRAFT
Active chemical (% Na_2O)	22	18	−18	17
Ratio hydroxide/sulfite, Na_2O	1.12	1.12	—	—
Total chemical (% Na_2O)	25.6	21	−18	21
	(45-14-41)	(45-14-41)	—	—
Anthraquinone, % on o.d. wood		0.1	—	—
Maximum temperature, °C	172	167	− 3	160
Time at temperature, min	65	65	0	65
Total yield, %	55.1	55.9	+ 1.5	55.5
Kappa No.	97.3	82.7	−15	85.5
Viscosity, mPa·s	32.4	47.5	+47	46.6
Refined freeness, CSF	500	500	0	500
Tear index, N·m²/kg	22.2	22.7	+ 2.3	24.1
Burst index, MN/kg	4.3	4.9	+14	4.2
Breaking length, km	5.8	7.5	+29	6.1
Brightness, ISO unbleached	20	20.9	+ 5	18.2

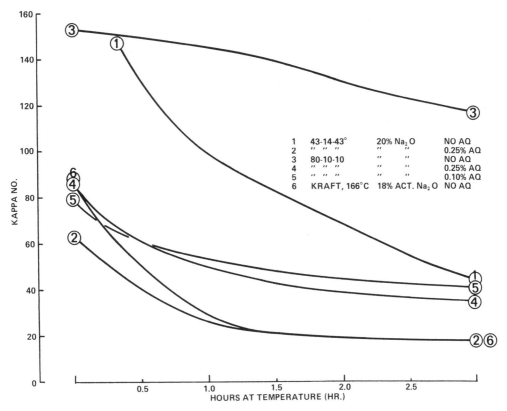

Fig. 44. Relationship between Kappa number and time at temperature of AS-AQ and kraft cooks of Northern softwood [38].

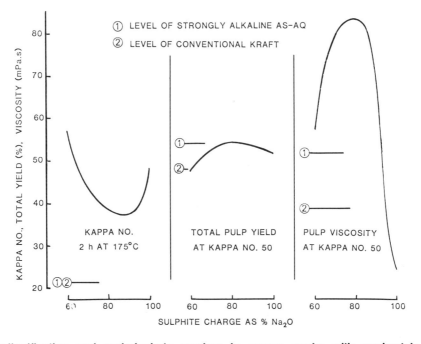

Fig. 45. Delignification and carbohydrate maxima in spruce cooks with moderately alkaline sulfite-anthraquinone liquor [38].

and higher strength of the bleached pulp. For packaging grades such as liner pulp for containers and bag paper (Table 8) the new process has the additional advantages of lower lignin content of the pulp at a given yield, one half the energy requirement for refining, higher burst and tensile strength which decrease little on raising the yield substantially, and higher brightness.

Economic evaluations of the new process and trials on the industrial scale are now in progress at various locations. The first application in industry has been reported from Finland [39] for the manufacture of strong semibleached softwood pulp for news furnish and market uses. Details are found in Chapter X, C.

One of the attractive features of the new process, by comparision with the conventional kraft process, besides the inherent advantage of being able to produce strong pulp without kraft odor, is its low requirement for caustic soda. In kraft pulping the charge of caustic soda is higher even than in the strongly alkaline sulfite pulping process and necessitates the operation and maintenance of large units for the re-causticization of the spent pulping liquor at considerable expense and fossil fuel consumption. With the moderately alkaline process, the necessary moderate alkalinity can be provided by soda (Na_2CO_3) which is available directly from the recovery process and, if needed, by a limited

Table 7. Comparison between the moderately alkaline sulfite-anthraquinone and kraft processes for bleached softwood pulp end use.

	AS-AQ (80-10-10)	Kraft	Δ% Kraft=100
Active chemical (% Na_2O)	22	18	+22
Total chemical (% Na_2O)	22	22	0
Anthraquinone, % on o.d. wood	0.1	—	—
Maximum temperature, °C	175	166	+5
Time at temperature, min	120	90	+33
Total yield, %	50.3	44.9	+12
Screened yield, %	50.2[1]	44.8[2]	+12
Kappa No.	31.2	29.5	+6
Viscosity mPa·s	54.2	23.4	+23
Bleach sequence, simplified	$C_D E(HD)$	$C_D E(HD)$	—
Chemical shrinkage	6.74	4.93	+16
Yield of bleached pulp, % o.w.	46.8	42.5	+10
Viscosity, mPa·s	43.0	17.7	+24
Brightness, ISO	91.1	89.6	+1.5
Reverted brightness, ISO	88.9	85.3	+4
Freeness, CSF	300	300	0
Burst index, MN/kg	11.9	11.1	+7
Tear index, N·m²/kg	11.3	10.7	+5
Breaking length, km	13.4	12.7	+5

1) Fiberized and screened
2) Blown and screened

Table 8. Comparison of the moderately alkaline sulfite-anthraquinone and kraft process for spruce-type liner board pulps (500 CSF).

	AS-AQ (80-10-10)		KRAFT	Δ% KRAFT=100	
Active chemical (% Na_2O)	20		18	+11	
Total chemical (% Na_2O)	20		22	−9	
Anthraquinone, % on o.d. wood	0.08		—	—	
Maximum temperature, °C	175		166	+5	
Time at temperature, min	100	50	30	+230	+67
Total yield, %	55.8	63.0	56.5	−1	+12
Kappa No.	56.1	85.5	83.5	−33	+2
Refining energy, GJ/o.d.t.	0.62	0.62	1.24	−50	−50
Tear index, N·m²/kg	13.7	13.2	14.2	−4	−7
Burst index, MN/kg	7.1	7.0	6.8	+4	+3
Breaking length, km	12.3	11.0	10.0	+23	+10
Brightness, ISO	34.0	32.8	19.0	+79	+73

amount of caustic soda (NaOH) used as make-up for losses in the process. Thereby the cauticization department can be completely eliminated, or if some free caustic is required elsewhere in the mill, for instance in the bleachery, a small causticization unit can be operated. The new process has advantages over the common kraft process in economy, much greater product flexibility and complete elimination of kraft odor, but it requires anthraquinone

catalyst. The kraft process produces now about 90 million tons of pulp per year worldwide and it seems possible that a major re-orientation in the field of alkaline pulping could be brought about. Such a change was anticipated after anthraquinone was first applied to soda pulping but was not realized because of various shortcomings of the soda process. Whether the alkaline sulfite-anthraquinone system will have more success remains to be seen.

REFERENCES

(CHAPTERS I AND II)

A. BOOKS OR BOOK CHAPTERS

Pulp wood and wood pulp in North America, by R.S. Kellog, McGraw Hill, New York, 1923.

Sulphite Pulping Technology, by H.F.J. Wenzl; Lockwood Trade Journal Company, New York, 1965.

Pulping Processes, by S.A. Rydholm; Interscience Publishers, division of John Wiley & Sons; London, 1965.

Kirk-Othmer Encyclopedia of Chemical Technology, Third Edition Volume 18, 1982 (Pulp).

Handbook of Chemistry and Physics, 61st Edition 1980-81; CRC Press.

Pulp and Paper Chemistry and Chemical Technology, edited by J.P. Casey; 3d Edition; John Wiley & Sons, New York, 1980.

Handbook for Pulp & Paper Technologists, Smook, G.A., Joint Textbook Committee of TAPPI & CPPA, 1982.

B. SELECTED ARTICLES AND ITEMS

1. Cotton, F.A. and Wilkinson, G., Advanced inorganic chemistry, p. 316. Interscience, 1962.
2. Ingruber, O.V., Temperature-dependent equilibria in sulfite solutions from room temperature to 160°C. *Svensk Papperstidn.* 65, 448-456 (1962).
3. Ingruber, O.V., High yield sulfite pulping, Part V, The direct measurement of pH at elevated temperature and pressure during sulfite pulping. *Pulp Paper Mag. Can.* 5, 124-131 (Sept. 1954).
4. Wenzl, H.F.J., Sulfite Pulping Technology. Lockwood Trade Journal, New York, 1965.
5. Engelhardt, G., Physical-chemical studies of bisulfite cooking acids. *Zellstoff Papier* 11, 43-50 (Feb. 1962).
6. Schwitzgebel, K. and Lowell, P.S., Thermodynamic basis for existing experimental data in Mg-SO_2-O_2 and Ca-SO_2-O_2 systems. *Environmental Sci. Techn.* 7, 1147-51 (Dec. 1973).
7. Hull, W.Q., Baker, R.E., and Rogers, C.E., Magnesium base sulfite pulping. *Ind. Eng. Chem.* 43, 2424 (Nov. 1951).
8. Kesler, R.B. and Han, S.T., Solubility and neutralization equilibria in the magnesium base sulfite acid system. *Tappi* 45, 534-8 (July 1962).
9. Markant, H.P. and coworkers, Physical and chemical properties of magnesium-base pulping solutions. *Tappi* 48, 648-53 (Nov. 1965).
10. Bryce, J.R.G. and Tomlinson, G.H., Modified magnefite process. *Pulp Paper Mag. Can.* 63,

T355-61 (July 1962).

11. Ingruber, O.V., Chemical equilibria in heated sulfite solutions. *Pulp Paper Mag. Can.* 66, T215-228 (Apr. 1965).

12. Morgan, R.S., A thermodynamic study of the system sodium sulfite-sodium bisulfite-water at 25°C. *Tappi* 43, 357-364 (1960).

13. Ingruber, O.V. and Allard, G.A., Alkaline sulfite pulping for kraft strength, *Pulp Paper Mag. Can.* 73, T354-T369 (Nov. 1973).

14. Bauer, T.W. and Dorland, R.M., Thermodynamics of the combustion of sodium-based pulping liquors. *Can. J. Techn.* 32, 91-101 (1954).

15. Magnusson, H. and Warnqvist, B., Properties of sodium sulfite-sodium carbonate melts. *Svensk Papperstidn,* 614-16 (1975).

16. Schöön, N.-H., Kinetics of the formation of thiosulfate, polythionates, and sulfate by the thermal decomposition of sulfite cooking liquors. *Svensk Papperstidn.* 65, 729-754 (1962).

17. Glasser, G.W. and Glasser, H.R., The evaluation of lignin's chemical structure by experimental and computer simulation techniques. *Paperi ja Puu* 63, 71-83 (Feb. 1981).

18. Gellerstedt, G., The reactions of lignin during sulfite pulping. *Svensk Papperstidning* 79, 537-43 (No. 16, 1976).

19. Rezanowich, A. and Goring, D.A.I., Poly electrolyte expansion of a lignin sulfonate microgel. *J. Coll. Science* 15, 452-471 (1960).

20. Rezanowich, A., Yean, N.Q. and Goring, D.A.I., High resolution electron microscopy of sodium lignin sulfonate. *J. Appl. Polymer Science* 8, 1801-1812 (1964).

21. Yean, W.Q. and Goring, D.A.I., Molecular properties of sodium lignosulphonates by a continuous flow bisulfite process. *Svensk Papperstidn.* 71, 739-43, (Oct. 1968).

22. Procter, A.R., Yean, W.Q., and Goring D.A.I., The topochemistry of delignification in kraft and sulfite pulping of sprucewood. *Pulp Paper Mag. Can.* 68, T445-53 (1967).

23. Ingruber, O.V. and Allard, G.A., The effect of hydrogen ion activity in the sulfite pulping of black spruce. *Tappi* 50, 597-614 (Dec. 1967).

24. Ingruber, O.V., Alkaline sulfite pulping. *Das Papier* 24, 711-27 (10A, 1970).

25. McLean, J.D., Hypo number. *Pulp Paper Mag. Can.* 66, Convention Issue, T103-6 (1965).

26. Kyrklund, B. and Strandell, G., Applicability of the chlorine number for evaluation of lignin content of pulp. *Papperi ja Puu* 51, 299-305 (4a, 1969).

27. Jutila, E., Passila, M., Uronen, P., Mathematical models for producing dissolving pulp. *Tappi* 64, 105-8 (Aug. 1981).

28. Kallonen, H. and Franzreb, J.P., Computer control of batch sulfite digesters. *Pulp & Paper Can.* 80, T437-9 (Dec. 1979).

29. Liebergott, N. and Yorston, F.H., The effects of temperature and liquor concentration on rate of sodium bisulfite pulping. *Pulp & Paper Can.* 77, T211-4 (1976).

30. Lemay, R., Batch digester computer cooking control. *Pulp Paper Can.* 80, T167-70 (June 1979).

31. Wilder, H.D. and Han, S.T., A composition of the kinetics of the neutral sulfite and kraft pulping processes. *Tappi* 45, 1-9 (Jan. 1962).

32. Jutila, E., Perron, M., Ahola, H., Hipeli, R., Computerized process control of a continuous digester for NSSC pulping. *Tappi* 63, 69-72 (Nov. 1980).

33. CPPA Useful Method G.10V, Chlorite delignification of cellulose material.

34. Strapp, R.K., Kerr, W.D. and Vroom, K.E., Comparison of bases used in sulfite cooking. *Pulp Paper Mag. Can.* 58, 277-283 (3, 1957).

35. Nomura, Y., Wakai, M., and Sato, H., Sulfite digestion of ligno-cellulose materials. Japanese Patent 112,903,1976; modified version, Process for producing pulp in the presence of a quinone compound. Canadian Patent 1 079 906, 1980.

36. Raubenheimer, S. and Eggers, S.H., 11th European ESPRA Meeting, Brussels, May 1979.

37. Virkola, N.-E., Pusa, R. and Kettunen, J., Neutral sulfite AQ pulping as an alternative to kraft pulping. *Tappi* 64, 103-7 (May 1981).

38. Ingruber, O.V., Stradal, M. and Histed, J.A., Alkaline sulphite-anthraquinone pulping of Eastern Canadian woods. *Pulp Paper Can.* 83, T342-49 (Dec. 1982).

39. Anon. Latest trends in pulping progress. *Pulp Paper Intl.* 24, 58 (June 1982).

40. Source: Adapted from J.K. Hamilton, *Pure Appl. Chem.,* 5(1-2):197-217 (1962) and reproduced by permission of the International Union of Pure and Applied Chemistry and Butterworth Scientific Publications, London W.C. 2, England.

41. Thompson, N.S., Peckham, J.R. and Thode, E.F. *Tappi* 45(6):433 (1962).

42. Stone, J.E., The penetrability of wood. *Pulp Paper Mag. Can.* 57, 139-45 (June 1956).

43. Alm, A. and Stockman, L., Chip impregnation with sulfite cooking liquor. *Svensk Papperstidn.* 61, 10-18 (Jan. 1958).

44. Smook, G.A., Handbook for Pulp & Paper Technologists. Joint Textbook Committee of the Paper Industry, 1982.

45. Adler, E., *Wood Sci. Technol.,* 11:169 (1977) in Parkam, R.A., Pulp and Paper Manufacture, Vol. 1, 3rd Ed. Joint Textbook Committee of the Paper Industry, 1983.

46. Rydholm, S.A., Pulping Processes, Interscience Publishers, Div. of John Wiley & Sons, London, 1965.

47. Pulp and Paper Chemistry and Chemical Technology, ed. by J.P. Casey; 3rd ed., John Wiley & Sons, New York, 1980.

48. Sjöstrom, E., Wood Chemistry, Academic Press, New York, 1981.

49. Falk, M. and Giguere, P.A., On the nature of sulfurous acid. *Can. J. Chem.* 36, 1121-25 (1958).

50. Scott, W.D. and McCarthy, J.L., The system sulfur dioxide-ammonia-water at 25°C. *Ind. Eng. Chem.* 6, 40-48 (Feb. 1967).

51. White, C.K., Vivian, J.E. and Whitney, R.P., The solubility of sulfur dioxide in calcium bisulphite solutions. Tech. Assoc. Papers 31, 143 (1948).

52. Schacht, W., Cooking liquor for manufacture of pulp from straw, esparto, wood and similar materials. German Patent 122 171, Oct. 10, 1900.

53. Sillen, L.G. and Anderson, T., Solid-gas equilibria of importance in burning concentrated calcium or magnesium sulfite waste liquor. *Svensk Papperstidn.* 55, 622-31 (1952).

54. Braun, C.A., See Schubert, M., Cellulose (Pulp) Manufacture, M. Krayn, Berlin, 1906, p. 183.

55. Stradal, M., Perrault, J., and Ingruber, O.V. 70% yield alkaline sulfite anthraquinone pulp for linerboard. *Tappi Journal* 66, 75-79 (Oct. 1983).

56. Ingruber, O.V., Development and status of alkaline sulfite anthraquinone pulping. Annual Meeting of the Technical Section, CPPA Jan./Feb. 1983, Preprints, B313.

III

PROCESS MEASUREMENTS AND CONTROL

O.V. INGRUBER

Retired Senior Research Associate from CIP, now an independent consultant

MANUSCRIPT REVIEWED BY
David Johnston, Manager Pulping
Nova Scotia Forest Industries
Port Hawkesbury, Nova Scotia

The most important control variable in pulping is the degree of delignification, expressed as Kappa No. or Roe No. The first is determined according to the TAPPI and CPPA Standard Methods by reacting the lignin in the pulp with measured potassium permanganate solution. Most foreign countries engaged in pulp and paper production have also adopted this standard. The Roe No. test (TAPPI and CPPA Standard Methods) measures the amount of chlorine gas consumed in the reaction with pulp lignin. This test is also widely used, but mainly for determining bleach chemical demand.

Next in importance as a cooking control variable is pulp viscosity for monitoring pulp degradation in dissolving pulp development and manufacture. In some cases other pulp properties, such as hemicellulose content or the amount of screening rejects, may be most significant for a process but their control will be indirect through establishing relations with the degree of delignification.

Pulp yield as such is not a primary control variable in industrial cooks because it cannot be determined directly from liquor or pulp samples. Values quoted by pulp mills are obtained by comparing wood intake with pulp production and estimated losses over long periods. Another way to estimate pulp yield is to establish yield-Kappa No. relations in laboratory cooked pulps and to assume that the relation will remain the same for pulps cooked in large-scale batch or continuous digesters. A discussion of pulp yield determination can be found in Chapter VI, H.

The subject of pulping process control can be divided into three aspects whose order also denotes their historical development: analysis, testing and measurement of significant variables, monitoring and manual control. The ultimate goal is for full operations and process control, utilizing all feasible and practical imputs into a computer model, including memory update.

A. ANALYSIS, MEASUREMENT, AND MONITORING OF PULPING VARIABLES, MANUAL CONTROL OF ENDPOINT

In acid bisulfite pulping the rate of the cooking reactions depends on pH, combined SO_2, temperature, and the amount of free SO_2 dissolved in the liquor and dissociated into ions. The pulping potential of the liquor is the resultant of the ionic equilibrium of the three interdependent chemical parameters at a given temperature level. Both cooking control variables are of technical importance in this range: Kappa No. or Roe No. of paper pulps, and viscosity of soft paper pulps and dissolving pulps of specified quality.

It is characteristic of acid bisulfite (and bisulfite) cooks to consume pulping chemical (bisulfite ion) in the reaction with lignin, while forming a strongly acid reaction product (ligno-sulfonic acid ion). The pH of the liquor decreases steadily with increasing degree of delignification and, consequently, the rate of delignification and carbohydrate hydrolysis and degradation greatly increases. Much greater precision and speed in the determination of the correct end point is therefore necessary to avoid the risk of over cooking or burning in extended acid bisulfite cooks than is necessary in short cooks to high yields or cooks at higher pH levels where the free SO_2 variable is absent.

Of the available tests and monitoring options, total SO_2 does not give useable information since it depends on chemical reactions and on the amount of top gas relief taken. Available combined SO_2 would be valuable but cannot be determined accurately. Hot pH is useful but existing equipment is difficult to maintain in a pulp mill routine. Conductivity which also indicates increase in acidity is less sensitive but still rarely used. What remains is the monitoring of the density change of liquor color, and this is indeed the most practical and widely used method for well-cooked acid bisulfite pulps. A comparison of some of these methods is made in Fig. 46 indicating their relative usefulness for intercepting an acid bisulfite dissolving pulp cook in its final stages at the right time. It is seen that cold pH and visible light absorption (500 nm) give by far the most sensitive indication for terminal viscosity change and lignin content, and that of the two, visible light absorption is not influenced by digester blow down.

1. pH and its effects

For most of its 100 year history, sulfite pulping has limited itself to the very acid range (cold pH 1.3-1.5, Ca base). After 1930, a narrow area closer to the top of the sulfite pH scale was developed for the chemical-mechanical high-yield pulping of hardwoods, primarily NSSC (cold pH 9.0, Na base). After 1950, sulfite technology moved to develop certain areas within the open field of 8 pH units: bisulfite at pH 4 with Na base (type: Arbiso) and Mg base (type: Magnefite); two-stage bisulfite-acid sulfite (type: Stora, Weyerhaeuser FB); two-stage bisulfite – neutral or alkaline (type: Weyer-haeuser HO, Sivola, Rauma). After their empirical discovery, the above sulfite processes were evaluated and developed individually for the production of pulps with different and desirable characteristics required by the market. In spite of

the obvious significance of pH in the sulfite pulping process, its effects were exposed in a coherent way and spanning the whole pH scale only in the late sixties [1]. The results from this work removed much of the confusion regarding the effects of temperature, base concentration, and free SO_2 on delignification, carbohydrate content, and the resultant pulp properties at different pH levels. Moreover, on account of these results, alkaline sulfite was finally and systematically added as the area of high strength pulp to complete the family of sulfite processes covering the entire pH scale.

2. pH measurement

The basis for these moves to bridge and expand the pH scale available for sulfite pulping was the development of a method (PAPRICAN) and of technical equipment (CIPRL) for the continuous measurement of pH at the temperature and pressure of cooking in the digester. To distinguish this pH value from the pH value measured in the sulfite liquor at room temperature (cold pH), and traditionally quoted for a given sulfite cook, the term "hot pH" or true pH is used in the following text and illustrations.

The cross-section of a hot pH flow cell and the actual stainless steel unit available for installation on liquor circulation lines of laboratory pilot plant and commercial cooking equipment are shown in Figs. 47 and 48. The assembly consists of a patented high-temperature, high-pressure reference electrode with connections for coolant and pressure compensation at the center, the high-pressure glass electrode adaptor at the left and a special double coil temperature

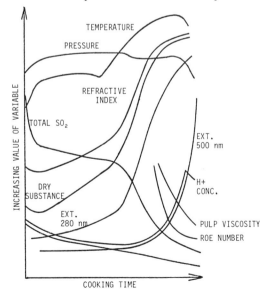

Fig. 46. Cooking end point methods [14].

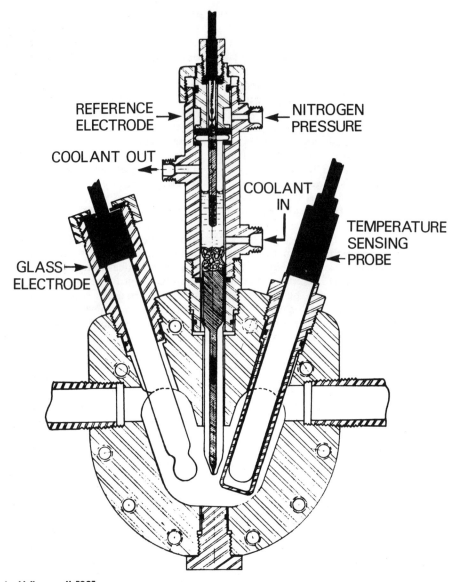

Fig. 47. Hot pH flow cell [23].

compensator in its well on the right. A pH meter with both EMF slope and pH-mV zero shift control is required.

For hot pH calibration of the equipment up to 95°C, the data supplied with most commercial buffer solutions may be used. For thorough calibrations from room temperature to 200°C, the following solutions provide chemical systems whose pH changes little over a wide range of temperature: HC1 of pH 2 or 3; magnesium or ammonium bisulfite solution, 0.30-0.35 N, pH 5-5.5; sodium sulfite solution, 0.31 N, pH 8 (cf Chap. I, D).

It must be emphasized that hot and cold pH are two very different values. Hot pH is an actual measure of the hydrogen ion activity as it exists at the temperature and pressure of the digester. The cooking process is influenced by this factor and not by the hydrogen ion activity measured in the cooled liquor sample.

Complete cooking data, as in Fig. 49, demonstrate the large difference between the pH of the cooled liquor sample and the effective pH at different stages of a calcium base acid bisulfite cook of a 75%-25% spruce-balsam wood blend. The cooking liquor contained 6.9% total SO_2 and 1.3% combined SO_2 and the pH was 1.3 at room temperature; the liquor (v) to wood (w) ratio was 5:1. As could be expected from the temperature effect on the pH of the plain liquor (Fig. 4),

Fig. 48. Hot pH flow cell [1].

heating did produce a substantial rise of pH, but in the presence of wood a number of factors determine its actual level. Hot pH is therefore a sensitive indicator of the cooking reactions taking place and also of the balance of base and free SO_2 in the system on the one hand, or, on the other hand, a sensitive monitor of induced changes of the system to control the rate of hydrolysis. The dashed portion at the beginning of the hot curve in Fig. 49 indicates a substantial increase of pH during the first 10 min from the time the digester was filled with 80°C acid from the accumulator. The pH of this liquor without wood rose from 1.3 to 1.85 at 80°C. The additional increase of almost 0.4 pH units was due to the uptake of free SO_2 by the wood. From the new level of about 2.24, the pH rose to 2.9 during 2.5 h of heating, then dropped to 2.7 during the following 2 h at constant temperature, and finally rose to 3.1 during the 1.5 h top relief period.

The sequence in the rise and fall of the pH curve is characteristic of conventional sulfite pulping operations. The three distinct sections can be explained as follows.

It was found that the temperature coefficient of pH in conventional cooking acids in the absence of wood, if taken over the range 25° to 150°C, is 0.011 to 0.012 pH per °C. Thus the effect of temperature is mainly responsible for

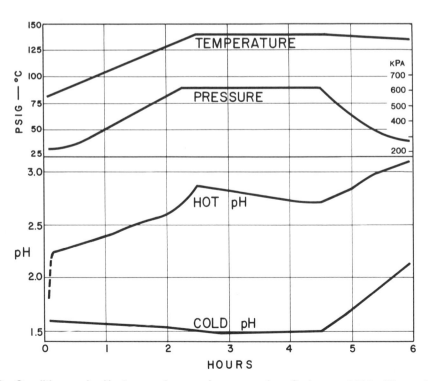

Fig. 49. Conditions of pH, temperature and pressure in a Ca base acid bisulfite cook [24].

the rise in the first section of the pH curve.

The second major factor influencing pH in cooking is introduced by the formation of strong lignosulfonic acid as explained earlier. This causes the fall in the second section of the pH curve.

Pressure determines the amount of SO_2 gas dissolved in the liquor. Top relief produces the pH rise in the third section of the curve.

Although each factor is dominant during a section of the cook, all three are active throughout to various degrees. The appearance of lignosulfonic acid in the liquor will slightly reduce the pH rise in the first section of the curve. The development of a peak between the first section and the second depends on the duration and rate of pressure relief taken towards the ends of the heating period. During cooking at temperature, volatile SO_2 becomes gradually displaced by stable lignosulfonic acid, therefore continuous pressure relief was taken in order to observe the prescribed 90 psig level. Without this continued relief of SO_2, a slightly steeper fall in the curve would have developed in the second section. In the final section of the cook, because the temperature was set to decrease by 5°C (augmenting acidity), the pressure relief had to be increased in order to raise the pH at the end to the predetermined level.

Examples for the course of the hot or true pH of strongly acid to strongly alkaline pulping liquors and actual wood cooks [2] are collected in Fig. 50. Starting at the acid bisulfite level it is seen that the strong sodium base ion produces a higher pH than the weaker calcium base ion under identical cooking conditions. This results in the higher yield, higher hemicellulose retention and better delignification observed in sodium base cooks. At the bisulfite level, the pH rises only about 0.5 unit during heating and decreases by about the same amount due to sulfonation in a cook to about 60% yield. Pressure relief at this level does cause a major increase in pH due to the dissociation of HSO_3 ion at the high temperature:

$$2HSO_3{}^- \xrightarrow{T} SO_3{}^{--} + SO_2 + H_2O.$$

The true pH of a sulfite (Na_2SO_3) solution drops about one unit upon heating to 175°C. With spruce wood present the drop is almost three pH units, the initial drop of one unit being due to the acid nature of the spruce wood (pH 4.5 at room temperature). With birch wood, and without the bicarbonate buffer used in technical sulfite cooks, the pH drop is 5 units due to the combined effect of wood acid and sulfonation of lignin.

The pH of a highly alkaline sulfite liquor, Fig. 50, top, decreased about 3 units due to heating to 175°C. With spruce wood present the pH of the cook decreased a further 0.7 pH units at the time when the maximum temperature of 175°C is reached, and another 0.75 pH units during cooking at 175°C. The concentration of hydroxyl ion, OH^-, which determines the rate of an alkaline cook is thus greatly reduced by cocurrent reactions with the wood components.

Note also that the iso-pH level, where cold and hot pH remain the same during heating, found to be at pH 8 for a pure sulfite solution (Fig. 24) is depressed by two units to pH 6 when spruce wood is present and the temperature of the cook is 160°C.

The relationship of hot and cold pH of solutions of some pulping chemicals and softwood cooks in Fig. 51 has been drawn from available data to describe the field in a coherent manner and for reference [3]. To date, data are insufficient to develop a pH diagram for hardwood over the complete range. A partial pH diagram in the alkaline sulfite range for eastern hardwood blend cooked at 175°C is available [3], however, and is shown in Fig. 52. The linear relation indicates a difference between cold and hot pH of about 3½ units. With such a relationship established for a given wood furnish, liquor chemistry and cooking temperature, hot pH in successive cooks can be controlled by monitoring the pH of cooled liquor samples and adjusting the charge of sodium hydroxide.

In alkaline pulping, the rate of both lignin and carbohydrate dissolution is governed by the sodium hydroxide concentration. The higher the alkali concentration, the lower is the selectivity of the cook, that is the ratio of lignin over carbohydrate removal. In practice this means that applying higher concentrations of NaOH to reduce cooking time will cause a corresponding loss of pulp yield.

The sodium hydroxide concentration of the initial cooking liquor decreases substantially upon contact with the wood due to neutralization of wood acids (formic acid, acetyl groups) and reactions with accessible carbohydrate components. It decreases further as the cooking reaction proceeds and its profile during the cook is the factor which determines the rate of the cook. Thus the total alkali charge required for a given degree of cooking (lignin removal) is the sum of the initial alkali demand of a particular wood furnish; the alkali consumed during the cook in chemical reactions with carbohydrates and lignin, including acid compounds present or formed; and the alkali concentration to be carried to the end of the cook to prevent its

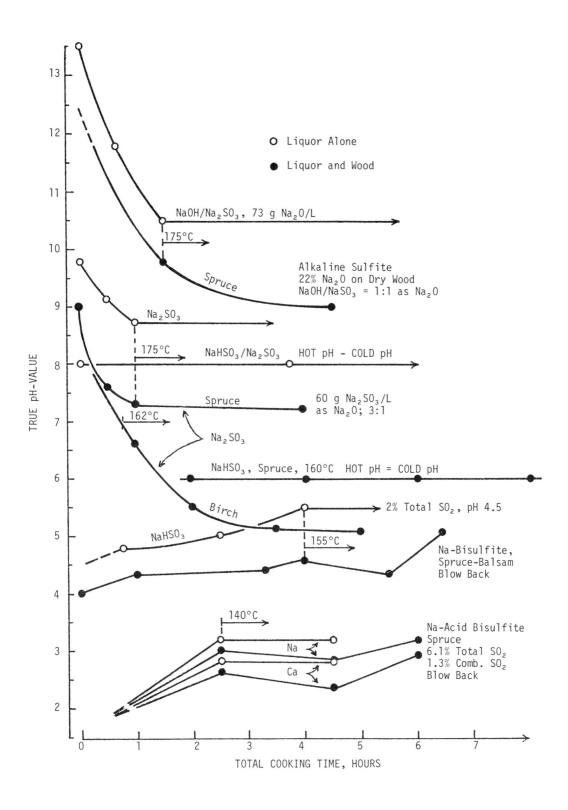

Fig. 50. Survey of true pH values in sulfite pulping solutions with and without wood [2].

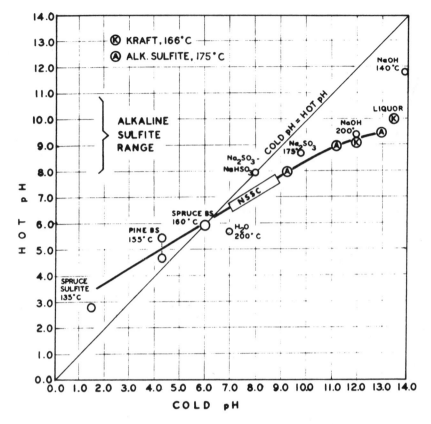

Fig. 51. Outline of hot and cold pH values in pulping solutions and softwood cooks [3].

premature slow-down or stoppage with possible redeposition of dissolved material.

It would appear from this that control of hydroxide concentration by injection during the cook would be a desirable feature of alkaline cooks. In kraft pulping, injection trials have been performed but the process has not been introduced into commercial practice because of more complicated operational and control demands.

3. pH control experiments

In alkaline sulfite pulping, pH control during the cook has been applied experimentally over the whole pH range and the response of wood components mapped as will be discussed below.

A brief exposure of the techniques used and of some results of the first complete survey of the sulfite field [1] follows. Published in 1967, it has remained the only one in existence, covering eastern spruce, the best all-around wood for paper making and sodium, the most versatile base. With the increasing utilization of hardwoods to alleviate wood shortages it would be desirable to add a sodium base survey of a common hardwood or hardwood blend. Since

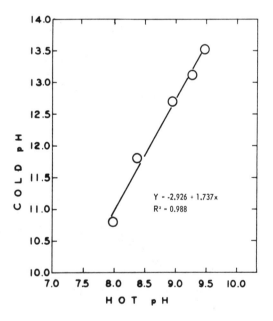

Fig. 52. Relation between hot and cold pH in alkaline sulphite cooks of Eastern hardwood blend: liquor to wood ratio 3 : 1 [3].

the spruce-sodium base field has been sufficiently well explored and mapped, substitution of other species or bases will require considerably less time and effort.

All cooks were made in a 70 L stainless steel experimental digester with indirect heating and forced circulation. The digester was steam-packed with 12 000 g o.d. black spruce chips and filled to the top with cooking liquor. The liquor:wood ratio was approximately 4.5:1.

The standard heating schedule used is shown in Fig. 53. Periods at cooking temperature were varied to obtain different yields, and digester pressure was relieved for 30 min to 345 kPa before blow.

Ideally, an isothermal schedule should have been used, but this was impractical because of limitations in heat exchanger capacity and, more pertinent, because of burning of chips in the acid bisulfite range if the temperature is raised too rapidly. The above cooking schedule is thus a compromise that can be used over the whole pH range. The final evaluation of all cooks indicated that rate of temperature rise was of little significance to cooking results.

It was attempted, however, to apply pH control for the whole cook, including the period to maximum temperature. Normally during heating, the pH rises in acid bisulfite cooks, remains rather level in an intermediate pH range, and drops in sulfite cooks in the alkaline range. Control of the pH level from the beginning of a cook eliminates inconsistencies that interfere with a clear separation of pH and temperature effects, particularly in short cooks. In the acid bisulfite range where pH control was effected by relief or injection of SO_2, the use of a liquid-filled digester assured that all SO_2 measured by liquor analysis was actually involved in the cook. The automatic top relief valve of the digester was

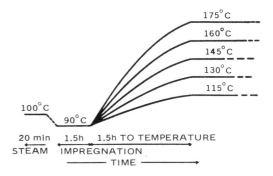

Fig. 53. Schematic cooking schedules used in the first survey of the sulfite field [1].

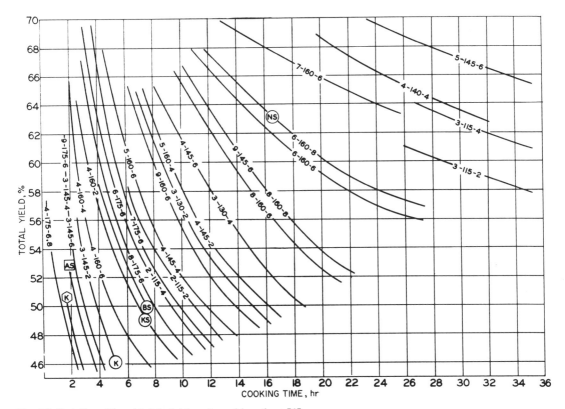

Fig. 54. Relationship of total yield and cooking time [1].

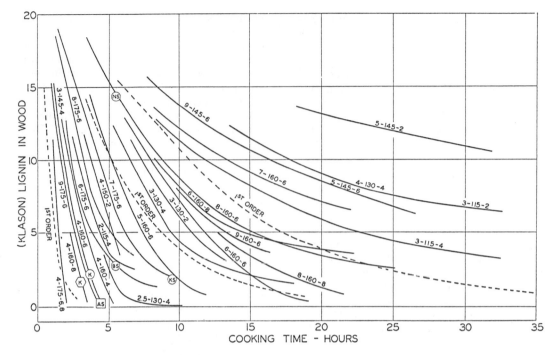

Fig. 55. Relationship of Klason lignin and cooking time [1].

normally closed, and pressure relief was taken only for the purpose of (upward) pH control in acid cooks. In the alkaline sulfite range, NaOH was injected to compensate for the normal pH drop. A record of chemicals injected and recovered from the digester was kept to give complete information on chemical consumption at the various pH levels.

The term "bound SO_2" is used here because it confers meaning over the whole range; it denotes the amount of SO_2 present as bisulfite and/or sulfite. (Refer to Analytical Considerations, Chap. I, B.) Sixty grams of bound SO_2 per liter corresponds to 3% combined SO_2 in acid bisulfite cooks and 6% total SO_2 above cold pH 4.2 ($NaHSO_3$ break at 25°C), or to 37.5 g NaOH per liter up to pH 4.2 and from 37.5 to 75 g NaOH per liter between (cold) pH 4.2 and 10, as the equilibrium shifts from $NaHSO_3$ to Na_2SO_3.

Preliminary work showed that cooks at 60 g bound SO_2 per liter were feasible throughout the pH range. In order to establish a relationship with common cooking processes, lower concentrations of bound SO_2 (2 and 4%) were added at low pH levels, and one higher concentration (8% bound SO_2) was added at the high pH levels.

Comprehensive rate relations in the sulfite field are found for total yield in Fig. 54, and for Klason Lignin in Fig. 55.

The rate curves of yield reduction and delignification bear some resemblance, but deviate in many ways. One reason for the deviations is the variable ratio between yield and lignin in comparable cooks over the whole pH range shown in Fig. 56. Another reason is the effect of bound SO_2 on both lignin left in the pulp and carbohydrate yield (=total pulp yield less lignin content).

Because of the clean separation between the effects of true pH and temperature, the two major kinetic factors in sulfite pulping, it was possible to characterize the optimum value of bound SO_2 at a given temperature, with slower cooking at both higher and lower levels of bound SO_2. In bisulfite cooks of 145 and 160°C the bound SO_2 optimum is around 4%. In acid bisulfite cooks at lower temperature it is about 2% (=1% combined SO_2) since a deficiency of HSO_3, furnished by the base system, can be compensated from the reservoir of aqueous free SO_2 (cf. Chap. I, A). Above pH 4 and up to pH 10 (cold) the optimum effect of bound SO_2 on the cook settles at about 6% and its rate effect amounts to about 45 min longer cooking time per unit increase between 6% and 8%. It appears that, provided that the sulfonating power of the liquor is adequate at a given pH level, higher sulfite (base) concentration causes a reduction of the hydrolytic power of the environment.

In sulfite cooks above pH 10 (cold) the requirement for sulfite as delignification agent is reduced and replaced more and more by that of the free alkali as the pH of the cooking liquor is raised. Its function in this range shifts from

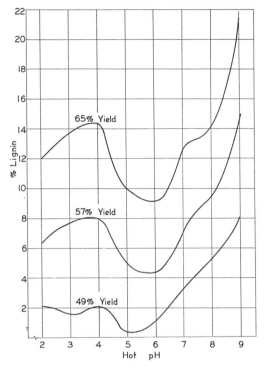

Fig. 56. Lignin profile at three yield levels [1].

delignification to protection of pulp quality in an increasingly alkaline environment. In a typical alkaline sulfite cook of pH 13.5 at the start, the chemical charge can be 22% Na_2O on oven dry wood, divided equally between Na_2SO_3 and NaOH. At a liquor to wood ratio of 4.5:1 exclusive of wood moisture, this would amount to a sulfite charge of 2.5% SO_2 in the liquor.

The most rapid cooks are made at pH 4 and 175°C with 6-8% bound SO_2 applied. The delignification rate of these cooks is greater than that of a standard kraft cook.

Alkaline sulfite cooks at 175°C and acid bisulfite cooks at 160°C with high bound SO_2 (4-160-8) are as fast as kraft cooks at 166°C. All sulfite cooks, if continued, proceed to lower lignin levels than kraft cooks. Points for the industrial sulfite processes, such as "neutral" sulfite, bisulfite and acid bisulfite, are found at or near their expected locations.

4. pH and delignification rate

There is no universal agreement on delignification kinetics in sulfite cooking, and approaches to first-order relationships between residual lignin and cooking time have been reported by various investigators studying confined areas in the field. First-order relationships are indicated by dotted lines in Fig. 55, and it appears that rapid cooks are approximately first order down to around 6% lignin and then

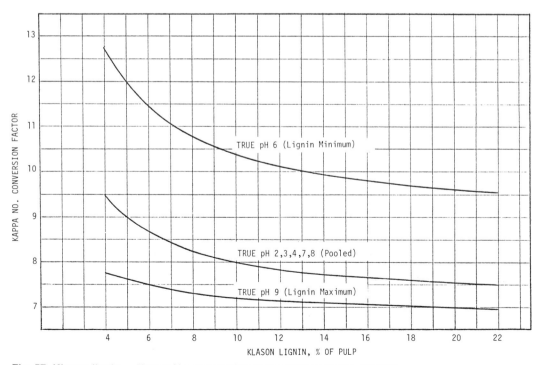

Fig. 57. Klason lignin — Kappa No. conversion in the field of sulfite pulping.

become lower (faster) than first order. Cooks of more than 8-10 h appear to be increasingly higher (slower) than first order, particularly below 6% lignin. There is also a shift of the point of maximum curvature (change to slower delignification) at lower lignin levels with increasing cooking rate. In the fastest cooks at pH 4 and 3, this point appears close to zero lignin.

The effect of bound SO_2 on delignification is positive at all pH levels except pH 6, where 8% bound SO_2 cooked more slowly than 6% bound SO_2.

The above discussion of delignification rate is based on Klason lignin data measured in the pulp. Figure 57 serves the purpose of handy conversion and to bring out two interesting variations of the relationship between data from lignin and Kappa No. tests. It is seen that the factor for converting Klason lignin to Kappa No. varies with the pH level in the sulfite cook and with the degree of delignification. From high levels of residual lignin in the pulp the factor rises increasingly as more and more lignin is removed. Also, the lowest and most consistent conversion factor of about 7 applies to strongly alkaline sulfite cooks in close agreement with kraft cooks, the pooled factor for all pH levels, except 6, runs somewhat higher, and for pH 6 cooks the factor runs exceptionally high.

Reason for this variability between the two lignin tests are losses of acid soluble lignin in the Klason test, decreasing accessibility and solubility of residual lignin in pulp, and oxidation of degraded carbohydrate by permanganate. It will be shown that cooking at pH 6 produces pulp with high hemicellulose content and maximum carbohydrate degradation as indicated by pulp viscosity.

To sum up the pH-time relationship in sulfite cooking, a contour graph for 49% and 65% total pulp yield was prepared (Fig. 58). The comparison of the effects of yield and temperature is based on 6% bound SO_2; some lower levels of bound SO_2 are given in the acid bisulfite range.

Minimum cooking rate to a given yield in this spruce system occurs at hot pH 7. Below and above this level the cooking rate increases rapidly in an almost symmetrical fashion, although the units on the pH scale are logarithmic. The clear exposure of hot pH 7 as the condition for minimum cooking rate in the 160-175°C range answers a question of long standing, at least for the spruce-sodium system. Systems containing other woods and bases can be expected to behave differently in this respect.

The coefficients of hot pH and temperature of

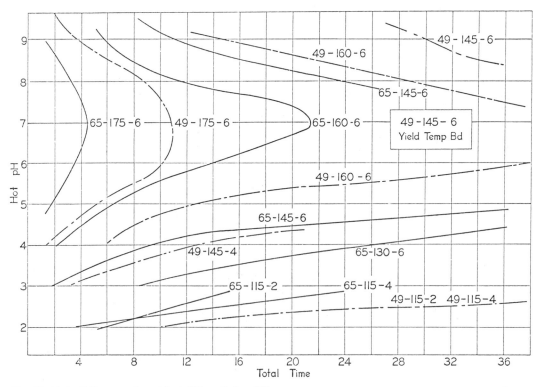

Fig. 58. Hot pH-time contours in sulfite pulping [1].

sulfite cooking time are evaluated in Table 9. The coefficients in this table were obtained from Fig. 58. The temperature coefficient of cooking time (min/°C) was obtained from the angle of the tangent of the yield line at any intersection with a pH level.

The values give the isolated effects of pH and temperature in sulfite pulping. They show the great variations of the temperature effect at different levels of temperature, and pH of the cook and the magnitude of the pH effect at these levels. Comparing similar temperature ranges for cooks reaching 65% yield, it is seen that the effect of temperature is greater at lower cooking temperature levels in alkaline cooks, becomes uniform over a wide temperature range at pH 7 where the effect of pH becomes zero, and becomes greater again in the acid range for lower cooking temperature levels, but much more so than on the alkaline side. The negative pH effect in pH 9 (alkaline sulfite) cooks is about the same at 160 or 175°C, but four times as great at 145°C. Below the pH 7 inversion the positive pH effect increases about six times between 175 and 160°C at pH 6 and about 3 times at pH 5 (bisulfite level). At pH 4 (cold pH about 3) the pH effect increases about 2.3 times for a temperature drop of 15°C.

At 49% total yield, several entries had to be left open because of excessive cooking time at some pH-temperature combinations. In these longer cooks, both the temperature and the pH coefficients are naturally greater than they are in high-yield cooks.

The accuracy of the present evaluation of rate coefficients, though limited by the available number of experimental data, gives a good idea of the magnitude of the individual effects of temperature and pH, and describes the sulfite field in terms that are of both basic and practical importance. Practically, it can be seen that the choice of comparing one centigrade unit with one tenth of a pH unit yielded coefficients of similar magnitude in some areas.

Of practical significance is also the amount of screening rejects to be expected from well delignified pulps cooked at the various hot pH levels. Figure 59 shows that rejects are low in the acid bisulfite and bisulfite range but rise above the neutral pH level. The great sensitivity to yield level and the steep rise at 50% yield show that alkaline sulfite pulps are similar to kraft pulps in this respect. However, the energy required to reduce alkaline sulfite screenings to usable pulp is considerably less than for kraft pulp which retains larger amounts of lignin which is not sulfonated.

Considering responses over the entire pH range, rejects increase with the cooking temperature and decrease with the amount of bound SO_2 charged up to pH 10 cold, and the amount of alkali charged above pH 10 cold.

In conclusion, this section sketches the rate of hot or true pH measurement and control in the recent expansion of the sulfite field to its complete pH range by adding the area of kraft-like (kraft = strong) pulps in the strongly alkaline range. It also describes the role of pH (hydrogen and hydroxyl ion acitivity) as the principal kinetic factor governing the varied delignification and carbohydrate reactions resulting in distinct pulp compositions and physical properties in different areas of the sulfite field.

With regard to the other three bases the present delignification results with sodium base can be compared in the acid bisulfite range which is common to all four.

Comparisons of the bases have been attempted by several authors (cf. Ref. 14, p. 568; Ref. 15, p. 70-80) with contradictory results which, however, permit the conclusion that the rate of delignification is somewhat increased by the use of soluble base, and particularly ammonium base which features a relatively high diffusion constant (cf. Chap. II, A, 2). Although the differences in rate and selectivity of delignification are not great [4], mill application has exposed several advantages, such as reduction in cooking temperature or time, and higher pulp yield and viscosity at a given lignin level combined with reduced rejects. All this indicates better penetration with cooking chemical and a more uniform cook when changing from calcium base to magnesium, ammonium, or sodium base. The use of ammonium base in particular has been advantageous for acid bisulfite pulping of hardwoods.

Another advantage of using soluble base is that the pH can be raised and thereby the softwood basis can be extended to include pines in general and jack pine in particular. Phenolic substances in the heartwood of pines prevent adequate sulfonation and cause high reject rates with the two bivalent bases, calcium and magnesium. Sodium and ammonium base liquor

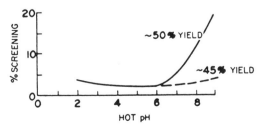

Fig. 59. Relationship of screenings and pH [1].

Table 9. Temperature and pH coefficients of sulfite cooking time [1].

Type (pH-Bound SO₂)	65% Total yield					49% Total yield				
	Temp. coefficient		pH coefficient		Total time, h:min	Temp. coefficient		pH coefficient		Total time, h:min
	min/°C	Temp. range, °C	min/0.1 pH	Temp., °C		min/°C	Temp. range, °C	min/0.1 pH	Temp., °C	
9-6	17.0	160-175	-14.0	175	1:10	43.3	160-175	-25.0	175	3:40
	20.8	145-160	-12.5	160	5:45	61.0	145-160	-82.5	160	14:40
			-44.0	145	11:05			(-48.5)	145	30:00
8-6	33.2	160-175	-10.6	175	3:20	80.0	160-175	-23.0	175	8:40
	47.9	145-160	-61.0	160	11:30	ND[a]	145-160	-82.5	160	28:40
			-79.0	145	23:30			>-100	145	38 +
7-6	67.5	160-175	0	175	4:30	236	160-175	3.1	175	10:50
	69.5	145-160	0	160	21:20	ND	145-160	77.5	160	43:00
			-82.5	145	38:20			ND	145	ND
6-6	42.0	160-175	8.5	175	3:45	112.0	160-175	14.5	175	10:10
	>300	145-160	52.08	160	14:00	ND	145-160	125.0	160	38:00
			>200	145	ND			ND	145	ND
5-6	19.4	160-175	11.5	175	1:45	30.0	160-175	26.5	175	5:15
	124.0	145-160	30.0	160	6:25	>130	145-160	105.0	160	12:40
			~330	145	44:00			~330	145	~45
4-6	6.0	160-175	12.5	175	<1:00	18.0	160-175	20.0	175	1:25
	28.4	145-160	21.0	160	2:10	20.5	145-160	21.0	160	5:50
	62.0	130-145	66.0	145	9:15	NE[b]	130-145	90.0	145	~11:00
3-6	26.6	130-145	150.0	130	26:30	NE	NE	NE	130	NE
			36.0	145	1:50	NE		NE		NE
			66.5	130	8:30	NE		NE		NE
3-4	NE	NE	NE	NE	NE	NE	NE	48.5[c]	145	3:05
3-4	NE	NE	150.0	115	27:00	NE	NE	~230	115	>50

[a] Not determined.
[b] Not explored.
[c] The time:pH ratio of 40 min:0.1 pH previously determined [24] in a small area of the acid-bisulfite region (hot pH 2.8-3.2, 140°C, 50% yield) shows good agreement with the present slope.

allow uniform digestion of pine heart-wood; magnesium base takes an intermediate position between sodium and calcium base.

At the level of bisulfite pulping the three suitable bases, ammonium, sodium and magnesium, produce pulps of similar yield and delignification at similar temperature and time. The only major difference is the lower brightness of ammonium base bisulfite pulp which has precluded this pulp from the high yield and ultra high yield developments taking place in the last twenty years.

At the level of semichemical sulfite pulping the remaining two bases, sodium and ammonium, give similar yield and product properties. In spite of its lower cost, only a few NSSC mills use ammonium base.

5. Light absorption techniques

Ultra violet absorption has been investigated since a pronounced peak at 280 nm is specific for dissolved lignin. However, due to interference from other organic substances formed during the cook (furfural, hydroxyl methyl furfural), bisulfite and sulfite ions, and variations of the absorptivity of dissolved lignin itself due to condensation, it lacks the necessary precision for delignification and point determination. Better results were obtained at 205 nm, but in both cases the absorbency changes much less in the last cooking phase than that of visible light which is therefore most suited.

Pure bisulfite solutions with or without free SO_2 are colorless with some absorption in the UV range. Fresh mill liquors are slightly yellow due to some used liquor being returned to the liquor making system from top and side relief. During the cook the color of the liquor gradually changes through yellow to orange in high yield cooks, to reddish brown in soft paper pulp cook, and finally to dark brown or black in dissolving pulp cooks. The chemistry of color formation is not known in detail. Color development is considered to be a function of the reactions consuming bisulfite ions and generating acidity, and therefore well suited to measure end points as the result of the rate effects of temperature and pH.

The method consists of shining unfiltered or filtered visible light through a controlled thickness of spent liquor onto a photo-electric cell whose output is amplified and processed. Application of the method varies widely from test tube comparisons to dip colorimeters with fiber optics (Fig. 60) and flow colorimeters (Fig. 61) for continuous monitoring and automatic control. Calibration of the instrument is necessary since the color density and hue of the spent

liquor varies with the base and with the wood species. Ammonium base develops much darker color than calcium, magnesium or sodium base, and birch spent liquors are darker and more reddish than spruce spent liquors, for example. The method is very flexible regarding adaptation to various color densities and hues by adjustments of the path length of the light penetrating the liquor, or of the amplification of the photo cell output, and by selecting the dominant wave length which gives optimum performance by filtering.

The method is useful also for end point determinations at the bisulfite level from low to high yields where applications have been reported. It is probably useful also at higher pH

Fig. 60. Industrial colorimeter with filter optics for end point control of sulfite operations.

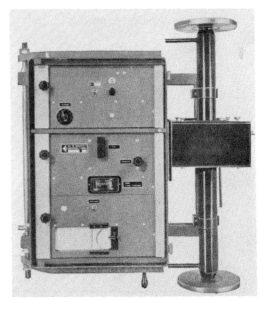

Fig. 61. Photometer for continuous color measurement in closed pulping systems (Courtesy of Sigvist Co.)

levels, although exploration and development is missing in these fields which are the natural domaine of integrated time-temperature cooking control.

6. Residual lignin measurements

The Kappa number test is useful for determining degree of cooking of bisulfite pulps over a wide range of yield. The test results become less precise above 65% yield and some advantage is claimed for the Hypo number test used in some high yield bisulfite mills in Canada [5]. Extensive comparisons made in Finland [6] have shown that the Hypo number (they use the term "Chlorine number") test is more accurate than the Kappa number test in the high yield range and that it has further advantages in linearity with lignin content and freedom from the influence of type of pulp and wood species. The relation of Hypo number and total lignin content is shown in Fig. 62. Unlike the Roe number test, the hypo number test is performed in aqueous solution using simple equipment and is therefore suitable as mill control test. Its accuracy approaches that of the Roe number test and the test values are identical. Generally, high yield bisulfite mills tend to control pulp quality more and more by paying attention to wood moisture, chemical charge, and to time-temperature schedules and less and less by frequent testing and monitoring of pulping variables.

In most NSSC mills and all ultra high yield bisulfite mills above 85% yield, wood moisture monitoring, liquor control and time-temperature control are the sole modes of process control (cf. Chap. VI, H).

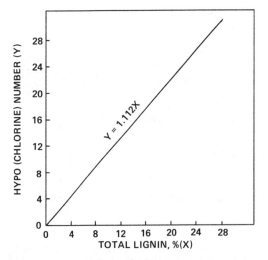

Fig. 62. Hypo (chlorine) number as a function of the total lignin content [6].

B. LIQUOR QUALITY CONTROL AND ANALYSIS

A constantly uniform liquor of good quality is obtainable only with proper regulation of the sulfur-burning process, adjustment of temperature, and the supply of the proper quantity of gas and water under general control by recording and indicating instruments and chemical analyses. Some of the methods in use will be considered.

1. SO$_2$ testing

A recording pyrometer is commonly used for measuring temperature of the gas before cooling. It is possible to use this type of pyrometer wherever it is desired to indicate or record temperatures up to 1650°C.

Sublimation. A simple test for sublimation of sulfur is to bring a cold glass rod in contact with the warm gases in the coolers. If the SO$_2$ gas contains sulfur vapors, these will be cooled down and will deposit as a yellow layer on the glass rod. To correct this, more air must be admitted for combustion. Acid makers sometimes examine the mercury bulb of the angle thermometer usually placed in the gas line after the coolers.

A glass peephole in the blank at the end of the gas header just before the cooler will enable the acid maker to check for sublimation of sulfur by the presence or absence of flame. A furnace-type peephole in the outlet from the combustion chamber is also used. With a spray cooling system, sublimation can be detected by looking at the primary cooler effluent.

Testing of SO$_2$ gas after cooling. Temperature is measured by means of an indicating or recording thermometer placed in the gas line. Vacuum or pressure conditions in the gas line are measured by means of U tubes made of glass. The vacuum before the fan is usually between 20 and 30 mm water column.

The efficiency of the sulfur burner is commonly controlled from determinations of the SO$_2$ content of the gas and adjustments of secondary air flow. Orsat or other absorption apparatus are used for manual analysis of gas samples; burner control is more and more based on on-line thermal conductivity equipment which measures the SO$_2$ concentration in the dry gas.

The content of SO$_3$ in the gas is determined by titrating a solution of the gas sample with sodium hydroxide for total acidity (SO$_3$ + SO$_2$) and with iodate for SO$_2$ content to obtain the SO$_3$ content by difference.

2. Chemical liquor tests

The actual chemical composition of sulfite liquors is accurately determined by chemical analysis. Essentially the methods use an iodo-metric (potassium iodate, KIO_3 or iodine, I_2) titration of SO_2 or sulfite compounds and an acid-base (NaOH or HCl) titration for determining base content. The iodate method is suitable for the acid range of sulfite liquor pH and the iodine method for the neutral and alkaline range.

The iodate method for acid bisulfite liquors was introduced by G.V. Palmrose in 1935 and was adopted as Standard by both TAPPI and CPPA. It is used for the analysis of bisulfite liquors as well. It gives clear and well-defined end-points also in colored liquors.

Procedure: Acid Bisulfite System

To an erlenmeyer flask containing 75 mL of distilled water add three drops of methyl red indicator solution. The solution should be yellow. Add 1 mL (20 drops) of starch-KI solution (do not use more).

Pipette 2.00 mL of the acid into the flask. The solution will be red.

Determination of total SO_2. Titrate with N/8 potassium iodate (KIO_3). Avoid shaking the flask until a blue color begins to appear. Titrate until one drop produces an intense blue. The milliliters of N/8 KIO_3 multiplied by 0.2 is the percent of total SO_2.

Determination of free SO_2. For accumulator acid. Titrate the contents of the flask with N/8 NaOH. The blue of the previous end point will disappear with the addition of the first few milliliters, and the solution in the flask will be red. Titrate until one drop turns the solution yellow. The milliliters of N/8 NaOH multiplied by 0.2 is the percent of free SO_2.

Determination of combined SO_2. Subtract the free SO_2 from the total SO_2.

Note that the values obtained by this method measure one half of the base bound sulfite, adding the other half to the actual free SO_2 dissolved in the liquor. The actual concentration of base bound SO_2 (bisulfite) is twice the value of "combined SO_2" (see Chap. I, B).

When using the method for raw acid bisulfite liquor from the absorber, the blue will not be discolored on titration with NaOH, so it is necessary to add one drop of 3% sodium thiosulfate to remove it. Do not use more. Titrate with N/8 NaOH as above. The difference between total and free SO_2 is the combined SO_2.

Standardization of the KIO_3. To 100 mL of distilled water in an erlenmeyer flask add 10 mL of a 30% KI solution and 10 mL of 1N H_2SO_4. Pipette 25 mL of KIO_3 into the mixture, and allow 3 min for complete reaction. Titrate with

standard sodium thiosulfate, using starch as an indicator when near the end point. It is necessary to run a blank on the reagents each time.

Procedure: Bisulfite Systems

When using this method for bisulfite liquor with a pH above 2.0 it is necessary to add N/8 HCl solution after the determination of total SO_2 until the color is distinctly red plus about 2 mL acid to the nearest unit. Note the mL of N/8 HCl required for acidification and subtract from the mL of N/8 NaOH used in the determination of free SO_2. In bisulfite solutions near pH 4 the total SO_2 is equal to twice the combined SO_2 or the actual base bound SO_2 concentration, that is equivalent to the bisulfite concentration.

The Palmrose method has been modified for electrometric end point determination by using a platinum electrode for the iodometric titration and a glass electrode for the acid base titration [4]. The advantages over the visual method are an improved acidimetric end point in used cooking liquor, adaptation to automatic titrators, and capability to record titrations.

In sulfite cooking liquors above pH 4 the total SO_2 alone is required for measuring the base bound SO_2 in the form of bisulfite and/or sulfite. Table 1B gives the proportion of bisulfite and sulfite over the range of pH 4-10 measured at room temperature.

Procedure: Neutral Sulfite System — Carbonate Present

Neutral sulfite solutions require different testing methods depending on whether they contain carbonate besides sulfite or only sulfite. If carbonate is present, the pH of the fresh solution varies linearly with the sulfite-carbonate ratio (S/C expressed as sulfite or, in many cases, as carbonate). A standard method has not been adopted so far. A suitable procedure follows.

1) Add 1 mL of solution from pipette into a flask with about 50 mL of cold water. Add a few drops of iodometric indicator (thydine) and titrate with N/10 iodine solution to a blue end point. Note mL consumed.

$$mL \ N/10 \ iodine \times 6.3 = g/L \ sulfite$$
$$\times 5.3 = g/L \ sulfite \ as \ carbonate$$

2) Dip the end of the sample pipette into the titrated solution to destroy the blue color and add a few drops of mixed (bromocresol red plus methyl red) indicator. Titrate the solution turned red with N/10 NaOH to a blue-grey end point.

mL N/10 iodine − mL N/10 NaOH × 5.3
= g/L carbonate
mL N/10 iodine − mL N/10 NaOH × 6.3
= g/L carbonate as sulfite

$$\text{Ratio S/C} = \frac{\text{mL iodine}}{\text{mL iodine} - \text{mL NaOH}}$$

If the solution after the first titration does not turn colorless when adding a trace of sample but instead light blue, the S/C ratio is smaller than one which indicates that sulfitation of carbonate by SO_2 was insufficient in liquor preparation. Another important consideration is that the pH of a neutral sulfite solution varies with time and temperature by up to 0.5 unit due to irreversible conversion of bicarbonate into carbonate and liberated CO_2 gas slowly escaping from the liquor. Although pH is an approximate indicator of the S/C ratio, only analytical determination can give precise liquor composition and characteristics.

TAPPI Useful Method 624 offers an alternate procedure, and is presented in Chapter VI, D, 3.

If the liquor is made from caustic and sulfur dioxide, carbonate is absent. The base concentration is then given by the amount of NaOH used and the pH is adjusted by adding SO_2 gas to the highly alkaline solution until the required level is reached. When preparing liquor in batches, the rate of SO_2 flow should be stopped as the pH change becomes more sensitive and stopped before reaching the required pH value to equalize the solution. Small amounts of SO_2 are then added to reach the specified pH. In continuous liquor making the flow rates of caustic and SO_2 are adjusted to obtain the specified pH from the mixing vessel.

For testing, only the first titration of the above testing procedure is required:

mL N/10 iodine × 6.3 = g/L sulfite.

In this type of liquor only pH and sulfite concentration need to be recorded.

Alkaline sulfite testing procedures differ again for liquors without and with carbonate. Testing methods for these liquors have not been published so far. Suitable procedures are given below.

Procedure: Alkaline sulfite liquor without carbonate (case of a liquor of high alkalinity)

To a 250 mL beaker containing 75-100 mL of distilled water add 2 mL sample from pipette. Place on magnetic stirrer and immerse pH electrodes.

1) Titrate with N/10 HCl to pH 9 and record mL spent (a). Add excess N/10 or N/1 HCl to pH 2.5. Do not record mL of acid.
2) Add starch indicator and titrate with N/8 KIO_3 solution to blue color, as for acid bisulfite liquors. Record mL of N/8 KIO_3 used (b).

If this procedure is followed, even the darkest spent cooking liquor gives a light yellow solution easy to titrate.

NaOH g/L = 2 × a
NaOH as Na_2O g/L = 1.55 × a
Na_2SO_3 g/L = 3.94 × b
Na_2SO_3 as Na_2O g/L = 1.94 × b

Example:
Used 23.48 mL N/10 HCl to pH 9.0
    23.48 × 2.00 = 47.0 g NaOH/L
    23.48 × 1.55 = 36.4 g Na_2O/L
Used 17.32 mL N/8 KIO_3 to starch end point
    17.32 × 3.94 = 68.2 g Na_2SO_3/L
    17.32 × 1.94 = 33.6 g Na_2O/L

Total Na_2O = 70.0 g/L

At 3:1 liquor-to-wood ratio this liquor would provide a chemical charge of 21% Na_2O on oven dry wood at a sulfite to caustic ratio, as Na_2O, of $\frac{36.4}{33.6} = 1.08$.

Procedure: Alkaline sulfite liquor with carbonate and caustic soda, as for AS-AQ cooks

This procedure is based on the test for neutral sulfite liquor with carbonate to which a separate determination of NaOH concentration is added.

1) Add 1 mL of sample into a flask with about 50 mL of distilled water. Add a few drops of iodometric indicator (thydine) and titrate with N/10 iodine solution to a blue end point (a).

mL N/10 iodine × 6.3 = g Na_2SO_3/L
        ”   × 3.1 = g Na_2O/L as sulfite

2) Dip end of sample pipette into above solution to destroy the blue color, then add a few drops of mixed indicator (bromocresol red plus methyl red). Titrate with N/10 NaOH to a blue-grey end point and note mL N/10 NaOH used (b). In alkaline sulfite liquor containing both caustic and carbonate this second titration will measure both:

(mL N/10 iodine − mL N/10 NaOH) × 3.1
= g Na_2O as NaOH + Na_2CO_3

A separate determination of NaOH after precipitating the carbonate is necessary for obtaining the concentration of carbonate.

3) To a 500 mL centrifuge bottle add 25 mL of liquor sample and 100 mL 10% BaCl₂ solution, dilute to 200 mL, mix and centrifuge for 5 min. Transfer 25 mL clear sample by pipette to a flask, add a few drops of phenol phthalein indicator and titrate with N/10 HCl to a sharp end point (pH 8) (c).

$$\text{mL N/10 HCl} \times 1.28 = \text{g NaOH/L}$$
$$\text{"} \times 0.992 = \text{g Na}_2\text{O/L}$$

To obtain carbonate as Na₂O subtract the Na₂O value of titration c from that of titration b,

$$\text{g Na}_2\text{O/L (NaOH + Na}_2\text{CO}_3) - \text{g Na}_2\text{O/L (NaOH)} = \text{g Na}_2\text{O/L (carbonate)}$$

To obtain the total Na₂O concentration of the liquor add Na₂O values a+b. The percent Na₂O ratio of the three chemicals is finally calculated from the total Na₂O concentration.

Example:
1) Used 18.52 mL N/10 iodine (a)
   18.52 × 6.3 = 116.68 g Na₂SO₃/L
   18.52 × 3.1 = 57.40 g Na₂O/L as Na₂SO₃
2) Used 14.44 mL N/10 NaOH (b)
   (18.52 − 14.44) × 3.1 = 12.6 g Na₂O as NaOH + Na₂CO₃
3) Used 5.65 mL N/10 HCl (c)
   5.65 × 1.28 = 7.22 g NaOH/L
   5.65 × 0.992 = 5.60 g Na₂O/L as NaOH
   Carbonate: 12.60 − 5.60 = 7.00 g Na₂O/L
   = 11.97 g Na₂CO₃/L
   Total Na₂O: 57.40 + 12.6 = 70 g/L
   Chemical ratio (% Na₂O): 82-10-8

The accuracy of titrations of 1 mL or 2 mL liquor samples can be greatly improved by adding a 10 mL sample from a pipette to a 100 mL volume in a flask, diluting with distilled water to 100 mL, and using 10 mL or 20 mL aliquotes for titrations. This will greatly reduce the pipette error.

Alkaline sulfite liquors and neutral sulfite liquors containing only sulfite and carbonate are similar and are tested by the same method. Some NSSC mills use sulfite with varying amounts of carbonate and caustic soda, that is three component liquors similar to the moderately

alkaline sulfite liquors. However, the pH of alkaline sulfite liquors and analysis of the alkaline liquor components are less critical for the high yield NSSC process than for the alkaline sulfite process producing fully delignified pulp.

Refer to Chap. VI,H and Chap. VII,A,4a for discussion and methods of digester yield determination.

C. WHOLE PROCESS CONTROL AND AUTOMATION OF SULFITE PULPING

An attempt to integrate the effect of cooking time and temperature of sulfite cooks was made in the thirties [16] at PAPRICAN. This early work was successful in expressing the rise-to-temperature portion of a batch-type acid bisulfite cooking curve in terms of equivalent cooking time at maximum temperature, though it did not accomplish the final step of expressing the whole cooking curve as a single factor. Based on this work, the first integration of complete alkaline (kraft) cooking cycles was produced in the fifties [17] but without having any impact on commercial process control practices. Again based on the early work at PAPRICAN, a practical method for integrated kraft cooking control by time-temperature integration was then published [18], the numerical value being given the name "H-Factor". This eventually unlocked the present era of automatic digester and cooking control in pulping.

Applications in sulfite pulping, though much less numerous than for kraft, developed contemporarily: first in bisulfite pulping, then in acid bisulfite pulping and finally in NSSC pulping; where no direct feedback of pulp quality data is available.

1. Batch digester plant control

The original direct manual operation of the batch cooking equipment functions has been gradually replaced by analogue control from a central instrument room. Control instrumentation found on individually controlled batch digesters includes temperature and pressure sensors and a controller/recorder, liquor level recorder and circulation flow recorder (if applicable) and remote valve actuation buttons, as shown in Fig. 90. The instruments are located in the control room on the cook's floor at the top of the digesters. A chart record of the schedule of two acid bisulfite cooks is shown in Fig. 63. Blow valves actuated from the control room are now almost standard. A terminal of liquor sampling lines from the individual digesters is usually found nearby.

Batch plants may contain from 4 to 40

individual digesters. Starting in the mid-sixties a considerable number of such plants has now been put under automatic control. The procedure for converting from manual or semi-manual operation to full control takes a number of steps, but always without interrupting normal production. The greater the number of individual digesters in the plant the easier it is to phase out one or more digesters for installation of automatic valves, sensors, etc. without affecting the level of production, since the productivity of the remaining digesters can be increased somewhat, for instance by raising the level of maximum cooking temperature. Along-side installations for automatic chip transport to and filling of the digesters as well as automatic liquor charging may be made. Existing analogue control instrumentation (recorders, controllers, indicators, etc.) may be improved and additional functions added. All pneumatic controls will be adapted to give electrical output. In general, preparing for automated systems presents a unique opportunity for updating and upgrading all functions of the pulping department, and to develop a complete operator-equipment interface. In some cases preparations for automatic control of pulp washing and bleaching are made at the same time. At this level of automation the entire system is run by the operator in the control room.

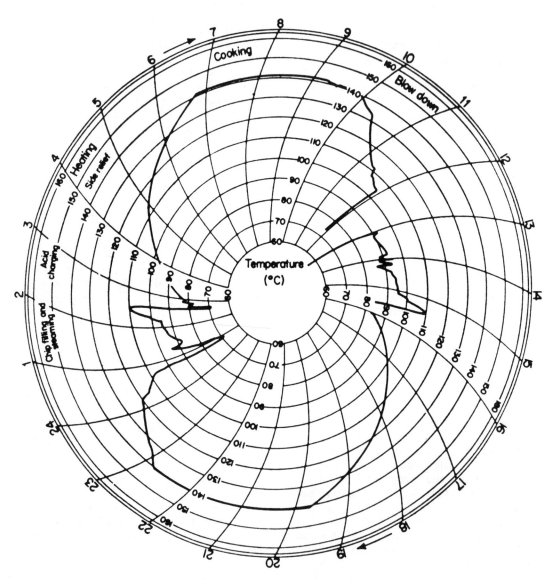

Fig. 63. Temperature diagram of two sulfite rayon pulp cooks [14].

The next step is to provide an analogue-digital interface where the analogue electrical signals are converted into numerical bits required by the computer. With this interface completed and a computer of sufficient capacity installed, the sequence of charging, cooking and discharging of any number of batch digesters can then be run by the computer program with complete supervisory power on process decisions by the operator as in Fig. 64. When necessary, individual digesters can be taken out of operation for repairs, or cooked manually for trials, or to correct for some malfunction in the automatic schedule. At this level scheduling of the individual cooks of the group is introduced, and the benefits of automatic operation are realized, such as improved yield and quality due to reduced Kappa number fluctuations, increased rate of production and energy savings due to "levelling of peaks in steam demand" [7].

The final step is the fully automatic computer control system where the supervisory function of the operator is exerted at the beginning of the charging cycle, at the start of blow back and at initiation of the blow; and concerns mainly a precise cooking end point (H-factor value), safety, warning signals by the system, and maintenance of the equipment.

2. Control kinetics

Any discussion of process control applications to pulping systems must start with the admission that all mathematical formulations relating degree of cooking with temperature and time are approximations based on empirical data and assumptions. Such expressions imply nothing about the mechanism of the cooking reactions involved or about strict order of rate (see Fig. 55), and numerical quantities are merely derived from empirical methods of accounting for variations. This is due to the complexity of any pulping reaction where the major and minor parameters as well as the composition of the heterogenous raw material change with time. Only in a system with well-defined reactions would the quantities in rate expressions assume some theoretical significance.

Nevertheless, and in spite of these theoretical shortcomings, the empirical approach to full

Fig. 64. Automatic sulfite pulping of five batch digesters. The operator controls every process in the entire digester house centrally at the process panel. All reports are written by the teletype. The display shows the state of each digester in the schedule [Courtesy Nokia Electronics].

cooking control has been remarkably successful in recent years, no doubt because of the high speeds and storage capacity of present computing devices supported by greatly improved control of digester functions from charging through heating and cooking to discharging of the product.

A simplified and idealized form of the delignification rate equation could be written:

$$r_L = kC^a L^b$$

where r_L is the time dependence of residual lignin, k is the rate constant, and a and b are exponential constants with respect to liquor concentration and residual lignin in the wood.

The rate constant k is theoretically related to the absolute temperature T (cooking temperature in °C plus 273) of the process by the Arrhenius equation:

$$k = Ae^{-E/RT}$$

where A is a constant, E is also a constant bearing the value of the Arrhenius activation energy, and R is the gas constant written in quantities which express E in the required energy units. The basic "H-Factor" is the integrated value of kdt.

Table 10 provides a listing of kinetic values of different pulping systems from literature sources. These are empirical data useful for making comparisons and for writing basic rate equations throughout the sulfite field.

As stated before, pulping reactions are a complex interaction between primary, second-ary and tertiary variables and pulping rate equations are neither theoretical nor accurate. Thus, to be of use in practical applications, extensive model research and development work was necessary to arrive at dynamic simulation models with modified constants which predict the course of the pulping process accurately.

3. Control strategies

From the development efforts made during the past decade the method now adopted [8] for putting a pulping process under complete control is:
1. To start with the assumption of first order reaction kinetics and Arrhenius-type temperature dependence as a basic structure;
2. To improve estimation of the effects or control of initial state conditions (wood and chip quality, wood moisture, liquor concentration, chip and liquor temperatures, time of temperature rise, etc.); and
3. To continuously adapt the general model to the demands of the actual process by dynamic simulations using mill data from all variables which are practical and can be reliably measured (material flows, amounts and concentration of chemicals, history of the cook at different stages – memory functions, temperatures, pressures, etc.) in a kind of evolutionary learning process.

The task of controlling a primary variable (e.g. lignin content of pulp) is understood from the simplified diagram in Fig. 65. Also indicated is actual measurement of the primary variable at some point, or points, during the cook, which would be most important for improving the

Table 10. Kinetic values of different pulping systems.

PROCESS	WOOD	RATE FUNCTION	ACTIVATION ENERGY kJ/mol	REACTION ORDER	TEMPERATURE COEFFICIENT per 10°C	REF
Acid bisulfite	Spruce	Delignification,	83.7-90.0	1 (assumed)	$\cong 2$	19, 20
Acid bisulfite		Hemicellulose hydrolysis	117			
		Cellulose hydrolysis	117-184			
Bisulfite	Spruce	Delignification	121.3		$\cong 2+$	
Bisulfite	Spruce	Delignification, Rate proportional to pH, no effect of HSO_3 concentration	150.6		3.5	21
NSSC	Aspen	Delignification			—	
		initial	150.6	2.5		
		main	122.2	1.2		
		Alpha cellulose	174.9	0 (main)		
		Pentosan	128.9	0 (main)		
Soda (NaOH)	Spruce	Delignification	121.3	1	$\cong 2$	22

accuracy of hitting the movable quality and time window. At present, however, reliable, easy-to-maintain samplers and analyzers for chips and liquor from the digester are not available. Several types of analyzers (colorimetric, spectrometric, titrometric, conductometric) have been tried to measure secondary process variables but have not been too successful so far in continuous industrial use because of sampling problems, low reliability, and investment and maintenance cost. Although primary and secondary analyzers are a most useful aid for arriving at the control target, they do not give process control because they do not provide information on the relation of variables needed for control purposes, and do not consider initial conditions.

4. Acid bisulfite control

In acid bisulfite pulping for paper and dissolving grades all early kinetic models proposed included some form of measuring free SO_2 (pressure) in order to account for variations or the dominant kinetic effects of the hydrogen ion concentration on delignification and carbohydrate depolymerization. A method for measuring hot pH was not available at that time and later, when it became available, its maintenance in an industrial environment was found to be difficult. The same applies to other methods for

the continuous measurement of primary and secondary state-of-the-process variables. It must be repeated also that even direct measurement of a primary process variable would not fulfill the requirement of taking the initial conditions of the process into account which can cause serious inaccuracy.

The acid bisulfite control models which did evolve and which are now in commercial use in various mills are modified H-Factor models resulting from extensive development work in Finland with the first computerized acid bisulfite digesting application in 1974. The second installation was at a modern calcium base acid bisulfite mill in Germany producing bleached paper and high quality dissolving pulp from beech and spruce [9]. Spent cooking liquor is recovered with 99% efficiency and lignin preparations, methanol and furfural are recovered as by-products.

Starting with a general sulfite control model, beech dissolving grade pulp was the first to be put under automatic control. During the "tuning" period of 5 months the cooking end point was determined by color measurements and the results were statistically evaluated and compared to the simulation model of the control system until the correct model was established. Separate simulation models were obtained for paper and dissolving pulps, i.e. for Kappa

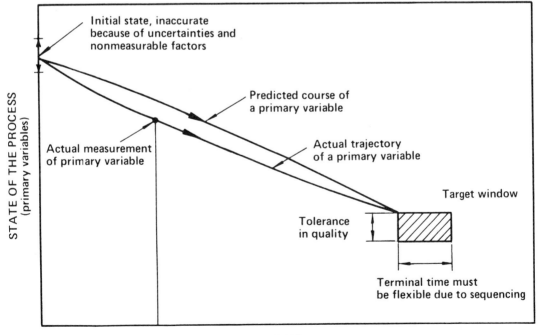

Fig. 65. The control of a batch process is a problem of controlling the course of primary variables [8].

number or viscosity end points. The target for dissolving pulp is high alpha cellulose content with a minimum of lignin and hemicellulose. This means that the Kappa number end point cannot be used. Dissolving pulp production also requires changes of the chemical charge and of the time-temperature relation. Moreover, as in this case, more than one wood species is involved and eight specified pulp grades are produced. Therefore, the simulation models used had to be adaptive and easily modifiable.

Of the range of control functions offered by the automatic system, only the following needed to be activated for accomplishing control:
1. monitoring and control of digester charging,
2. steam and cooking temperature control,
3. control of pulp quality by the cooking degree model,
4. control of digester blow back and liquor displacement,
5. monitoring of blow, and
6. process monitoring and recording.

The benefits reported by this mill due to the introduction of automatic cooking control were increased production, more uniform pulp quality, and better delignification which, in turn have significantly lowered both the consumption of bleach chemicals and the water pollution load.

5. Bisulfite control

Attempts to control the bisulfite pulping process began in the late fifties as a consequence of a major growth of interest in this process earlier in the decade. This new development in the pulping field was due to the tendency of raising the yield of sulfite pulp for disposable news and other publication grades. Calcium acid bisulfite liquors used earlier for this purpose cannot be prepared at a pH level higher than 2 where the effect of lignin and carbohydrate hydrolysis is still strong and pulp yields and strength are relatively low. Meanwhile almost all production facilities of this kind have replaced calcium base by sodium or magnesium base and are operating at pH levels from about 3.5 to 4.5. In Canada, where the idea of higher sulfite yield was put originally into practice (PAPRICAN, Abitibi) strong preference was given from the outset to sodium base which offers the advantage of complete flexibility in pH, stability and ease of cooking liquor preparation. This contrasts with Sweden, for example, where preference has been given to magnesium base.

The tendency to produce more pulp from a given weight of wood, once the low pH barrier was overcome, did not halt at a similar degree of cooking, but swiftly took advantage of the newly discovered possibility characteristic of the sulfite process to expand into the high yield range at 60-70% yield from wood with the aid of mechanical fiberization. Most recently, the yield range was extended to its practical limit of 90% ("ultra high yield") with the help of sodium base liquors at pH 4 and at pH 8 (refer to Chap. VII).

Although bisulfite pulps can be made at any yield level, the field is dominated by high yield pulping; and control systems have developed in this range.

Early studies were aimed at finding the temperature coefficient of the rate of delignification and the magnitude of the effects of bisulfite concentration and pH level. Conflicting results for the activation energy were obtained in these experiments, ranging from 96 to 150.7 kJ. It was found [10] that the reaction rate changed little even with substantial changes of bisulfite concentration and that pH was the controlling factor.

The following is an example of a full process control application to a high yield bisulfite batch operation using mainly Eastern spruce wood [11]. The mill was faced with a major problem of variable pulp quality when the yield of sodium bisulfite pulp was raised from 52% to 65%.

When producing high yield pulp it is important that uniform pulp quality is obtained in the cooking stage in order to minimize quality fluctuations in the refining stage. The power required for refining to a given freeness is highly dependent on the degree of cooking, measured as Hypo number. If both the Hypo number fluctuations and the blow frequency at the cooking stage are high, it is practically impossible to compensate in the refining stage for non-uniformity.

Initial attempts to improve the reproducibility of cooks included better control of SO_2 content and pH of the cooking liquor and monitoring of cooking progress by measurement of SO_2 consumption. The results were still unsatisfactory until it was discovered that a modified H-factor model for sodium bisulfite pulping correlated better with the Hypo number data than with SO_2 consumption (Fig. 66.) At this stage it seemed promising to calculate the H-factor every 2 min and to blow the digester down as soon as a predetermined value was reached. It soon became evident, however, that only moderate improvements could be obtained as long as the digesters were under manual control.

The final decision was then made to develop, in cooperation with a supplier, a computerized batch digester system providing the following

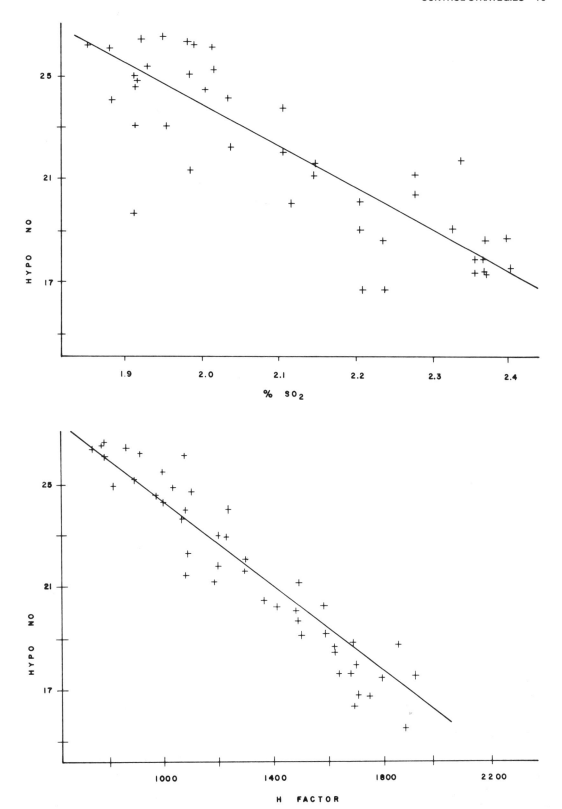

Fig. 66. H-factor and SO₂ consumption vs Hypo No. of a high-yield sodium bisulfite pulping operation [11].

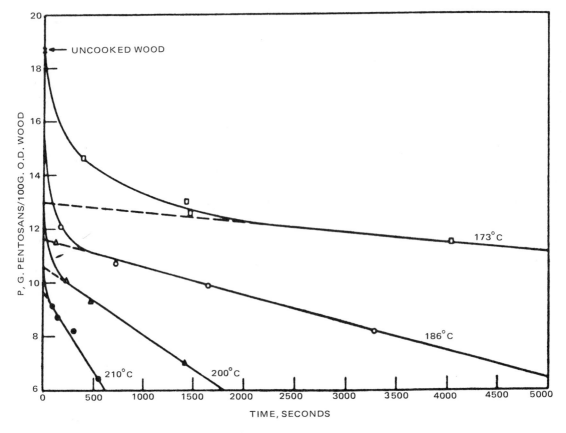

Fig. 67. Pentosan content as a function of time during neutral sulfite pulping [12].

functions:
- Digester monitoring
- Temperature and pressure control, inclusive of temperature rise and blow down
- Steaming rate control
- Blow time prediction
- Steam levelling by enforcing the production schedule
- Hypo number control with a feedback adaptive feature to update the mathematical model in the computer.

As a result of automatic control the range of Hypo number variation was reduced 30%, of freeness variations 55% (helped also by a continuous freeness analyzer) and the variations of steam demand flow substantially reduced.

6. Neutral sulfite control

Detailed information on the kinetics of neutral sulfite pulping is available for the case of aspen wood [12]. This work did consider cellulose and pentosan removal besides delignification during cooks at temperatures from 166°C to 210°C. Delignification is usually the major concern in pulping; the rate of cellulose

degradation is associated with losses in pulp yield and strength; and pentosan content is the most important pulp component determining NSSC strength properties. Knowledge of the relative rate of removal of these components is essential for process design, particularly with regard to the right cooking temperature for optimum production rate and pulp quality.

It is found that removal rates of the three pulp components are high during the first 10 min of the cook and then follow approximately a first order reaction for lignin removal and steady decline (zero order reaction) for the carbohydrates during the cook. This is illustrated in Fig. 67 for xylan (pentosan) content of the pulp. The temperature dependence of removal rate of the three components is found to differ significantly. At higher cooking temperature the initial, more rapid delignification reaction becomes more important and the degree of delignification increases. Alpha cellulose, on the other hand, suffers an initial loss of about 8% (based on the original content of 56% in oven dry wood) quite independent of the level of temperature. However, the steady alpha cellulose loss during the cook is more temperature dependent than delignification and trying to decrease cooking time by increasing the temperature can become counter productive with regard to pulp yield and quality.

Unlike alpha cellulose, xylan is removed more rapidly in the initial period at higher temperature of the cook, and up to one half of the pentosan content may be lost during this short period at temperatures above 200°C (Fig. 68).

The conclusions drawn from this kinetic analysis are that the neutral sulfite process is most suitable for producing a high yield, relatively low lignin pulp. Compared with the kraft process it has the advantage of being considerably more selective in short cooks, removing a larger portion of lignin at the same yield. The production rate can be increased by raising the cooking temperature but pulp quality is compromised.

Complete process control of a NSSC plant is discussed by virtue of a recent application in France [13]. The original installation consists of a 350 t/d M&D continuous digester (see Fig. 69) pneumatically supplied with hardwood chips from two piles. Facilities are provided for washing the chips and for presteaming and impregnation with ammonium base cooking liquor with a pH of 5-5.5. Cooking at 170-175°C takes about 25 min and the product is fiberized in two disk refiners using 145 kWh/t of pulp. The yield of pulp is 79% calculated on dry wood. Processing steps include 3-stage pressurized drum washing, high density storage, waste paper addition, screening and paper (corrugated medium) making. The most important objective in this type of process control application is to develop a practical mathematical model for digesting and refining where no direct feedback measurement related to pulp quality is available.

The immediate product after cooking is still wood-like and no Kappa number, chlorine number or similar test is feasible. Even after refining, such analyses do not give useful results. Refining has a profound influence on pulp quality and it is very difficult to decide whether an observed change is caused by chemical treatment or changes of the refiner disk gap or wear. There are no pulp tests available in NSSC technology which allow one to predict runnability of the pulp on the corrugator and its final quality.

Two mistakes are often made initially: (1) to choose variables too difficult to measure or to design formulas too complicated for realistic practical mode; (2) to assume that a formula as

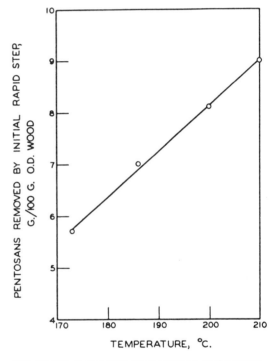

Fig. 68. Effect of pulping temperature on the initial removal of pentosans during neutral sulfite pulping [12].

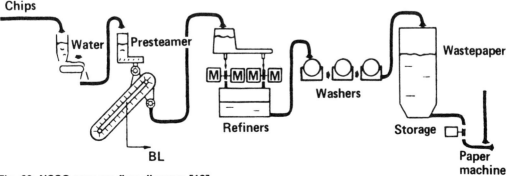

Fig. 69. NSSC process flow diagram [13].

such is capable of controlling the process. The time for progressing from this stage to a real applicable model can be long.

In this particular case a way was sought: (1) to simulate the process so that it would be possible to maintain uniform pulp quality with constant and varying values of production rate and other primary variables in the process; (2) to control pulp quality in a desired way through changing the primary variables of the process.

A major improvement of digester instrumentation was made by adding four temperature sensing locations to the two of the original installations. The result was a realistic temperature profile as never seen before (Fig. 70).

Several months were spent on planned mill-scale experimentation without loss of production, collecting data of measurements and laboratory analyses and giving special consideration to specific time constants of the process, standard energy demands, pulp properties and cooking liquor variations. Formulas were also developed on the basis of past experience, to represent an H-factor-like component expressing dependance of reaction rate in the final model.

Finally, the simplified, practical model was obtained for programming the computer. It used as input: (1) the cooking time and temperature profile; (2) amount of chemical charged; (3) liquor-to-wood ratio; and (4) production rate. The pH of the spent liquor is used as the "fast" feedback and the specific refining energy demand per tonne of fiberized pulp as the "long-term" feedback. The target used for the model is "specific energy", that is the amount of kWh the operator wants to spend for refining a given amount of pulp. This, in turn, represents a primary pulp quality factor, termed "hardness" which replaces the Kappa No. used for softer pulp grades. Based on this target setting, the cooking model controls the required cooking temperature, and the "fast" pH feedback helps to do this correctly and to compensate for input fluctuations such as moisture content and cooking liquor composition.

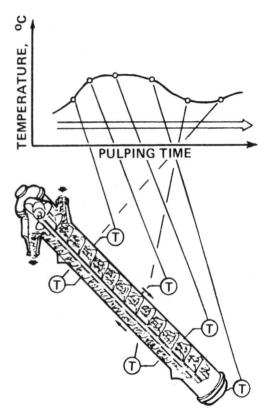

Fig. 70. Actual M&D digester temperature profile [13].

Process control has brought major operational changes. For instance, when changing production rate, only the new value is fed to the computer which in turn controls the presteamer feeding screw, inlet valve, digester chip conveyor chain, outlet valve, chemical dosage, temperature, and more.

Desired changes in pulp quality can be made by giving a new target for the "specific energy". But the main advantage of the system is that it maintains a given pulp quality when changing production rates. Although not one of the original goals, a 10% increase in production has been reported.

REFERENCES

1. Ingruber, O.V. and Allard, G.A., The effect of hydrogen ion activity in the sulfite pulping of black spruce. *Tappi* 50, 597-614 (Dec. 1967).
2. Ingruber, O.V., Alkaline sulfite pulping. *Das Papier* 24, 711-27 (10A, 1970).
3. Ingruber, O.V. and Allard, G.A., Alkaline sulfite pulping for kraft strength, *Pulp Paper Mag. Can.* 73, T354-T369 (Nov. 1973).
4. Tanny, B., and Ingruber, O.V., The application of potentiometry to the determination of total and combined SO_2 in sulfite liquors. *Pulp Paper Mag. Can.* 67, T407-13 (Sept. 1966).
5. McLean, J.D., Hypo number. *Pulp Paper Mag. Can.* 66, Convention Issue, T103-6 (1965).
6. Kyrklund, B. and Strandell, G., Applicability of the chlorine number for evaluation of lignin content of pulp. *Paperi ja Puu* 51, 299-305 (4a, 1969).
7. Paasila, M. and Reynolds, M., Batch digester control of sulfite pulping is improved via computer. *Pulp Paper,* 130-1 (Mar. 1977).
8. Jutila, E., Paasila, M., Uronen, P., Mathematical models for producing dissolving pulp. *Tappi* 64, 105-8 (Aug. 1981).
9. Kallonen, H. and Franzreb, J.P., Computer control of batch sulfite digesters. *Pulp Paper Can.* 80, T437-9 (Dec. 1979).
10. Liebergott, N. and Yorston, F.H., The effects of temperature and liquor concentration on rate of sodium bisulfite pulping. *Pulp Paper Can.* 77, T211-4 (1976).
11. Lemay, R., Batch digester computer cooking control. *Pulp Paper Can.* 80, T167-70 (June 1979).
12. Wilder, H.D. and Han, S.T., A composition of the kinetics of the neutral sulfite and kraft pulping processes. *Tappi* 45, 1-9 (Jan. 1962).
13. Jutila, E., Perron, M., Ahola, H., Hipeli, R., Computerized process control of a continuous digester for NSSC pulping. *Tappi* 63, 69-72 (Nov. 1980).
14. Rydholm, S.A., Pulping Processes, Interstate Publishers, Div. of John Wiley & Sons, London, 1965.
15. Wenzl, H.F.J., Sulphite Pulping Technology, Lockwood Trade Journal Company, New York, 1965.
16. Calhoun, J.M., Yorston, F.H., and Maas, O., A study of the mechanism and kinetics of the sulfite process. *Can J. Research,* B 15, 457-74 (1937).
17. Felton, E.W., A study of the boiling process and a mill application. Worlds Paper Trade Review 140, 1905-6.
18. Vroom, K.E., The "H" Factor: A means of expressing cooking times and temperatures as a single variable. *Pulp Paper Mag. Can.* 58, 228-31 (Convention Issue, 1957).
19. Yorston, F.H., Studies in sulfite pulping. *Dominion Forest Service Bull.* 97, 20-29 (1942).
20. Yorston, F.H., and Liebergott, N., Correlation of the rate of sulfite pulping with temperature and pressure. Further studies. *Pulp Paper Mag. Can.* 66, T272-78 (1965).
21. Dorland, R.M., Leask, R.A., and McKinney, J.W., Pulp production with sodium bisulfite. I. The cooking of spruce. *Pulp Paper Mag. Can.* 59, 236 (May 1958).
22. Larocque, G.L. and Maas, O., The mechanism of the alkaline delignification of wood. *Can. J. Research,* B19, 1-16 (1941).
23. Ingruber, O.V. Flow cell containing clustered electrodes. Can. Pat. 794,495, Sept. 10, 1968. U.S. Pat. 3,471,383, Oct. 7, 1969.
24. Ingruber, O.V. pH control in sulfite pulping. The effect of pH level on cooking rate. *Pulp Paper Mag. Can.* T344-52 (Nov. 1959).

PART TWO

SULFITE PULPING TECHNOLOGY

IV

RAW MATERIAL CONSIDERA-TIONS AND COOKING LIQUOR PREPARATION

O.V. INGRUBER

Retired Senior Research Associate from CIP, now an independent consultant

MANUSCRIPT REVIEWED BY
David Johnston, Manager Pulping
Nova Scotia Forest Industries
Port Hawkesbury, Nova Scotia

A. WOOD SPECIES AND PROPERTIES

When sulfite pulping took its rise in the last century it did so in the form of the acid bisulfite process in North America, Northern Europe and in the alpine region of Middle Europe where spruce, balsam fir and hemlock are the predominant species. Spruce makes choice timber and also the best paper, and fir and hemlock follow closely in papermaking properties. All three species, often called "spruce-type", cook without difficulty in the acid bisulfite process. The pines, larches, and Douglas fir, which make a large part of available softwoods (Table 11) are less suitable due to the presence of phenolic substances in the heartwood which tend to condense with the lignin content and cause high levels of screening rejects and high lignin content in the pulp. Sapwood of these species, such as saw mill slabs, is pulped as well as sprucewood.

Among the pines, red pine shows least resistance to acid bisulfite pulping due to lower heartwood content. Solvent extraction of wood from difficult species with acetone or methanol reduces the phenolic content to very low levels, and could be employed to render them suitable for the acid bisulfite process.

High resin content is also undesirable since it impedes penetration of the wood with calcium base cooking liquor and generates pitch problems in pulp processing equipment. Modifications of the acid bisulfite process which favor sulfonation, such as less acidity, lower cooking temperature, and higher concentration of bisulfite with soluble base, have made it possible to use less desirable softwood species to some extent.

Certain softwood and hardwood species have structural peculiarities which act as penetration barriers. See Volume 1, "Raw Materials", of this series for more detail.

Hardwoods contain 80-90% by weight libriform fibers useful for papermaking which measure 1 mm or less in length compared to 2-5 mm for softwood fibers and therefore give inferior pulp strength properties. Nevertheless, dwindling supply of spruce-type pulp wood and

Table 11. Major wood species containing phenolic inhibitors to acid bisulfite pulping.

Species	Generic Name	Phenolic Substance
Pines	Pinus L.	Pinosylvin
Jack Pine	Pinus banksiana	Pinobanksin
Larches	Larix L.	Flavanones
Douglas Fir	Pseudotsuga	Taxifolin
Spruce bark	Picea L.	Tannins
Oak	Quercus L.	Tannins
Chestnut	Castanea	Tannins

oversupply of hardwoods in pulp wood producing areas have helped to develop methods for hardwood utilization which also uncovered a number of advantages. Some, such as eucalyptus and poplar, grow faster than softwoods; hardwoods are denser than softwoods and therefore cost less per ton on a volume basis; and their lower lignin content permits shorter cooking schedules or lower temperature. Direct advantages for hardwoods were found in dissolving pulp manufacture, particularly in ammonium acid bisulfite cooks, and for improving opacity in fine papers. A list of the suitability of Eastern North American hardwoods for acid bisulfite pulping is given in Table 12.

Aspen, poplar, beech, birch and maple are used predominantly in the northeastern part of North America and also in Europe for acid bisulfite, bisulfite, and neutral sulfite pulping. Hardwood utilization follows the decreasing quantity order: kraft, NSSC, acid bisulfite and bisulfite soda. The reason for the large share of kraft hardwood pulp is paper strength, which is approximately on the same level as that of softwood acid bisulfite pulp.

Wood species tolerance in the field of sulfite pulping is lowest with acid bisulfite and somewhat improved with bisulfite from pH 4 up. Pines, in particular, can be accommodated on this level in a single-stage process with adequate sulfonation and even better in two- and three-stage processes. At pH levels above neutral such as NSSC and alkaline sulfite, all wood species and annual plants suitable for pulping can be used, though with some natural individual differences in process and product characteristics, depending mainly on density, lignin content and fiber dimensions.

Annual plants (straw) provided the raw material for the first alkaline sulfite cooks. Since industrial chemical pulping started with soda cooks of straw in continuation of age-old pre-industrial practices, it was only natural to try

soda in combination with sulfite after the latter became known. According to a patent issued in Germany at the turn of the century [15], the alkaline sulfite liquor is made by reacting recovered soda solution with sulfur dioxide gas to form sulfite, and causticizing the excess soda with lime to form caustic soda. Compared with soda pulp, the early alkaline sulfite straw pulp had 28% higher yield, and was brighter, stronger and more easily bleached than soda pulp.

Annual plant species which now find large-scale industrial use for kraft, alkaline sulfite, NSSC and soda pulp grades include sugarcane (bagasse), bamboo, cereal straw, esparto grass, kenaf, and reeds. Several others are used on a smaller scale. The acid bisulfite process is not suitable for pulping these materials. Pulp production from these lignocellulosic resources has greatly increased since the fifties and now supplies about 10% of world pulp.

Kenaf, in particular, is receiving increasing attention as a fiber crop in the United States. The plant originates in Africa and is widely cultivated also in Asia, including Russia. Extensive investigations which began in 1945 indicate [1] that kenaf can give the highest yield of fiber per hectare compared to all other lingo-cellulosic raw materials. The bast fibers of kenaf are found to have excellent strength and can be used to replace long fiber softwood pulp. Kenaf paper is stronger than paper from hardwood and, except for tearing resistance, similar to paper from softwood.

B. WOOD CHIPS

In the process of converting wood or other ligno-cellulosic raw materials into useful fiber the quality of the chips entering the process is of primary importance for the efficiency and cost of cooking, bleaching or mechanical refining, and the quality of the resulting pulp. Uniform chip size is the basis of uniform pulp quality; and also of optimum pulp quality because the greater the range of chip dimensions, the greater will be the compromise in the effectiveness of chemical or mechanical treatment.

Requirements are as stringent for acid pulping processes as they are for alkaline processes, only the chip dimensions differ depending on the difference of the mechanism of impregnation. These were outlined in the section on penetration of cooking liquor and impregnation (see Chap. II, A). Acid liquors penetrate mainly through the cut ends and chip length is therefore the important dimension. With conventional chippers, length can be controlled but not chip thickness which increases with the length setting. Wood fibers at and near the cut ends are

Table 12. Suitability of North American hardwoods to acid bisulfite pulping [18].

Diffuse-porous species	Suit-ability	Ring-porous species	Suit-ability
Poplars	+	Hackberry	+
Gums	+	Elm	+
Willows	+	Ash	+
Magnolias	+	Hickory	+
Maples	+	Catalpa	−
Beeches	+	Sassafras	−
Sycamore	+	Red oak	−
Birches	+	White oak	−
Butternut	+	Chestnut	−
Basswood	−	Slippery elm	−
Red gum	−	Black locust	−

damaged by the cut and the impact of the chipper knife (compression damage), and these effects cause a deterioration of pulp strength when the chips are cut short. Blunt chipper knives greatly increase the fiber damage. Shortening of the average fiber length was detectable when the chips were cut below 19 mm in length; but pulp strength did not suffer down to between 6 and 13 mm of chip length. For industrial chips of this type, the optimum thickness is between 5 and 8 mm. Compression damage parallel to the grain has a more severe effect on pulp strength than compression at right angle to the grain. Strength degradation due to compression of the chip is greatest in acid bisulfite cooks and least in alkaline cooks.

In alkaline cooking liquor, penetration proceeds at nearly equal rate in all three chip dimensions but the gradient of chemical concentration increases with chip thickness. This is therefore the important dimension in alkaline pulping; and its control has received increasing attention in recent years through the introduction of slotted screen chip classifiers which, of course, measure only the compact thickness but not the amount of fissuring of the chip. The major effect of chip thickness on screening rejects from an alkaline cook, and the almost ideal behavior of the 3 mm chip [2] are shown in Fig. 71.

In the sixties, several new chipper designs were introduced industrially for the purpose of minimizing wood damage, and to produce thin chips of uniform thickness (Anglo drum chiper, 1960; Souderham H-P chipper; Swedish CCL chipper, 1963). See Volume 1 for a more complete discussion of chipping.

Although the above chip quality requirements for producing uniform pulp are well known, they are often not met in industrial operations for practical reasons such as saving labor costs by less frequent knife replacement and sharpening; or decreasing chipping cost and increasing output by making larger chips. There is also a range of tolerance depending on the quality of available wood and pulp quality specifications. However, increased production of pulp and a more favorable overall income balance of a mill, when due to lowering chip quality specifications, will always be at the cost of higher consumption of chemicals and energy and at a compromise with pulp quality.

Investigation of the material collected in knotter screens from pulp blown from the digester has shown that it consists predominantly of oversized, poorly cooked chips with only a small fraction of actual knots. There are three ways to dispose of this material: as land fill, as fuel for steam generation, or by re-cooking. The latter option is the most profitable since it produces saleable pulp though at the expense of somewhat reduced digester space and some additional steam.

Constant wood moisture facilitates the operation of the pulping process and allows more uniform pulp to be produced. However, most mills have some variations in wood moisture and several have to make pulp with wood moistures ranging from 25 to 60%.

Variations in wood moisture affect the degree of chip packing in successive digesters, because wet wood chips pack very densely. In the same digester, they cause some of the charge to be either undercooked or overcooked; thus producing pulp of varying bleach demand from a single digester, and high rejects.

Dry wood inhibits chip packing considerably and can cause a loss in daily production by the loss in individual digester yields. Generally speaking, large variations in wood moisture increase pulping rejects considerably. Ideal wood moisture is estimated to be about 45% to satisfy both penetration and production. The problem of wood moisture variation is greatly relieved by presteaming which tends to adjust the moisture level towards 50% from either below or above.

C. SULFUR

Sulfur dioxide gas for preparing sulfite cooking liquor is obtained by burning pure sulfur in special burners with air. Sulfur is received at the mill as powder or in the molten form in heated tank cars or trucks. Sulfur

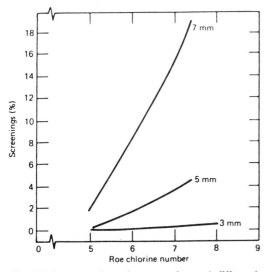

Fig. 71. Increase in pulp screenings at different Roe chlorine number levels for handmade chips of three different thicknesses [16].

obtained by the Frasch mining process is available in the United States and Canada mainly from Texas and Louisiana. Sulfur recovered from "sour" natural gas is available in both the United States and Canada from western portions of the countries. Other sources of sulfur are Sicily, Japan, and Mexico.

Sulfur is a brittle, pale yellow solid. It melts at 112.8°C to form a brownish yellow transparent liquid. When the temperature is raised, the liquid becomes more fluid and then suddenly becomes dark red. A rapid increase of viscosity occurs above 159.9°C. At 230°C the liquid is black and viscous. Beyond 230°C the viscosity decreases but the color remains dark, and the liquid boils at 444.6°C. If the boiling sulfur is allowed to cool slowly, it passes through this series of changes of color and viscosity in reverse order. Ordinarily it will be found most convenient to keep molten sulfur at 140 to 160°C and to avoid the temperature 200°C, at which it becomes extremely viscous.

The chemical symbol for sulfur is S; the atomic weight is 32.07; and one atom of sulfur will combine with two atoms of hydrogen or a monovalent metal or with one atom of a bivalent metal to form sulfides.

For many years iron sulfide, or pyrite, has been used by sulfite mills to produce SO_2 by a roasting process. This practice has been discontinued and, instead the use of liquid SO_2 has increased to supplement sulfur burner capacity. This is particularly attractive in Canada where liquid SO_2 is available at relatively low cost from base metal refining operations. The liquid is shipped in pressurized and insulated tank cars, since it develops an equilibrium pressure of about 300 kPa at normal temperature. Unless the price is kept exceptionally low, the cost of the liquid SO_2 is high relative to raw sulfur, and freight charges must be paid on oxygen as well as on the sulfur content of the gas. However, operation of the liquor making plant on liquid SO_2 is extremely simple.

1. Sulfur handling and storage

Dry sulfur can be fed by a screw conveyor directly into a rotary sulfur burner. A majority of mills, however, melt the sulfur and supply it in liquid form to the sulfur burner. Molten sulfur is practically a necessity if a spray-type burner is used, although some spray burner designs are said to be capable of using dry sulfur. Figure 72 shows the melter designed by Canadian Industries Limited (Canadian Patent 382,014) for melting dry sulfur. It consists of a vertical cylinder set over a concave bottom section; it is trouble-free and low in maintenance costs.

Cylinder A and bottom pan E are steam-jacketed, steam entering at B and G and emerging partly or wholly condensed at C and H. The lugs J support the cylinder.

In operation, the sulfur is fed at D, normally from an overhead storage bin containing at least 8 h supply, which in turn is supplied by bucket elevator or belt or screw conveyors or combinations of these. All the moving parts of the conveyors, elevators, etc., for handling dry sulfur are constructed of nonsparking materials to reduce the fire hazard. Molten sulfur flows down the inner side of the melter to the collecting pan, from which it flows through the outlet pipe F to a molten-sulfur storage tank. Debris which finds its way into the sulfur during shipment and storage is for the most part carried by the liquid and must be screened or decanted out. The melter operates at any pressure of steam from 50 to 100 psig with the optimum at 75 psig. The melting time can be varied by adjusting the steam pressure or the sulfur feed as long as the steam pressure is not raised so high that the temperature of the sulfur at the melter shell gets into the high-viscosity range which slows down the melting. The melter can be brought up to production in 10 to 15 min, or shut down in approximately the same time. The capacity per melter is normally 50 000 lb of dry sulfur per 24 h.

Sulfur is also melted and stored in several types of storage tanks or pits where heating is provided by steam coils. Most melting pits,

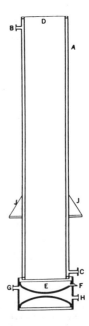

Fig. 72. Tubular sulfur melter (CIL).

whether of concrete or steel, have a brick lining to reduce the tank corrosion at the molten sulfur-air interface and to insulate the tank. Various means are used for supplying dry sulfur to the melting pit or tank, and a steam-jacketed sump-type pump is commonly used for transferring sulfur from the melting-storage pit to the burner.

The availability of liquid sulfur in tank cars or trucks has led to the installation of relatively large storage tanks by sulfite mills for storing the liquid sulfur. Such an installation consists of steam-jacketed (or traced) unloading piping to transfer the sulfur from car or truck to the tank. A syphon type of unloading system with compressed air as a padding medium is commonly used, and gravity flow or a steam-jacketed transfer pump can also be used. The molten-sulfur storage tank is generally of mild-steel construction with insulation on the outside. Steam-heating units or coils are provided in the tank, and instrumentation may be provided to measure and record sulfur level and temperatures. Sulfur is transferred from the tank to the process by gravity flow or steam-jacketed pump.

Sulfite mills with Stora-type sodium base recovery feature complete recycle systems for handling and storing liquid sulfur produced from spent cooking liquor components in Claus reactors and converted back to SO_2 in the sulfur burner (refer to Chap. IX, D, 1a and case 1 of up-to-date sulfate process applications).

2. Sulfur burning

When sulfur or a sulfide burns, it combines with the oxygen of the air to form oxidation products of sulfur, mainly sulfur dioxide (SO_2). The nature of the gas, i.e., the proportion of the sulfur dioxide to the other combinations of sulfur and oxygen (SO_3, etc.), depends upon the prevailing conditions; such as temperatures to which the gases are exposed, and the proportion of sulfur gases to oxygen or air. In other words, the quality of the gas depends upon the construction and the operation of the sulfur burner.

(a) Rotary burner

The rotary sulfur burner consists of a steel plate cylindrical sheel with conical ends. The burner rests on rollers and is turned at approximately 1 to 1.5 rpm. Burners are manufactured in various sizes from 50 cm up to 1.5 m in diameter and from 1 to 5 m long. Modern units are lined with firebrick in the cylindrical section.

Rotary burners are supplied with dry or

molten sulfur (120-140°C) and air supply is regulated by dampers in the front head. Only part of the sulfur is burned to SO_2; the rest is merely vaporized at 310-370°C. For complete oxidation of this gaseous sulfur to sulfur dioxide at 850-950°C it is absolutely essential to have adequate combustion volume and adequately sized air ports between the rotary burner and the combustion chamber which can be regulated to produce the proper gas strength. The combustion chamber is usually a steel plate cylinder erected vertically with firebrick lining and baffle wall. In Fig. 73 the gases enter from the rotary burner at E, pass over the baffle wall, and leave by outlet F. In small installations the combustion chamber is sometimes of cast-iron material with firebrick lining. In larger installations two combustion chambers are often placed in series and the baffle wall is eliminated, since the path of the gas between the primary and secondary chambers acts as a baffle.

(b) Jet burner

A different principle of adding sulfur is employed in the spray-type burner. Liquid sulfur is atomized under pressure and combined with the proper amount of air to complete combustion. Various types of burners are in use with different spray nozzles, or jets, in which the sulfur is atomized by pressure, atomizing air, or steam. The basic principle remains the same, however. The combustion is completed with secondary air admitted through entry ports in a

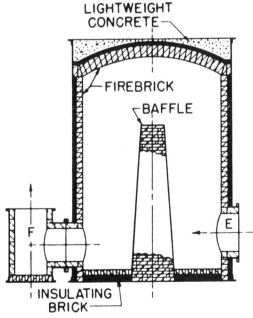

Fig. 73. Combustion chamber of rotary sulfur burner.

combustion chamber which may be similar to that used with rotary burners but more often is a horizontally placed cylinder with refractory insulating brick lining, as shown in Fig. 74A. A substantial reduction in length of the horizontal sulfur combustion chamber has been achieved by arranging the air entry ports so that the flame rotates (cyclone) as it advances as in Fig. 74B, thus increasing the length of the combustion path several times. This is the same principle as found in the Loddby combustion chambers for burning concentrated spent sulfite liquor (see Chap. IX, C, 1). Here the combustion tempera-

ture is 650-850°C. Heat of combustion is used to preheat the primary and secondary air.

The advantages of a jet-type burner are quick start-up, instantaneous shutdown, almost complete elimination of SO_3 formation, and ability to produce a high concentration of SO_2 gas. There are other sulfur burners which are commercially available, the principal requirement being a method of mixing sulfur and air to promote combustion and providing sufficient volume so that combustion is completed. However, the majority of installations in sulfite pulp mills on the North American continent consist of the types mentioned above.

Since the incoming air used for combustion of sulfur contains 21% oxygen by volume, it is theoretically possible to obtain a burner gas containing 21% sulfur dioxide. Actually, rotary burners produce a concentration of about 15-17% of sulfur dioxide and spray burners about 17-19%. It is desirable to produce as high a sulfur dioxide gas strength as possible, not only because of the increased absorption efficiency for most acid-making requirements but also because a smaller total volume of gas will have to be handled. However, if the SO_2 gas strength is carried too near the theoretical limit, occasional sublimed sulfur particles (unburned sulfur) are carried into the balance of the system, causing plugging and inefficient operation.

Low sulfur dioxide concentration in the burner gas, and consequently a high level of excess air, are conducive to the formation of

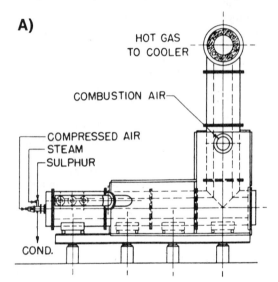

A)

HOT GAS TO COOLER

COMBUSTION AIR

COMPRESSED AIR
STEAM
SULPHUR

COND.

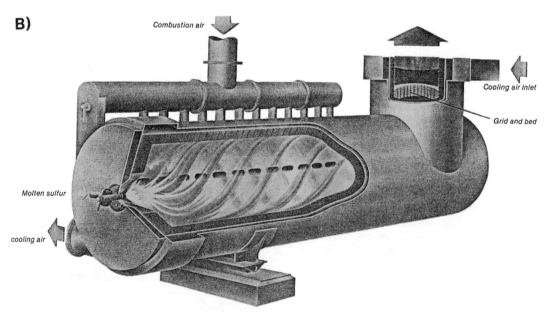

B)

Combustion air

Cooling air inlet

Grid and bed

Molten sulfur

cooling air

Fig. 74A. Jet-type sulfur burner (Chemipulp Division, The Stebbins Engineering & Mfg. Co.) B. Jet-type sulfur burner (Cellchem).

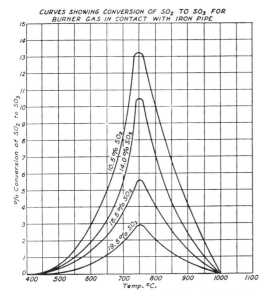

Fig. 75. Conversion of SO₂ to SO₃ in contact with iron.

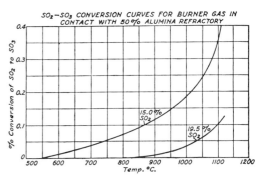

Fig. 76. Conversion of SO₂ to SO₃ in contact with alumina brick.

sulfur trioxide, the anhydride of sulfuric acid: $SO_2 + 1/2 O_2 \rightarrow SO_3$. Sulfur trioxide formation causes loss of sulfur and base from the pulping system, and corrosion and scaling of equipment. Some oxygen is always present, and this unwanted conversion can be assisted by the presence of catalysts. The only catalyst present in the conventional sulfur-burning equipment is iron, and research has indicated that the use of cast iron or steel piping, particularly in the critical temperature range, can promote the formation of sulfur trioxide (Fig. 75). The conversion of SO_2 to SO_3 can be inhibited by the operation of the burner at the highest possible gas strength and corresponding high temperature; use of refractory-lined piping (Fig. 76) instead of cast iron or other metal exposed to hot gases; and use of a minimum amount of piping between combustion system and burner. It is also important to cool the gas rapidly through the critical range of maximum SO_3 formation between 600 and 900°C.

(c) Waste-heat boiler

Interest in the use of steam generators for utilizing the sensible heat of sulfur burner gas has been increasing in recent years as part of measures to improve the heat balance of the pulping process within the general context of energy conservation. A modern spray-type sulfur burner with the steam generator replacing the first gas cooling stage is shown in Fig. 77. Steam generators are standard equipment on the S/H₂S burners of state-of-the-art sodium base

sulfite recovery systems (cf. Chap. IX, D, 1b,c).

The heat generated by burning to SO_2 1 kg of sulfur is 9.27 MJ and of 1 kg of H₂S 31.1 MJ. One tonne of sulfur can produce about 2.7 tons of steam at 80% boiler efficiency.

The gas leaving the burner has a temperature close to 1000°C and the operation of the waste heat boiler is usually so regulated that the range of SO_3 formation is avoided.

3. Cooling the SO₂ gas

In view of the volatile nature of SO_2 gas explained previously, it is necessary in the preparation of acid bisulfite liquor in atmospheric systems to cool the gas to 25-30°C. For bisulfite liquor preparation, cooling of the gas to 50-70°C is sufficient and even higher temperatures are possible for neutral sulfite and alkaline sulfite liquors. Generally, any system which involves handling of SO_2 gas can be made simpler, more effective and more economical by keeping it under a slight positive pressure to counteract the volatile tendency of SO_2.

The chief requirement for the design of the cooling system is to lower the temperature of the gas rapidly to below 600°C in order to avoid SO_3 formation. There are basically two types of coolers in use in the sulfite pulp industry. In the indirect type, the sulfur dioxide gas is inside pipes of corrosion-resistant material (normally lead) and is cooled by water on the outside of the pipes, either in a pond or flowing over the pipes. Of the indirect types, one of the more common consists of a combination of horizontal pipes that are submerged in water and larger vertical pipes with water running down the outside.

In the spray-type cooler, the gas is cooled by direct contact with a water spray, as shown in Fig. 78. In the two-tower system the hot gas enters the first tower and is quickly cooled by the water spray to below the critical range for SO_3 formation. The wet gas at saturation temperature then passes into the bottom of the secondary tower and rises against two sprays which complete the cooling. Some water is sewered from both the primary and the secondary towers through the weir box, and this is a means of getting rid of SO_3 in the system, since much of the SO_3, if present in the burner gas, will be washed out in the primary tower. Because of the recirculation of the cooling solution over the secondary tower through the heat exchanger, very little SO_2 is lost in the system; the recirculated solution builds up to an equilibrium concentration of SO_2. Make-up water is added at the weir box or may be added in a final spray in the secondary cooling tower.

The towers are made of stainless steel or, in older systems, of steel lined with lead and acid resistant brick. The secondary tower is fitted with ceramic packing rings to provide intimate contact of recirculating solution and gas.

The spray system offers several advantages; among them are that it takes up less space, wastes no water (since the warm water from the heat exchanger can be utilized in other mill locations), reduces SO_3 content of the gas by rapidly dropping the temperature of the gas and by scrubbing action in the primary tower. Under present conditions, the indirect type of cooler can be justified for relatively small installations where space is not at a premium and where there is little use for small quantities of warm water from the heat exchanger.

D. BASE MATERIALS
1. Limestone

The appearance of a limestone is an indication of the purity of the stone. In its purest form, the stone is white and crystalline and consists essentially of calcium carbonate ($CaCO_3$); an example is marble. Commercial stone usually contains impurities, such as oxides of iron and aluminum and insoluble silica, which affect its appearance. This varies from the highly crystalline variety to the amorphous, porous stone, while the color may vary from white to yellow and gray. The specific gravity varies, being highest with the highly crystalline stone. The solubility of the limestone in SO_2 water increases with the crystalline character and, of course,

Fig. 77. Sulfur burning with waste heat boiler (Cellchem Co.).

with the strength and temperature of the acid. Usually, the limestone contains magnesium carbonate ($MgCO_3$), and the properties of the stone vary according to the amount of magnesia.

Other forms in which calcium carbonate can be used in the sulfite mill are finely ground marble and sludge recovered from the causticizing process in alkaline pulp mills. In both cases the carbonate is used as a slurry in a milk-of-lime type of acid system.

2. Dolomite

Dolomite is a calcium-magnesium carbonate, normally containing 54.00% calcium carbonate and 46.00% magnesium carbonate. The stone containing one part of $MgCO_3$ to two parts of $CaCO_3$ is also called dolomite. Limestone and dolomite are found in many places in the United States and Canada and are of many different qualities.

Calcium oxide (CaO) and magnesium oxide (MgO) are obtained by burning limestone or

dolomite at high temperature:

$$CaCO_3 \rightarrow CaO + CO_2$$
$$MgCO_3 \rightarrow MgO + CO_2$$

When slurried in water, milk of lime (magnesia) results:

$$CaO + H_2O \rightarrow Ca(OH)_2$$
$$MgO + H_2O \rightarrow Mg(OH)_2$$

While a low magnesia content, preferably not above 5%, is considered the most suitable for the tower system, a high-magnesia or dolomite lime is preferred in the milk of lime system.

3. Magnesia

Magnesium base is received in the form of dry or slurried magnesium hydroxide, $Mg(OH)_2$, or magnesium oxide (MgO) powder. The slurry is shipped in tank cars, trucks or ships and stored in slowly agitated tanks at the mill. Natural magnesium carbonate or magnesite ($MgCO_3$) does not react with sulfur dioxide at a sufficient rate to be of use in common liquor making systems. Dolomite, $CaMg(CO_3)_2$ and brucite, $Mg(OH)_2 \cdot CaCO_3$ contain too much calcium to

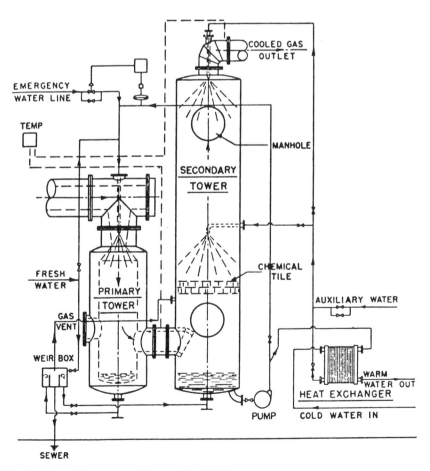

Fig. 78. Chemipulp — KC spray cooling system [16].

be used in true magnesium base pulping operations. For mills with magnesia recovery systems producing MgO, make-up chemical in the form of magnesium sulfate, $MgSO_4$, can be added to the spent cooking liquor before the recovery furnace. The sulfur dioxide formed from the decomposition of the sulfate then provides part of the sulfur make-up of the system (see Chap. IX, C, 3, 8).

4. Ammonia

Ammonia (NH_3) gas is prepared as a by-product of the production of coke by the Haber process, or by the cyanamide process. It is shipped in liquid form in high-pressure, insulated tank cars or trucks, or as aqueous ammonia at about 30% concentration. Liquid ammonia systems are common for permanent installations and require facilities for unloading, dissolving to 15-20% concentration, and storage. Special precautions are necessary to confine the ammonia gas.

5. Soda ash

Sodium carbonate, or soda ash, is obtained in a very pure state by means of the Solvay process, and it is used in the preparation of soda-base cooking acids and neutral and moderately alkaline sulfite liquors. The commercial standard for soda ash is expressed as equivalent sodium oxide (Na_2O) content. Hence a 55% soda ash contains sodium carbonate equivalent of 55% Na_2O (93.5% pure).

Two standard grades are commercially available; they are known as light soda ash and dense soda ash. These grades differ only in physical characteristics such as size and shape of particles and bulk or apparent density, which influences the flow characteristics. The soda ash used in pulp mills is ordinarily shipped in railroad cars and unloaded with mechanical or pneumatic conveyors. Storage at the mill may be in the powdered form or as a slurry in which the supernatant liquid, at approximately 30% sodium carbonate by weight, is drawn off to the process.

6. Sodium hydroxide

Sodium hydroxide, NaOH or caustic soda, is the raw material source of soda for sodium-base and strongly alkaline sulfite pulping liquors. Caustic soda may be received in solid form and dissolved in water at the mill, but for pulping requirements it is normally supplied in liquid form. Caustic soda solution is produced and shipped at or slightly above 50% concentration by weight (765 g NaOH/L).

Specific gravity and concentration data are found in Table 13. These data are for solutions of

Table 13. Specific gravity — concentration data for pure caustic soda solutions at 15.6°C.

Based on tables from International Critical Tables, Vol. 3, p. 79.

% NaOH	Specific Gravity 15.6°C	Baume Am. Std.	% Na_2O	g NaOH/L
1	1.012	1.72	0.78	10.11
2	1.023	3.26	1.55	20.44
3	1.034	4.77	2.32	30.99
4	1.045	6.25	3.10	41.77
5	1.056	7.69	3.87	52.77
6	1.067	9.10	4.65	63.99
7	1.079	10.65	5.42	75.44
8	1.090	11.97	6.20	87.10
9	1.101	13.30	6.98	98.99
10	1.112	14.60	7.75	111.1
11	1.123	15.88	8.52	123.5
12	1.134	17.13	9.30	135.9
13	1.145	18.36	10.08	148.9
14	1.156	19.57	10.85	161.7
15	1.167	20.75	11.62	175.0
16	1.178	21.91	12.40	188.4
17	1.190	23.15	13.17	202.1
18	1.201	24.27	13.95	215.9
19	1.212	25.36	14.73	229.1
20	1.223	26.45	15.50	224.3
21	1.234	27.50	16.28	258.9
22	1.245	28.53	17.05	273.6
23	1.256	29.55	17.83	288.6
24	1.267	30.56	18.60	303.7
25	1.278	31.57	19.38	319.2
26	1.289	32.51	20.15	334.8
27	1.300	33.46	20.92	350.7
28	1.310	34.31	21.70	366.6
29	1.321	35.23	22.48	382.8
30	1.332	36.14	23.25	399.2
31	1.343	37.03	24.02	416.0
32	1.353	37.83	24.80	432.6
33	1.363	38.62	25.58	449.7
34	1.374	39.47	26.35	466.7
35	1.384	40.23	27.12	484.4
36	1.394	40.98	27.90	501.5
37	1.404	41.72	28.68	519.4
38	1.415	42.53	29.45	537.1
39	1.425	43.24	30.22	555.3
40	1.434	43.88	31.00	573.3
41	1.444	44.58	31.78	591.6
42	1.454	45.28	32.55	610.2
43	1.463	45.89	33.32	629.5
44	1.473	46.56	34.10	647.6
45	1.483	47.22	34.88	667.8
46	1.492	47.82	35.65	685.9
47	1.502	48.46	36.42	705.9
48	1.511	49.04	37.20	724.8
49	1.521	49.67	37.98	744.8
50	1.530	50.23	38.75	764.5
51	1.540	50.84	39.52	784.3
52	1.549	51.39	40.30	804.4

pure NaOH in water and may be somewhat different for some industrial grades because of the presence of variable amounts of sodium chloride (NaCl).

E. COOKING LIQUOR PREPARATION
1. Absorption

The next step in the process of acid making is the absorption of the cooled sulfur dioxide gas in water in the presence of calcium, magnesium, sodium, or ammonium compounds with which it forms the corresponding sulfite, bisulfite, or acid bisulfite solutions used in pulping. The absorption process requires a means of providing intimate contact between the gas and liquid in order to accomplish the transfer of the sulfur dioxide from the gas phase to the liquid phase.

Two factors are involved: (1) the equilibrium solubility of sulfur dioxide in the absorbing liquid; and (2) the rate at which equilibrium is reached, that is, absorption kinetics. As shown earlier, the solubility is a function of temperature, pressure, and pH at a given stage of absorption, while the rate depends on base concentration, temperature and turbulence (reaction surface) of gas and liquor, the latter varying with the design and flow rates of the absorption system. Solubility relations for sulfur dioxide and water without and with base present, discussed previously, have shown that the partial pressure of SO_2 over the solution containing base becomes very small at pH 4 and disappears at higher pH levels. Consequently, the rate and efficiency of SO_2 absorption increase with rising pH above 4. Conversely, in acid bisulfite liquors below pH 4, to increase the concentration of free SO_2 gas in solution, increasingly high SO_2 concentration in the gas, low liquor temperature, and high excess pressure are required as the pH decreases.

(a) Packed tower.

One of the principal types of industrial absorption equipment is the packed tower. Packed towers can be so arranged that the gas-liquid flow is either concurrent or countercurrent; the latter is most common for sulfur dioxide absorption. The absorbing liquid is introduced at the top of the column and flows down through a bed of packing material that is chemically inert to the system. The sulfur dioxide burner gas is forced through the tower from the bottom to the top by means of a differential pressure which may be either an induced-draft fan or a pressure blower on the inlet of the absorption tower. The packing material may be spheres, coke, manufactured grid shapes, Raschig rings, cross partition rings, or various types of saddles. The latter are available in ceramic or plastic materials.

(b) Jenssen packed tower

Traditionally the Jenssen packed tower system has been used in North America for the preparation of calcium base acid bisulfite cooking liquor. In this special case the packing consists of limestone or dolomite, ranging in size from 25-38 cm supplying the cooking base. A schematic of the system and its operation is shown in Fig. 79. Normally two large towers, 2-4 m in diameter and 20-33 m in height, are combined in a unit. Four-way gas valves and six-way liquor valves permit continuous operation while one tower (the "weak tower") is replenished or "charged" with stone daily. Under favorable conditions the first or "strong" tower can absorb as much as 97% of the SO_2 in the burner gas. The second or "weak" tower completes the SO_2 absorption and the remaining gas is vented. The Jenssen tower system has the advantage of simple operation but the disadvantage of the high cost of charging the stone and lack of control over the composition of the acid cooking liquor. Since the base is present in excess, base concentration and pH can be controlled only to the extent that the raw water temperature and the SO_2 gas strength can be maintained at the required levels. In fact,

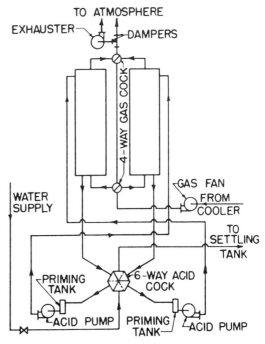

Fig. 79. Two-tower acid system (Jenssen Division, The Stebbins Engineering & Mfg. Co.).

seasonal variations are common with this system, liquor with low free and high combined SO_2 in summer and with low combined SO_2 in winter.

(c) Bubble cap/plate towers

Another type of absorption apparatus consists of fixed plates in a tower which maintain a static fluid level through which the upcoming gas must rise. Bubble caps, impingement plates, or other means are used to force the gas to go through the liquid head and to provide numerous gas inlets to ensure intimate liquid-gas contact. These types of absorption towers have been used in the sulfite industry primarily for milk of lime systems in which burnt lime, usually containing a high percentage of magnesia (dolomite), is slaked in warm water to form hydroxide of lime-magnesia. These hydroxides are used as the absorbing medium in the absorption tower.

(d) Milk of lime bubble cap system

With powdered limestone or milk of lime, the final pH of the liquor can be regulated by controlling the input of the base. The flow diagram of a milk of lime system in Fig. 80 indicates the supply of burnt lime being delivered to the lime slaker B, from which it is pumped to the slurry tank C. From C the lime slurry flows to the limewater proportioning pumps D, which can be operated by either steam or electric power. The discharge from these plunger-type pumps can be regulated by a flow of water to produce a predetermined quantity of acid per day and to deliver a regular flow of limewater to the absorption towers E. The towers are usually constructed of chrome-nickel steel and are arranged in series. Each tower contains four active bubble compartments, gas distribution in each compartment being obtained through the use of bubble caps over the apertures in the plates. The towers are so equipped that they can be reversed by the operation of four-way cocks F. By reversal, any lime or other incrustation "make-up" in either tower during its period of functioning as the weak tower is removed by solution during the period of operation as the strong tower. From the cooling system the burner gas is conducted through the gas compressor G and then through a hard-lead pipe H to the absorption towers E. The gas mixes with a downward counterflow of limewater, which enters the tower 15 cm above the top compartment and crosses this compartment to an overflow section opposite the inlet, the slurry being in contact with the gas in the system and the resultant mixture being delivered to the next compartment through the weir overflow. The flow through succeeding compartments will be the same as the top compartment. It will be evident from the diagram that the settings shown for valves F will put fresh limewater in tower E_2, weak liquor from E_2 to E_1, strong liquor from E_1 to storage I, strong gas to E_1, weak gas to E_2, and

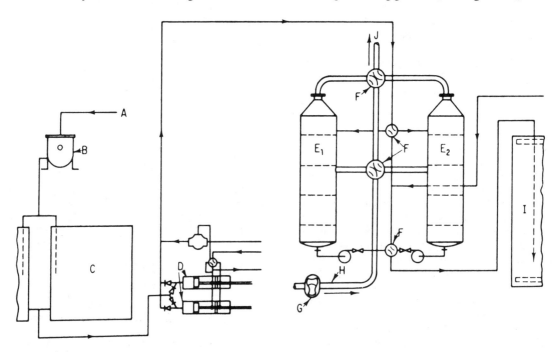

Fig. 80. Milk of lime system (Chemipulp Division, The Stebbins Engineering & Mfg. Co.).

spent gas to atmosphere through vent J. The capacity of the systems is regulated by the amount of limewater admitted, and since there must be a certain relation between the limewater and the gas, the number or size of the perforations in the partitions must be regulated to meet these requirements. The mill combined SO_2 is regulated by the strength of the limewater.

This system still suffers from poor control of liquor concentration by seasonal temperature variations. These effects can be eliminated and acid strength (SO_2 concentration) raised by using an absorber operating under slight pressure up to 140 kPa above atmospheric. It is necessary to use as a gas pump a compressor of acid-resistant construction and another compressor for adding milk of lime. A throttling valve on the tail gas outlet at the top permits operating at any desired pressure up to the design maximum.

Heat is developed in both systems because of the chemical reactions, which increase the temperature of the acid usually about 8°C, but slightly greater with the milk of lime system than with the tower system. Less power is required in the tower system, and it is also claimed that less labor is needed to operate it than the milk of lime system. There is a difference in the finished acid, owing to the fact that lime with a high magnesia content is advantageous in the milk of lime system, which is, therefore, generally employed. In the tower system, a limestone with low magnesia content is preferable, although not necessary, for a satisfactory operation. As a consequence, the acid made by the milk of lime system generally has a much higher magnesia content, and some advantages have been claimed for this acid in the cooking process. The magnesium salts are more soluble than the corresponding calcium salts and therefore do not decompose so easily; for that reason there is a more gradual liberation of the SO_2 in the cooking process when an acid with high magnesia content is used. This permits, it is held, a higher temperature in the process, which is necessary in quick cooking. There is no doubt that the pulp resulting from cooking with high-magnesia liquor has a more pliable fiber than that resulting from pure calcium bisulfite liquor, and it has a tendency to be whiter in the unbleached condition.

(e) Other systems

With the change to soluble base bisulfite pulping, several mills continued to use their Jenssen towers by packing them with inexpensive, inert material. Initially, even packing with jack pine logs was practiced which, incidentally, pulped the jack pine wood to pure white cellulose

within a few years.

Packed towers for the preparation of acid bisulfite and bisulfite cooking liquor with slurried or dissolved base materials such as milk of lime, milk of magnesia, ammonia, soda or caustic soda, are in general only about half as wide and high than those required for limestone or dolomite rock, and one tower is sufficient for most sulfite mills. For bisulfite, neutral sulfite and alkaline sulfite absorption systems packing made of heat resistant material should be used. In order to achieve efficient absorption of sulfur dioxide and to avoid deposits plugging the tower, particularly with slurried bases, weir-type distributors are normally installed at the top of the tower. Mist eliminators at the gas outlet are used to minimize the discharge of liquor droplets into the atmosphere. Sulfur dioxide absorption efficiencies in excess of 99% can be achieved with this kind of equipment.

(f) Turbulent Contact Absorber

The Turbulent Contact Absorber (TCA) is a relatively new development in gas-liquid absorption equipment that has found application in the preparation of sulfite liquor. This tower contains a packing of low density, hollow plastic spheres placed between retaining grids sufficiently far apart so that the high gas and liquid velocities maintain the spheres in a turbulent and random motion. This turbulence contributes to intimate contact between the gas and liquid phases, resulting in a high degree of mass transfer and allowing efficient absorption to take place in towers much smaller than would be needed if static packing were used. The tower is essentially nonplugging and is particularly useful when a solid phase is involved, either in the feed material or the product [3]. A TCA tower for a typical sulfite mill is about 1 m in diameter and 7 to 10 m in height, with the configuration shown in Fig. 81. Four stages are usually used, each partially filled with hollow polypropylene spheres 40 mm in diameter. To achieve turbulent motion in the bed, the static volume of the packing should fill less than 50% of the total absorption zone volume. The tower should be sized so that the gas velocity is in the range of 200 to 400 m/min to maintain the proper bed action. The gas is admitted to the bottom of the tower and the base is introduced above the top bed. Alternatively, the base may be added to the third bed and fresh water above the top bed to minimize liquor carryover and to recover heat from the exit gas. An expanded section in the base of the tower (sump) collects the liquor. Product acid and any liquor required for recirculation are taken off at this point. The ratio of recirculation to product is around 10:1.

(g) Venturi scrubber

Multiple stage Venturi scrubbers are in use for preparing magnesium base bisulfite liquors in mills with chemical recovery. Several stages are necessary for achieving efficient absorption of SO_2 gas. Control is accomplished by adding $Mg(OH)_2$ base at each stage while maintaining the pH at a level high enough for absorption, but low enough to prevent monosulfite ($MgSO_3$) formation. In the classical system three or four Venturi absorption stages are arranged horizontally in series (Fig. 260) whereas in more recent designs vertical arrangement in a slim and high tower is chosen. The stages in this tower combine TCA absorption with multiple-Venturi absorption (see Chap. IX,C,8c).

(h) Liquor quality

Control over bisulfite liquor quality is much easier to achieve than with acid bisulfite liquor.

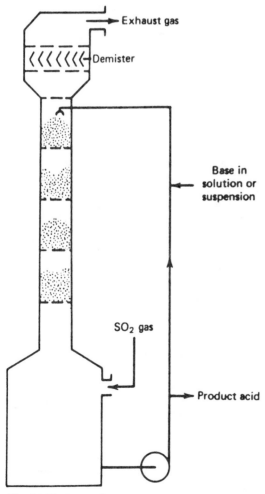

Fig. 81. Turbulent Contact Absorber (TCA) tower [16].

The production rate is controlled by the operating rate of the sulfur burner, that is the rate of feeding sulfur and air. The strength of the liquor is controlled by the total water consumed by the system. The pH, or ratio of SO_2 concentration to base concentration leaving the tower, is determined by the flow rate of base, which should be controlled so that an excess of monosulfite is maintained in the system to ensure efficient SO_2 absorption, while avoiding sulfite deposits.

The final cooking liquor produced for acid bisulfite and bisulfite operations is obtained by combining the raw liquor stream from the absorption system with the highly acid contents of the accumulator system for the recovery of liquor and gas relieved from the digester during the cook.

2. Recovery of chemical and heat from the digester

(a) Top relief

Complete impregnation of the chips with cooking liquor is of first importance in acid bisulfite and bisulfite cooks. To accomplish this in the batch digester, sufficient cooking liquor is charged to cover the chips or, preferably, the digester is filled with liquor to the top and pressurized to 500-650 kPa (depending on digester safety limits) with a suitable pump. Liquid expansion produced by slight temperature rise can also be used to produce hydraulic pressure for impregnation. When the charge is then heated in order to reach the temperature required for complete cooking, the pressure increases rapidly (Fig. 82) and both liquor and free SO_2 gas need to be relieved through an automatic top valve set to meet digester pressure specifications. This top relief is reclaimed as completely as possible to utilize its content of chemical and heat for subsequent cooks and to meet environmental standards.

(b) Side and blow-down relief

During the period of heating to cooking temperature it is common practice to withdraw liquor from the impregnated chip charge to provide vapor space at the top for gas relief; to reclaim liquor chemical not consumed in impregnation; and, in cooks with direct steam heating, to provide space for condensate formed. This side relief forms the second stream recovered during a cook. The third stream, blow down relief, is taken during a sufficiently long period before discharging the digester to reduce pressure and consists of SO_2 gas and steam with entrained liquor.

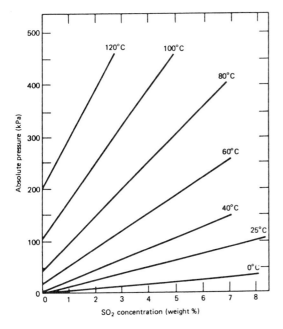

Fig. 82. Vapor pressure of sulfur dioxide solutions at various concentrations and temperatures [16].

(c) Pressurized acid system — fortification

In acid bisulfite pulping, the three relief flows combined contain 30-70% of the sulfur dioxide charged to the digester. They are most useful for building up the required concentration of free SO₂ in the fresh liquor since they contain a higher concentration of SO₂ than burner gas.

In bisulfite pulping, only about 10% of the sulfur dioxide charged to the cook appear in the digester relief streams, but the heat economy of reclaiming these streams is substantial, considering the high cooking temperature used.

The basic system for preparing fresh hot cooking liquor by combining the raw liquor from the absorbers with the above digester relief streams is shown in Fig. 83. Some variant of it is in use by most acid bisulfite and bisulfite mills. The principle involved is countercurrent gas recovery with the fresh liquor flowing from the acid plant (SO₂ burner and absorber) to the digester and the relief gas returning to the acid plant, whereby the strong gas is brought in contact with the strong acid. At the same time a large portion of the heat content and unused chemical in the cooking liquor is transferred to freshly prepared cooking liquor. In the schematic diagram of the recovery plant, the hot side relief liquor and high pressure top relief from the

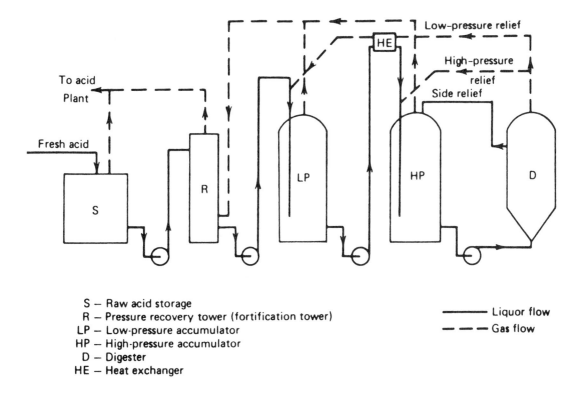

S – Raw acid storage
R – Pressure recovery tower (fortification tower)
LP – Low-pressure accumulator
HP – High-pressure accumulator
D – Digester
HE – Heat exchanger

—— Liquor flow
– – – Gas flow

Fig. 83. Chemipulp hot-acid system [16].

digester are returned to the high pressure accumulator operating at 50-80°C, and the low pressure top relief to the low pressure accumulator. In both cases eductors (jet mixers) with acid pumps developing 500 kPa pressure are used for affecting intimate contact between top gas relief and onflowing fresh cooking liquor. The low pressure relief from the latter part of digester blow down is cooled by fresh cooking liquor in order to facilitate SO_2 uptake in the low pressure accumulator and pressure recovery tower. Top relief SO_2 from both accumulators is piped to the pressure recovery tower where additional absorption and recovery in fresh liquor of relatively low temperature takes place, and a high degree of fortification is achieved.

Typical acid strengths for acid bisulfite are shown below:

	Fortified Raw Acid	Cooking Acid
% Free SO_2	2.5-3.0	5.0-7.0
% Combined	1.0-1.7	1.0-1.5
% Total	4.0-4.5	6.0-8.0

Accumulator capacity required is from 1.5 to 3 digester liquor charges. In acid bisulfite systems the high pressure accumulator operates at about 275 kPa and the low pressure accumulator at about 70-100 kPa pressure, whereas in bisulfite operations the system runs under very low or no over pressure. Since the pH of bisulfite liquor is highly sensitive to changes in SO_2 concentration, it is common practice to provide liquor circulation loops on the high pressure accumulator for mixing and pH control of the liquor to the digester by injection of base. The hot acid system can also be operated with a single accumulator and, since they are of similar construction, idle digesters can serve as accumulators.

(d) Neutral and alkaline sulfite systems

For neutral sulfite pulping the equipment and procedure for cooking liquor production can be simplified further since there is no free SO_2 and no need to contain or recover gaseous chemical. Liquors for the original NSSC process are made by treating soda ash (sodium carbonate) solution with burner SO_2 gas in a simple absorption tower as a continuous process or in agitated tanks as a batch process. The pH of the fresh liquor is 8.0-8.5 and therefore contains two buffer systems, bicarbonate-carbonate and bisulfite-sulfite, which help to increase the pulp yield and reduce corrosion of equipment by neutralizing acids released from the wood. The presence of the bicarbonate-carbonate buffer is now found to be considerably less important than thought originally, and NSSC pulp of good quality is produced with little or no carbonate in the liquor. In mills with no or only partial recovery of chemicals, where the cooking liquor is made with purchased sodium chemical, the operation can be greatly simplified and the control over liquor composition and pH much improved by using caustic soda at 50% concentration from tank cars. In this case the liquor contains only bisulfite-sulfite, and the end pH of the cook is controlled by the pH level of the fresh liquor. Sodium base NSSC mills with complete chemical recovery necessarily use carbonate produced in the recovery process for liquor making.

In liquid phase NSSC cooks, up to one half of the cooking liquor charged can be reclaimed and its heat and residual chemical utilized in subsequent cooks. In vapor phase NSSC cooks which have become common because of substantial savings in energy and chemicals, only the amount of cooking liquor taken up by the chips during impregnation enters the cook.

Alkaline sulfite liquors are basically "neutral" sulfite liquors whose pH has been raised to the moderately or strong alkaline level by adding sodium carbonate and/or caustic soda. The production of the sulfite component involves the same principles and methods as used for the production of "neutral" sulfite liquor from carbonate or caustic soda and sulfur dioxide, with the difference that SO_2 addition is taken only far enough to retain the desired concentration of unreacted alkali. Make-up caustic soda can be used to raise the pH above that of the sulfite or the carbonate as the case may be.

F. LIQUOR QUALITY CONTROL

Refer to Chapter III,B for a discussion of liquor analysis. Procedures are covered for all sulfite liquors.

REFERENCES

See end of Chapter V for references for both Chapters IV and V.

V

PULPING DIGESTER SYSTEMS AND OPERATION

O.V. INGRUBER

Retired Senior Research Associate from CIP, now an independent consultant

H.E. WORSTER

Senior Research Associate, Special Projects
MacMillan Bloedel Research, Vancouver, B.C.

MANUSCRIPT REVIEWED BY
David Johnston, Manager Pulping
Nova Scotia Forest Industries
Port Hawkesbury, Nova Scotia

Digesters or cookers for producing sulfite pulps of different descriptions are available in several variants which can be divided into two groups: discontinuous or batch type and continuous type. The original industrial sulfite cooks used batch type pressure vessels of both horizontal and vertical design. The vertical batch digester succeeded in the long run and far out numbers continuous equipment in installations and tons of pulp produced. The first continuous pulping equipment was the Asplund-Defibrator, introduced in 1934 in Sweden and in 1937 in North America for the production of mechanical pulp. Its modern variant is a standard for producing high yield NSSC, and sulfite chemimechanical pulp from hardwoods and softwoods. Continuous digesters are of the vertical, horizontal and inclined type.

A. BATCH DIGESTERS AND THEIR OPERATION

The common upright, cylindrical batch digester is shown in Fig. 84. The cylindrical shell carries a hemispherical or conical top with a large flanged opening for chip filling and access. The neck of the top fitting made of cast steel carries a pipe for pressure relief with a strainer on the inside wall. The conical bottom, which is usually given an angle of 70° to assist in the clean blowing of the cooked material, carries the bottom fitting with flanged connections for the blow valve, steam line, liquor and drain line and liquor circulation line in the case of indirect heated digesters.

1. Shell

The shell of the digester is made of carbon steel plate. For many years, and still common in older installations, the construction method consisted of riveting together preformed plate of 28.5 mm thickness by double butt straps for rigidity. Later the method of construction changed to completely welded shells made of 51 mm steel plate and sometimes reinforced by using 76 mm plate for bottom sections. This has been the standard for most new digesters for some time. Batch digester closures in use range from manually tightened steel plate covers, bolted to the neck piece and the handwheel operated blow valve at the bottom on older digesters; to hydraulically or electrically operated gate — or yoke type closures and blow valves, actuated remotely from the control room on the cook's floor. For some time automated ball valves have been installed in increasing numbers, first as blow valves and more recently, and of much larger size, for top closures, particularly in batch digester plants with auto-

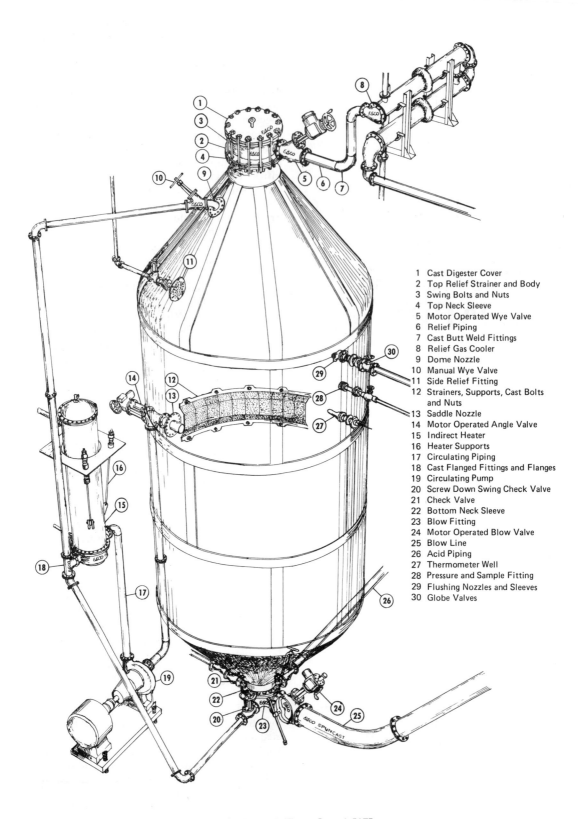

1   Cast Digester Cover
2   Top Relief Strainer and Body
3   Swing Bolts and Nuts
4   Top Neck Sleeve
5   Motor Operated Wye Valve
6   Relief Piping
7   Cast Butt Weld Fittings
8   Relief Gas Cooler
9   Dome Nozzle
10 Manual Wye Valve
11 Side Relief Fitting
12 Strainers, Supports, Cast Bolts
    and Nuts
13 Saddle Nozzle
14 Motor Operated Angle Valve
15 Indirect Heater
16 Heater Supports
17 Circulating Piping
18 Cast Flanged Fittings and Flanges
19 Circulating Pump
20 Screw Down Swing Check Valve
21 Check Valve
22 Bottom Neck Sleeve
23 Blow Fitting
24 Motor Operated Blow Valve
25 Blow Line
26 Acid Piping
27 Thermometer Well
28 Pressure and Sample Fitting
29 Flushing Nozzles and Sleeves
30 Globe Valves

Fig. 84. Batch digester with ancillary equipment (Esco Corp.) [17].

matic charging, cooking and dumping control and production and steam scheduling (Fig. 89).

The size of batch digesters ranges from about 70 to 350 m³ capacity and a production of 5 to 25 tons of unbleached pulp per blow, requiring a charge of wood on oven dry basis of 11-54 tons at a yield of 46%. High yield cooking produces proportionally more pulp from the same wood charge.

2. Materials of construction

(a) Digesters

Sulfite cooking liquors in the acid pH range are highly corrosive to carbon steel and proper protection of the inner wall of the steel shell and all fixtures exposed to liquor contact is a necessity. The inner lining consists of ceramic or carbon tile, or of stainless steel. The transition from tile to stainless steel started in the fifties under Scandinavian leadership, and has been completed there in 1980 with the last brick-lined digester being taken out of commission. In North America steel-lined sulfite digesters are in use, but brick lined digesters are still the most common type (Fig. 85).

With regard to brick linings, it is of considerable importance whether the cooking acid used contains a soluble or insoluble base. The acid-proof ceramic brick normally used in digesters have a certain porosity. With calcium bisulfite this porosity is slowly eliminated due to the formation of insoluble deposits within the bricks, and complete protection of the digester shell results. With liquors of high SO_2 content or with soluble bases (magnesium, sodium, and ammonium), little or no deposition of insoluble material takes place. To achieve protection, two layers of brick are recommended.

Before the digester is lined, it is exposed to high temperature and pressure with water and the inside of the shell is thoroughly cleaned for the removal of oil and grease. The lining consists of a grout layer 38 mm to 75 mm thick of cement and crushed quartz mixed with a dilute solution of sodium silicate, followed by one or two courses of brick. For handling low-acid-strength calcium base cooking liquors, a single course of brick 3 in thick has generally been adequate. These bricks in recent years have been laid up with a furane resin cement mortar. For handling liquors of high acid strength or for soluble bases, two layers of brick are employed. The inner layer or brick, usually 50 mm thick, is laid up in a litharge-glycerine mortar, and the lining face brick, usually 75 mm thick, is laid with furane cement mortar. Carbon brick is used only for the facing layer in two-layer linings.

The quality of brick for digester linings is important. These bricks are specially designed to combine chemical and spalling resistance. They are molded to fit the contour of the vessel, and special shapes are required for cones, corners, necks etc. All fittings projecting through the lining are made of 316 stainless steel.

Stainless steel sleeves are employed at all openings, and the joint between these sleeves and the brick is arranged to prevent acid seepage between the lining and the shell. Brick linings are designed to remain in compression at all times to assure a tight seal between the steel shell and the lining. A certain minimum compressive stress in the brickwork is necessary to overcome the form-changing factors that tend to cause a separation of the lining from the shell or of the components of the lining.

Form-changing factors which contribute to the danger of cracking of the brick lining or of separation of the lining from the shell include (1) the rapid cooling of the brick lining, especially its face, during blowing and during washing of the digester contents, (2) the operating pressure, which tends to expand both lining and shell, and (3) the differential expansion between the digester shell and the brickwork during the steaming period.

With welded digester shells, rigidity is less than with riveted butt straps of the older models and service time for brick linings has decreased from 10 years to 6 years, in general.

Initial cost for ceramic bricks is lower than for carbon bricks, the latter having lower maintenance and longer life and being superior with soluble bases. Both types of lining have the disadvantage of occupying 15-18% of digester space; continually introduce dirt into the pulp

Fig. 85. Strainer being installed in a brick-lined digester (ESCO) [17].

due to brick erosion and spalling; and require frequent inspections and minor repairs with cooling periods up to 4 h.

Two types of stainless steel protection are in use: stainless steel lining installed in existing digesters, and stainless steel cladding bonded to carbon steel plate in the rolling mill. Of the two, the latter is superior due to smooth and uniform surface and the strong continuous bond to the shell with low susceptibility to thermal shock and consequently less downtime for repairs, no break-in period and good for wash down with cold water (cf Chapter X, A, 3).

Carbon steel digesters are common with alkaline pulping processes with an average life span of 20 years, but suffer also from corrosion, particularly at the upper and middle sections, at an average rate of 0.76 mm per year. Sodium hydroxide concentration is a critical factor. In critical areas the corrosion rate may vary from 1 mm to 2 mm per year. The corrosion rates for Inconel or stainless steel clad alkaline digesters are about 10% of the above figures.

While the corrosivity of strongly alkaline sulfite liquor is not greater than that of kraft or soda liquor, data for the new moderately alkaline may not be available immediately. Neither are data available as yet on the corrosivity of alkaline cooks containing the anthraquinone redox system. At any rate, the conclusion to be drawn from the above data is that it is always safe and perhaps even economically justified to use stainless steel clad carbon steel digesters for extended life at reduced maintenance for flexibility of performing cooks with liquors of all descriptions.

(b) Liquor and pulp processing equipment

Modern design usually provides stainless steel, higher alloys and nonmetallics at strategic points in all equipment that comes into contact with the spent liquor, usually up to and including the first washer. Most mills now use plastic wires for improved operation and reduced corrosion.

There are many commercial stainless steels and specialty alloys but only a few find extensive use in pulp mills. Stainless steels with an "L" in their designation are now carbon grades recommended for welding. In addition, the "L" grades have lower impurity levels and are thus usually more resistant to localized corrosion.

Due to problems with stress chloride cracking, higher alloys such as ASTM B625, Hastelloy G, Incollay 800 and Carpenter 20C63 are being used.

Typical materials of construction are as follows:

– Digesters
 • 316/316L
 • 317/317L
 • ASTM B625
– Evaporator Tubes
 • 304/304L
 • 316/316L
 • ASTM B625
 • Sandvik 3RE60
 • Inconel 600
– Evaporator Body
 • 304/304L
 • 316/316L
– Washers
 • 304/304L
 • 316/316L
– Piping
 • 304/316L
– Refiner Bodies
 • Stainless Steel
– Refiner Discs
 • 17-4 pH Stainless Steel
– Refiner Plates
 • White Cast Iron (WCI)
 • Ni-Hard
 • Highly Alloyed WCI

For extremely abrasive conditions, as found in digester chip feeders, screw presses and refiner plates, corrosion and wear resistant alloys are used.

In other parts of the mill where operations are similar to those in mills using other pulping processes, materials of construction are the same. Plastics are replacing many of the older materials of construction (lead, ceramic, terra cotta, etc). Fiberglass reinforced plastics have shown exceptionally good service in many areas.

(c) Corrosion prevention and repair
Thermal spray coatings

The life of plain carbon steel digesters has been successfully extended by thermal spray coatings. Thermal spraying is a process in which metallic and nonmetallic materials are deposited in a semimolten or plastic condition to form a corrosion and/or erosion resistant coating on the base metal. There are three major processes: flame spraying, electric arc, and plasma spraying. Coating thicknesses vary with the process and materials being sprayed but thicknesses in excess of 0.030″ can be applied. These processes must be used with caution since even in the most dense coatings (plasma spray) have some porosity and can disbond due to undercutting corrosion.

Weld overlay

A second and much better method for

refurbishing (or on new construction) digesters is weld overlay. Both manual and automatic processes are used. The automatic submerged arc process has been used extensively for extending the service life of sulfite and kraft digesters with excellent success (Fig. 86).

The submerged arc process is more versatile than manual, since an infinite range of alloy compositions can be produced by using steel tubing filled with metal powders for the electrode. Thus a specific composition, e.g. 316SS, can be specified for the surface of submerged arc overlay. In addition, submerged arc overlays are much more uniform in composition (reduces chance of galvanic corrosion) and have fewer flaws (sites for initiation of corrosion undercutting).

Stainless steel clad steel

Roll-bonded or explosion-bonded clad steel can be used for new construction to reduce the cost of solid stainless steel or higher alloy equipment. Loose clad or bag liner type construction should not be considered due to the possibility of perforation of the cladding and resultant corrosion of the base metal. Weld procedures for clad plate construction must be closely followed to prevent corrosion of the welds.

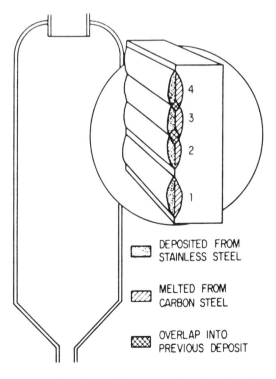

DEPOSITED FROM
STAINLESS STEEL

MELTED FROM
CARBON STEEL

OVERLAP INTO
PREVIOUS DEPOSIT

Fig. 86. Overlay welding procedure for digester repair.

Equipment failures

Failures of stainless steel equipment are frequently due to either misapplication (e.g. incorrect alloy) or intergranular corrosion. Intergranular corrosion of austenitic stainless steels is normally the result of "sensitization" which involves the intergranular precipitation of chromium carbides when the metal is exposed to temperatures in the approximate range of 425-815°C. Precipitation increases with increasing carbon content and exposure time. The commonly accepted mechanism attributes intergranular corrosion to a difference in corrosion potential between a chromium depleted zone at the grain boundaries and the chromium-rich central zone of the grains.

"Weld decay" is an example of intergranular attack that occurs in bonds adjacent to welds which were exposed to the critical temperature range.

Sensitization of stainless steels

Sensitization can be prevented or reduced significantly by reducing the carbon content to values such that carbide formation is inconsequential, less than 0.02% or 0.03%, depending on time and temperature; or by using carbide former alloy additions such as titanium (321SS) or columbium/tantalum (347SS) to "stabilize" the alloy.

"Desensitization", redissolving carbides, can be carried out by annealing the stainless steel at 1040-1150°C and then rapidly cooling to prevent reprecipitation. This procedure is not practical in the mill or for anything but the simplest construction and thus is very seldom used.

Stress corrosion

Stress corrosion cracking (SCC) is a form of corrosion that occurs in certain environments when a metal is under the combined influence of stress and corrosion. Neither the stress nor corrosion, when operating independently, can cause failure.

One major drawback in the use of austenitic stainless steels has been the sudden and quite frequently catastrophic failure of equipment due to chloride stress corrosion cracking (stress chloride cracking).

SCC of stainless steel occurs most frequently in mildly corrosive environments containing chlorides (pH 4-9, temperature >50°C).

In all industries this problem of stress chloride cracking of stainless steels necessitates design limitations such that materials with superior corrosion resistance, or more expensive materials must be used where stainless steels would otherwise have been satisfactory.

The stresses that cause cracking are most frequently residual that result from welding and fabrication, e.g. rolling. These stresses can be reduced by thermal treatments, e.g. annealing or low temperature stress relief (538-649°C). The low temperature treatment cannot be carried out on materials susceptible to sensitization and intergranular corrosion. Even the low carbon and stabilized grades will suffer sensitization when exposed to the 538-649°C treatment for the required time.

Localized corrosion

Localized corrosion, of which pitting and crevice corrosion are two forms, can be defined as the selective removal of metal by corrosion at small areas or zones on a metal surface in contact with an aqueous environment.

Only alloys such as stainless steels, which depend on a passive film for corrosion protection, are susceptible to localized corrosion. Attack occurs either as pitting of exposed metal or crevice corrosion in areas partially shielded from the environment.

Localized corrosion can only occur in an environment which contains an oxidizing agent and aggressive anions, e.g. chlorides.

Attack is usually confined to a small fraction of the total metal surface (hence the term "localized") where passive film breakdown results in propagation of corrosion.

Deposits on metallic surfaces should be

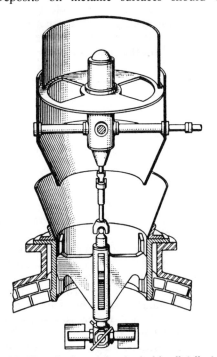

Fig. 87. Chemipulp mechanical chip distributor.

removed periodically to reduce the possibility of localized corrosion due to oxygen depletion.

Corrosion-erosion

Good design will avoid abrasion by suspended solids or impingement of liquor, since this scours off the oxide layer that gives stainless steel its corrosion resistant qualities. In stationary digesters, the vapor zone is especially susceptible because of condensation, as is the side wall where liquor and chips wash the surface and tend to wipe off the protective coating.

3. Chip packing

For producing uniform pulp at high yield per digester it is necessary to accommodate as much chip weight as possible and to distribute the material uniformly throughout the whole volume of the digester. Free falling chips form a cone in the digester, packing density is non-uniform, and the least charge results. To obtain uniform chip distribution and greater packing density, chip packers have been in use for many years. Using either mechanical action or steam, these devices impart a tangential motion to the falling chips and produce a horizontal surface of the charge. A mechanical distributor is shown in Fig. 87. Packing and density can be influenced to some degree by adjusting the tilt of the paddles.

A steam packer assembly is shown in Fig. 88. Steam chip packers consist of a steam pipe with nozzles slanted downward below the lower peripheral edge of the top sleeve. Both stationary or movable models are available, although recently these packers have been built into the neck of the digester. Since the noncondensible gases and air have to be removed simultaneously, an evacuating system is used on the digester circulation lines. This prevents blowback of chips onto the digester floor and enables increased steaming of the chips. Presteaming may be completed by adding steam through the bottom fitting.

It is customary when charging by either method to admit steam to the bottom of the digester to remove air and to increase chip packing. This also presteams the chips and allows for better penetration of liquor. The digester charge can then be raised to maximum cooking temperature at a faster rate.

Mechanical or steam packers are used in both acid and alkaline pulping to achieve high chip packing density. In alkaline pulping a common method is also to begin liquor filling through the bottom as soon as the first chips are charged from the top. Due to alkaline swelling the chip surfaces become slippery, and the chips slide into a more density packed position under the weight

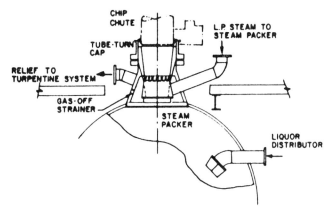

Fig. 88. Rader steam packer assembly [17].

of the charge. This method gives good packing in minimum time. Liquor filling while chip charging precludes presteaming.

(a) Degree of packing.

The degree of packing is the ratio of dry wood substance in a digester to the theoretical weight of dry wood substance that the digester could hold assuming the total digester volume to be solid wood. It is expressed as a percent. This is also known as a wood charging factor (see CPPA Data Sheet C-00.) The following is an example.

Digester volume, m^3	131
Wood density, g/cm^3	0.405
Wood moisture, %	46
Dry substance, %	54
Digester wood charge, wet weight, t	50

Degree of packing:

$$\frac{\text{dry weight of chips}}{\text{weight of dry solid wood volume of digester}}$$

$$\frac{50 \times 0.54 \times 100}{131 \times 0.405} = 50.9\%$$

When the digester is full of chips, the top is sealed either by hand or automatically (Fig. 89) and liquor is filled from the bottom by pumping from liquor storage or, for acid bisulfite cooks, from the high pressure hot accumulator. During pumping the air still entrained in chips is vented through a vent line at the top either to the atmosphere or, for acid bisulfite, back to the absorption unit for reclaiming SO_2 gas.

It is customary for acid bisulfite cooks to circulate the hot liquor back to the accumulator

Fig. 89. Automated ball valves for chip filling. (Kamyr Inc.)

for 15 to 45 min depending on time available and cooking schedule. This ensures minimum variation of the strength in the digester, owing to variations in wood moisture, and allows a stronger liquor to be present when steaming of the digester charge is started. During circulation, penetration of the acid into the wood is also proceeding. If circulation is eliminated, the digester is usually steamed at a slower rate to provide proper penetration of the chips by the liquor.

A)

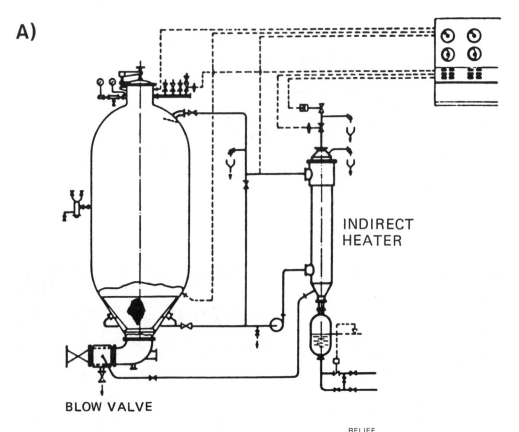

INDIRECT HEATER

BLOW VALVE

B)

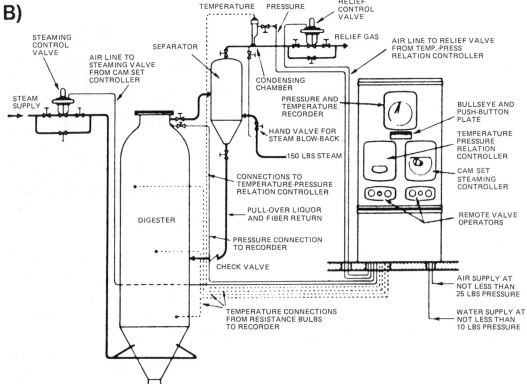

Fig. 90. Batch digesters for (a) indirect and (b) direct heating [18].

(b) Liquor-to-wood ratio

The volume of liquor charged on dry wood in the digester is known as the liquor-to-wood ratio. When calculating the ratio, moisture present in the wood must be considered with the quantity of liquor charged to the digester to obtain the true ratio. This ratio can vary from 3:1 to 6:1 depending upon the degree of chip packing and the moisture content of the wood. When dealing with liquor-to-wood ratio data, it should always be verified whether wood moisture has been included or excluded, the latter often being the case in established routines. The true ratio is always required for process calculations.

4. Heating systems

Basically there are two types of batch digesters, with indirect steam heating and with direct steam heating. The first type is equipped with forced liquor circulation and a heat exchanger in the circulation line (Fig. 90a), the second type features direct steam admission into the cooking liquor. The simplest variant of direct steam application is into the bottom cone, either as shown in Fig. 90b or through the bottom fitting. A variant giving better control of the heating schedule and uniformity of cooking features forced liquor circulation and direct steam admission into the circulation line by

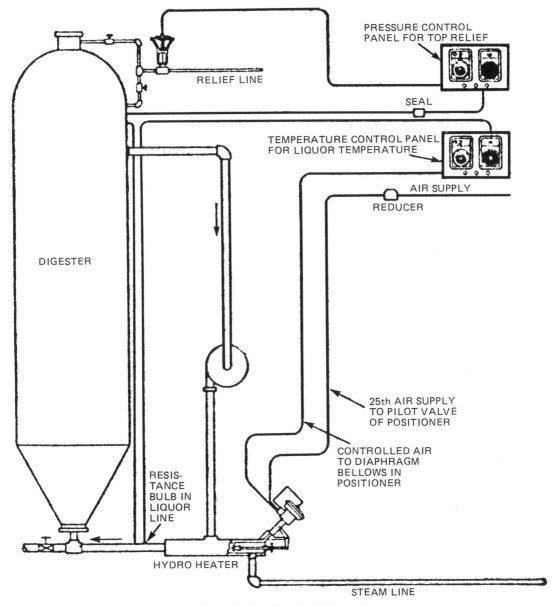

Fig. 91. (a) Hydroheater installation in batch digester circulation system.

means of a device called a hydroheater (Fig. 91a, b).

(a) Direct steaming without and with circulation

Direct steaming without circulation is seldom practiced in sulfite pulping because of its low efficiency relative to modern systems and practice. Without forced circulation, the heat from the direct steam at the bottom of the digester rises through the central portion and, upon reaching the top, starts to work down and cook the peripheral volume. Digesters so cooked produce pulp containing large numbers of slivers, shives, and uncooked chips. It is not possible to have uniform cooking with natural convection of steam heat admitted into the bottom of the digester, and non-uniformity of temperature and liquor concentration obviously must increase with the size of the digester. Also, the steam contacting the cooking liquor at the bottom causes dilution through condensate and displacement of free SO_2 from acid bisulfite by thermal decomposition explained earlier. Consequently, the chips in this area darken and cannot be delignified because of lignin condensation. In high yield bisulfite cooks where lignin content is high, both brightness and strength, the chief characteristics of high yield pulps for publication papers, are damaged by direct steaming. Damage is less on the neutral sulfite and alkaline pH level but non-uniformity and non-efficiency prevail. Liquor circulation is necessary for mills making high quality pulps, such as paper grades and dissolving pulps; for mills with recovery systems so as to avoid unnecessary dilution of spent cooking liquor; and for general pulp quality reflected in operating costs and market value.

Forced liquor circulating systems use pumps, and differ only in the location where the cooking liquor leaves the digester and is returned to it. Liquor may be withdrawn from the digester at a higher or lower level and, after passing through the pump and a heat exchanger or a steam injector, return to the bottom of the digester. Systems with reversed flow are also in use, withdrawing the liquor from the conical end and feeding it to the top of the digester. Many pulp mills use a system where the acid is withdrawn from the middle of the digester and fed to both ends. Some mills also have provisions for reversing the direction of liquor flow. Too slow circulation will result in a noticeable temperature gradient in the digester from the inlets to the strainers. Too rapid circulation may result in clogging of the strainers when the wood in the digester reaches the point of fiber liberation in soft cooks.

The results from a survey of the performance of circulation systems indicate that split circulation, from the middle to top and bottom, maintained liquor flow during the later part of extended soft cooks where circulation may stop due to plugging of the strainer.

The hydroheater (Fig. 91) steam injector is usually located after the discharge of the circulating pump to prevent uncondensed steam from binding the pump and lowering the rate of acid circulation. The temperature of the acid should be measured before the hydroheater section; otherwise, readings will be influenced by

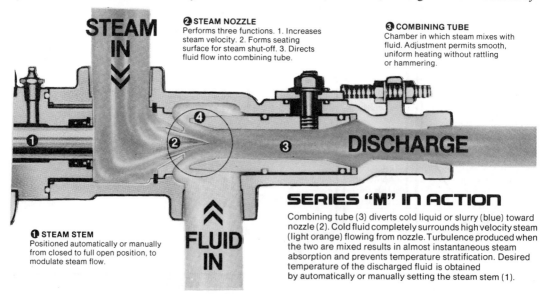

Fig. 91. (b) Hydroheater operation (Hydro-Thermal Corp.).

variations in rate of steaming. Steaming can be carried out at a much faster rate with circulation, the limiting requirement being complete acid penetration of chips before temperatures above 100°C are reached. The initial problem of steam hammering has been largely solved by design improvements [5].

(b) Circulation and indirect steam

A heat exchanger in place of the hydroheater allows the condensate to be recovered and reused as make-up in the boiler house, thereby reducing overall steam costs. Acid-to-wood ratio is held more uniform throughout the cook, and better pulping conditions exist. Solids content of the spent liquor is increased, a real advantage in chemical-recovery systems. These benefits and economies can usually justify the capital cost of the heat exchanger installation.

When such a circulating system is designed, the heat exchange surface must be calculated from the differential between maximum cooking temperature and the steam temperature. Undersized heat exchangers limit the rate of ascent of the digester temperature when approaching maximum cooking temperature. Calcium-base acid sulfite has a tendency to deposit insoluble calcium monosulfite and sulfate, which hampers heat exchange and reduces flow through the heat exchangers. Therefore, the design should also include provision for cleaning. Change of process to soluble base solves this scaling problem.

Circulation is usually started after liquor filling is complete. The cooking liquor is typically recirculated once every 10 to 15 min. Therefore, the time lag between the bottom and top temperatures cannot exceed 5 to 7.5 min at any point in the temperature curves, whereas in the old systems the bottom and top temperatures never matched at all. Circulating systems have thus resulted in more uniformity in penetration, temperature, and acid strength throughout the whole digester. This improvement, in turn, has permitted increased chip charge, improved wood and digester yields, produced cleaner, more uniform pulp, and has reduced steam cost. However, maximum chip packing may impede circulation even with the best of systems, and the right balance is found only by experience. Wide variation in initial moisture content of chips is still detrimental and is to be avoided if possible.

Failure of the circulating system before maximum temperature has been attained presents many problems, especially with a high degree of chip packing. The pulp will be dirty and "shivier". The possible actions to be considered are:

1. Steam the digester at a slower rate so that as much natural convection as possible can be obtained, especially if failure of the system occurs at the beginning of the cook. The cooking time will be longer, but at least the pulp produced will be closer to normal.
2. Steam the digester up to about 100°C and then shut the steam off for 30 min or longer if time is available. This will allow the top and bottom temperatures to equalize after penetration has occurred.
3. If the failure is not mechanical, but due to chips plugging the line, it may be possible to get the system back into operation after the temperature has risen to 100°C. This should always be tried by stopping and starting the circulating pump.

Although pumps are designed for a 10 to 15 min circulating time, as the temperature rises to near maximum, the chips become soft and rate of circulation is reduced to 50% or less. To save on maintenance costs, some mills shut down the circulating pumps about an hour after maximum temperature has been reached. Pumping rates are reduced earlier if the free SO_2 content of the cooking acid is greater than 6.00%. This is especially true after the side relief has been removed from the digester if the suction of the circulating pump is close to the level of liquid in the digester. This is probably due to foaming of the acid, with high separation of SO_2 gas.

Circulation in high yield cooks usually remains at a high rate during the cook because of little deformation of the chip charge to reduce void space.

5. Liquor relief during temperature rise

Cooks in the acid sulfite range, whether for high grade dissolving pulps or strong, uniform high yield pulps, require complete impregnation with cooking liquor. This is easily accomplished by the best available method consisting of presteaming the chips, liquor pumping to the top of the digester and application of hydraulic pressure at 550-700 kPa during steaming to about 110-125°C with liquor circulation.

After complete penetration it is no longer necessary to maintain the digester full of liquor. Hence a quantity of the cooking acid is removed through a side relief fitting or through the circulation strainer and returned to heat the liquor in the hot accumulator and to maintain a temperature of about 80°C in this vessel. Excess chemicals are thus recovered to improve the economy of the process. Liquor relief is removed by volume determined as a compromise between chemical and heat recovery, cooking time and pulp quality. Depending on the design and size of the strainer, the time required for side relief is from 15 to 45 min. An efficient way of removing liquor relief is to pump the excess liquor through the circulation strainer or collector from the digester and deliver it through the circulation pump to the liquor-making system for re-use. The large circulation strainers do not impede the flow and the digester will be on dry gas relief longer and the accumulator liquor further fortified with SO_2.

If the side relief is removed too early, chips left above the liquid level in the digester will show signs of improper penetration by the bisulfite. These chips show up in the screen room with black centers from long acid bisulfite cooks and with white, uncooked centers from short high yield bisulfite cooks, being cooked only on the outside where some penetration has occurred. Taking side relief too early also reduces the temperature, and causes low-temperature cooking liquor for the succeeding cook. If the side relief is removed too late, the heat value of the liquor will be high and the time remaining on dry gas will be shorter, thus providing a cooking acid of low strength for succeeding cooks.

If the quantity of side relief removed is too great, the heat value to the accumulator liquor will be high. Also, the liquid level in the digester may be too low for good circulation. Insufficient side relief causes low temperature in the hot accumulator and increased chemical consumption. Also, if condensate from direct steam has to be added to the digester to obtain maximum temperature, it may occupy the vapor space and thereby cause difficulty in obtaining dry-gas relief. Wet wood requires a later side relief and also a stronger initial cooking liquor.

The amount of liquor which can be removed from the digester after impregnation is limited by the location of the strainer or collector through which the liquor is withdrawn. It is advantageous in high yield bisulfite cooks to withdraw as much of the impregnating liquor as possible after adequate impregnation, and to cook the charge more or less in the vapor phase. Mills having digesters with strainers in the cone and bottom-to-top circulation can make the best use of this method. The top circulation return in the dome can be equipped with liquor distributors or sprays, and the small volume of liquor remaining in the digester circulated to produce what is called a trickle phase cook. This increases the effectiveness and uniformity of the vapor phase cook and produces a spent liquor of high solids content for recovery. In the case of sodium base high-yield bisulfite mill, the liquor strainers were lowered from the upper half of the digester to the top of the cone resulting in significant improvement in pulp uniformity, and savings in chemicals, total cooking time, and steam.

6. Digester top relief

Top relief in sulfite cooks up to pH 5 consists of the liquor relief removed automatically by a pressure controller before the side relief is taken; and of vapor or dry gas relief after the side relief has been taken, containing H_2O, SO_2, CO_2, O_2, N_2, and organic vapors in small quantities. During side relief there is no top relief, because any pressure buildup is used to drive the required quantity of side relief to the accumulator. Cymene may be recovered in a separator from the top relief of acid cooks. Top relief is one of the most important components of acid bisulfite liquor preparation because SO_2 gas at 70-90% concentration is recovered and returned to liquor preparation together with a substantial amount of heat.

Control of the digester pressure is of importance for the cooking liquor composition and therefore the rate of cooking. A high pressure maintains a high sulfur dioxide concentration and results in a rapid acid bisulfite cook. A high pressure in the neutral sulfite cook, buffered with sodium bicarbonate, leads to somewhat higher acidity and inferior cooking result than when the carbon dioxide formed is allowed to escape. The gases of the alkaline cook pass through a cooler to yield hot water and turpentine, and those of the sulfite cook, after cooling and cymene separation, pass to the acid system to be absorbed for re-use.

7. Digester blowdown

At the end of the cooking period, at a predetermined time, heating is discontinued and the digester pressure is relieved during a blowdown period, varying from 30 to 90 min depending on the pH level and the size of the digester. Free SO_2 boils off rapidly from heated acid bisulfite solutions, but with the large dimension of a digester densely packed with cooked chips and pulp it takes much longer for the gas to be freed and to reach the top of the liquid. As the pressure and level of free SO_2 become lower, the rate of pulping slows to an insignificant level.

High pressure blowdown is rate controlled allowing a consistent flow of hot vapor to the hot accumulator. It is normally terminated when the back pressure from the hot-acid accumulator starts to reduce the flow of dry gas from the digester, i.e., when the preset cam in the pressure controller is no longer the controlling factor. For example, if the digester pressure is 550 kPa and the pressure on the accumulator is at 275 kPa, high-pressure blowdown is normally terminated at 350 kPa. Other factors are the liquid level in the accumulator, the size of the connecting pipelines, and flow from other digesters connected to the common relief header (i.e., the overall net pressure differential in the system).

On low-pressure blowdown controlled manually, or with a precut cam, a weaker dry gas is sent either to a low-pressure accumulator or to a packed pressure-absorption tower. The gas in this case is still being used to fortify cold storage acid which eventually is pumped into the high-pressure hot-acid accumulator. This low-pressure blowdown is continued until the predetermined pressure is reached, depending upon whether the digester is to be pressure-emptied (i.e., the stock is to be blown into blowpits), which usually requires 175 to 275 kPa; or is to be emptied by diluting the pulp in the digester with spent liquor from preceding cooks and then pumping it into the blowpit.

Acid bisulfite spent liquors used for producing by-products by biological processes need to be stripped of free SO_2. Much of this can be done by low pressure blowdown in the digester or tank.

If the blow pressure is too high, SO_2 gas and heat are lost; if it is too low incomplete blow results and the remaining material has to be removed by re-blowing or washing. Many mills have flushing systems to obtain clean blow and this is necessary when pulping to high yield. Some mills practice complete pressure and liquor blowdown and pre-washing the pulp in the digester, thus avoiding blow pits or tanks.

To help clean blows, a common practice has been to blowback to a given pressure and admit steam for 15 min through the bottom. This shake-up steaming is effective, but causes some damage to pulp strength, particularly tear. It is not known whether this effect is due to the mechanical agitation of the pulp while hot, or whether it is caused by the rise in temperature due to the steam. Probably it is a combination of both.

Blowing has been improved considerably by replacing the older 25 or 30 cm seated or Y-type valves with high friction by gate valves, or preferably by teflon-seated ball valves. If the blowpits are installed below the digester level, spent liquor can be pumped through the bottom connection into the bottom cone of the digester, thus reducing the consistency of the pulp. In this manner, pulp can be sluiced or pumped into the blowpits and the number of washouts or nonclean blows can be reduced. Shake-up steam, with its attendant ill effects to the pulp, is then

not necessary. This method is known as a cold blow to differentiate it from the steam shake-up methods.

Systems for pumping spent liquor or hot water into the digester for emptying require a pump with a minimum capacity of 4 m^3 per min at a head of 60 m and a tank of sufficient size for storing spent liquor at 65-90°C. Usually the flushing solution is pumped into the bottom cone, but some mills pump into the dome or back through the circulation strainers. In the absence of a recovery system for chemicals and heat the purpose is to reduce the consistency in the digester. In mills with recovery, this method is most useful also for increasing the solids content of the spent liquor and thereby to reduce the load to the evaporators.

8. Steam requirements

When cooking digesters with direct steam, the steam is invariably added either to the bottom digester fitting, or into the circulation line through a hydroheater. If the digester pressure is 550 kPa at the top of the digester, the minimum steam pressure required to balance the pressure at the bottom of the digester is 550 kPa plus the pressure due to the level of the acid in the digester – say, an additional 170 kPa – to make a total of 725 kPa. This is why most batch digester plants use reduced mill steam at 950 kPa to ensure adequate steam flow. Saturated steam at this pressure has a temperature of 185°C. The benefit of using superheated steam in direct-steam cooking is less liquor dilution by condensate. However the gain in heat content is relatively small and offset by serious side effects, such as deactivation of the cooking liquor and black uncooked chips. Black uncooked chips can also result from prolonged steaming with regular saturated steam.

If indirect cooking is to be used, whereby heat exchangers are installed on the forced-circulation lines, heat flow must be calculated between steam temperature and maximum cooking temperature. This enables the designer to estimate the maximum demand for the heat exchanger. From this temperature differential and the heat flow required, the steam pressure required can be estimated. However, in case the heat exchanger system fails, it is still necessary to have steam at a minimum pressure of 950 kPa for direct steaming of the digester.

Therefore it is a necessity that a desuperheater system be provided for all sulfite mills. Generally, all mills operating in the acid range practice presteaming. The system may be used, or not, at the discretion of the operating personnel depending on the wood species and type of cook.

To compare steam consumption of various mills the quantities reported should be from and at 100°C. If the mill used direct steam, the energy reported to cook the digester should include energy to heat cold river water to 100°C plus that required to cook the digester. With indirect steam, where there is condensate return, steam consumption should be only the steam required to cook the digester calculated on the basis from and at 100°C. This shows the benefits of heat exchangers and also establishes a basis on which all mills could exchange data.

The following factors contribute to high steam consumption:

1. High maximum cooking temperatures.
2. Low temperature of initial cooking acid.
3. Cold wood chips in wintertime.
4. Low degree of chip packing.
5. Low production: standard system heat loss divided by lower unit production increases steam consumption per ton.
6. Low-density wood.
7. Very wet wood.
8. Poor recovery of heat from low-pressure blowdown.
9. Side relief taken too late.
10. Side relief taken in too small a quantity.

A major portion of heat applied to the cook is lost to the environment when digesters are blown. In alkaline cooks where temperatures are high and the contents are often blown without pressure relief, these losses are particularly high and have led to the development and industrial use of blow-heat recovery systems. The principle is based on heat exchange for generating steam and/or preheating the cooking liquor charged to a subsequent cook. Considering the high heat content of the aqueous system this practice results in major steam savings. It also permits higher cooking temperature and therefore production rate if agreeable with pulp quality. Due to increasing energy cost, the efficiency of energy usage of pulp mills has been increasing in recent years, and blow-heat recovery is a major step towards the goal of energy self-sufficiency.

9. Batch digester plant control

Refer to the section on Process Control in Chapter III, C, 1 for a discussion of batch digester control.

B. CONTINUOUS DIGESTERS AND THEIR OPERATION

The era of continuous production of cooked ligno-cellulosic material for the manufacture of chemical and semichemical pulp started in 1948 with the installation of the first 50 t/d unit by Kamyr in Sweden [2]. It can safely be said that Sweden is the country where continuous industrial pulping machines, other than the stone grinder, originated since the Asplund defibrator [7] for mechanical pulping of chips was invented there in 1930. This, however, was originally a horizontal, screw operated device where the treatment time was much too short for cooking action to take place.

It was only in 1958 that the vertical model of the defibrator digester appeared [7] which allowed a residence time of the wood material of 20 or more minutes with provisions for steaming and impregnating the chips so that at least NSSC pulp could be produced. To this category belongs also the Pandia digester [8] produced by the Black Clawson Company which represents a further development of the horizontal mechanical defibrator by adding several screw operated tube lengths to provide longer cooking time to make semichemical pulps.

Another successful design which departs radically from the above models is the Bauer M&D continuous digester [9] which is typically mounted at an angle of 45° and where the charge is positively moved through the cooking medium by an internal chain-driven conveyor.

The present status of the two remaining upright continuous digesters which have reached industrial application is briefly mentioned: The upflow continuous digester (Improved Machinery Co. Inc.) described in the second edition of this textbook has been used to cook kraft and NSSC pulp but is no longer produced. The design of the ESCO upright continuous digester, which represents a simplified version less sensitive to wood quality and chip size than the Kamyr digester, was modified by Rauma Repola in Finland (Fig. 92) for the cooking of high yield bisulfite news furnish [10]. The digester was used for some period and then shut down. Installed ESCO digesters are used for kraft and NSSC pulping in various countries. The exclusive rights to design and manufacturing were acquired in 1982 by Ingersoll Rand.

In parallel with the development of continuous digesters went the development of continuous feeders for adding the material to be cooked against the pressure maintained in the digester.

Again, typically, there are two devices for accomplishing this:

1. The screw plug feeder (Fig. 93) is used for higher yield pulping applications, particularly of hardwoods and annual plants where the material is forced by a tapered screw into a tapered housing and compressed to form a

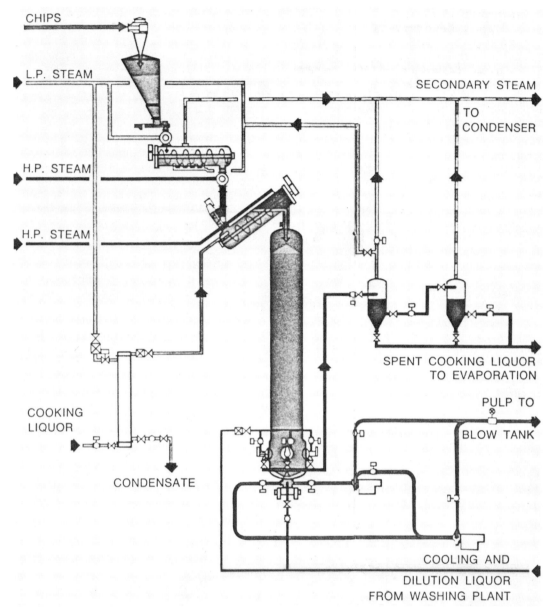

Fig. 92. Modified ESCO continuous digester for sodium base high-yield bisulfite pulping (Rauma Repola).

tight plug which advances into the cooker. Screw feeding devices were developed in Europe in 1938 [7]. It is an advantage of this mode of feeding that the material is thoroughly impregnated when the plug expands in cooking liquor inside the digester. A disadvantage is compression damage to the fibers which may cause inferior pulp quality with chemical pulp grades.

2. The rotary pocket valve feeder (Fig. 94) is used for softwood and is necessary for high quality pulping applications. The material is

fed from the top into the pockets of a pressure-sealed rotating plug and empties from the pockets into the vessel below. Extra

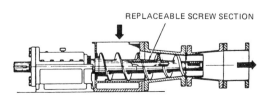

Fig. 93. Screw feeder (Black Clawson).

arrangements are required for balancing the rotating plug against the digester pressure by distribution of steam or liquor, and to eliminate expansion problems so that the clearances between rotor and housing can be minimized without metal-to-metal wear.

1. Kamyr sulfite hydraulic digesters

The development of the continuous digester was initially aimed at kraft cooking applications where impregnation of the wood chips with cooking liquor takes place quite rapidly. For several years modifications of the original model installed in 1948 were made in order to improve the reliability of the system and, more important, to improve the mechanical properties of the pulp [12]. The mechanical outlet device for discharging the hot pulp from the bottom of the digester was identified as the source of most of the damage done to kraft, acid bisulfite and even neutral sulfite pulps. The problem was alleviated if not eliminated by introducing "cold blow", that is circulating cool weak liquor through the bottom section to cool the pulp and introducing a blow valve which directly reduced the digester pressure.

The next modification leading to the present digester design was the attachment of a counter-current washing section allowing up to 4 h of residence time to remove spent liquor from the softened but undisintegrated chips or other material while still in the digester. With this, the original and most widely used hydraulic Kamyr digester was complete. Its first sulfite version is shown with the auxiliary equipment in Fig. 95. "Hydraulic" implies that the digester operates

completely filled with cooking liquor, without gas space.

This equipment was used for the first two continuous sulfite installations, one in Switzerland in 1963 [11] and one in Canada in 1964 [12], producing high yield bisulfite news furnish with sodium or magnesium base liquor, respectively. The following description concerns the first Mg base bisulfite application [12].

Chip quality with respect to size class and uniformity is of greater importance with this type of continuous equipment than with batch equipment. Somewhat smaller chips are preferred than is common for the industry and thin chips are recommended in view of penetration and diffusion. Sawdust or fines fractions are avoided as much as practical to prevent plugging of circulation strainers.

The chips are fed from a volumetric chip meter by the low pressure rotary feeder into the presteaming vessel which operates at 100-130 kPa steam pressure.

In acid sulfite pulping, proper presteaming of the wood chips is especially important since this is the simplest method available to prepare the chips for the ensuing satisfactory penetration with cooking liquor. Considerable thought has been given to this aspect and the standard horizontal presteamer used for alkaline cooks was mounted on a slight angle with the outlet raised to increase the quantity of chips, to lengthen residence time, and to assist the run-off of condensate. Even so, the presteaming time is only 4-5 min. The steam is taken from a low pressure mill supply and admitted through three bottom strainers. Air and uncondensed steam are vented by the top through a controllable exhaust system. No flashed vapors from the digester are returned to presteaming. In neutral and alkaline operations these flashed vapors are returned for steaming.

Transfer of the chips from the chute to the high-pressure zone of the system is accomplished with the high-pressure feeder, aided by cooking liquor circulating through the chip chute and the feeder. At this point the high pressure impregnation period of the chips begins, which extends through 1 min of transport into the top of the digester, and down movement of the chip plug inside the digester until the temperature becomes high enough to initiate the digestion reaction.

The digester pressure is maintained in the 1275-1380 kPa range, which is well above the vapor pressure of the cooking liquor at the maximum cooking temperature, thereby preventing boiling and flashing. Kamyr digester shells are made of carbon steel to withstand 1750 kPa pressure, and all welded 10% clad with

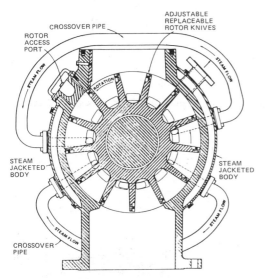

Fig. 94. Rotary pocket feeder with steam balancing and emptying (ESCO).

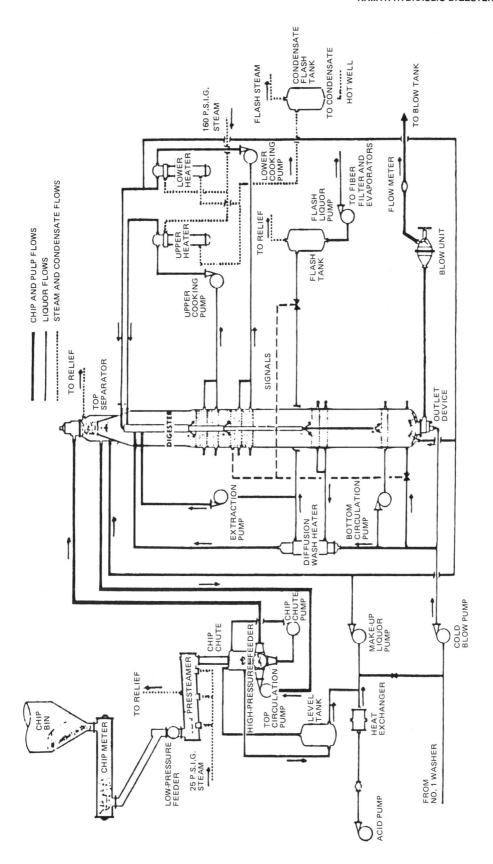

Fig. 95. Flow sheet of the hydraulic Kamyr continuous Mg base sulphite digester [12].

stainless steel. Concentric liquor inlet pipes and strainers for the extraction of liquor, all made of 316SS, are spaced down the length of the vessel to provide 45 min of impregnation, 2 h of cooking and 1 h of countercurrent washing. A slotted strainer is provided at the top for separating chips and recycled liquor.

The fresh cooking liquor for the process is stored in an outside tank where it is preheated to 82°C to assure constant cooking conditions. The liquor is then forced into the pressure system by a make-up liquor pump. A level tank, with controls as indicated in the diagram, serves to maintain a constant liquor level in the chip chute on the high pressure feeder and discharges also to the suction side of the make-up liquor pump.

Pressure of the "hydraulically full" digesters is maintained with the aid of two automatic liquor valves actuated from a common pressure signal. The fresh cooking liquor enters the digester via the jacket around the top separator. In the case of the magnesium base bisulfite application, this liquor is made from recovered chemicals from the process and some make-up chemicals, and does not contain any relief liquor as is common in conventional sulfite processes with the hot-acid system. It contains 6-7% total sulfur dioxide content and has a pH around 3.7 to 4.

The liquor flow rate is mainly determined in accordance with the throughput rate of chips, but can be further adjusted to accommodate variations to maintain a desired liquor-to-wood ratio of 3.25:1.

The digester has the usual two heating zones, a cooking zone, and a section for countercurrent washing or "Hi-Heat Washing".

The plug of chips and liquor descends from the impregnation zone to the upper heating zone where liquor is extracted through two complete sets of strainers. The extracted liquor is heated indirectly by 1100 kPa (160 psi) steam and returned to the area where it was extracted through the outer one of four concentric liquor inlet pipes. The chips then pass through a dwell zone for the purpose of diffusing the heat throughout the individual chips. From the dwell zone the mass descends into the lower heating zone which is similar in design as the upper heating zone.

The liquor temperature is raised at this point to near the maximum cooking temperature of 170°C. (Since the digestion reactions are exo-thermic, a further rise in temperature will take place.) Only a little scaling is encountered with Magnefite liquor and hence no spare heater has been included in the system.

The lower heating zone is followed by the cooking zone where the digestion reactions proceed until they are quenched by a rapid cooling of the ambient liquor.

As digestion takes place, the total volume of chips and liquor contracts, due to solid sub-stances going into solution. This phenomenon, plus the recirculation of cooking liquor, sets up currents relative to the movement of the chips, a condition which required careful thought in the process design, as such currents unchecked may lead to a build-up of thiosulfate, a compound harmful to proper cooking in acid bisulfite or bisulfite liquor. The digestion is stopped by rapidly extracting hot liquor from the system and replacing it by cooler liquid. Liquor is extracted through the double set of upper extraction strainers, then cooled on a pass through the shell side of the diffusion-wash-heater, a tube-type heat exchanger, after which about half of this flow is returned to the upper extraction zone via a third concentric tube, and the other half is bled under the digester pressure to a flash tank whence it is pumped to weak liquor storage for the recovery of chemicals.

The discrepancy in the upper extraction zone, amounting to the sum of the liquor sent to storage and the volume contraction of the cook on its way down to this zone, is made up by the inflow of a weak wash liquor from below, and thus counter-current washing takes place in the washing zone.

Below the washing zone, liquor is removed through the lower extraction strainer by the bottom circulation pump, and returned together with very weak liquor from external pulp washers via the tubes of the diffusion-wash heater through the innermost pipe. Even with the make-up from the external washers, most of the discrepancy mentioned in the previous para-graph still exists and most of the liquor that entered through the innermost tube diffuses to the upper extraction strainers through the plug of chips.

Further make-up flow from the external washers is pumped into the flow zone by a "cold blow" pump, and thus provides some more counter-current washing with a weaker liquor.

The function of the diffusion-wash heater is to provide a fairly high temperature in the washing zone for optimum washing conditions. It is the task of the digester operator to maintain liquor extraction and pulp washing conditions which achieve a full stop of the reactions, and washing at maximum efficiency to gain the highest possible solids content in the liquor sent to the chemical recovery plant. This plant, in turn, will then operate under the best conditions. An outlet device at the bottom of the digester breaks up the mass of chips, cooled and diluted from 19% to 8% consistency by weak wash liquor from the washers in the screenroom, by paddle arms rotating at 5-6 rpm. The chips leave the digester under pressure through a valve and the Kamyr blow unit, to be blown tangentially into the blow tank.

Pilot-plant studies of continuous sulfite cooking were made at the Mo & Domsjo Sulfite Mill in Sweden during the years 1950-1953 [13]. The result of the trials indicated that the equipment developed for kraft pulping was not quite appropriate to acid sulfite cooking liquors. The calcium base gave scaling difficulties in the heat exchangers and on the extraction strainers. The high SO_2 vapor pressure of the cooking liquor caused certain problems in the chip feeding system. It was concluded that continuous pulping with cooking liquor rich in excess free SO_2 would demand a redesign of the digester system. The introduction of acid bisulfite pulping with the soluble bases eliminated the problem of calcium base, but applications of the hydraulic digester were successful only for unbleached bisulfite pulp at higher yield; but not for bleachable pulp grades requiring longer cooking times, because of liquor breakdown.

2. Kamyr steam phase digester

Experiments conducted at the Billerud experimental pulp mill identified the problem as being due to excessive thiosulfate formation in the hydraulic digester, by intermixing of hot used liquor from the cooking zone with fresh cool liquor in the impregnation zone. For extended cooks to bleachable pulp grades, therefore, it was necessary to find a modified digester system which eliminated the thermal flows and the liquor intermixing. The answer to this problem proved to be a digester system which made it possible to carry liquor at full cooking temperature in the top of the digester [13].

The first model built on these principles is shown enlarged in Fig. 96. The top separator inside the standard hydraulic model has been removed and instead an inclined vessel (Mumin), supplied with direct steam and combining the functions of impregnation and chip-liquor separation, has been placed on the outside of digester's top. This design permits maintaining a vapor space atop the digester while retaining most of the standard equipment, and the high pressure chip and liquor feeding system.

Kamyr digesters carrying this modification were supplied between 1965 and 1968 when the final modification of the modern steam-liquor phase digester became available. Adhering to the above principles, but in order to circumvent certain problems encountered with the external model, the inverted top impregnator-separator was placed inside the top of the digester as shown in Fig. 97.

The inverted separator is kept full of cooking liquor through which the chips are moved upward by action of a screw and decant over the rim. Thus chip impregnation time from the pressure feeder is increased to about 2 min. Impregnated chips and overflowing cooking liquor are heated in the top with direct steam and are brought uniformly to maximum cooking temperature.

(a) Flow ratio vs liquor-to-wood ratio

At this point it will be useful to enlarge on the meaning of liquor-to-wood ratio and degree of packing in continuous cooking, since their relation differs greatly from that common in batch cooking [2]. In a batch digester, the chips are packed with suitable devices and then liquor is added into the digester to cover the chips. Depending on the degree of packing, the wood density, etc., the liquor ratio is about 4-5:1 including wood moisture. In a continuous digester, there is about the same degree of chip packing and consequently a similar liquor ratio when the liquor has been introduced to cover the chips. However, this liquor ratio is not of the same interest as in batch cooking regarding steam consumption and spent liquor concentration. Instead, what is of interest to continuous cooking is the flow ratio between liquor and wood chips, and this is to be called liquor ratio in continuous cooking. When this flow ratio is 2.2,

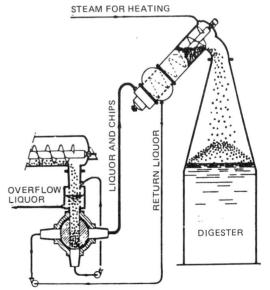

Fig. 96. Steam phase Kamyr digester with inclined top impregnator-decanter (Kamyr).

but the liquor ratio to cover the chips in the digester is 4.4, this means that the linear flow of the chips through the digester is about twice as fast as the linear flow of the liquor, and that the retention time of the liquor is twice that of the chips.

(b) Pulping conditions

In a typical bisulfite cook, fresh bisulfite liquor of pH 4-5 at 16% total SO_2 on oven dry wood is introduced into the feeding circulation, where the presteamed chips are impregnated at a pressure of 1000 kPa and at a temperature of 105°C. The retention time of chips in the impregnation stage is about 1.5 h. After impregnation the chips are instantaneously heated by direct steam to full cooking temperature of about 160°C. The liquor level in the digester is the same as the chip level which means that the cooking is conducted entirely in the liquor phase. Retention time in the cooking zone is about 3 h. After cooking, the pulp is

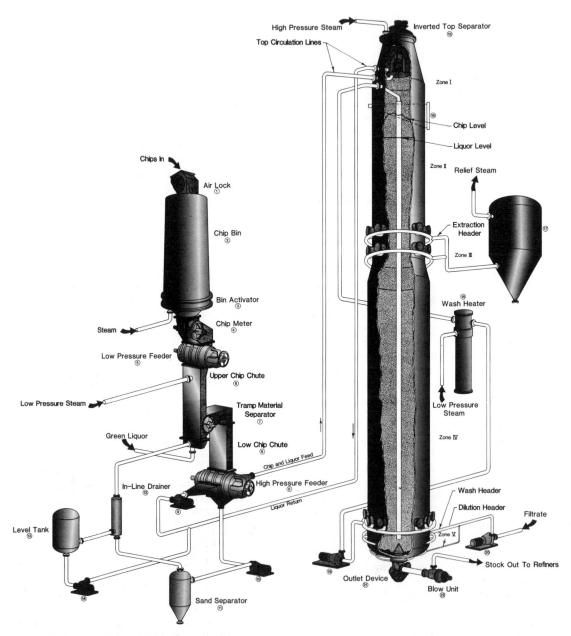

Fig. 97. Kamyr semichemical steam phase digester.

countercurrently washed in the digester for about 1.5 h. In this manner, bisulfite pulps have been cooked to different yields varying from 75% down to 53%. The raw material used was mainly spruce but also birch and pine.

For continuous cooking of acid bisulfite dissolving and specialty paper pulp, the pre-steamed wood is impregnated in the feeding circulation with sodium or magnesium or ammonium base bisulfite liquor at pH 4. The acid sulfite conditions are obtained by injection of liquid SO_2 at the top of the digester together with the steam. The quantity of liquid SO_2 injected is determined by the desired partial SO_2 pressure and concentration of free SO_2 in the liquor. The digester contents are heated instantaneously with direct steam to a temperature of 135-145°C. Trim circulation and indirect heating are used in all cases to establish uniform liquor and temperature distribution. The cooking time is about 3 h. The liquor to wood ratio in the digester can be kept much lower than what is usual in batch cooking, and consequently the continuous system offers a better heat economy in both cooking and recovery. Dissolving and paper pulps of good quality are obtained. Free SO_2 is recovered from the extracted spent liquor by flashing to atmospheric pressure and condensing the SO_2 gas to liquid SO_2. Magnesium and sodium base are recovered by their respective systems, as described later. Ammonium base is commonly burnt with the concentrated spent liquor.

The Kamyr steam phase digester is also well suited for producing liquid phase NSSC pulp. Carbon dioxide liberated in the cook from the decomposition of sodium bicarbonate can be relieved continuously from the vapor space and does not upset the system as it would in the standard hydraulic digester. Similar results are obtained with sodium or ammonium neutral sulfite liquors. Vapor phase cooking by lowering the liquor level in the cooking zone of the digester has not been found to give any particular advantage, but rather produces somewhat less uniform pulp.

The digester is also suitable for two-stage sulfite pulping of the neutral-acid type [2,13]. Two-stage acid sulfite pulping for higher yield of carbohydrates is done so that the first stage mainly de-acetylates the hemicellulose, and the second stage mainly takes cares of the necessary delignification. De-acetylation stabilizes glucomannan and increases pulp yield, as explained earlier. The treatment can be done in a neutral stage at high temperatures and a comparatively long time, or in a strong alkaline stage with lower temperature and a short time. In the latter case, the conditions must be balanced so that the delignification is not retarded due to condensation reactions.

In the case of neutral-acid two stage pulping, the presteamed chips are impregnated in the feeding line with neutral sulfite liquor of pH 7. The first-stage cooking temperature is obtained by direct heating at the top of the digester. The acid stage is obtained by adding liquid SO_2 to a liquor circulation. This circulation level also serves for adjustment of the second-stage cooking temperature. After cooking, the pulp is washed countercurrently in the conventional manner.

At low liquor-to-wood ratios and with lower temperature in the first stage than in the second one, the border line between the two stages was somewhat diffuse, indicating an upflow of acidic second-stage cooking liquor. By simply increasing the temperature of the first stage, however, it is possible to obtain well-defined and sharp zones. Pulps made from both spruce and pine give low rejects, high brightness, good bleachability and, as indicated by carbohydrate analyses, a yield increase of about 5% calculated on the wood. The strength properties correspond well with same grades of laboratory-cooked pulps.

A simpler version of two-stage pulping was carried out by impregnating the chips with sodium hydroxide in the feeding circulation, and adding liquid SO_2 and direct steam to the top of the digester. The temperature in the impregnating circulation was about 100°C and the retention time approximately 1.5 min. This short treatment was sufficient to give a de-acetylation corresponding to a yield increase of about 3% calculated on the wood. The delignification posed no difficulties, and the pulp characteristics were normal except for reduced brightness. Higher sulfite content gave better brightness of the unbleached pulp while the yield was somewhat reduced. Sulfite was obtained in the feeding circulation liquor with no special arrangements due to absorption of some SO_2 gas. The bleachability of the pulp was good, and the strength properties about the same as for the neutral/acid sulfite two-stage pulp. There is, however, no doubt that optimum cooking results in any kind of pulping depend on optimum amount and distribution of chemical at the reaction sites in the material to be cooked. Therefore adaptability and control of the impregnation stage is always desirable, particularly with the sulfite processes.

3. Semichemical Kamyr steam phase digester (Fig. 97)

The semichemical Kamyr steam phase down-flow digester is a modification of the steam phase digester described in Section 2. The major differences between the two types are that the semichemical units are equipped with a vapor zone and have no forced circulation flow circuits in the pulping zone. The chip feed systems are the same except that the semichemical digesters do not have a pressurized steaming vessel. Steaming is done at atmospheric pressure in the chip bin. Both systems incorporate countercurrent hi-heat washing and cold blowing auxiliaries. Normal blow temperatures are approximately 88°C (190°F).

The impregnation, heating and pulping zones, due to inherent design, do not have distinct separations in a semichemical digester. There are no heaters in the pulping zone, direct or indirect, nor are there any circulation flows. All heating is done in the vapor zone of the digester where a chip level is maintained approximately four feet above the liquor level. Relatively mild pulping of short duration (approximately 35 min) takes place in the liquid phase and continues down to the extraction zone.

Figure 97 shows a typical installation. Chips, after screening, enter the chip bin (2), through a constant rpm rotary airlock feeder (1). In the chip bin, the chips are steamed to remove air. The bin activator (3), ensures uniform chip movement into the chip meter (4). The production rate is controlled by setting the speed of rotation of the chip meter rotor. From the chip meter, the chips pass through the low pressure feeder (5), into the upper chip chute (6).

The upper chip chute pressure is controlled using low pressure steam at 104-138 kPa (15 to 20 psig). This operating pressure maintains the required pressure balance in the feeding system to avoid boiling in the lower chip chute. From the upper chip chute, the chips pass through the tramp material separator (7) and cascade into the liquor phase in the lower chip chute. The liquor in the lower chip chute comes from two sources, pulping liquor is added at the tramp material separator and cascades with the chips into the lower chip chute. The balance is controlled leakage around the high pressure feeder rotor.

From the chip chute, the chips are pulled into the high pressure feeder (8), by the chip chute pump (10). There is a grid on the bottom of the high pressure feeder which prevents chips from entering the chip chute circulation.

The chip chute liquor is circulated by the chip chute pump through a sand separator (11), which removes sand from the liquor, then through an in-line drainer system (12) and back to the tramp material separator (7). Subsequently, it flows down through the lower chip chute with the chips into the high pressure feeder. To maintain liquor level control in the chip chute, excess chip chute liquor is removed at the in-line drainer. This liquor flows into a level tank (13) and is then pumped to the digester by the make-up liquor pump (14), via the top circulation lines. A screen in the in-line drainer keeps pin chips from entering the make-up liquor pump.

The high pressure feeder (8), is a rotary valve which transfers the chips, with no mechanical action, from the 104-138 kPa (15 to 20 psig) chip chute pressure, to the full digester pressure. When one of the four pockets in the rotor of the high pressure feeder lines up with the top circulation pump (9), the chips and pulping liquor are flushed up the top circulation lines to the top of the digester, where the chips and liquor enter the inverted top separator (15). Most of the liquor is extracted through a screen in the inverted top separator and returned to the suction of the top circulation pump.

The impregnated chips are elevated by means of the separator screw and discharge from the basket top into the vapor zone I, of the digester. A gamma type measuring device (16), with an eight-foot range is used to detect the chip level in the vapor zone. The vapor zone is maintained at the desired pressure, e.g. 794 kPa (115 psig) with make-up process steam which heats the chips to full pulping temperature. The liquor level is maintained approximately 1.2 m (4 ft) below the chip level.

Chips and liquor move down the pulping zone II, to the extraction zone III, where the hot spent liquor is extracted to the flash tank (17). Released steam vapors are recovered from this vessel, reducing the net energy consumption of the system.

After the digested chips pass through the extraction zone they enter the hi-heat washing zone IV of the digester. Recycled filtrate from the subsequent external washing system is pumped by the cold blow pump (20), through dilution nozzles located around the bottom of the digester. This filtrate then moves upward through the downward moving chips, providing a countercurrent wash flow. In the wash zone, dissolved solids diffuse out of the chips into the upward moving wash liquor.

At the bottom of the wash zone, the wash liquor is extracted through a set of circumferential screens to the wash heater (18) back into the digester via the wash central pipe. This pipe extends back down to the bottom of the wash zone and reintroduces the heated filtrate into the center of the chip column at an elevation just above the wash extraction screens. This heating of the wash liquor improves the efficiency of the digester hi-heat washing phase.

At the extraction zone III, the combined flows of wash filtrate and spent liquor are removed to the flash tank. After flashing (17), the liquor is pumped to the evaporators and recovery boiler to recycle the pulping chemicals.

The digested chips are discharged from the bottom of the digester with the aid of an outlet device (21). The outlet device is a scraper assembly which removes the chips uniformly from the entire cross-section of the digester. Due to the cool filtrate entering the bottom of the digester, the pulp is blown through the blow unit (22), at 88°C (190°F) or less.

From the blow unit, the digested chips pass through a blow valve, operated on flow control, to one or more refiners. Some defibering of the chips occurs in the blow valve as the pressure is abruptly reduced under highly turbulent conditions. Final defibering occurs in the refiners. After refining, the pulp can be blown to a blow tank and then pumped to an external washing system where it will receive the final wash treatment. Alternatively, the pulp may pass directly through a diffusion washer upon leaving the refiners, after which it is discharged to the blow tank.

Much of the instrumentation, piping and valve arrangement, and mechanical equipment is the same as used on the hydraulic continuous digester.

NSSC pulping in a Kamyr semichemical digester lends itself well to computer control. The digester instrumentation is supervised by the computer in order to produce uniform pulp during steady state conditions and during production changes. The computer system controls, for example, the chemical-to-wood and liquor-to-wood ratios, taking into account the moisture content of the ingoing chips and the chemical composition of the pulping liquor. By controlling all flows to and from and the digester, including steam flow, the important conditions in the pulping and washing zones and at the bottom of the digester can be supervised efficiently. During a production rate change, the computer schedules the temperature change in the pulping zone. In this manner, production of the off-grade pulp is avoided. Constant pulp properties and yield are very important for smooth operation of the entire mill.

4. Kamyr two-vessel system

To meet these requirements and to improve the flexibility and capability of the original steam phase concept the two-vessel system was eventually introduced. By this is meant the combination, shown in Fig. 98, of the steam phase continuous digester, with a continuous pre-impregnation vessel which operates at the pressure of the digester plus the hydrostatic head resulting from the difference in height. The vessel thus operates hydraulically full, and is equipped with a standard top separator. The purpose of the impregnation is to provide the best possible impregnation of various chip size fractions at high pressure and a relatively low temperature of 99-115°C. These conditions are known to result in higher yields, lower rejects and more uniform pulp with any type of cook, from acid to alkaline and from high yield to low yield.

Even with the inverted top separator added, the total time available for high pressure impregnation of the presteamed chips is not more than 2 min. While the best results would naturally be obtained with alkaline processes, there were always some reservations whether presteaming and impregnation time provided by the common Kamyr system would be sufficient for best results with acid bisulfite and bisulfite liquors, and even neutral sulfite liquors. Many years of industrial experience with Kamyr digesters have proven that, for most applications, chip impregnation is adequate for producing acceptable pulp quality; although provisions have been made in several cases for extended presteaming and, of course, for extended impregnation, by adding the inverted top separator to the feeding system. With this modification, the amount of sulfite chemical entering the wood under the impregnation pressure of some 1000 kPa is obviously sufficient for the initial cooking reaction, plus later diffusion in the liquor phase to complete the cook.

The two-vessel system is now common for sulfite pulping installations and all digester installations made previously for acid bisulfite and bisulfite cooking have been retro-fitted with the high pressure impregnation vessel.

In existing acid bisulfite and bisulfite operations, the liquor-to-wood ratio is usually held in the range of 3:1 to 4:1. The initial digestion chemical charge is 13 to 18% sulfur dioxide based on oven-dry wood, depending on the desired yield. Duration of the cook varies from 2.5 to 4 h, with maximum temperature in the range of 157 to 165°C. Screened, unbleached yields of the pulp products have generally been in the range of 51 up to 63%, the latter still easily fiberized.

Moreover, in order to ensure minimum thiosulfate formation, there is no reuse of any reclaimed or spent liquor in cooking acid fortification.

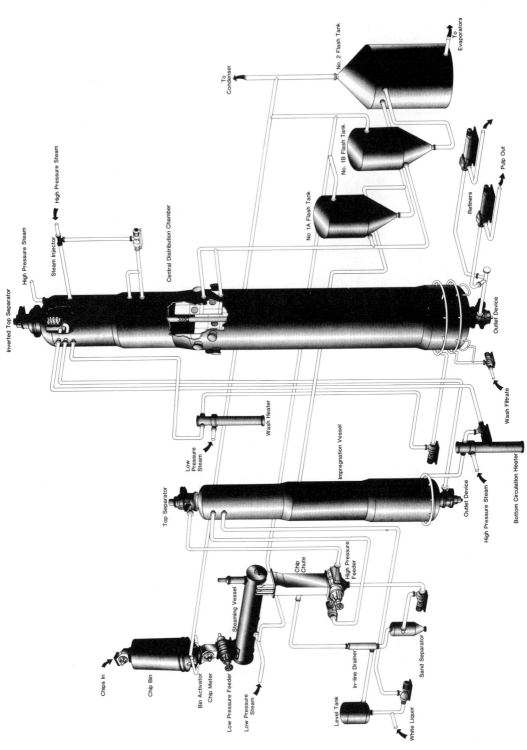

Fig. 98. Kamyr continuous sulfite pulping system with high-pressure impregnation vessel and steam phase digester. Refiners are optional for high-yield pulping.

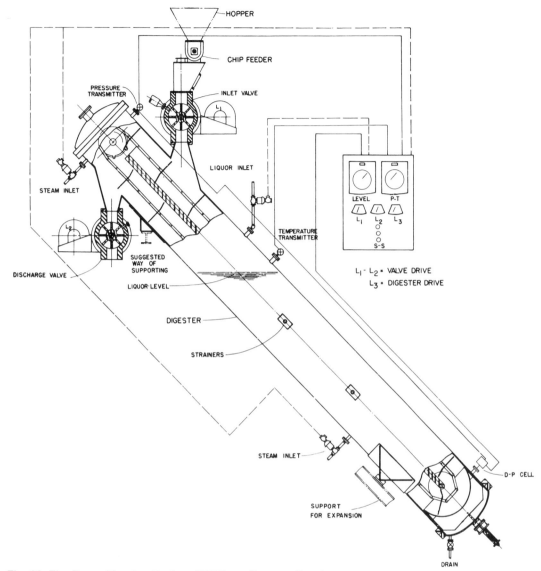

Fig. 99. The Bauer Messing-Durkee (M&D) continuous digester.

Recent Kamyr digesters used for the production of high yield linerboard pulp have been equipped with blow line refiners also shown in Fig. 98. This facility will be useful not only for the production of the common kraft packaging grade pulp but also for AS-AQ pulp which can be obtained at considerably higher yield without loss of mechanical strength (see Chapter II,G).

5. The Bauer M&D inclined continuous digester

The first modern version of this versatile continuous digester was installed in 1959. The "M&D" stands for the inventors, H.S. Messing and C.L. Durkee, of Bauer Bros. Co. [9].

Principles of design and operation are explained in Fig. 99. The digester consists of a large tube with dome-shaped ends made of carbon steel for alkaline pulping, and stainless steel for acid pulping, installed at an angle of 45°C. The tube is divided into an upper and lower compartment by a midfeather, and equipped with a chain conveyor which moves the material to be cooked by means of flights down the upper half and up the lower half of the tube. Connections for charging and discharging are located near the top end in the upper and lower wall. The material is fed with or without presteaming into the tube by a high pressure rotary pocket valve. The pocket feeder is shown in Fig. 100. For comparison, a screw feeder designed by Black Clawson is shown in Fig. 101. The pockets move on while being filled, and submerge the chips into the pool of cooking liquor. By sectional forced liquor circulation, with or without heat exchangers, one M&D tube may be divided into four treatment areas as far as liquor temperature or chemical concentration are concerned. After travelling for the specified time down and up the tube, the chips are then lifted out (decanted) of the cooking liquor and dropped into the rotary discharge valve. Several industrial installations feature two parallel inclined tubes connected by a rotary pressure valve. This so-called piggyback arrangement permits cooking with a great variety of treatment stages, be it temperature, chemical composition, or steam-liquid phases.

The M&D digester is the only continuous digester where the cooking material is positively moved through the liquor, and also the most versatile of all regarding furnish quality and pulp grade. The range of pulp grades produced extends from soft, bleachable pulps to ultra high yield pulps for news furnish.

One disadvantage which in some cases limits applicability is a relatively high liquor-to-wood ratio in the cooking stage (but not in feeding and discharge), which causes the cooking of the chips to proceed in a medium consisting of fresh and exhausted cooking liquor. Caution is also necessary against over-filling the flight space at maximum production rate, because the drive mechanism can get stuck and to free it may not be easy with a full digester. Since the retention time is given by the speed of the chain drive, the production rate at a given degree of cooking is adjusted by the amount of wood charged.

Of over 70 installations, nearly 30 are for sulfite operations, ranging from magnesium base paper pulp, through NSSC pulps from hard-woods, to high and ultra high yield news furnish from softwood.

A modified inclined digester for vapor phase pulping has just been introduced for producing high yield chemimechanical pulp. The operational structure is shown in Fig. 102A and B. In this first installation long, thin softwood fragments, so-called pin chips, are presteamed to remove air, pressure-impregnated in the digester with bisulfite solution, and exposed to a short vapor phase cook at 170°C to give a yield of up to 90%.

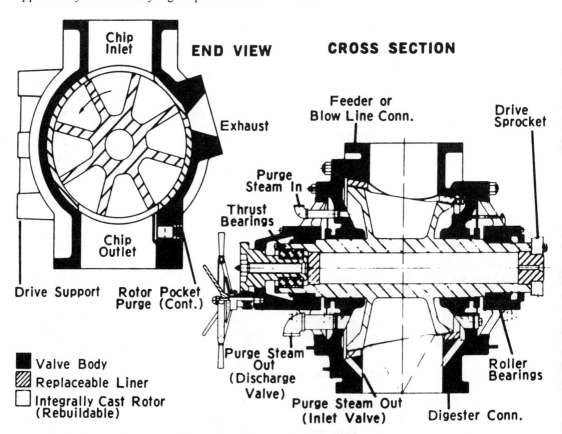

Fig. 100. Pocket-type feeder for continuous digester (Bauer Bros. Co.).

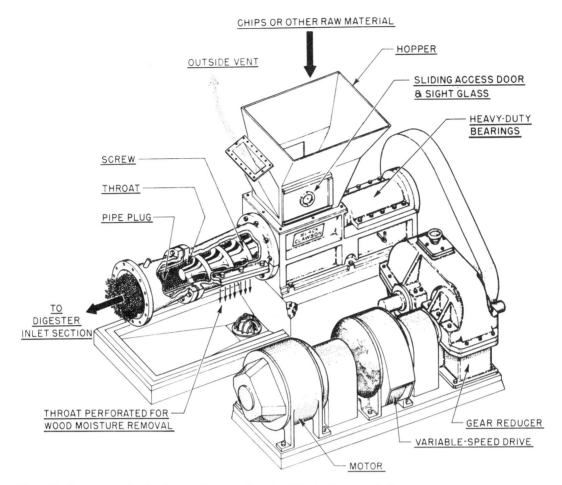

CHIPS OR OTHER RAW MATERIAL

OUTSIDE VENT

HOPPER

SLIDING ACCESS DOOR
& SIGHT GLASS

HEAVY-DUTY
BEARINGS

SCREW

THROAT

PIPE PLUG

TO
DIGESTER
INLET SECTION

THROAT PERFORATED FOR
WOOD MOISTURE REMOVAL

GEAR REDUCER

VARIABLE-SPEED DRIVE

MOTOR

Fig. 101. Screw-type feeder for continuous digester (Black Clawson Co.).

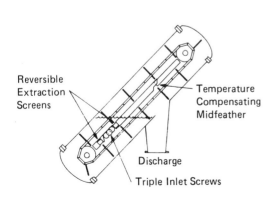

Reversible
Extraction
Screens

Temperature
Compensating
Midfeather

Discharge

Triple Inlet Screws

Fig. 102A. Main features of M&D vapor-phase digester.

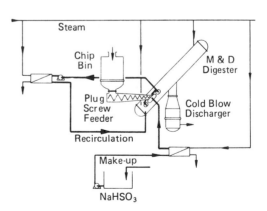

Steam

Chip
Bin

M & D
Digester

Plug
Screw
Feeder

Cold Blow
Discharger

Recirculation

Make-up

$NaHSO_3$

Fig. 102B. Partial process diagram for bisulphite CMP.

Three changes are made to the design of the original M&D digester: the locations for chip intake and discharge are lowered to the bottom quarter of the digester tube in order to provide vapor space; the movement of the internal chip conveyor is reversed so that chips forced into the lower division of the tube by means of a screw plug feeder (see Fig. 102B) are first submerged into the cooking liquor where they travel down and up into the upper division, and then up and down in the vapor phase to the cold blow discharge; a new midfeather design with tele-scopic joint near its middle to allow for temperature expansion when fastened to the vessel at both ends. The latter is necessary for closing the gap at the sprocket shaft and preventing chips from falling through as they are carried over the upper end of the conveyor. The midfeather also accommodated the chip inlet from the screw feeder and the liquor extraction screens in the liquid phase region.

In this particular application, maximum retention time in the cooking liquor is 12 min and in the vapor phase 60 min. The "cold" blow discharger eliminates steam loss, quality loss experienced with refiner discharge, and provides preliminary washing of the cooked chips. The material can be discharged at 3% consistency

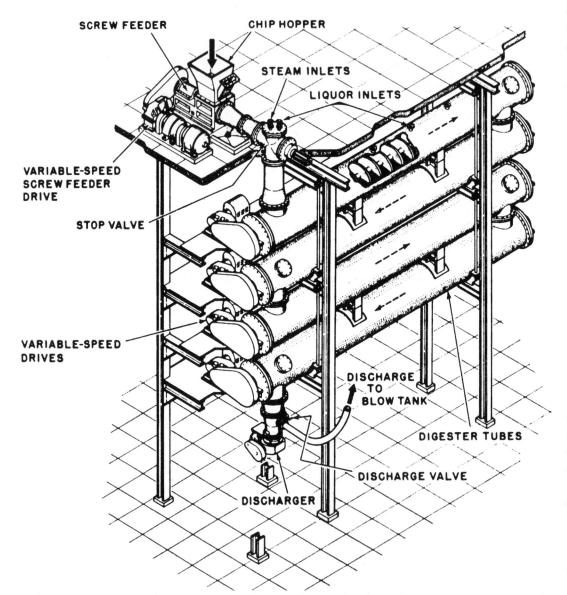

Fig. 103. Pandia horizontal multiple-tube continuous digester (Black Clawson).

below 100°C, or at 5% consistency above 100°C if flash steam is used in the presteamer to replace fresh process steam.

6. Black Clawson Pandia Digester

The Pandia digester consists of one or more horizontal tubes stacked one above the other. A typical, multiple horizontal tube digester is shown in Fig. 103. Two tube units are more common than four tube units. A separate chamber may be provided for hydrostatic impregnation of the chips or liquor may be sprayed on the chips compressed in a screw feeder (Fig. 101) as they enter the top tube. As another alternative, the top tube may be used to steam the chips before the liquor is added. However, these preimpregnation treatments are less effective with woods containing tyloses in their heartwood. Slowly revolving screws move the chips from the one end of each tube to the other, where they fall to the next tube. In modern installations, a disc mill, usually referred to as refiner, is attached to the end of the last tube to fiberize the chips under pressure before they are discharged. This procedure is known as "blow line" or "in-line" refining. The energy application made in blow line refining varies from mill to mill depending on wood species, chemical application, pulp yield and the degree of fiberization desired. For blow line refining, the Sunds Defibrator RGP and the Sprout-Waldron 42-ICP are being used in many mills.

As already stated, in modern installations the digester content is discharged into a pressurized blow line refiner. In some older mills, the pulped material may be discharged directly into a leach caster, live bottom bin or an ordinary blow tank.

Details of one type of discharge valve are shown in Figs. 104 and 105. Continuous digesters typically discharge to a cyclone, where steam and gas are separated from the pulp liquor.

7. Sunds Defibrator

Equipment suitable for neutral sulfite semichemical pulping was initially developed by the Defibrator Company in Sweden and a chemimechanical pulping process, designed for the manufacture of pulps suitable for printing grades of paper, was developed jointly in Sweden by Defibrator AB and Stora Kopparberg AB. The first commercial unit, utilizing birch as the raw material, went into operation at the Kvarnsveden mill of Stora Kopparberg in 1958, and the vapor phase technique is now used extensively on a wide range of wood species, the impregnation and digestion equipment representing integral components of Sunds Defibrator equipment (refer to Chap. VII,E).

For the production of good quality, high-yield chemimechanical pulp, efficient chemical impregnation represents the critical phase of the process and is achieved by first replacing air present in the wood with water vapor by means of atmospheric presteaming. The softened chips are then pressed in a compression screw which feeds them directly into the impregnating vessel, the moisture lost during the compression being replaced by the chemical solution. In at least one mill, the pulping liquor is preheated with a steam heat exchange before it is pumped to the impregnation vessel. During the impregnation phase, the temperature differential between the steamed chips and the cooler impregnating liquor causes a reduction in the gas volume in the

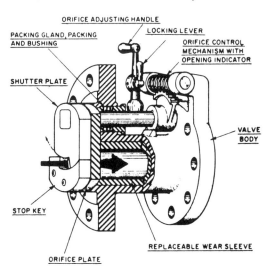

Fig. 104. Details of manually operated discharge valve (Black Clawson Co.).

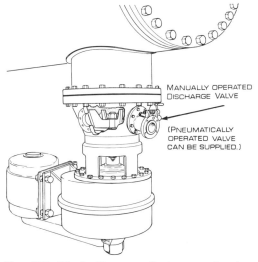

Fig. 105. Black Clawson discharge valve for continuous digester.

chips and the partial vacuum thus created draws chemical solution into the fiber structure. After impregnation of the chips with pulping liquor of relatively high concentration followed by removal of excess superficial liquor, vapor phase digestion is carried out under time and temperature conditions suitable for attaining the level of sulfonation required.

A typical schematic diagram for chemimechanical pulping operation is shown in Fig. 106. After atmospheric presteaming and chip washing, the excess water is removed in the inclined drainage section of the washer and the chips fall into a small buffer bin where additional presteaming can be carried out. The chips are then removed from the bin and conveyed through a screw feeder into the bottom of the impregnating vessel which is equipped with vertical lift screws. The air-free chips travel up through the impregnating liquor and, after drainage of excess superficial liquor in the upper section of the impregnating vessel, the chemically impregnated chips are transferred into the vapor phase digester. The equipment is designed for retention times in the range 10-45 min and digestion temperatures in the range 140°-175°C, the actual conditions depending on the wood species used and on the level of sulfonation required. The pulped chips are removed from the bottom of the digester by means of two variable speed screws feeding into a Defibrator discharger with a variable orifice blow valve. Further details of the equipment, showing the inlet feeder and the discharge arrangement are shown in Fig. 107. The steam chip mixture after digestion is blown to a cyclone where steam flashes off and the digested chips, which discharge from the lower end of the cyclone, are fed into a press where excess pulping liquor is removed and recycled to the process. After the liquor extraction phase, the digested chips are refined in either a single- or two-stage refining process. For pulps in the lower yield range where the specific energy requirements are low, the digested chips can be transferred directly into the refiner.

8. Escher-Wyss

A downflow digester developed by Escher-Wyss has been installed in several European mills. The chips are evacuated while immersed in pulping liquor, allowed to stay for a certain time in pulping liquor at atmospheric pressure, and then introduced into the top of a vertical vessel where they cook in the vapor phase without free liquor. Apparently, this digester works best with relatively dry low density wood, for example European poplar or aspen.

A rotary valve is most commonly used to feed the chips into a digester operating at high pressure; it consists of a solid rotating cylinder in which a series of pockets consecutively fill with chips at atmospheric pressure, rotate through half a revolution, and discharge into the digester (Fig. 100). The empty pockets returning to atmospheric pressure vent the inert gases along with some steam. Another device in frequent use consists of a screw flight in which the chips are progressively compressed until the plug is sufficiently dense to resist the digester pressure (Fig. 101).

Too vigorous mechanical treatment of the chips before pulping or while they are at pulping temperature can result in sufficient fiber damage to weaken the pulp. For this reason the compacting ratio of the screw-type feeder is limited. Moderate wringing action on the wood is beneficial in as much as it opens the chips along the natural lines of cleavage and facilitates liquor penetration. Where cooked chips are fed to a directly connected fiberizing mill operated at digester pressure, provision may be made to cool the chips before they pass between the discs.

9. Batch versus continuous digester

From a modest beginning in 1952, the number of continuous digester installations has risen to just over 300 units for the leading Kamyr model and its modifications. The average capacity from all Kamyr installations is about 400 air dry tons of pulp per day, but units with up to 1350 t/d are in operation. About 40-45% of world pulp is presently produced by continuous cooking equipment. In this discussion it is useful to consider not only the magnitude of this stupendous growth but also its distribution on the time scale and the advances made in the batch cooking field. The curve has a pronounced sigmoid shape with a steep section between 1964 and 1974 and a tendency to flatten out in the last few years. During the same period, batch cooking technology has added stainless steel clad equipment, remote control of all operating functions from charging to discharging, and finally complete automation of multiple-unit plants. Production scheduling and levelling of steam demand has resulted in a substantial increase of production, higher yield with better uniformity of product, reduced peak steam demand and drastic reduction in labor.

Presently, the primary objectives in designing a new mill or the expansion of an existing facility are to produce quality at reduced cost and to meet the demands of environmental protection. Different approaches are taken by different mills to meet these objectives, and evaluations to

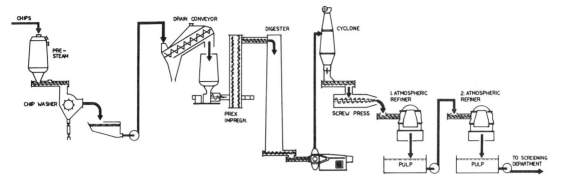

Fig. 106. Defibrator semichemical pulping system with vapor phase digester.

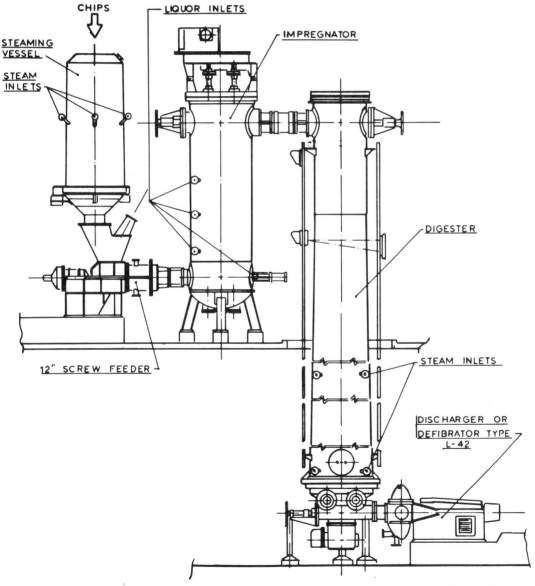

Fig. 107. Steaming, impregnation and digester vessels of Defibrator semichemical pulping system.

chose the proper digester for a specific mill cannot be generalized [14].

There is no difference between batch and continuous digesters in the processing of any type or quality of wood furnish, or in the pulp quality which can be produced. However, pulp strength can be affected by the mechanical action of the bottom discharge plough of continous digesters, and a cooling section is often provided to alleviate this problem. Sulfite batch digesters may be blown under pressure, or diffusion washed in the vessel and emptied by gravity to preserve pulp quality.

In order to lower capital expenditures which are high for continuous equipment, the tendency has been to minimize the number of pulping lines installed while increasing the capacity of the unit. This naturally reduces flexibility in the acceptance of different type of wood supply and in the pulp grades produced. Batch digester plants are naturally more flexible and specific in this respect and can produce any ratio for optimum economy or paper quality specifications. The other important aspect of the number of pulping lines installed is plant availability. With a single contiuous digester, as is the case in many mills, any malfunction which cannot be repaired within the time allowed by unbleached pulp storage capacity will shut down the whole mill. This is aggravated by the more complex mechanical equipment requiring systematic preventive maintenance. With a batch plant or a mixed batch/continuous plant (a very practical choice) production will never be interrupted or even curtailed unless scheduled.

The liquor-to-wood ratio in a continuous upright digester is naturally lower than in a liquid-filled batch digester, and the solids content of the spent liquor therefore some 2% points higher. This is not the case in sulfite digester schedules with side relief, where the ratio of cooking liquor withdrawn after impregnation can be varied within wide margins, and the solids content of the spent liquor adjusted for minimum steam consumption in the evaporation plant of the recovery system.

The continuous digester uses less steam for cooking due to uninterrupted operation. On the other hand, its consumption of electrical energy is higher than for the batch digester. In choosing a cooking system for a pulp mill with cogeneration of power the correct balance between electrical energy and steam energy must be considered.

When developing comparisons for deciding on the choice of cooking plant for a given pulping installation it is always necessary to consider the interdependence of digester operations and overall mill operations, particularly

any source of heat which can be recovered and used. For example, the recovery of blow heat from batch digesters should be taken into account, which is used most efficiently for replacing low pressure steam (400 kPa), which in turn should replace as much higher pressure steam (1000 kPa) as possible, as for multiple effect evaporation.

One final item concerns safety. Batch digesters have been considered safe for a long time and the same was true for the Kamyr digester. However, very recently two serious cases of continuous digester blowout failure were reported which brought a major revision of the attitude towards operations safety. Since these accidents happened, many digesters were inspected and weakening of the shell due to corrosion induced cracks of welded seams, and also hair cracks in adjacent wall portions and even in wall plates away from seams identified. A certain reluctance against frequent inspection and repair of continuous production equipment, particularly of the size of large Kamyr digesters, is understandable, but it is now obvious that this is necessary.

It can be concluded that the decision for selecting either cooking equipment should be based on a careful valuation of all possible factors. Quite generally, the continuous digester offers high productivity and the batch digester plant high flexibility.

Factors favoring each system are summarized in Table 14.

Table 14. Advantages: batch vs. continuous digesters.

Factors Favoring Continuous Digesters

1. Lower steam requirement (less energy)
2. More constant steam demand
3. More compact; less space requirement
4. Lower capacity of ancillary components because of constant loading (e.g. chip conveyor belt, heat recovery system)
5. Easier treatment of non-condensibles because of uniform flow
6. Includes diffusion washing stage (Kamyr only!)
7. Adaptable for digestion of all wood subdivisions

Factors Favoring Batch Digesters

1. Production reliability
2. Operating flexibility
   - ability to change grades
   - ability to cook softwood and hardwood concurrently
   - ease of start-up and shutdown
3. More efficient turpentine recovery
4. Lower maintenance requirement

Factors Which May Favor Either Type of Digester

1. Capital cost
2. Operating manpower.

REFERENCES

(CHAPTERS IV AND V)

1. Dempsey, J.M., Fiber crops. The University Presses of Florida, Gainsville, 1975.
2. Rydholm, S., Continuous pulping processes. TAPPI, Special Publication No. 7, 1970.
3. Douglas, W.J.M., Heat and mass transfer in a turbulent bed contactor. *Chem. Eng. Progr.* 60, 66-71 (7-1964).
4. Tanny, G. and Ingruber, O.V., The application of potentiometry to the determination of total and combined SO_2 in sulphite liquors. *Pulp Paper Mag. Can.* 67, T407-13 (Sept. 1966).
5. Inderdohnen, J.F., Control of direct heated sulphite digesters. *Paper Trade J.,* 35-37 (July 30, 1942).
6. Paasila, M. and Reynolds, M., Batch digester control of sulfite pulping is improved via computer. *Pulp Paper* 130-1 (Mar. 1977).
7. Lowgren, V., Development of defibrator continuous digesters. *Tappi* 45, 210A-215A (July 1962).
8. Herbert, W. The modern continuous digester. *Tappi,* T207A-210A (July 1962).
9. Goodwin, R.G., Continuous digester specifications, applications and operations. *Tappi* 45, 214A-219A (July 1962).
10. Hassinen, I., Continuous digester for bisul-phite pulp is success at Finnish Mill. *Pulp Paper* 92-4 (Nov. 1975).
11. Herzog, P., Semichemical Pulp. *Wochenblatt Papierfabr.* 101, 802-4 (No. 21, 1975).
12. VanEyken, H.K., Continuous magnefite for newsprint. *Pulp Paper Mag. Can.* 65, 103-16 (Sept. 1964).
13. Annergren, G. and Backlund, A., Continuous sulfite cooking. *Pulp Paper Mag. Can. 67,* T220-T224 (Apr. 1966).
14. Patton, W.R., Batch vs. continuous. *Pulp Paper,* 70-2 (May 1977).
15. Schact, W., Cooking liquor for manufacture of pulp from straw, esparto, wood and similar materials. German Patent 122 171, Oct. 10, 1900.
16. Pulp and Paper Chemistry and Chemical Technology, ed. by J.P. Casey; 3rd ed., John Wiley & Sons, New York, 1980.
17. Smook, G.A., Handbook for Pulp & Paper Technologists. Joint Textbook Committee of the Paper Industry, 1982.
18. Rydholm, S.A., Pulping Processes, Interscience Publishers, Div. of John Wiley & Sons, London, 1965.

VI

SEMICHEMICAL PULPING FOR CORRUGATING GRADES

H.E. WORSTER

Hans E. Worster holds M.Sc. and Ph.D. degrees from the Technical University, Darmstadt, Germany. He has 25 years experience in the North American pulp and paper industry in research, engineering and technical positions. Work areas have included dissolving grade pulps, bleached market pulp, linerboard and corrugating medium. Dr. Worster holds 10 patents and is the author of more than 30 papers on wood chemistry, pulping, bleaching and associated areas. He is a member of TAPPI and the Technical Section, Canadian Pulp and Paper Association and, in the past, was active on the Alkaline Pulping and Education Committees of the Technical Section, CPPA.

MANUSCRIPT REVIEWED BY
C. Fred Mills, Technical Director
Packaging Corp. of America
Filer City, Michigan

William Smith, Director Technical Services,
Paper Division Sonoco Products Co.
Hartsville, South Carolina

Semichemical pulping combines chemical and mechanical methods. Conceptually, it can encompass the entire intermediate range of pulp yields between pure chemical and pure mechanical pulping, i.e. 55% to 92% yield. In practice, it includes yields in the 70% to 85% range, to distinguish this process from other high and ultra high yield processes.

This chapter will discuss the semichemical pulping of hardwoods, and includes the Neutral Sulfite Semichemical (NSSC), Green liquor nonsulfur, and other related processes. High yield, and ultra high yield pulping of softwoods and hardwoods will be discussed in Chapter VII. Taken together, these two chapters represent processes covering the 55% to 92% yield range.

Semichemical pulping is a two-stage process. The first stage is a mild chemical treatment of wood chips, sawdust or other lignocellulosic materials. The objective of the chemical treatment is to weaken inter-fiber bonding by removal of some hemicelluloses and a portion of the lignin. The second stage is a mechanical treatment which converts the chemically treated material into individual fibers, bundles of fibers called shives and slivers. Normally, little breaking and splitting of fibers occurs. Microscopic inspection of the fibers of semichemical pulps shows much more resemblance with chemical pulps than with mechanical pulps such as groundwood.

Semichemical pulps are obtained with approximately 70% to 85% yield depending on the type of raw material and the nature and intensity of the chemical treatment. In contrast to chemical pulps, especially bleachable pulps which contain only small amounts of lignin, semichemical hardwood pulps may contain more than half of the 18% to 27% ligning found in the original wood.

Semichemical pulps can be prepared by any of the established commercial pulping processes and their modifications by reducing the chemical application, pulping time and temperature or combinations of these parameters.

The continuous growth of the semichemical pulping processes during the last few decades is to a large extent a result of their adaptability to low cost abundant hardwoods to produce high yield pulps for corrugating medium and similar products with good strength properties, especially compression strength required in corrugated containers.

Bleaching of semichemical pulps is not practical due to their high lignin contents. The chemical requirements would be extremely high. Also effluent treatment would be very expensive.

However, the pulps are well suited for a variety of products, for example light-colored NSSC pulps can be used to some extent in transparent papers.

A. HISTORY

Semichemical pulping dates back to 1874, when A. Mitscherlich proposed that chips be treated with sulfurous acid or bisulfite followed by grinding or rubbing to make a pulp; and to 1880, when C.F. Cross described the advantages of cooking in neutral or slightly alkaline solution with sodium sulfite. Exploitation of the process was delayed by lack of demonstrated use for the pulp and lack of suitable equipment, particularly for fiberizing.

In investigating corrosion at a mill operating on the Keebra process for fully cooked sodium sulfite pulp, workers at the U.S. Forest Products Laboratory found it advantageous to add a small amount of an alkaline agent to the sodium sulfite in the liquor to neutralize the acids released by wood. They also found that full pulping was not necessary; partially cooked chips at 60% to 75% yield could be mechanically pulped with no appreciable damage to the wood fibers. An opportunity to introduce this semichemical process commercially occurred in 1924, when the laboratory was asked to find a profitable outlet for chestnut chips from which the tannin had been extracted. The first neutral sulfite semichemical (NSSC) mill started operation in 1925 for the manufacture of corrugating medium. The product found ready acceptance and, following the first publication [1] on the process, by Rue, Wells, Rawling and Staidl of the Forest Products Laboratory in 1926, several similar mills were soon in operation.

During the 1930's, the daily capacity remained close to 350 tons, but rapid expansion began in 1944 following conversion of a kraft mill to NSSC with a daily production of 200 tons. The same year marked the first preparation of NSSC pulp by a continuous process. Four years later, fully bleached pulp went into commercial production.

Ammonium sulfite was resorted to by several mills temporarily in 1950 during a shortage of soda ash. Cooking was appreciably faster, but the pulps were darker and in some respects a little weaker. In the middle 1960's, ammonia base again came into use because it leaves no residual that would require additional processing if the spent liquor is disposed of by burning. However, due to a variety of factors such as corrosion problems, especially with copper, increased cost of ammonia and the complexity of ammonia recovery, very few ammonium base

NSSC mills were in operation by 1980.

In the late 1950's, forty-seven mills were operating in the United States, and in 1963, about 2.5 million tons of NSSC pulp were produced. As paper and paperboard requirements continue to increase, further growth in semichemical pulping is to be expected, although it is unlikely that the proportional increase will approach that of the past. Its growth is assured by the unrivaled compression strength that hardwood imparts to corrugated containers, the demand for which is expected to increase steadily. However, since the mid-seventies, many NSSC mills in the United States and Canada were converted to nonsulfur (sodium carbonate and sodium hydroxide) semichemical pulping to reduce environmental problems. Also the NSSC operation of many integrated kraft-NSSC mills were switched over to kraft green liquor semichemical pulping. Several small NSSC mills were shut down because pollution problems could not be solved at acceptable cost. Chemical recovery would reduce significantly water pollution caused by small mills. However, capital costs of recovery for small mills are usually very high.

B. WOOD SPECIES

Nearly all important hardwoods of the temperate zone have been evaluated and found to be suitable to NSSC pulping. Semichemical pulps with excellent strength properties can be prepared from birch, gums and poplar. Many other species also give good quality semichemical pulps, for example eucalyptus. Some tropical species of exceptionally high density and lignin or extractive content are more difficult to process. North American hardwoods grow in mixed stands and are frequently harvested as they occur and cooked in mixtures of a dozen or more species.

Semichemical pulping has been limited almost entirely to the hardwoods, although excellent pulps can be made from some softwoods, especially the finer fibered species such as western red cedar. In addition to higher chemical applications, softwoods require considerably more energy for fiberizing and refining. Softwoods having a high resin content are likely to cause pitch troubles in the paper mill unless precautions are taken. Despite these handicaps, some softwoods such as western hemlock, cedar and Douglas fir are pulped by the NSSC process on the west coast of North America where little hardwood is available.

Due to the differences in pulping rate between softwoods and hardwoods, these two types of wood should be pulped separately for optimum results. Accidental contamination of the birch-aspen pile with pine in a Swedish corrugating medium mill [1A] caused considerable problems such as high shive content of the pulp.

C. WOOD PREPARATION

The preparation of wood chips for semichemical pulping is the same as for full chemical pulping. Woodroom equipment and layout are similar. Since some hardwood species are difficult to bark, they are pulped without barking in some mills. However, bark consumes chemicals and persists in the pulp as dark specs, which can be difficult to remove. Bark containing a high proportion of bast fiber can contribute to the yield of pulp without lowering the strength appreciably. Barks without useful fiber add little to yield and reduce strength even if present in only moderate amounts [2,3].

Chips should be thinner than softwood chips, for example pine chips used for kraft pulping. Thick chips are difficult to penetrate with semichemical pulping liquor because either the chemical concentration is lower or less liquor by volume is applied than in chemical pulping. To minimize the portion of thick chips, chipper knives may be set to cut chips as short as ⅜″ to ½″ in grain direction without significantly reducing the fiber length of the pulp.

Thick chips can be removed efficiently from the digester furnish with new disc screens [4]. In 1981, at least one corrugating medium mill installed a disc screen [4] to produce uniformly sized chips between ⁵/₃₂″ and ⁵/₁₆″ thickness.

As previously mentioned, the NSSC process was originally developed to utilize waste chestnut chips from which the tannin had been extracted. In the meantime, semichemical pulping has been applied to a variety wood residues including sawmill chips, furniture mill residue, hogged wood, veneer cores, sawdust, etc. Much of this material has given usable pulps provided the particle size was not so small that the fiber length was greatly shortened.

D. NSSC PROCESS

In the NSSC process, the lignocellulosic material is pulped with a solution of sodium sulfite containing a small amount of an alkaline agent such as sodium carbonate, bicarbonate or hydroxide as a buffer. In mills producing both kraft pulp and NSSC pulp, frequently kraft white liquor or kraft green liquor is added as buffer to the sodium sulfite solution. In some mills, the white liquor is oxidized to convert sodium sulfite to thiosulfate. The reason for the oxidation is to eliminate possible reactions of sodium sulfide with sodium sulfite.

1. Reactions in the chemical treatment

As in the acid sulfite process, neutral sulfite pulping depends on sulfonation and hydrolysis of the lignin to soften the rigid matrix in which the wood fibers are bound together. Since sulfonation takes place slowly and to a limited extent, involving "A" groups (least etherfied, fastest reacting groups) of lignin only, high temperatures in the range of 160° to 190°C are required to complete the reaction within a reasonable time. Although the dissolution of the lignin is described as a two-stage process – sulfonation followed by hydrolysis – solubilization begins quite soon, since some of the lignin molecules are small enough to dissolve immediately on sulfonation, in contrast to the large molecules which must hydrolyze before they can pass into solution. Refer to Chapter II, 1 for additional information on lignin reactions.

In the neutral sulfite pulping of hardwoods, the lignin is sulfonated only to the extent of about 0.07 sulfur atoms per methoxyl group and partially depolymerized. After a fast initial rate, delignification slows down and the main portion of the lignin does not dissolve from the chips; only after mechanical fiberization of the chips does a substantial portion of the lignin disperse in the hot water applied during washing. The heterogeneous nature and lower molecular weight of the native lignin in hardwoods partly account for the adaptability of the species to neutral sulfite pulping.

Although softwood lignin is sulfonated more than hardwood lignin by neutral sulfite liquor, coniferous wood reacts slower and requires more chemical for a given degree of delignification. This is largely due to the greater lignin content of softwood, its inaccessability in the cell wall, and its higher molecular weight. The resins in coniferous woods also resist dissolution in neutral sulfite liquor, and consequently pulps from resinous woods may contain objectionable quantities of pitch.

During the neutral sulfite cooking of hardwoods, hemicelluloses of increasing molecular weight are dissolved and continuously degraded. Part of these polysaccharides survive the digestion and can be recovered from the spent liquor. No simple sugars or di- or trisaccharides are found in the liquor, indicating that the low molecular weight oligosaccharides are easily degraded to aldonic and sulfocarboxylic acids in the presence of high concentrations of sulfite ions.

An aqueous solution of sodium sulfite is definitely alkaline, having a pH above 9. When heated with wood chips, however, the solution soon becomes acidic, unless the sodium sulfite solution is quite concentrated or heavily buf-fered, owing to saponification or hydrolysis of acetyl groups associated with the hemicellulose. Typical hardwoods release acetyl groups equiv-alent to from 3% to 5% of the wood, and typical softwoods from 1% to 1.5%. A much smaller amount of formic acid is also formed. Early digestion with sodium sulfite alone was attended by pronounced corrosion until the cause was recognized and partially corrected by the addi-tion of alkali. The alkalinity of the liquor is a variable of the process. Higher pH increases pulp strength and rate of cooking, but darkens the pulp and reduces its drainage rate. Although the neutral character of the liquor is a handicap in sulfonation, it is beneficial in so far as it reduces loss of carbohydrates by hydrolysis. This con-tributes to the exceptionally high yield and strength of NSSC pulps.

2. NSSC liquor preparation

Most NSSC pulping liquor is prepared at the mill by burning sulfur to sulfur doxide and absorbing the gas in a solution of soda ash or sodium hydroxide. Some of the smaller mills dissolve purchased sodium sulfite and soda ash in water if inexpensive sodium sulfite is conven-iently available, e.g. as a by-product from a plant making phenol by the benzene sulfonate process. Various methods for recovering base and sulfur from the spent cooking liquors in a form suitable for reuse exist and will be reviewed later.

Batch and continuous systems for preparing liquor from purchased soda ash and sulfur are shown in Figs. 108 and 109. The sulfur burning operation is similar to that described for the manufacture of acid sulfite liquor (Chapter IV) except that the heat in the gases is an asset instead of a liability. It helps drive off the carbon dioxide formed in the sulfiting operating of sodium carbonate, increases the solubility of the soda ash, and reduces the amount of steam subsequently needed to heat the liquor to cooking temperature. The absorption tower (Fig. 109), in which the solution of soda ash or sodium hydroxide introduced at the top reacts with the sulfur dioxide gas admitted at the bottom, is relatively small, since no excess sulfur dioxide must be held in simple solution. The tower is packed with an inert filler such as ceramic rings or plastic balls to provide a large surface for absorption of the gas. Other types of absorbers such as scrubbers or ventures are applicable but are not in general use.

The soda ash or sodium hydroxide may be sulfited and additional soda ash or other compounds for buffering, such as green liquor from a kraft operation added later, or the sulfiting may be stopped at a point correspond-ing to the required alkalinity. Residual soda ash is converted to sodium bicarbonate by the carbon dioxide generated. Preparation of NSSC pulping liquor of precise composition or pH is difficult when sodium carbonate is used as base because of dissolved carbon dioxide. The use of sodium hydroxide is much easier.

A typical strong liquor contains between 120 and 200 g/L of sodium sulfite and 30 to 50 g/L of sodium carbonate or an equivalent amount of sodium hydroxide. In charging digesters, the concentrated liquor may be diluted with hot spent liquor from a preceding digestion to conserve heat and increase the solids content of spent liquor, which may later require evapora-tion preparatory to diposal.

The TAPPI standards do not provide specifi-cally for the analysis of NSSC liquor. However, the TAPPI "Useful Method" 624 contributed by the Forest Products Laboratory, Madison, Wis-consin, can serve as a rapid method for analyzing neutral sulfite chemical liquor. The procedure is based on oxidation of reduced sulfur compounds such as sodium sulfite with iodine and the neutralization of alkali with sulfuric acid. This method cannot be applied to pulping liquors containing significant amounts of reducing compounds other than sodium sulfite, for example sodium thiosulfate.

Useful Method 624 consists of the following steps:

— Pipette 25 mL of 0.1N iodine solution into an Erlenmeyer flask containing a small amount of distilled water. Measure a few drops of 0.1N sulfuric acid into the flask to acidify the iodine. Pipette in the 2 mL of the liquor to be analyzed. Back titrate with 0.1N sodium thiosulfate, using a starch indicator.

— Add enough 0.1N sulfuric acid so that the total volume added equals 5 mL. Boil gently for a few minutes using a 2½′ air condenser; then cool the flask in cold water. Add 2 to 3 drops of phenolphthalein indicator and with a buret and sufficient carbonate-free 0.1N sodium hydroxide to give a pink coloration. Be sure to measure the amount of sodium hydroxide used. Back titrate with 0.1N sulfuric acid until the pink color disap-pears.

The results are calculated as follows:

— g/L (Na_2SO_3) = net milliliters of 0.1N iodine × 3.15.

— Alkalinity (g/L of $NaHCO_3$) = 4.2 (net volume of 0.1N iodine + total volume of 0.2N sulfuric acid − volume of 0.1N sodium hydroxide). A blank should be titrated to obtain the net volume of the 0.1N iodine.

An alternate procedure is reviewed in Chapter III, B, 2.

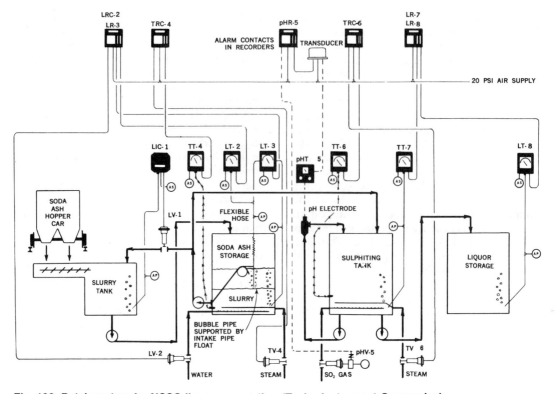

Fig. 108. Batch system for NSSC liquor preparation (Taylor Instrument Companies).

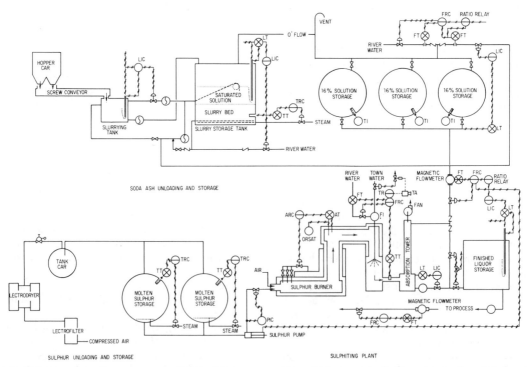

Fig. 109. Continuous system for NSSC liquor preparation. LIC, level indicating controller; LT, level transmitter; TT, temperature transmitter; FT, flow transmitter; TRC, temperature recorder controller; FRC, flow recorder controller. (Pulp and Paper Magazine of Canada).

The results of a more complete analysis of NSSC liquor prepared from purchased sodium hydroxide and sulfur dioxide are listed below in Table 15.

The results show that approximately 2% of the sulfur is in the form of sodium sulfate, which is an inert compound, and 96% is in the form of sodium sulfite. This leaves 2% of the sulfur as unaccounted for, which is well within the sensitivity of the analytical techniques. For all practical purposes, sodium sulfide and thiosulfate are not present.

The 2% sulfur, which is in the form of sodium sulfate, could well be due to a minor quantity of sulfur trioxide contained in the sulfur dioxide gas used to prepare the NSSC liquor.

3. Pulping

In batch pulping, chips are charged to the digester with or without the addition of steam. Chip packing and liquor penetration are usually enhanced by steam addition. Subsequently, the pulping liquor is introduced, and the charge is heated to maximum pulping temperature. The heat-up period may extend over 45 min to 2 h. The length of the heat-up period is usually determined by the capability of the steam generating equipment. The 45 min to 2 h heat-up usually allows uniform chip penetration with the pulping liquor and also uniform pulping. This is a prerequisite for producing satisfactory corrugating medium.

In continuous pulping, the chips are normally presteamed either in the chip bin ahead of the digester, in a presteaming vessel, or in the first tube of a multitube digester. Thorough presteaming is very desirable to assist chip penetration with pulping chemicals, especially in view of the shorter retention times and higher temperatures usually chosen in continuous pulping compared to batch pulping. In continuous digesters, the charge is exposed to full digester pressure before the maximum pulping temperature is reached. The early exposure to relatively high pressure also enhances chip impregnation.

Pulping rate is controlled by the temperature, with the rate roughly doubling for an increase of 10°C. The selectivity of the process, i.e. the relative amounts of lignin and carbohydrate removed from the wood, decreases as the temperature is raised (Table 16), but experiments have shown that the cellulose loss is not excessive. The chief obstacles to the use of higher temperatures lie in securing sufficiently rapid liquor penetration and chemical distribution and in the increase in pressure, which is nearly 690 kPa (100 psig) when the temperature is raised from 190° to 210°C. Tube-type continuous digesters, which are of limited volume, require high temperatures to give a satisfactory rate of production. Retention time for corrugating grade pulps is usually 15 to 35 min. Continuous vertical digesters permit longer pulping at moderate temperature. Typical batch digesters usually operate on a 2 to 4 h cover-to-cover cycle, allowing for charging and discharging.

The ratio of liquor to wood in the digester ranges from 1.5:1 to 4:1. Low liquor ratio is practiced by impregnating the chips under high pressure in an excess of liquor at temperatures ranging up to maximum pulp temperature. This is followed by draining the free liquor and completing the digestion with steam admitted directly to the chips. High liquor ratios are necessary if the digester is heated by circulating the liquor through an external heat exchanger. Low liquor ratio pulping obviously saves steam, since less is required to heat the digester charge to maximum pulping temperature and to evaporate the spent liquor.

The concentration of chemical in the digester is fixed by the ratio of liquor to wood on the one hand and the amount of chemical charged on the other. Over the commercial range, the liquor concentration has little influence on the rate of pulping or on the strength properties of the pulp unless the amount of chemical applied is insufficient.

Table 15. Chemical composition of typical NSSC liquor.

Chemical Compound	Concentration, GPL as Chemical
Sodium Sulfite	133
Sodium Hydroxide	5.8
Sodium Sulfate	3.2
Sodium Thiosulfate	<0.1
Sodium Sulfide	<0.1
Total Sodium	53.0
Total Sulfur	35.1

Table 16. Effect of cooking temperature on selectivity in NSSC Pulping (mixed eastern hardwoods, 70% yield).

Cooking temp, °C	Lignin[a] retained in pulp, g per 100 g wood	Pentosans retained in pulp, g per 100 g wood
165	9.4	12.6
175	9.3	11.9
185	9.6	10.9
195	11.3	9.8

[a]Klason-type (TAPPI) lignin. No allowance made for acid-soluble lignin.
Source: Data from U.S. Forest Products Laboratory.

For the manufacture of corrugating medium, from 8% to 14% of sodium sulfite, calculated on the weight of moisture-free wood, is charged, depending on the species and the yield desired, usually between 75% and 85%. The amount of sodium sulfite should be sufficient to give a residual of from 5 to 10 g/L in the spent liquor. If the chemical is completely exhausted before the chips are blown, the pulp darkens.

From 2.5% to 5% (wood basis) of soda ash, or an amount of sodium bicarbonate, caustic soda, or kraft green liquor sufficient of comparable terminal pH, is included in the pulping liquor to ensure that it will remain slightly alkaline. The precise amount depends on the wood species and the preference of the individual mill. A minimum amount, sufficient for a spent liquor pH slightly over 7, gives the brightest pulp. However, advantages accompany a pH of 8 to 9, including faster pulping, some improvement in pulp strength, and reduced corrosion of the equipment. Typically, the pH measured at room temperature on the spent liquor sample drawn from the digester will range between 7.2 and 9. However, some mills pulp to a spent liquor pH as low as 5.5 to save buffer chemicals and to enhance pulp drainage.

The alkalinity of the pulping liquor can be adjusted to some extent by controlling the amount of carbon dioxide gas that is relieved from the digester. Formed by reaction of the wood acids with the carbonate and bicarbonate in the liquor, the gas buffers the solution at a lower alkalinity than would be the case in its absence. Practically, the carbon dioxide and other noncondensable gases generated are released as a matter of course to lower the digester pressure, to facilitate temperature control and reduce the amount of alkali needed for the desired pH.

Figure 110 shows the relationships between yield and chemical requirement, pulping time, and lignin content of the pulp from several species. The lignin values do not include the "acid soluble" lignin. Pulp yield after mild digestion depends on the amount of hot water soluble and readily hydrolyzable material originally present in the wood, as well as on the amount of lignin removed. The proportion of this material varies widely with species and the history of the wood. It includes a number of diverse compounds such as tannins, starches and other short chain carbohydrates, inorganic salts, acetyl groups and many compounds unique to a single species or group of species. Oak (Fig. 110) illustrates the response to pulping of a wood high in extractives, showing a rapid decrease in yield with the application of small amounts of sodium

sulfite.

The effects of some variables in NSSC pulping of Australian eucalyptus for corrugating medium in a Kamyr continuous digester on bursting strength and concora crush strength at 400 mL Cs freeness are presented in Table 17 [5]. The data show that a presteaming temperature of 121°C (250°F) instead of 99°C (210°F), a low height of the chip level above the liquor level, and a Kappa number 100 instead of 130 gives better quality. A blow temperature of 104°C (220°F) is only marginally better than a blow temperature of 127°C (260°F).

4. Digesters
(a) Batch digesters

Batch digesters include globe rotary vessels in older mills, some of which are originally used to

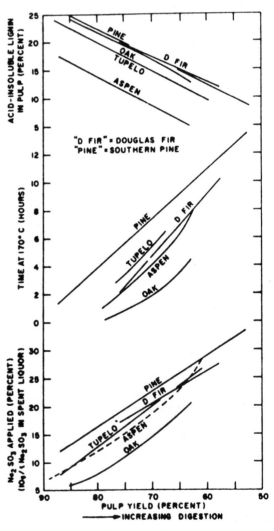

Fig. 110. NSSC pulping of aspen, white oak, black tupelo, short leaf pine, and Douglas fir.

Table 17. Effect of pulping variables on pulp quality.

(a) Effect of degree of cooking on pulp quality Huon "C" Pulp

Kappa Number	Burst Factor	Concora Crush
100	38	80
110	36	80
120	33	78
130	30	75

(b) Effect of presteaming vessel temperature on pulp quality
Huon "C" pulp cooked to 110 Kappa number

Temperature °F	Burst Factor	Concora Crush
250	36	80
230	34	79
210	32	76

(c) Effect of chip height above liquor on pulp quality
Huon "C" pulp cooked to 110 Kappa number

Chip Height Above Liquor	Burst Factor	Concora Crush
91 cm (3 ft)	36	80
6 m (20 ft) approx	22	62

(d) Effect of blowing temperature on pulp quality
Huon "C" pulp cooked to 110 Kappa number

Blowing Temperature °C °F	Burst Factor	Concora Crush
104 220	36	80
116 240	35.5	78
127 260	35	79

pulp straw, and vertical stationary digesters similar to those used for chemical pulping. The early vessels were built of mild steel and gave satisfactory service, but changes in wood species and pulping conditions have increased corrosion to a degree that the use of stainless steel, usually type 316, is mandatory. Older digesters can be periodically spray coated with stainless steel, or an overlay may be deposited. Newer vessels are fabricated of a stainless steel clad material or its equivalent. Similar material is used in the construction of continuous digesters, which are used by both old and new mills to take advantage of the reduced labor, steam and space requirements, and possibly increased uniformity of pulping.

The first NSSC mills used spherical rotary digesters, 14′ to 16′ in diameter, which held 6 to 10 cords of wood. The spherical digester consists of a simple shell mounted on trunnions through which steam for heating can be introduced and the noncondensable gases vented. Since the digester slowly rotates during operation, uniform pulping can be obtained with low ratios of liquor to wood. After the chips are impregnated, all of the free liquor can be removed and digestion completed in the so-called vapor phase. Both kraft and acid sulfite vertical digesters have been converted to NSSC pulping. The digesters can be heated by direct steam or indirectly with an external heat exchanger and pump to circulate the cooking liquor. Stationary digesters intended for semichemical pulping have blow lines and blow valves larger than usual because softened chips are more difficult than slush pulp to blow from the digester. To ensure a clean blow, the digester may be filled completely with spent liquor from a proceeding digestion at the end of pulping, or spent liquor may be injected continuously into the bottom cone during the blow.

See Chapter V for a detailed discussion of batch digester operations.

(b) Continuous digesters

Several types of continuous digesters are used for semichemical pulping. They differ considerably in design and can be divided more or less into two groups, namely digesters with "liquid phase" pulping and those with "vapor phase" pulping.

In liquid phase pulping, liquor in addition to that absorbed by the chips is present, whereas in vapor phase pulping the free liquor is removed from the impregnated chips before direct heating with steam.

The digesters with a liquid phase include the Pandia, Bauer M&D and Kamyr digesters. Vapor phase units include the Sunds Defibrator, and modified liquid phase digesters. They are discussed in some detail in the section on Continuous Digesters in Chapter V.

Vapor phase digestion represents an alternative approach to liquid phase digestion, and the technique is now a well-established commercial reality. The technique involves the impregnation of wood chips with a pulping liquor of higher than normal concentration followed by removal of excess superficial liquor prior to direct heating of the chips with steam in a continuous vertical digester.

Digestion in the vapor phase offers considerable savings over liquid phase digestion in terms of steam requirement. The technique increases the solids content of the waste liquor and renders rapid cooking possible. A prerequisite for vapor phase digestion is that the pulping chemicals must be introduced into the chips during the impregnation phase of the process. In order that this can be achieved efficiently, air present in the incoming chips is removed by atmospheric steaming prior to impregnation.

In modern installations the digester content is discharged into a pressurized blowline refiner. In some older mills, the pulped material may be discharged directly into a leach caster, live bottom bin or an ordinary blow tank. Continuous digesters typically discharge to a cyclone, where steam and gas are separated from the pulp liquor.

5. Fiberizing and initial refining

A number of mills have introduced one or more stages of screw pressing before the digested chips are sent to the disc (refiner) mill for fiberizing. The Bauer screw press, the French Oil Mills screw press, the Impco twin roll press, and the Thune screw press are being used successfully. A schematic of the Bauer screw press is shown in Fig. 111. Advantages include the recovery of a maximum amount of the spent cooking liquor and partial washing with minimum dilution. The energy required is in the order of 0.14 GJ/t (2 hp day/ton) of pulp per stage, but this is more than made up in energy saved in subsequent treatment and in greater uniformity of the pulp discharged from the disc mill.

The Impco twin roll roll press is shown in Fig. 112. This press consists of two horizontal porous rolls mounted in a sealed vat and rotating at the same speed towards each other. One roll is fixed, while the other is movable and is loaded by air pressure to maintain constant nip loadings under varying openings. Slurry is fed to the vat under pressure and is dispersed within the vat to maintain uniform consistency. A dewatered mat forms on the rolls under vat pressure, and this mat is further dewatered in the nip. The dewatered cake is shredded and conveyed away from the press by a serrated conveyor positioned in the nip above the rolls. Pressate passes through the perforated roll face and is discharged from the roll ends. It then flows to an outlet connection located on the vat bottom.

With or without screw pressing, the disc refiner is the basic unit in the semichemical process for fiberizing and refining the cooked chips. It consists essentially of two heavy steel discs (Fig. 113) to which are bolted segmental plates of various configurations between which passes the material (chips, screenings or pulp) to be treated. In the disc refiner, the softened chips are fiberized by passing between two closely spaced ridged plates one or both of which may be rotating. Figure 114 illustrates the classic double rotating disc refiner in which chips or pulp is fed to the center of the refiner and carried through the grooves to discharge by centrifugal force.

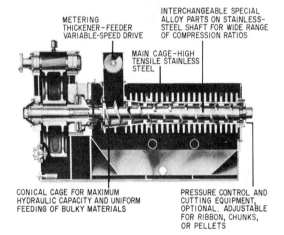

METERING THICKENER-FEEDER VARIABLE-SPEED DRIVE

INTERCHANGEABLE SPECIAL ALLOY PARTS ON STAINLESS-STEEL SHAFT FOR WIDE RANGE OF COMPRESSION RATIOS

MAIN CAGE-HIGH TENSILE STAINLESS STEEL

CONICAL CAGE FOR MAXIMUM HYDRAULIC CAPACITY AND UNIFORM FEEDING OF BULKY MATERIALS

PRESSURE CONTROL AND CUTTING EQUIPMENT, OPTIONAL. ADJUSTABLE FOR RIBBON, CHUNKS, OR PELLETS

Fig. 111. Single-stage screw press (Bauer Bros. Co.).

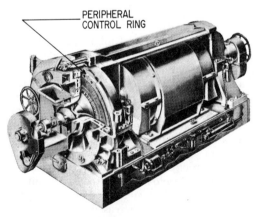

PERIPHERAL CONTROL RING

Fig. 113. Single-rotating-disk refiner (Sprout, Waldron & Co., Inc.).

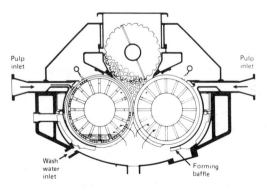

Pulp inlet

Pulp inlet

Wash water inlet

Forming baffle

Fig. 112. Impco twin roll press.

Construction is heavy to avoid mechanical strain, since the plates must remain parallel under all conditions while separated by a few thousandths of an inch. Commercial equipment operates at speed from 600 to 1800 rpm, with connected horsepower of 150 to 3700 kW (200 to 5000 hp) and discs from 61 to 132 cm (24 to 52 inches) in diameter.

Plate clearance may be maintained mechanically with a spring, or hydraulically to keep constant pressure or loading. The plates are segmented for quick replacement and bolted to the heavy disc, which is an integral part of the machine. A set of plates may last 500 to 2000 h or more, depending on operating conditions and material of construction. Plates are available in many compositions, included hard white iron, steel, stainless steel, and nickel steels. The plates also come in numerous patterns. To protect refiners and other processing equipment, powerful magnets are mounted over chip conveyors to remove tramp metal.

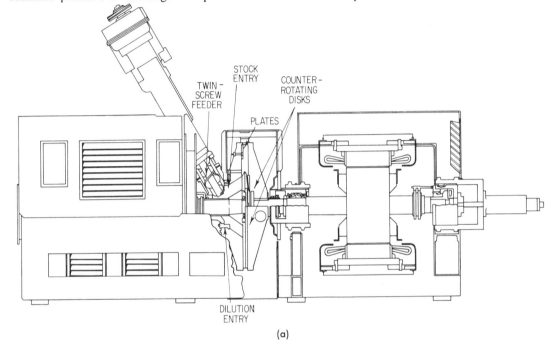

(a)

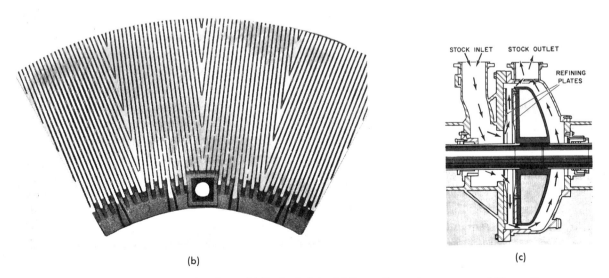

(b)   (c)

Fig. 114. (a) Double-rotating-disk refiner. (b) Typical plate. (c) Pump-through arrangement (Bauer Bros. Co.).

Refer to Volume 2 on Mechanical Pulping for illustrations and discussions of refiners.

Before entering the refiner, the pulped chips and hot spent liquor may be separated in a pit or on a conveyor belt and recombined at the inlet of the refiner to give a consistency of from 10% to 15% between the plates. The fiberizing is done with discharge temperatures close to 93°C (200°F) to take advantage of the savings in energy afforded by the softening effect of heat on the lignin that bonds the fibers together. For maximum efficiency, the motors driving the discs should be operating at full capacity. The loading can be increased by increasing the consistency and the feed rate. Positive pressure by the screw feeders forces the plates apart and, more or less, determines the clearance against a fixed plate loading. The total energy required for fiberizing and secondary refining of the pulp in subsequent treatment is typically between 0.71 and 2.13 GJ/t (10 and 30 hp day/t) of air dry pulp, depending chiefly on the yield to which the chips are cooked and the ultimate freeness required. Softwoods require much more energy than hardwoods.

6. Washing and cleaning

In many mills, the material discharged to the blow tank is still in chip form, and the spent liquor retained within the chips cannot be readily separated in a blow pit or diffuser. Since the liquor is frequently corrosive, stainless steel is used liberally up to and including the first washer stage. The use of the screw press for washing was referred to in the proceeding section. More than one stage may be used, with the cake emerging from one stage being sprayed with water before it enters the next stage.

More commonly, the pulped chips as they come from the digester are fed to a disc refiner and then to one or more multistage vacuum washers. The pulp drains somewhat slowly and the capacity per square foot of wire is less than with kraft or sulfite softwood pulp. The drainage rate can be improved by increasing the freeness of the stock from the fiberizer or by including a small amount of softwood fiber. Screening is usually omitted in semichemical pulp mills. However, in principle, centrifugal or flat screens may be used before or after washing.

Especially in some older mills the pulp is frequently passed through batteries of centrifugal cleaners before or after the second stage of refining to remove shives and dirt. In more modern mills, stock cleaning is usually conducted immediately ahead of the paper machine.

7. Secondary refining

Fiberizing and refining to the freeness necessary to develop the required pulp qualities can often be accomplished with a single pass through a disc refiner, especially if the pulped chips have received a preliminary breakdown in a screw press. In two-stage refining, however, usually the first pass is primarily to fiberize the chips, with a subsequent pass through a refiner to develop strength. Such refining can be accomplished either in a double rotating disc refiner (Fig. 114) fitted with appropriate plates or in a disc refiner with one rotating and one stationary plate (Fig. 113). With only one rotating plate the relative speed differential between the plates is halved. The disc refiners are often the pump-through type, instead of the free discharge type, so that retention time of the pulp suspension between the plates can be controlled. In secondary refining, consistency is lower than the first stage. Secondary refining in more than one stage is becoming more common to upgrade pulp quality and uniformity.

In newer semichemical pulp mills, a refining step is frequently conducted immediately after washing at the washer discharge consistency in the pulp mill. A high speed refiner with low unit power loading per active plate area is usually installed, for example a Sprout-Waldron 45/50-1B, a Sunds Defibrator RG or a CE Bauer Double Disc refiner. At the washer discharge consistency of 10% and higher, a high fiber-

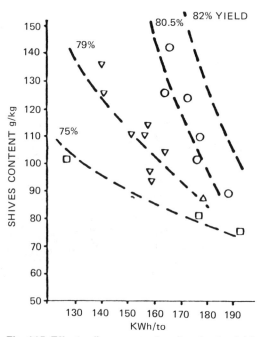

Fig. 115. Effect refiner energy input and pulp yield of shive content of birch NSSC pulp.

to-fiber rubbing action occurs between the refiner plates. This action converts shives and fiber bundles to individual fibers without significant freeness reduction. This procedure is referred to as "deshive refining".

Large shives are undesirable in semichemical pulp because they affect adversely concora crush strength, which is one of the most important end use properties of corrugating medium [7]. The quantity of shives contained in semichemical pulp depends mainly on pulp yield and the refining energy application. The effects of these two parameters on the shive content of birch-aspen NSSC pulp is shown in Fig. 115 [1].

At least one Scandinavian corrugating medium mill [8] is controlling the pulp mill refiners accurately by a computer to operate at constant specific refiner energy (kWh/t). The feed to each refiner is calculated and the plate distance is set for constant energy application per ton of pulp. The set point of the refiner load is adjusted by the computer every hour based on a shive analysis. It was found that the specific energy required for constant shive content is a very good indication of pulp yield.

8. Flow sheets

The equipment just described may be assembled in a variety of ways and there is, in fact, considerable difference among semichemical mills. Figure 116 is a flow sheet showing continuous horizontal tube digesters discharging into a live bottom bin. Steam from the flow is condensed by white water from the paper machine, to furnish hot water to the washers that follow the primary refiners. The filtrate from the washers, which contains a good deal of the spent liquor, goes to the kraft mill. Thickened stock from the washers is diluted with white water, passes through the secondary refiners, and is then mixed with kraft pulp and broke before it enters the chest of the paper machine. Storage chests between the operations take care of surges and brief interruptions in the flow of stock.

The second flow sheet (Fig. 117) shows a rotary batch digester operated with a leach caster from which cooked chips are fed at a constant rate to the horizontal fiber presses. Rotary digesters date back many years, and were used to cook straw and rags before semichemical pulping was practiced.

This type of digester is no longer installed, and

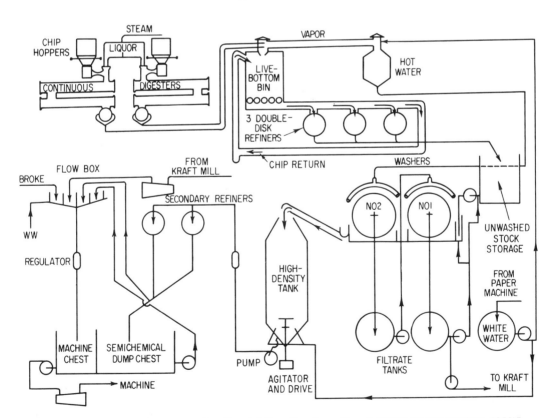

Fig. 116. Flow sheet for NSSC mill [Continental Can Co., Inc.; *Tappi*, 41(12):58 (December, 1958)].

many of the old units have been retired. In some mills they serve for liquor storage. The sidehill decker screens fibers from the press effluent, and the effluent (consisting of strong spent cooking liquor) is pumped to the recovery plant. A conveyor takes the pressed chips from the presses and distributes them among the primary refiners, each of which is fitted with a magnetic separator in the entrance chute to remove tramp iron that would otherwise damage the equipment. Such separators are commonly found just before the feeders that supply chips to the digesters. As in the first flow sheet, the pulp is then put through vacuum washers and given a second refining before it is blended with long fibered stock.

A third, more comprehensive, flow sheet is shown in Fig. 118.

Refer to Chapter III,C,6 for NSSC process control and example of an automated NSSC production line.

9. Materials of construction

NSSC spent liquors are more corrosive than kraft liquors, although considerably less than acid sulfite liquors. First applications of the process showed no serious corrosion of the mild steel digesters used to cook the extracted chestnut chip. But changes in wood species, alkalinity, digester operation and design, and the introduction of corrosive sulfur-containing com-

pounds, such as thiosulfate, sulfides and polysulfides, in regenerated cooking liquors, have contributed toward the widespread application of stainless steels.

Refer to Chapter V for a detailed discussion of materials and corrosion.

10. Properties of NSSC spent liquors

The odor of NSSC spent liquors is relatively weak compared to conventional sulfite and kraft black liquors. However, NSSC spent liquors are intensely colored and may contain small amounts of toxic materials depending mainly on wood species. The spent liquors have a significant biological oxygen demand, which needs to be taken into account before discharging weak spent liquor or more concentrated liquor spills into bodies of fresh water, especially if the flow rate of the stream is limited. The complex organic compounds in the liquor account for a relatively small part of the oxygen consumption. One half or more of the total biochemical oxygen demand (BOD) is due to sodium salts of the lower fatty acids, principally acetic acid.

Some analytical test results established for NSSC spent liquors are presented in Table 18.

(a) Evaporation and burning

Spent liquor from typical batch digestion may contain 12% to 14% solids. Using spent liquor from one digestion to dilute the strong fresh

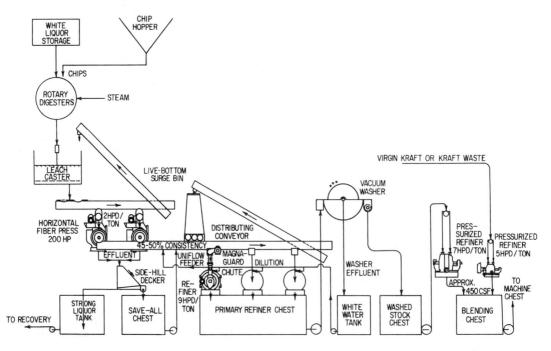

Fig. 117. Flow sheet for NSSC mill showing application of rotary digesters (Sprout, Waldron & Co., Inc.).

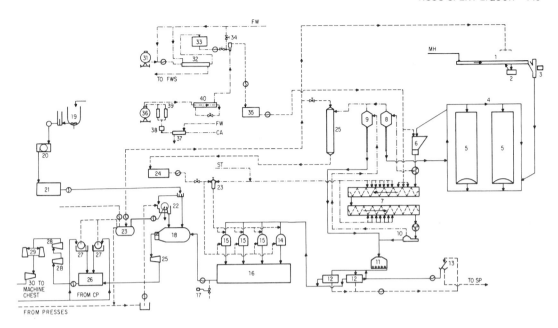

Fig. 118. 130-ton neutral sulfite semichemical mill. MH, mixed hardwoods; SP, storage pond; FW, fresh water; S, steam; CA, compressed air; FWS, fresh-water storage; HW, hot water ("Pulp and Paper Manual of Canada, 1962," page 117).

1. Log conveyor
2. Splitter
3. 96-inc. 12-knife chipper
4. Conveyor
05. Chip silo
6. Chip-feeding hopper
7. Continuous digester
8. Dry-chip cyclone
9. Wet-chip cyclone
10. Expansion box
11. Live-bottom bin
12. Pressafiner
13. Save-all
14. Two Sprout Waldron disk refiners

15. Bauer disk refiners
16. Pulp chest
17. Consistency controller
18. "Tugboat" thick stock chest
19. Hydrapulper
20. Thickener
21. Stock chest
22. Disk-type vacuum filter
23. No. 1 hot-water tank
24. No. 2 hot-water tank
25. Refiner
26. Beater chest
27. Gravity save-alls
28. Cyclifiners

29. Jordans
30. Jordan
31. 73% NaOH
32. Heat exchanger
33. 50% NaOH storage
34. Reaction chamber
35. Na_2SO_3 storage
36. Liquid SO_2
37. Aftercooler
38. Lectrofilter
39. Lectrodryer
40. SO_2 cooling chamber

Table 18. NSSC spent liquor characteristics from analyses made at Virginia Polytechnic Institute.

	Range		Average	
pH	6.5-8.5		7.5	
Total solids, %	8-22		12	
Volatile solids,[a] %	43-52	based on total solids	47.9	based on total solids
5-day, 20°C, BOD, ppm	16 000- 50 000		25 000	
Acetate, g/L	12-20		18	
Wood sugars (mostly pentoses), g/L	5-10		7	
Oxygen consumption				
From $KMnO_4$, ppm	55 000-142 000		65 000	
From Ag-catalyzed dichromate, ppm	83 000-235 000		100 000	
Lignin, ppm	25 000- 85 000		45 000	

[a]Loss on igniting 1 h at 600°C after previous drying at 103°C.
Source: *Tech. Bull.* 83, National Council for Stream Improvement, Inc.

liquor for the next may increase this to 18% to 20%. However, continuous digester operations may discharge the dissolved material at 7% to 9% solids in the washer filtrate. Many of the disposal methods proposed for mills that can not sewer their spent liquor require that it be evaporated to a syrupy consistency and burned. Because of scaling in the tubes, the most suitable for the conventional evaporator forms are the vertical long tube and the forced circulation. The scale has its origin in calcium dissolved from the wood, as well as a neutral lignosulfonate susceptible to precipitation, fine fibers and short chain carbohydrate material that may approach the colloidal in dimensions. Analysis of such a scale showed calcium sulfite, sulfate and carbonate, plus 15% to 20% of organic matter. To remove this scale, the tubes are boiled out at intervals with water or caustic solutions, with occasional acid treatment if necessary. Removal of fine fibers and other suspended material from the weak spent liquor by filtration over a fine mesh wire reduces the scale formation. Some mills have installed weak liquor filters for this purpose.

Problems in burning concentrated NSSC

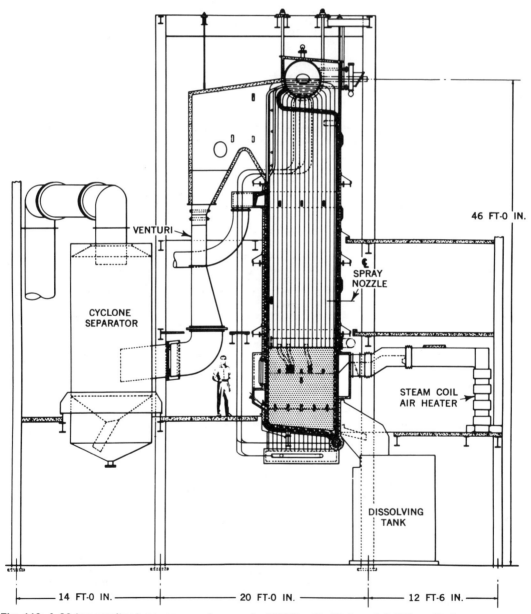

Fig. 119. A 30-ton smelter-type recovery furnace for NSSC mills (Babcock & Wilcox Co.).

liquor, compared with kraft, arise from a generally lower heat value per pounds of solids, which makes it more difficult to maintain combustion without auxiliary fuel. The small production of many NSSC mills in conjunction with the low heat potential per ton of pulp produced 6.3 to 12.7 GJ (6 to 12 million Btu) for NSSC vs 22 GJ (21 million) for kraft does not justify the high capital investment required by the standard kraft-type furnace. However, a simpler smelter-type furnace is available at a price of somewhat greater operating costs (Fig. 119). Refer to Chapter IX for additional details.

Technical data of interest in processing spent liquor are shown in Figs. 120 and 121. NSSC liquors obtained from four mills were more viscous than representative kraft or acid sulfite liquors of equal solids content, but they showed considerable difference between mills. The mills that pulped to 5% to 15% greater yield with one half the chemical charge furnished the more viscous liquors. However, the four liquors were similar and comparable with kraft black liquor in boiling point rise. The high boiling point rise, compared with that of acid sulfite liquor, is probably associated with the greater inorganic content. NSSC liquor is highly thixotropic, tending to gel on standing but regaining fluidity after moderate stirring.

11. Chemical recovery and effluent disposal (Refer also to Chapter IX)

The chemical recovery methods developed for soda-base sulfite and bisulfite mills are also applicable to the NSSC process. Owing to the relatively small size of the older mills and the low organic content and heat value of the spent liquor, as well as the low cost of the cooking chemicals, there has been little economic incentive for high capital expenditure for recovery plants.

An economical and effective method adopted by a number of mills, in which both NSSC and kraft operations are conducted at the same site, is "cross-recovery". Since both linerboard (kraft) and corrugating medium (semichemical) are required in the manufacture of boxboard, such joint operation is desirable. The sulfur and soda obtained from the semichemical spent liquors, when introduced in the kraft recovery process, replace the salt cake and equivalent chemicals used to make up chemical losses in the kraft mill. To utilize all of the chemical from the NSSC spent liquor, the sulfur and sodium input from the NSSC spent liquor must not exceed the make-up of these two components for the entire mill complex.

During the last two or three decades, several different sodium-based NSSC chemical recovery systems have been proposed and operated. Many of them were abandoned due to various problems, e.g. poor economics or air pollution problems. The common elements of the recovery processes currently practiced are spent liquor collection, concentration by evaporation, incineration of the concentrated liquor, and chemical conversion to pulping liquor. In operation are the Sivola-Lurgi, Tampella, SCA-Billerud and Sonoco processes. Refer to Chapter IX for

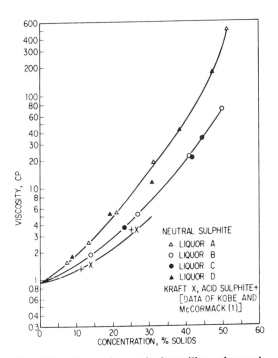

Fig. 120. Comparison of viscosities of spent liquors at 76.5°C [*Tappi*, 40:921 (1957)].

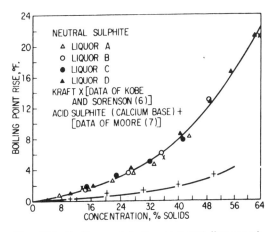

Fig. 121. Boiling point rise of spent liquors at atmospheric pressure [*Tappi*, 40:921 (1957)].

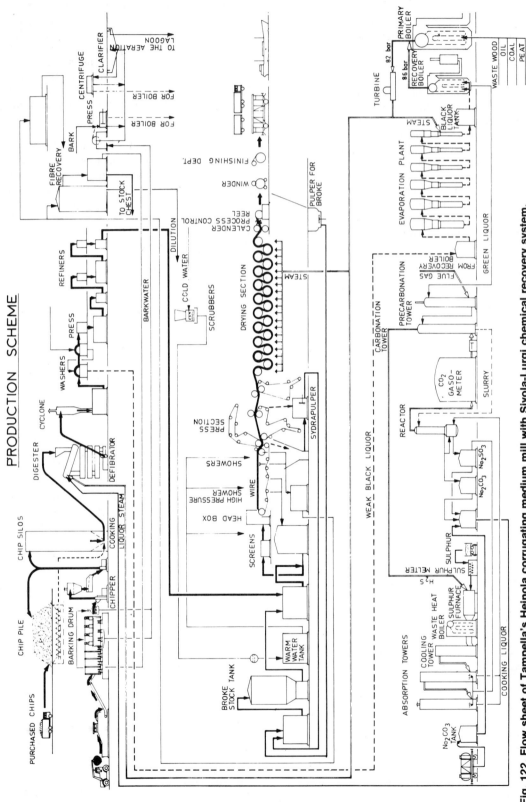

Fig. 122. Flow sheet of Tampella's Heinola corrugating medium mill with Sivola-Lurgi chemical recovery system.

further discussion of these systems.

Since 1962, the Sivola-Lurgi process [9] has been used successfully at the Heinola corrugating medium of Tampella Oy, Finland. A flow sheet of the entire mill is shown in Fig. 122.

More important than economic factors in recovering the pulping chemicals for many NSSC mills is the urgent need of disposing effluents without stream pollution. In addition to the chemical recovery methods described earlier, most of the practices developed by acid sulfite mills for controlling stream pollution, such as ponding, soil filtration, aeration and the use of the liquor as road binder, are also applicable to NSSC effluent.

Materials other than soda and sulfur can be recovered from NSSC spent liquors. A previously described hardwood contains from 3% to 5% of acetyl groups, which are saponified during digestion. Consequently, spent liquors typically contain the equivalent of 10 to 13 g/L of acetic acid in the form of sodium acetate, as well as a small amount of sodium formate. The process of the Sonoco Products Company [12] recovers the acids by acidifying the concentrated liquor with sulfuric acid and extracting the organic acids with methylethyl ketone. The acetic and formic acids are separated by azeotropic distillation using ethylene dichloride. The residual is sold as a replacement for salt cake.

No practical recovery system for ammonia-base spent liquor has been developed. In principle, this type of spent liquor can be disposed of by burning, leaving as solid residue only a very small amount of inorganic matter which originates mainly from wood. Sulfur dioxide recovery from the combustion gases is possible by scrubbing.

12. Chemical properties of NSSC pulps

Semichemical pulps for corrugating medium contain 15% to 20% lignin depending on wood species and pulping conditions. Bleaching of semichemical pulps is not practical due to high bleach chemical requirements.

Hardwood NSSC pulps contain more pentosans than pulps cooked by any other common process. This is largely due to the neutral nature of the pulping liquor, which restrains hydrolysis and dissolution of the hemicelluloses during pulping. Both yield and strength benefit from high retention of hemicelluloses. With light mechanical processing, the pentosans on the surface of the fiber swell rapidly and furnish additional bonding. The chemical composition of selected pulps is shown in Table 19. It might

Table 19. The chemical composition of selected pulps.

Constituent	Birch		Spruce	
	Wood, %	Pulp, %	Wood, %	Pulp, %
Analyses of wood and unbleached neutral sulphite pulps				
Yield		77.0		81.0
Pentosan	22.5	23.3	7.5	7.9
Uronic acids	4.7	2.8	2.3	1.1
Lignin	20.0	10.8	28.0	20.2
Methoxyl	6.1	4.4	4.9	3.5
Roe number		15.0		25.0
Carbohydrates, total	75.5	84.0	70.5	75.0
Polysaccharides in wood and neutral sulphite pulps[a]				
Galactan	1.5	trace	3.0	trace
Glucan	59.5	67.5	66.0	69.0
Mannan	2.5	1.0	17.0	19.0
Araban	2.5	1.0	1.5	1.0
Xylan	27.5	27.5	9.0	9.5
Rhamnose	0.4	0	0	0
Uronic acid	6.1	3.3	3.5	1.5
Losses in neutral sulphite pulping[b]				
Pentosans	20.5		19.0	
Uronic acids	54.0		61.0	
Lignin	58.5		41.5	
Galactan	100.0		100.0	
Glucan	3.0		10.5	
Mannan	66.0		3.0	
Araban	65.0		33.0	
Xylan	14.0		10.0	
Rhamnose	100.0			

[a] Relative amounts expressed in percent.
[b] Expressed in percent of the amount in wood.

Source: Data from Bjorkvist, Gustafsson, and Jorgensen, *Pulp Paper Mag. Can.*, 55(2):68-72 (1954).

be mentioned that birch wood and birch pulps contain an exceptionally large amount of pentosans compared to other species. Table 19 shows 23% (see also Chap. II,F).

13. Physical properties of NSSC pulps

Semichemical fibers are heavier than those that are fully pulped because of the large amount of lignin retained. Consequently, a sheet of standard weight contains fewer fibers. The semichemical fibers are also stiffer than comparable chemical pulps prepared from the same species. As the yield is reduced by extended pulping, higher chemical application or higher pulping temperature, the increasing number of fibers per gram and increasing flexibility contribute to greater bonding within the sheet. As would be expected, sheet density, when pressed under similar conditions, also increases as the fibers become less rigid.

Strength varies with species, pulp yield and degree of refining. Species such as poplar give pulp of excellent quality, light in color, low in lignin content and high in strength. Birch is a much denser species that likewise gives a light, strong pulp. Oaks and certain other dense woods give products of lower strength and greater bulk, although some oak species give a notably better pulp than others. In spite of relatively low strength and high lignin content, large amounts of oak are pulped for corrugating medium. Because of their bulk, these pulps impart adequate rigidity to corrugating medium. The properties of a number of corrugating mediums are shown in Table 20.

In the literature, a number of studies on the effects of NSSC pulp yield are reported. Although a certain degree of disagreement on the magnitude of the yield effects on strength properties appears to exist, the general conclusion is that the lower the yield, the better the average strength. The results of a systematic research study [13] on the effects of NSSC pulp yield are presented in Fig. 123 for 26 lb/Msf handsheets prepared from pulp refined to 450 mL Cs freeness. The figure shows a continuous decrease in almost all strength properties as yield increases from 70% to 90%. In view of the importance of concora strength (flat crush), the effect of NSSC pulp yield on this property is shown in more detail in Fig. 124. Closer inspection of the data reveals that the concora values of the pulps obtained with 72% to 81% yield are similar. This yield range corresponds to standard operating conditions in many corrugating medium mills.

NSSC hardwood pulps respond to refining similar to other pulps except that they increase somewhat in tear as freeness is reduced. Other properties such as density, flat crush (concora), ring crush and tensile strengths also benefit from freeness reduction, while caliper, bulk and stock drainage rate decrease. Some typical data established for eucalyptus NSSC pulp are presented in Table 21 [14].

14. Use of NSSC pulps

Almost all NSSC pulp produced is converted to corrugating medium. A small amount of NSSC pulp is used in the base sheet duplex linerboard as low cost furnish component, which imparts good compression strength.

Table 20. Properties of nominal 26-lb corrugating boards made from neutral sulphite semichemical pulps.

| Wood species | Thick-ness, mils | Bursting strength points | Tearing resis-tance, g | Tensile strength, lb per in. width | Ring crush,[a] lb | | Flat crush, lb |
					Machine direction	Cross machine direction	Concora medium test
			Commercial boards				
Mixed northern hardwoods	8.8	48	93	36	57	43	79
East central hardwoods	9.4	34	77	29	59	42	74
Red alder and Douglas fir (3:1)	8.1	41	—	—	—	49	55
			Experimental boards				
Hardwoods							
White and red oaks (2:1)	9.1	47	74	36	63	46	68
Sweet gum	8.2	51	108	40	60	46	85
Mixed northern hardwoods	8.1	54	74	39	69	56	76
Mixed Wisconsin hardwoods	8.2	60	81	45	76	60	90

[a]6.0 in. specimen, 0.5 in. wide.   Source: Data from U.S. Forest Products Laboratory.

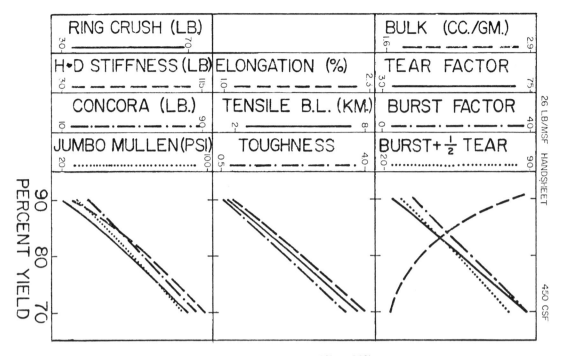

Fig. 123. Properties of refined NSSC pulps with yields of 70% to 90%.

The quality of NSSC pulp has to be high enough to meet the requirements in three areas. They are paper machine runnability, corrugator runnability and end use as component of corrugated containers.

On the paper machine, the fully refined machine furnish has to drain sufficiently fast and wet web strength, i.e. tensile and stretch, has to be high enough that the machine can be run at acceptable speed. To meet these requirements, usually 10% to 25% long fibers such as softwood kraft pulp have to be added to NSSC pulp. As alternatives, secondary fibers prepared by recycling corrugated container clippings or old corrugated container waste may be used. However, due to the hardwood fiber content of these two products, more secondary fiber is required

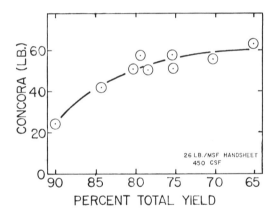

Fig. 124. Effect of yield on concora crash strength of NSSC pulps.

Table 21. Valley beater evaluation of NSSC corrugating pulp.

Beating Time min	Csf	Tear Factor	Breaking Length km	Stretch %	Burst Factor	Bulk	Air Res sec/ 100 mL	Concora Crush	Drainage Time s
0	660	46	2.5	1.0	11.0	2.49	1.0	21	4.5
15	480	65	5.2	1.7	25.0	2.12	4.0	52	5.5
23	315	79	6.1	1.9	36.0	1.88	21.0	67	7.5
31	150	97	7.9	2.9	49.0	1.70	231.0	80	21
37	70	105	8.7	3.2	58.0	1.52	—	78	132
32	150*	98	7.7	2.6	50.0	1.65	230	78	23

*Properties extracted at 150 Csf.
Substance of evaluation sheets 120 g/m².

than virgin kraft pulp to achieve the same improvement in stock drainage and wet web strength. The better the quality of the NSSC pulp, the less long fiber is required and the lower are the manufacturing costs. Top quality NSSC pulp does not need the addition of long fibers nor do low machine speeds, e.g. 304 m/min (1000 fpm) or less. One Finnish corrugating medium has for many years produced a good quality product at speeds of 396 to 488 m/min (1300 to 1600 fpm) without using long fiber. Major reasons for the ability to produce corrugating medium only from NSSC pulp are the high quality of the birch pulp and the availability of a vacuum pick-up on the machine.

Runnability of corrugating medium has been defined as the ability of medium to withstand the stresses and strains imposed by the single facing operation without rupture of flutes. Increased bonding results in higher elongation at rupture. Therefore, low yield pulps elongate more. Runnability depends also on tearing strength, which is again influenced by pulp yield, since in tearing, the work to rupture will depend partly on the number of fibers involved in rupture. Tensile strength depends mainly on bonding. The higher fiber flexibility at lower yields is conducive to more bonding which increases the tensile strength of paper.

Refining also affects favorably the various properties related to corrugator runnability. Thus, NSSC pulps with yields at the high end of the commercial range can give medium with acceptable runnability if refining is severe enough. However, this approach is energy intensive and also results in slow stock drainage and a less porous sheet. Finally, tearing and tensile strengths required for good corrugator runnability are also enhanced by the addition of long fibers to semichemical pulp.

At normal corrugator speeds and temperatures, increasing amounts of lignin due to pulping to higher yields cause incipient cracking and finally breakage of the flutes. This effect can be at least partially offset by raising the preheater and single-facer temperature as can be seen from Fig. 125. The explanation for this phenomenon is the fact that corrugating medium behaves as a thermoplastic material [13]. The strength of the flutes is highly dependent on the temperature of the corrugating operation. The thermoplastic behavior is most likely due to the ability of lignin and hemicelluloses to soften at elevated temperature in the presence of sufficient moisture, thus minimizing the damage suffered by the fibers due to the bending and tensile forces at the top of the flutes and shear forces at the shank. The thermosoftening components set on cooling and impart additional strength to the flutes.

The important function of corrugating medium in corrugated containers is to enhance compression strength. This end use property is affected favorably by the high lignin content of semichemical pulps, refining and machine parameters such as good sheet formation and absence of weak spots in the sheets, such as flaws caused by large shives.

E. KRAFT SEMICHEMICAL PULPING

Corrugating medium with good properties can also be produced from hardwoods pulped to approximately 70% yield by the kraft process. For some specialty end uses such as wax reinforcement, kraft corrugating medium is preferred over other types of medium because it does not become as brittle in the heat treatment involved in waxing.

Due to the relatively low yield and the need to process the pulping liquor through the entire kraft chemical cycle, kraft semichemical hardwood pulps are more expensive to produce than other types of semichemical pulps. Probably, for this reason, very little kraft semichemical hardwood pulp is being produced.

F. GREEN LIQUOR SEMICHEMICAL PULPING

Semichemical pulping of hardwoods with kraft green liquor for corrugating medium and similar products was adopted by the North

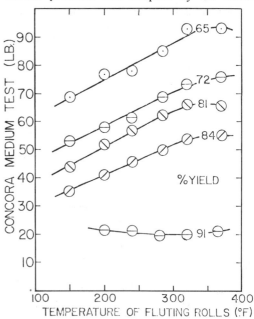

Fig. 125. Effect of fluting temperature on concora crush strength of NSSC pulp.

American industry in the mid seventies. The major advantage over the NSSC process and nonsulfur pulping is the ease of the chemical preparation and of chemical recovery in a mill, which produces both kraft pulp, e.g. for linerboard, and semichemical pulp. As a result, capital and operating costs may be reduced significantly.

The first trials documented in the literature in 1967 were conducted in Sweden [15] with birch and a mixture of beech and birch. Both laboratory and full-scale mill trials indicated no major differences in the important pulp properties between NSSC and green liquor pulps at a given yield. The major difference is the very dark color of green liquor pulp. However, this difference is of no significance in corrugating medium nor in the base sheet of duplex linerboard.

Extensive investigations conducted in North America [16-21] have essentially confirmed the Swedish conclusion that green liquor semichemical pulp can be equivalent in mechanical strength properties to NSSC pulp prepared from the same raw materials. In the manufacture of green liquor corrugating medium, a wetting agent has to be added to the machine furnish to meet the desired specifications on water drop

test, which is an indicator for the rate of glue absorption on the corrugator. Corrugator plant trials showed that green liquor corrugating medium is equal to NSSC medium in handling, runnability, and quality [17, 20].

Up to 10% green liquor semichemical hardwood pulp can be used in linerboard as a kraft pulp substitute without significant quality losses [24]. Due to the higher yield of the semichemical pulp (approximately 70% compared to 53% to 54%), raw material and chemical costs are reduced significantly by this substitution.

Regular kraft mill green liquor or synthetic green liquor prepared from purchased sodium carbonate and sodium sulfide are being used for pulping. Mill liquor should be well clarified because suspended material may interfere with liquor penetration of the chips. The chemical composition of typical kraft green liquor is given in Table 22.

A system for synthetic green liquor preparation is shown in Fig. 126. It also allows adjustments of the chemical composition of mill green liquors, e.g. addition of caustic soda or sodium carbonate. In a similar manner, sodium hydrosulfide or sodium sulfite can be added to kraft green liquor [16, 18] within the limits of the sulfur balance of the total mill.

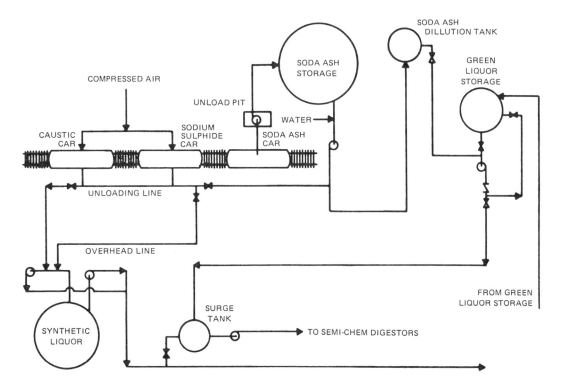

Fig. 126. System for preparation of green liquor from purchased chemicals [20].

Similar to NSSC pulping, relatively thin chips are desirable to facilitate liquor penetration, and the bark content should be as low as possible to minimize chemical requirements and product quality problems.

In principle, all commercial digester systems are suitable for green liquor pulping. In North America, the Bauer Messing-Derkee, Defibrator, Kamyr and Pandia digesters have been used successfully. In some mills, which converted from NSSC to green liquor pulping, digester pressure and temperature could be reduced significantly. This made possible steam savings.

No unusual corrosion problems have been observed in green liquor pulping [20]. Due to the higher pH level than common in NSSC pulping, the metallurgical requirements are less stringent, and are similar to those of the kraft process.

Spent liquor from green liquor pulping can be unstable. Precipitation of organic material can occur when temperature or pH are allowed to decrease significantly. The addition of sodium sulfite to green liquor [16] or sodium hydroxide to spent liquor for pH control [22] reduces or even eliminates the precipitation problem. In integrated kraft pulp, green liquor semichemical pulp mills, the weak spent green liquor is blended with kraft black liquor before evaporation. This procedure stabilizes the spent liquor and reduces scaling and heat transfer problems.

While the green liquor process can be applied successfully to hardwoods, softwoods do not respond well [16]. Even with large chemical applications, Kappa numbers are very high [23]. Mill trials failed to produce pulp with acceptable quality. Therefore, in green liquor semichemical pulp mills, contamination of hardwood with softwood is avoided as much as practically possible.

In 1980, approximately five United States corrugating medium mills were pulping with green liquor and three other companies had announced plans for green liquor semichemical pulping. A 660 ton/day mill that started operation in 1982 in the southern USA received considerable publicity in trade journals [51-54] due to its "state-of-the-art" equipment, operating procedures and integration with an existing linerboard kraft mill.

G. NONSULFUR SEMICHEMICAL PULPING

Due to concerns about pollution problems, especially air pollution with sulfur compounds such as sulfur dioxide emissions caused by NSSC pulp mills (and kraft odor caused by green liquor pulping due to its content of sulfide), and due to high capital and operating costs of chemical recovery systems for small stand-alone mills, considerable efforts have been made since the mid sixties to develop sulfur-free semichemical pulping processes. The efforts led to pulping with sodium carbonate only and with mixtures of sodium carbonate and sodium hydroxide.

1. Sodium carbonate semichemical pulping

In 1965, laboratory studies demonstrated that semichemical pulping of hardwoods with only sodium carbonate was feasible [27]. A Canadian patent [28] covers a specific range of process conditions. Mixed hardwoods were pulped with a sodium carbonate solution containing 65 to 120 g/L expressed as Na_2O at a temperature between 170° and 200°C to a pulp yield of about 70% to 85%. After separation of the spent liquor, the digested chips were converted to pulp by primary refining to a freeness not greater than 480 mL Cs. This was followed by secondary refining to a freeness of 180 to 350 mL Cs prior to feeding the pulp to the paper machine system. Corrugating medium with good flat crush (concora) and stiffness was apparently produced under these conditions. The patent also spells out the possibility of burning the spent liquor under oxidative conditions in a fluidized bed recovery furnace to recover sodium carbonate. An alternative to chemical recovery in a fluidized bed system would be the Zimpro wet combustion process [29], which has been used successfully in some nonsulfur pulp mills.

The Zimpro process involves aqueous phase oxidation of organic matter by molecular oxygen. Autothermal reaction takes place at an elevated temperature and pressure using air as

Table 22. Chemical composition of typical kraft green liquor.

Chemical	Concentration g/L
Sodium Carbonate	130
Sodium Sulfide	42
Sodium Hydroxide*	12
Sodium Sulfate	10
Total Sodium	92
Total Titratable Alkali as Na_2O	119
Sulfidity, % of Total Titratable Alkali	28
Density at 20°C	1.18

* Sodium hydroxide is introduced into green liquor with wash water from recausticizing used for dissolving the smelt from the recovery boiler.

the normal source of oxygen. The conventional evaporation, incineration and dissolving steps in the recovery circuit can be replaced by the Zimpro process which has proved to be clean and efficient. There is no fume loss and only negligible chemical loss. The recovery of thermal values as usable steam is apparently significantly higher than conventional practice. The major process steps and equipment layout of the Zimpro system operated at the Burnie mill of Associated Pulp and Paper Mills, Australia, which produced soda pulp, are shown in Fig. 127.

At least one North American corrugating medium mill is pulping with sodium carbonate.

2. Sodium carbonate-hydroxide semichemical pulping

Owens-Illinois, one of the leading North American manufacturers of corrugating medium, was granted a U.S. patent [30] which covers pulping of hardwood in a sulfur-free aqueous solution containing from about 15% to 50% of the total chemical as sodium hydroxide and 50% to 85% of the total chemical as sodium carbonate both expressed as sodium oxide. The chemical application based on wood is 3% to 10% by weight expressed as sodium oxide. The pulping cycle can be very short, for example, 4 min at a temperature of 191°C (376°F), or much longer, for example, heating to a temperature of 171°C (340°F) in 50 min and maintaining this temperature for 45 min. The resulting corrugating medium compares favorably in quality to that of NSSC medium.

In 1978, at least six Northern American corrugating medium mills were pulping hardwood with mixtures of sodium carbonate and sodium hydroxide [31]. They included two Owens-Illinois mills. A flow sheet of the Owens-Illinois mill, located at Tomahawk, Wisconsin, is shown in Fig. 128 [32]. Semichemical pulp is being produced in four six-tube 24″ diameter Pandia digesters. The total Na_2O on wood is about 4.5%, which is significantly less than the 6%-8% usually used for NSSC pulping. The pulping time is only approximately 5 min at

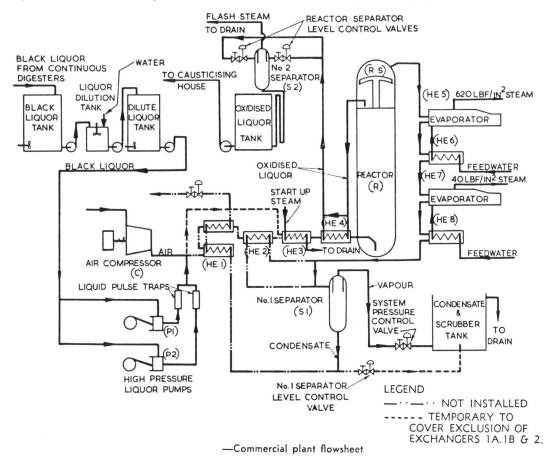

—Commercial plant flowsheet

Fig. 127. Zimpro wet combustion process.

190°C. The steam consumption is 5% to 10% less than for NSSC pulping in the same Pandia digesters. The yield in the low eighties is several percent higher than was the NSSC yield.

The pulp is discharged from the digesters through defibrators to the blow tank. Blow steam is used to heat wash water. The blown stock is given primary refining in Sutherland refiners and then goes to the washers. Two lines of 8' × 10' and one line of 9.5' × 12', three drum vacuum washers, are being used for 900 tons of pulp. The washed pulp is further refined in disc refiners, combined with about 15% recycled fibers prepared from corrugated clippings and then passed through tickler refiners to the paper machines. The total refining power is about 1.1 GJ/t (15 hpd/ton) of medium, about the same as that for the NSSC pulp produced previously.

The filtrate from the washers is evaporated in multiple effect evaporators, with forced circulation in the first effects. The strong liquor is then burned in a 150 ton kraft recovery furnace, followed by an electrostatic precipitator. The liquor solids have a heat of combustion of about 14 GJ/t (6000 Btu/lb), which is about 15% higher than the NSSC liquor solids. The smelt from the furnace as well as the precipitator catch are dissolved in the smelt dissolving tank to form a sodium carbonate green liquor. The quantity of dregs at this mill is very small, so that a clarifier and dregs washer are not required. However, dregs do settle in the liquor tank, which is cleaned out once or twice a year.

No new capital investment was required in the two Owens-Illinois mills for pulping chemical recovery. The mills already had washers, evaporators, and either a kraft type recovery furnace or smelters [33]. The recovered smelt consisting primarily of sodium carbonate is clarified and returned to the pulping system. The required sodium hydroxide is added as make-up chemical. Chemical recovery is 93% on the average so that sodium hydroxide presents 7% of the pulping liquor.

The Owens-Illinois nonsulfur pulping process also makes possible savings in capital expenditures, due to simplified chemical unloading and chemical recovery systems compared to the NSSC process.

With the removal of sulfur from the system, the fusion temperature of the inorganic combustion residue has increased. A higher fusion temperature has permitted operation at higher combustion chamber temperatures without slagging of screen tubes and heat transfer surfaces in kraft type furnaces. However, at the same time, it has made smelting more difficult. Initially, more problems with plugged smelt spouts occurred after conversion to nonsulfur pulping. This problem has been solved. A major factor in this improved operation, at the Owens-Illinois Tomahawk mill, is the maintenance of approximately 20 mol percent potassium carbonate in the smelt by occasional additions of KOH instead of NaOH make-up [34].

Papier Cascades (Cabano) Inc. at Cabano,

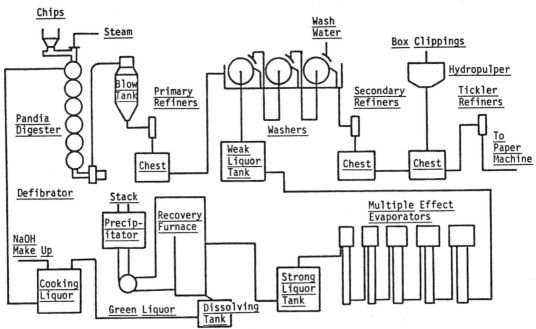

Fig. 128. Owens-Illinois Tomahawk mill pulp and liquor system.

Quebec [35] and Stone Container, Coshocton, Ohio [36] use Copeland fluidized bed reactors instead of recovery furnaces for combustion of the spent cooking liquor. The fluid bed product is recycled in a manner similar to the green liquor at the Owens-Illinois Tomahawk mill.

H. PULP YIELD DETERMINATION IN SEMICHEMICAL PULPING

Wood is one of the major cost items in the manufacture of semichemical pulp. Therefore, exact knowledge of the quantity of wood consumed and of the quantity of pulp produced is very important. Usually, the ratio of the oven dry weight of pulp to the oven dry weight of wood is referred to as pulp yield.

Temler and Bryce [37] recommended that pulp mills should consider two definitions of yield. The first is the total tonnage of usable pulp produced over an extended time period as percentage of the total oven dry wood consumed during this period for the manufacture of this pulp. The percentage should be called "mill yield". It is the average of many individual measurements. Wood losses incurred during barking, chip storage, screening and conveying and pulp losses, e.g. fiber losses to the sewer, should be included in the mill yield. Reliable data on mill yield are very important for profit maximization, mill and sales planning.

The second type of yield is that obtained in the pulp mill. It is referred to as "digester yield". Reliable data on digester yield are required for the evaluation of the effects of changes in wood supply to the pulp mill and in digester operating conditions.

In semichemical pulp mills, mainly two methods are being used to determine pulp yield. They are:
1. Gravimetric determination of the weight of wood used for pulping and of the weight of pulp produced during a given time period.
2. Determination of the amount of dissolved organic solids present in a sample representing the total product of digestion and subsequent mass balance calculation.

Very few semichemical pulp mills determine the residual lignin content of their pulp by an indirect method (e.g. Kappa number common in bleachable pulp mills [38]) as a measure of pulp yield. Obviously, laboratory data are needed to establish the relationship between pulp yield and lignin content measured by the indirect method for the wood furnish and simulated pulping conditions applied at the particular mill.

(a) Yield determination based on weights of wood and pulp

In most mills with continuous digesters, chips are weighed before being charged into a storage bin immediately ahead of the digester. In many mills with batch digesters, the chips are weighed either while being fed to the digester or to a storage bin. The number of digester fillings from the bin is estimated. However, this estimate can be a source of error.

A major uncertainty in determining dry chip weight is sampling and measurement of wood moisture content. Samples are best taken from the feed belts to the chip bins or to the digester. The frequency of sampling is dictated by the variations of chip moisture content. Wide variations experienced with chips from different sources require more frequent sampling than chips that originate from similar supplies, or that have been blended in pile storage. The samples are oven dried in the laboratory. Continuous moisture meters are available from several suppliers. However, for a variety of reasons, the meters have not found wide acceptance.

For measuring the weight of chips, weight-ometers are widely used. However, incorrect information may be obtained due to factors such as poor calibration, overloading or chip surges. Frequent calibration, use of overload alarm at the weightometer, and levelling out the chip flow, assist in obtaining more meaningful data on chip weight.

In semichemical pulp mills, the quantity of pulp produced may be measured as slush stock. Only in those few mills, where the finished product is manufactured only from semichemical pulp, the quantity of product off the machine can serve as a measure of pulp production if the fiber loss from the mill is known. However, most semichemical pulp is blended with other pulps, e.g. secondary fiber, in stock preparation. The ratio of these two components needs to be established accurately if the quantity of pulp produced is to be estimated from the finished product.

The quantity of pulp in slush form is calculated from pulp flow and consistency measurements. The accuracy of this method is limited by the ability to measure and control pulp flow rate and consistency. Flow meters and consistency measuring devices need to be calibrated frequently to ensure accurate results.

(b) Yield determination based on dissolved organic solids

This method [38] is based on the fact that the product of semichemical pulping is a mixture of the following components:
1. pulp suspended in the spent liquor,
2. organic wood components such as lignin and

hemicelluloses dissolved in the spent liquor and

3. chemicals from the pulping liquor, also dissolved in the spent liquor.

The pulp yield can be determined by measuring the percentage of each of the three components. To this end, the pulp is separated from the spent liquor, washed with hot water and oven dried. The weight of the two dissolved components is derived from the residue obtained after calcination of the spent liquor. The percentage of inorganic material can also be calculated from the sodium content of the sample.

Temler and Bryce [37] tested the accuracy of this method by determining the yield of a laboratory pulping experiment by this dissolved solids technique, and gravimetrically from the weight of oven dry wood used for pulping and the weight of pulp produced. The results differed from each other only by 1%. The higher value was obtained with the dissolved solids method. Temler and Bryce stated that the 1% difference is probably related to chemical changes that may occur during pulping. They include increase of weight of lignin left in pulp due to chemical modifications, formation and loss of volatile organic compounds and formation of water due to condensation reactions.

Sampling for this method has to be representative for the total digester charge. Sampling from the blow line of continuous digesters other than the Kamyr digester is very suitable. In the case of batch digesters, a close approximation of a representative sample can apparently be arrived at by volumetrically sampling both the digested chips and the spent liquor [38]. Much care would have to be taken when applying this procedure in a mill with a Kamyr digester due to removal of the spent liquor in the extraction zone of the digester and due to in-digester washing. Some of the important steps would include obtaining a spent liquor sample that is not diluted with wash liquor and a sample of digested chips that corresponds to the liquor sample.

According to Temler, Bryce and Robertson [37, 38], this method is simple, rapid, and accurate within ±1%.

(c) Yield determination based on residual lignin content of pulp

As already stated above, determinations of the residual lignin content of semichemical pulps using the Kappa number [39], hypochlorite number or chlorine number [40] tests are conducted only in very few semichemical pulp mills. The main reason is probably the fact that the reliability of these methods as a measure of

pulp yield tends to diminish in mill operation as the yield increases. Raw material variations may be a major factor.

The literature on this subject is rather limited. Only three publications [41-43] report linear relationships between NSSC pulp yield and Kappa number or chlorine number at high yield ranges, i.e. 70% to 85% [41] and 72% to 80% [43], for a given raw material. The error is smaller for the Kappa number (0.9%) than for the chlorine number (1.5%).

An extensive study with eucalyptus NSSC pulps cooked to approximately 60% to 75% yield in the laboratory under well-controlled conditions [44] led to the conclusion that the relationship between Kappa number and pulp yield is markedly influenced by wood properties and pulping conditions. A modification of the Kappa number test by including hot water extraction before the Kappa number determination improves the correlation with yield, but does not eliminate the effects of wood properties and pulping conditions. It was also found that the sulfated ash content of spent liquor gives a smooth plot against pulp yield for an individual wood sample, but samples of different basic density give different plots.

I. NEW DEVELOPMENTS IN NSSC PULPING

Anthraquinone, generally referred to as AQ, catalyzes the neutral sulfite delignification of softwoods [45]. The addition of 0.1% AQ based on the dry weight of wood accelerated pulping of *Pinus radiata* in laboratory experiments and a mill trial. For example, 0.1% AQ gave a pulp with a yield of 65.7% and Kappa number 80 after 1.5 h pulping at a temperature of 180°C. Without AQ, the same time-temperature conditions produced a pulp with 65.1% yield and Kappa number 113. At 600 mL Csf, the AQ pulp had 20% higher tearing strength, 8% higher breaking length and 18% higher bursting strength than the pulp prepared without AQ.

Note, however, that the pulps termed "neutral sulfite" in the two Finnish and Australian studies were cooked with sulfite-carbonate liquors of initial pH above 11, whereas the initial pH range for common neutral sulfite pulp is between 7 and 9.

Finnish researchers [46] reported that neutral sulfite pulping with AQ makes it possible to produce linerboard grade pulp from softwood with "17 to 22 percentage points" higher yield than conventional kraft pulping.

In contrast to the favorable response of softwood to AQ, hardwood NSSC pulping is enhanced only moderately or not at all. For example, AQ NSSC pulping of low density, young eucalyptus gave pulp with only marginally lower yield and Kappa number, and old dense eucalyptus did not respond to AQ [45].

NSSC pulps prepared from western softwoods with Kappa numbers of 140 to 170 and yields of 67% to 81% can be further delignified with oxygen under pressure and in alkaline medium [47, 48] similar to oxygen delignification of chemical pulps. In laboratory experiments, the oxygen-treated NSSC pulps were obtained with 10% to 17% higher yields based on wood than conventional kraft pulps with the same Kappa number and gave simulated duplex linerboard sheets with the same or better mullen, ring crush and brightness than comparable kraft pulps.

ACKNOWLEDGEMENTS

This chapter is an update of a similar chapter prepared for a previous edition of this textbook by G.H. Chidester, E.L. Keller and N. Sanyer. Therefore, information that is still considered valid in the previous edition has been utilized.

Several suppliers of pulp and paper mill equipment, e.g. Combustion Engineering Co., Ingersoll Rand, Kamyr Inc. and Sunds Defibrator, have contributed information to this updated edition. Their assistance is appreciated.

REFERENCES

1. U.S. Patent 1,673,089 June 12, 1928.
1A. Granehult, L., et al., *Pulp Paper International*, p. 80-82 (May 1981).
2. Branch, B.A., *Pulp Paper Mag. Can.* 72 (4):84-87 (Apr. 1971).
3. Keller, E.L., *Tappi* 33 (11):556-560 (Nov. 1950).
4. Porter, J.B., *Pulp and Paper*, p. 76-79 (Jun. 1981).
5. Reid, R.B. and Rodd, C.J., *Appita* 18 (4):139-150 (Jan. 1965).
6. Alsholm, O. and Petterson, B., Transactions of Symposium "Papermaking Systems and Their Control", Technical Section of the British Paper and Board Association, London, Vol. 1, p. 209-220 (1970).
7. Jayme, G. and Schwartzkopff, U., *Das Papier* 17 (12):697-702 (Dec. 1963).
8. Raij, U., Highly Automated Medium Manufacture at Billerud, preprint book of 1974 TAPPI Annual Meeting.
9. Ek, K.E., Proceedings of the 1968 Symposium on Recovery of Pulping Chemicals, Helsinki, p. 611-620.
10. Horntvedt, E. and Edmarks, S., *Pulp Paper*, p. 54-55 (Nov. 1977).
11. Cook, W.R., *Tappi* 57 (9):94-96 (1974); 60 (4):106-108 (1977).
12. Copenhaver, J.E., Briggs, W.A., Jr., Baxley, W.H. and Wise, J.T., U.S. Patent 2,714,118 (Jul. 26, 1955).

13. Gartaganis, P.A. and Ostrowksi, H.J., Variables Affecting the Converting Efficiency of Corrugated Combined Board and Their Measurement, paper presented at TAPPI 18th Testing Conference, Chicago, Illinois (Sept. 26-29, 1967).
14. Murphy, D.C., *Appita* 18 (2):69-77 (Sept. 1964).
15. Vardheim, S., *Paperi ja Puu* (9):613-619 (1967).
16. Worster, H.E. and McCandless, D.L., *Tappi* 57 (10):75-79 (Oct. 1974).
17. Dawson, R.L., *Tappi* 57 (12):113-116 (Dec. 1974).
18. Battan, R.R., Ahlquist, Q.J. and Snyder, E.J., *Tappi* 59 (6):130-133 (Jun. 1976).
19. Bublitz, W.J. and Fang, Y.P., *Tappi* 60 (4):90-93 (Apr. 1977).
20. Bellefortaine, W.F., Continental Forest Industries Experience with Green Liquor Pulping, paper presented at TAPPI Alkaline Pulping, Secondary Fiber Conference 1977, preprint book, p. 395-399.
21. Bublitz, W.J. and Hull, J.T., 1979 TAPPI Pulping Conference Proceedings, p. 89-95.
22. Ringley, M.B., U.S. Patent 3,811,995 (May 21, 1974).
23. Charbonnier, H.Y., Rushton, J.D. and Schwalbe, H.C., *Tappi* 57 (12):109-112 (Dec. 1974).

24. Worster, H.E. and Bartels, M.E., *South Pulp and Paper* 39 (2):32-36 (Feb. 1976).
25. Stein, R.B., *Tappi* 64 (7):71-75 (Jul. 1981).
26. Hammond, J.A. and Karter, E.M., U.S. Patent 4,073,678 (Feb. 14, 1978).
27. Winczakiewics, A. and Kasynska, J., *Papier, Carton, Cellulose* 14 (1):96-98 (1965).
28. Temler, J.S., Canadian Patent 1,042,159 (Nov. 14, 1978).
29. Morgan, J.E. and Saul, C.M., *Appita* 22 (3):60-75; and Gommi, J.V., *Pulp Paper Mag. Can.* 74 (5):20 (May 1973).
30. Dillard, B.M., Gilmer, R.J. and Kennedy, J.D., U.S. Patent 3,954,553.
31. Hanson, J.P., *Pulp Paper* 52 (3):116-127 (Mar. 1978).
32. Shick, P.E., Owens-Illinois non-sulfur pulping process. TAPPI Alkaline Pulping/Secondary Fibers Conf. Nov. 1977. Papers: 391-2.
33. Shick, P.E., *South Pulp and Paper* 40 (2):13-16 (Feb. 1977).
34. Fisher, W., U.S. Patent 3,873,413 (Mar. 25, 1975).
35. Micheals, W., *Can. Pulp Paper Ind.* (3):13-19 (Apr. 5, 1977).
36. Patrick, K.L., *Pulp & Paper* 53:119-123 (May 1979).
37. Temler, J., and Bryce, J.R.G., *Pulp & Paper Can.* 76 (2):92-95 (Feb. 1975).
38. Robertson, J.D., *Tappi* 43 (3):192A-193A (Mar. 1960).
39. Berzins, V. and Tasman, J.E., *Pulp Paper Mag. Can.* 58 (10):145 (1957).
40. Chlorine Number of Pulp, TAPPI Official Standard T202 OS-69.
41. Surewiz, W., and Mudrzejewski, K., *Przeglad Papier* 34 (9):347-348 (Sept. 1978).
42. Surewiz, W., Modrzejweski, K., Mroz, G., and Gizowski, M., *Przeglad Papier* 35 (1):38-39 (Jan. 1979).
43. Nedeva, S., Toshkova, B., and Genev, I., *Tseluloza Khartiya* 4 (2):5-8 (Mar./Apr. 1973).
44. Elloy, E.H.C., Nelson, P.F., and Smith, J.G., *Appita* 28 (2):21-28.
45. Cameron, D.W., Jessup, B., Nelson, P.F., Raverty, W.D., and Vanderhock, N., *Appita* 35 (4):307-315 (1982).
46. Kettunen, J., Pusa, R., and Virkola, N.E., TAPPI Pulp Conf. Preprints, p. 425, Atlanta (1980).
47. Worster, H.E., and Pudek, M.F., *Pulp Paper Mag. Can.* 75:T2-T6 (Jan. 1974).
48. Worster, H.E., and Pudek, M.F., *Tappi* 57 (3):138-141 (Mar. 1974).
49. Anon., Pine Hill investment pays off for MacMillan Bloedel, *Southern Pulp Paper* 47 (9):43-46 (Sept. 1984).
50. Smith, K.E., *Pulp Paper* 58 (9):118-140.
51. Johnson, K.A., *Paper Age* 100 (9):28-33, 42-46, 69 (Sept. 1984).
52. Koncel, J.A., *Paper Trade J.* 168 (15):44-51 (Sept. 1984).

VII

HIGH YIELD AND VERY HIGH YIELD PULPING

A.M. AYROUD

Dr. Ayroud received his B.Sc. and Chemical Engineering degrees from Lille University, and a Doctor of Engineering from Grenoble University in France. He began his business career at the French Papermaking School in Grenoble in 1953, as Head of the Pulping and Wood Chemistry Section, Assistant to the Director. In 1957, he came to Canada and worked as Chemical Engineer with the Pulp and Paper Research Institute of Canada. He joined Consolidated-Bathurst Inc. (then Consolidated Paper Corp.) in 1958 as Senior Research Chemist, and advanced through various levels to his present position as Director of Research. He has published many papers in the fields of wood chemistry, pulping, pollution and chemical analysis of pulps and liquors.

MANUSCRIPT REVIEWED BY
Anthony Binotto, Technical Assistant to the Mill Manager, Great Northern Paper Co., Millinocket, Maine

Nomenclature and definitions of pulping processes are constantly evolving, as modifications in the processes occur. In order to clarify the position of the high yield processes discussed in this chapter, the following guidelines will be used:

Semichemical — Methods which encompass the entire intermediate range of pulp yields between pure chemical and pure mechanical pulping, i.e. 55% to 92%, although yields of 70% to 85% is the accepted range to avoid confusion with other specific processes.

HYS — High yield sulfite, 60-80% yield.

VHY — Very high yield sulfite, 80-92% yield.

CMP — Chemimechanical pulping, involving chemical pretreatment of chips before atmospheric refining to produce a pulp with yields above 80%.

CTMP — Chemithermomechanical pulping, involving chemical additions prior to or during the pressure steaming of chips, followed by a first-stage pressurized refining, and subsequent refining above or at atmospheric pressure. Yields are in the plus 90% range.

Figures 129 and 130 show typical conditions to distinguish between CTMP and CMP.

This chapter will discuss softwood and hardwood High Yield Sulfite (HYS) & Bisulfite pulping in the 60-80% yield range, followed by Very High Yield (VHY) Sulfite/Bisulfite pulping in the 80-92% yield range.

Other related processes are covered elsewhere. Semichemical pulping, emphasizing Neutral Sulfite Semichemical (NSSC) and related processes for hardwoods, is covered in the previous chapter. CTMP pulping is discussed in the volume on Mechanical Pulping, and also in the very high yield section (D) of this chapter, as a member of the chemimechanical pulping family.

Interest in semichemical sulfite/bisulfite pulps for newsprint furnish was intensified in late 50's [1-3, 17]. This was due to increased consciousness in preserving forest resources, reducing wood costs and pollution loads per ton of pulp and to meet increased competition for improved product quality. The results of advances in improved disc refiner design and newly acquired knowledge in process application helped in the implementation of this challenging and growing technology.

At the early stage, conversion from low (48%) yield Ca-base acid sulfite to higher (60-65%) yield was practiced at some Canadian newsprint mills. However, the superior performance of soluble bases such as Na "Arbiso" process [4, 25-27] and Mg "Magnefite" process [5, 28, 80, 82, 86, 87] and to a much lesser extent, ammonium [18], introduced in the early 60's put a stop to the use of Ca base liquor for the production of high yield pulps. Since then, the subject was extensively covered in the literature [6-10, 15, 16].

Encouraging mill application of soluble base in high yield (60-75%) sulfite (HYS) pulp production for newsprint furnish started early in the 60's particularly in Eastern Canada [21, 25]. However, the discovery of thermomechanical pulp (TMP) and its modified forms in the early 70's increased the pressure on the sulfite industry to further reduce the pollution load. This spurred development to increase yields of HYS pulps to ultimate limits of 80%-92%. Properties were maintained closer to chemical than mechanical pulps.

A. HIGH YIELD SULFITE/BISULFITE PULPING (60% < YIELD < 80%) OF SOFTWOODS

1. Liquor preparation and characteristics (For details, refer to Chap. IV)

In this category, sodium and magnesium base liquors in the pH range of 3.5 to 5.0 (20°C) are commonly used for the production of the so called "HY Bisulfite Pulps". The risk of $MgSO_3$ precipitation at a pH of about 5.9 puts some restriction on the use of the magnesium base. This makes the more flexible Na base desirable for mill operation when higher pH liquors are required. Refer to Chapter I,C,D.

In both cases, the cooking liquors are prepared by the absorption of SO_2 (sulfur burner gas) in sodium hydroxide or carbonate solutions or magnesium hydroxide slurries. Relationships of pH/SO_2 content for both the Na and Mg bases were discussed in the earlier chapter, and are presented here in greater detail.

These relationships are extremely helpful not only for liquor preparation but have also been

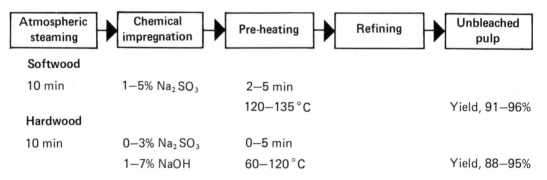

Softwood

10 min   1–5% Na_2SO_3 2–5 min
                                          120–135 °C   Yield, 91–96%

Hardwood

10 min   0–3% Na_2SO_3 0–5 min
                1–7% NaOH   60–120 °C   Yield, 88–95%

Fig. 129. CTMP, typical pulping conditions.

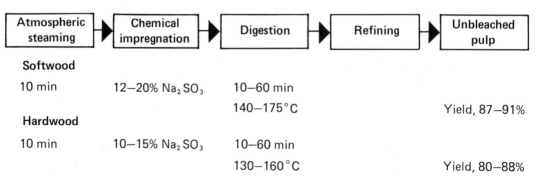

Softwood

10 min   12–20% Na_2SO_3 10–60 min
                                            140–175°C   Yield, 87–91%

Hardwood

10 min   10–15% Na_2SO_3 10–60 min
                                            130–160°C   Yield, 80–88%

Fig. 130. CMP, typical pulping conditions.

proven practical in mill operations for adjusting pH and concentration of the cooking liquor to the desired range, as explained below.

(a) Adjustment of cooking liquor pH

It is possible to calculate the amount and effect of SO_2 or NaOH addition on a sodium bisulfite solution. This can be done by using the relation between pH and the molar ratio of SO_2/base, as expressed in Figs. 131 and 132. Two simple equations have been developed to permit adjustment of the final SO_2 concentration (X_2); as well as the pH of the liquor by addition of SO_2 (Y) or NaOH as required.

Equation 1 allows the calculation of the change in liquor concentration resulting from a pH adjustment. The equation illustrates the relationship between pH and molar ratios of chemicals.

$$X_2 = X_1 \left(1 + \frac{M_2 - M_1}{M_2 M_1} \right) \qquad (1)$$

Where $X_1 =$ the initial liquor concentration as % total SO_2
$X_2 =$ the final liquor concentration as % total SO_2
$M_1 =$ the molar ratio SO_2/NaOH corresponding to the initial pH
$M_2 =$ the molar ratio SO_2/NaOH corresponding to the final pH

The second equation allows the calculation of the amount of SO_2 or NaOH (a negative value corresponds to the addition of base to neutralize an equivalent amount of SO_2) required to modify the system's molar ratio.

$$Y = \frac{QX_1}{100} \left(\frac{M_2 - M_1}{M_2 M_1} \right) \qquad (2)$$

Where Q = the total amount of liquor (in lb or kg) to be adjusted and

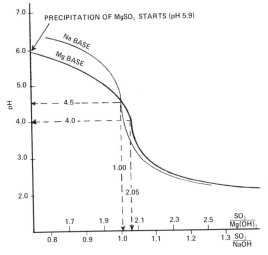

Fig. 131. Variation of pH of bisulfite liquors with SO_2/base ratio.

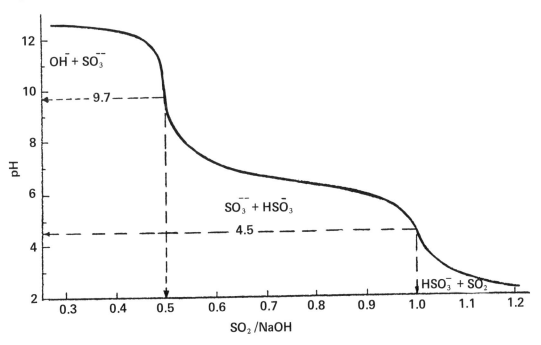

Fig. 132. Variation of pH with SO_2/NaOH molar ratio.

Y = the amount of SO_2 to be added or neutralized, in lb or kg.

To better illustrate these equations, the following examples are given:

EXAMPLE 1

The initial liquor existing in the storage tank (LST), Q = 450 000 lb has a pH of 5.5, and a concentration of 2.96% total SO_2.

It is required to adjust the pH to a value of 4.5 and have the total SO_2 concentration in the order of 3.1% (this corresponds to 5.0% $NaHSO_3$).

What amounts of SO_2 should be added to the LST in order to lower the pH to 4.5?

To solve the problem, it is necessary to know the initial and required molar ratio M_1 and M_2 of $SO_2/NaOH$. This can be obtained from Fig. 132 at pH 5.5 and 4.5. The values are:

$M_1 = 0.96$ at pH = 5.5
$M_2 = 1.00$ at pH = 4.5

By using Eq. 2, the amount of SO_2 (Y) to be added to the LST can be easily calculated as follows:

$$Y = \frac{450\,000 \times 2.96}{100} \left(\frac{1.00 - 0.96}{0.96} \right)$$
$$= 554 \text{ lb } SO_2$$

To calculate the final SO_2 concentration which will result from this addition, we may use Eq. 1.

$$X_2 = 2.96 (1 + 0.0416) = 3.08\% \ SO_2$$

Since this SO_2 concentration is almost exactly what is required, no concentration adjustment is necessary in this case.

EXAMPLE 2

The liquor in the LST is as follows:
Q   = 450 000 lb
x   = 3.12% SO_2
pH = 3.5

It is required to bring the pH up to 4.5. Using Eq. 2 once more, after obtaining $M_1 = 1.03$ and $M_2 = 1.00$ from Fig. 131, we can calculate a value for Y of -421.2 lb. Since this quantity is negative it means that 409 lb of SO_2 must be neutralized by an appropriate amount of NaOH. Since 1 mol of NaOH will react with 1 mol of SO_2 to produce 1 mol of sodium bisulfite:

$$421.2 \text{ lb } SO_2 \left[\frac{1 \text{ mol } SO_2}{64 \text{ lb } SO_2} \times \frac{1 \text{ mol NaOH}}{1 \text{ mol } SO_2} \right.$$
$$\left. \times \frac{40 \text{ lb NaOH}}{1 \text{ mol NaOH}} \right] = 263.2 \text{ lb NaOH}$$

Since a small volume of concentrated caustic solution would be added, in this case, to a very large volume of cooking liquor, the SO_2 concentration will remain within the required value of 3.12% and no further adjustments would be necessary.

In general, if the pH of the liquor in the LST is kept between 4 and 5, the volume changing resulting from such adjustments will be negligibly small.

(b) Adjustment of the liquor concentration

The examples were chosen such that, once the pH had been adjusted, the SO_2 concentration in the LST was in the acceptable range. Therefore, no further adjustments were necessary. In practice, adjustment of the liquor concentration might also be required. Two cases could result:

1. After adjustment of the pH to 4.5, the liquor concentration was higher than the required value. Here the amount of dilution water (W, in lb or kg) to be added to Q (lb or kg) of liquor can be calculated from following equation:

$$W, \text{ lb} = \frac{Q(c_1 - c_2)}{c_2} \qquad (3)$$

where c_1 is the initial liquor concentration, %
      c_2 is the final liquor concentration, %

EXAMPLE : Q = 450 000 lb
            c_1 = 3.20% Total SO_2
            c_2 = 3.08 Total SO_2

The amount of water to be added to the LST can be calculated from Eq. 3 as being 17 532 lb.

2. After adjustment of the pH to 4.5, the liquor concentration was lower than the required values. In this case, SO_2 and NaOH should be added in equimolecular quantities to bring up the liquor concentration to the right range. The amount of SO_2 and NaOH in pounds can be calculated as follows:

$$SO_2, \text{ lb} = \frac{Q(c_2 - c_1)}{100} \qquad (4)$$

$$NaOH, \text{ lb} = \frac{0.625(c_2 - c_1)}{100} \qquad (5)$$

EXAMPLE : Q = 450 000 lb
            c_1 = 2.58% Total SO_2
            c_2 = 3.08% Total SO_2

The amount of SO_2 to be absorbed can be calculated from Eq. 4 as being 2250 lb, and the amount of NaOH can be calculated from Eq. 5 as being 1406 lb.

(c) Liquor stability (refer for details to Chap. I, D, 5)

Bisulfite liquors are subject to reduced activity through autocatalytic decomposition of the bisulfite ions [31]. Such reactions involve the formation of thiosulfate in the presence of reducing agents such as wood sugars [12] and formic acid [13]. These are produced during cooking, according to the following reactions:

$$2HSO_3^- + 2C_6H_{12}O6 \rightarrow$$
bisulfite   sugar
ions

$$S_2O_3^= + 2C_6H_{12}O7 + H_2O$$
thiosulfate   aldonic   water
ions   acid

$$2HSO_3^- + 2HCOOH \rightarrow$$
formic acid

$$S_2O_3^= + 2CO_2 + H_2O$$

$$5S_2O_3^= + 4HSO_3^- + 6H^+ \rightarrow$$
$$6S_2O_3^= + 2SO_4^= + 8H^+ + H_2O$$

This latter reaction [14] illustrates that the formation of thiosulfate from bisulfite is autocatalytic. Once a certain amount is formed, there is further decomposition of the bisulfite that would normally be available for pulping. With increasing acidity of the cook, through depletion of the base, further decomposition of the thiosulfate into elemental sulfur and bisulfite results:

$$S_2O_3^= + H^+ \rightarrow HSO_3^- + S$$

The above indicate that special precautions will have to be taken to insure that the thiosulfate and other undesirable side reactions remain within acceptable limits when considering liquor recycling.

2. Digesters

Batch and continuous stainless steel digesters [88, 89] are both used and each offers certain advantages depending on application. In all cases it should be remembered that, under certain unfavorable operating conditions, formation of the very corrosive sulfuric acid could occur, especially in the upper space "steam phase" of the digester. This means that higher quality stainless steel, with a higher molybdenum content such as the 904L than is present in 316 or 317L for digester construction, would be recommended.

In the case of the tile-lined batch digesters, conversion from low yield to high yield sulfite pulping would necessitate some changes within the digester. For instance, the installation of high pressure nozzles to assist in blowing the semi-cooked chips with spent liquor will avoid chip hang-up.

See Chapter V for a discussion of Digester Pulping systems

3. Wood species (for details refer to Chap. IV, A)

Softwoods such as black and white spruce, balsam and other firs are the preferred species for producing HY bisulfite pulps for newsprint applications. However, pine such as jack pine and slash pine [11, 4, 19] can also be successfully pulped with the bisulfite liquor. Worldwide, many softwood and hardwood species are being converted to pulps of various yields by cooking with bisulfite-sulfite liquors.

4. Pulp production and effects of process parameters

The process consists of producing partially cooked chips with sulfite/bisulfite liquors, followed by disc refining, washing, screening, cleaning and thickening. Post refining is also used in special cases to minimize the pulp shive content.

Cooking of chips can be conducted in liquid phase or in vapor phase. This latter mode is faster, more economical, but also more critical for the control of pulping operation [80, 86].

Good chip penetration by the cooking liquor should be accomplished to insure uniformity of liquor concentration and penetration prior to reaching maximum cooking temperature [81, 85]. This results in evenly softened chips for better defibration by refining. Failing to achieve this results in heterogeneous pulps with higher rejects, higher sliver contents and lower strength properties.

In batch operations, the concentration of the liquor entering the digester depends on whether direct or indirect heating and chip packing are applied, as well as the desired yield range. In general, it varies between 3-4% of total SO_2 (when spent liquor recovery is not practiced) at a pH of 4-5 for pulp yields ranging from 65-75%. In most cases the liquor to wood ratios fluctuate between 4.5/1 to 6/1.

In the early development and application stages of the bisulfite process, information on softwood pulp properties did not conform to cooking parameters and yields. Difficulty in accurately determining yields of HY pulps, the lack of accurate refining energy data, and the freeness/drainage characteristics may have caused these discrepancies.

For this reason, an extensive study was undertaken [11] to develop a practical method for yield determination, to define process parameters for mill applications and their effects on the pulp quality. The main findings are discussed below.

(a) Yield determination

It is relatively simple to determine pulp yield in the lab. Accurate data on chip weight and moisture, and resulting HY pulp can be obtained. However, at the mill, such determination is very difficult. Measurements of chip weight and moisture entering the digester, as well as the weight of pulp resulting from the cook, are not exact. Generally, the yield is estimated based on tonnage of paper produced and the proportion of component pulps. A number of methods were investigated. A modified permanganate No. (Kj) developed originally by Jayme [20], and further simplified [11], was found very useful for mill application. This correlated well with lab

studies in the 60-75% range using reasonably constant chip (say spruce/balsam) supply. The correlation of Kj with pulp yield is shown in Fig. 133. This correlation was verified in mill cooks where special baskets containing known amounts of wood were suspended in the digester and resulting pulps recovered quantitatively for yield and Kj No. determination.

There are also other useful correlations for mill application, such as the one between pulp yield and the consistency of free-drained whole cooked chips after washing (Fig. 134), as well as between the pulp yield and sodium bisulfite consumed (Fig. 135). In all cases the amount of chemicals applied to the digester should exceed the amount consumed to protect the strength and optical properties of the resulting pulp. The amount of residual chemicals will be dependent on whether chemical recovery is practiced and to a lesser degree on the required pulp yield and brightness levels.

In HY bisulfite cooking as in LY pulping, the yield is related to the cooking time and temperature as illustrated in Fig. 136. The pulps resulted from pilot cooks where the liquor concentration, expressed in total SO_2, varied from 1.80 to 2.50% (measured when the digester temperature reached 100°C) and cooking tem-

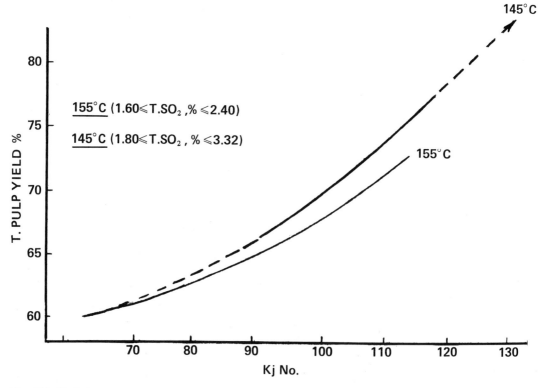

Fig. 133. Variation of pulp yield with Kj No.

peratures from 145 to 155°C. An equation to calculate pulp yield is as follows:

$$Y_{(t)} = A (10)^{-bt}$$

Where $Y_{(t)}$ = pulp yield at time, t
A = pulp yield when the cooking temp. reaches maximum
b = a constant, whose value depends on the temperature and initial liquor concentration. At fixed temperature and at a given liquor concentration, b can be calculated as follows:

$$b = \frac{\log A - \log Y_t}{\log t}$$

t = the cooking time (in hours) at maximum temperature. Results show that, at fixed liquor concentration, the pulping rate at 155°C is approximately double than at 145°C.

(b) Effects of liquor pH

The most commonly used pH range of cooking liquor for bisulfite pulping is from 3.5 to 5.5

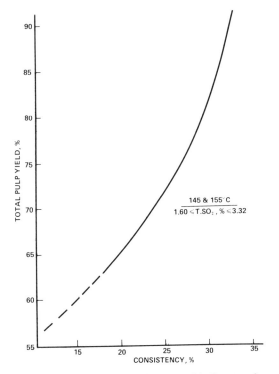

Fig. 134. Variation of pulp yield with the consistency of the free-drained cooked chips.

measured at room temperature. An increase in pH of bisulfite solutions from 4.3 to 5.4 doubles the cooking time of softwood for the production of similar yield pulps, as shown in Fig. 137. This means that an increase of one pH unit would have the same effect as decreasing the cooking temperature by 10°C, i.e. from 155 to 145°C

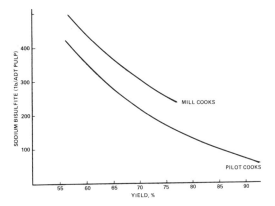

Fig. 135. Variation of sodium bisulfite consumption with pulp yield.

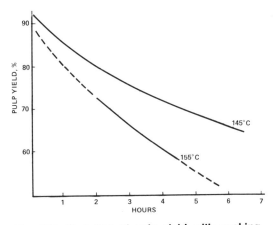

Fig. 136. Variation of pulp yield with cooking time.

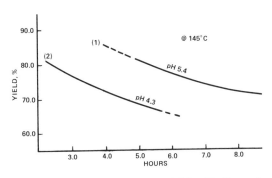

Fig. 137. Variation of pulp yield with time at maximum temperature.

(refer to Chap. III, A, 4).

Pulps produced at pH 5.4 contain less lignin than pulps produced at pH 4.3 at corresponding yields. Compared at the same refining energy and yield levels, the former pulps have lower rejects, bulk and freeness than the latter pulps. They exhibit comparable brightness but somewhat lower tearing and bursting strength at comparable breaking length.

It is worth remembering that the pulping cost at higher pH would be greater due to the increased consumption of applied base.

(c) Effects of cooking liquor concentration

The effect of chemical concentration on the rate of pulping is a well-known factor in the production of low- and medium-yield chemical pulps. This effect was not clearly defined for high and very high yield pulps.

In all cases, a minimum amount of chemicals should be left at the end of the cook to avoid degradation of pulp and lowering of its quality. An optimum amount of chemical should be used initially from an economical quality standpoint.

It was found [11] that harder pulps would result by reducing the initial concentration of the cooking liquor. For example, a 74% yield pulp produced with a cooking liquor concentration of 3.3% (expressed as total SO_2) had lower rejects than a pulp of corresponding yield produced with a lower liquor concentration of 2.4% at the same refining energy consumption. Compared at the same rejects and physical properties levels, the latter pulp would consume about 22% more refining energy than the former pulp. When the initial cooking liquor concentration drops further, say to about 1.6%, the difference in energy consumption would be still higher.

The pulp yield has a tendency to increase when the liquor concentration decreases (all other cooking variables are being kept constant). However, by a proper increase of cooking time, temperature, or both, a fixed pulp yield would result.

In general, initial liquor concentrations of 3.0 to 4.0% total SO_2 are used in commercial batch operations with direct steam heating and where spent sulfite liquor recovery is not practiced.

(d) Effects of wood species

Spruce, balsam and jack pine (main Eastern Canadian softwood species) and a wood mix were cooked under the same bisulfite cooking conditions (20% $NaHSO_3$ based on wood, L/W = 5/1, 3.75 h at 155°C). Figure 138a-f illustrates the effect of refining energy on pulp properties.

Comparison of refined pulps derived from different species at similar yields (63 ± 1%) showed the following:

1. The lignin content of the species are as follows:
   Spruce < mixed wood < balsam < jack pine
2. At the same degree of refining, the physical characteristics of the species are as follows:

Rejects:	mixed wood < balsam < spruce < jack pine
Burst factor:	jack pine < balsam < mixed wood < spruce
Tear factor:	balsam < spruce < mixed wood < jack pine
Strength factor:	jack pine < balsam < mixed wood < spruce
Opacity:	spruce < mixed wood < balsam < jack pine

It should be added here that the brightness was higher for spruce and lowest for jack pine. This latter pulp was yellowish and relatively sensitive to light, which explains its high opacity.

(e) Effects of cooking temperature

Although the cooking temperature has a strong effect on the pulping rate, which practically doubles for each 10°C increase in temperature, its effects on the pulp quality is much less noticeable, at least between 145°C and 155°C, as practiced by most HYS mills.

It was found that 73% yield pulps prepared at 155°C and at 145°C have similar physical characteristics at comparable refining energy input. Similar findings were obtained when 66% yield pulps were prepared at 155°C and at 145°C.

It can be concluded that variations of the cooking temperature, within practical limits, do not affect the pulp strength significantly, when pulps are compared at similar yields and refining energy levels.

If proper penetration of the cooking liquor prior to reaching maximum temperature cannot be achieved, pulping at lower cooking temperature would be recommended. This should avoid significantly differential gradients in chemical concentration and heat that would prevail within the chips at higher temperatures, resulting in heterogeneous pulps.

(f) Effects of single- or multi-stage refining

It is generally considered necessary that the refining of high yield chip be carried out progressively, that is, the amount of required

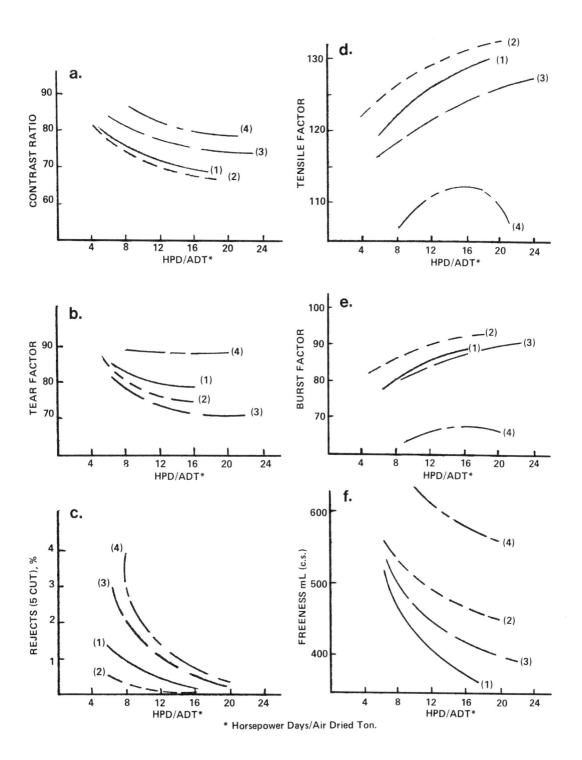

Fig. 138. Variation of pulp properties with refining energy. (1) Mixed Wood; (2) Spruce; (3) Balsam; (4) Jack pine.

energy should be distributed in multi-passes rather than a single pass. This practice is widely used in the industry.

It was not clear whether multi-stage refining, which is more elaborate than the one-stage operation, improves the physical characteristics of pulp. To establish this, refining of 65 to 75% yield bisulfite pulps prepared at 155°C and at 145°C in one, two and three stages was carried out at different refining energies [11].

The physical characteristics of the refined pulp were compared to the refining energy. The following is a summary of the results.

At constant refining energy, no significant differences in the amount of rejects, freeness, bulk, bursting, tearing and tensile strengths are apparent. However, the preferred mill practice is based on refining the high yield chips in two stages, with an added inter-stage washing. This enables the mill to have greater flexibility in refining; and improved washing operation to eliminate the residual amounts of corrosive spent liquor from the pulp.

(g) Effects of recycling spent liquors

Recycling of the spent liquor still containing active chemicals would be desirable from the economic standpoint of heat and chemical consumption. The presence of deleterious substances such as thiosulfate, sulfur, and sulfate must be kept within tolerable limits (see Chap. I, D, 5).

Trials were made [11] aimed at the production of 75% yield pulp. About 18% and 25% of sodium bisulfite (on O.D. wood basis) were used in the first and second series respectively. The spent liquor from the first cook was drained from the digester before blowing. The cooking liquor for the following cook was prepared by mixing fresh liquor with spent liquor. The pH of the cooking liquor was then adjusted to the same initial value (about 4) with caustic. Initial liquor concentration was kept constant by adding the required amounts of sodium metabisulphite. The mixed liquor was formed with approximately equal parts of spent liquor (similar to the kraft cooking liquor). The spent liquors were used twice and three times in the first and second series, respectively. In all cases, pulping proceeded by maintaining the other cooking variables constant. The resulting blown chips were refined under the same conditions at different energy levels.

Properties of the refined pulps from the two trial series were plotted against refining energy. A summary of the results follows:
1. The pulp yield increases with repetitive recycling of spent liquor. This results, when

compared at the same refining energy, in increased rejects (5 cut) and generally higher freeness of the accepted stock.
2. The physical strengths of the pulps were not significantly affected by recycling of the spent liquors.
3. The brightness of the pulps was adversely affected, about one point loss in brightness for each recycle.

It would appear that savings in cooking chemicals and steam can be realized by recycling the spent liquor. However, some adjustment of replenishing cooking chemicals and refining conditions should be made to compensate for the slight increase in pulp yield. This will produce an end product of comparable quality.

(h) Effects of hot refining

Cold water at 15°C was generally added to the refiner to keep the refined stock consistency close to 10%. It was of interest to see whether hot refining in the presence of hot water or hot spent liquor had any effect on pulp quality and refining energy consumption.

Studies were carried out using hot water at about 75 and 60°C and hot spent liquor at about 57°C. The resulting refined pulps were tested for the quantity of rejects and physical properties.

It was found that, at comparable energy levels, no significant differences could be noticed in the physical characteristics of pulp whether refining was done in the presence of hot spent liquor or hot water. This implies that washing may not be necessary before refining; the refiners could then be located in the blow line system, with washing at a later point.

(i) Effects of refining energy on rejects level

With increasing expenditure of refining energy, the amount of screen rejects (expressed as % of pulp weight) drop sharply at first to the approximate 1% level (5 cut screen:0.127 mm = 0.005 in slits) and then decrease very slowly. From an economic standpoint, this means that at an optimum refining energy (depending on the desired pulp yield, quality and cleanliness), rejects should be eliminated from the refined stock by screening rather than by further refining.

There are strong indications from mill experience that the higher yield (75% +) pulps if properly refined and processed contain less shives and debris than lower yield (70%) pulps.

Figure 139 represents variations of refining energy with pulp yield at two practical rejects levels (i.e. 1 & 3%).

Refer to Volume 2 on Mechanical Pulping for illustrations and discussions of refiners.

5. Comparison of Na-base pulps at different yield

Familiarity with mill facilities and operator habits are important in facilitating the implementation of development studies into mill practice. In HY pulping, variations in cooking (e.g. pH, liquor concentration, etc. . .) refining (e.g. energy applied, type of refiner, disc pattern, etc. . .) and processing (e.g. screening and cleaning) have significant influences on pulp characteristics even when they are produced at similar yields. These are some of the reasons for conflicting information and lack of specific data on pulp treatment found in the literature.

A practical way to treat and present test data for proper comparison of HY pulps, proven successful in mill application, is highlighted below.

For illustration, let us consider the variation of the tear factor with refining energy and pulp yield. For this purpose, sodium bisulfite cooks of spruce and balsam of known yield levels were produced. The cooked chips were refined individually with different plate settings (i.e. different energy input to reduce their

freeness levels) and then tested for their physical and optical properties.

Tear factor vs refining energy plots can thus be made for each pulp yield as shown in Fig. 140a. By interpolating these constant yield curves at fixed energy levels, it is then possible to plot the tear factor vs yield at constant energy levels as shown in Fig. 140b. By selecting the necessary energy levels needed to obtain pulps of acceptable quality for use in newsprint, one can define areas of practical application (shaded areas). It is then possible to plot, by joining the centre of these shaded areas, an "optimum" curve representing the tear factor variation as related to pulp yield; this is shown as the heavy curve in Fig. 140b. Similar plots can be made for other pulp properties as shown on Figs. 141 and 142.

Practical experience has shown that, within the 60-80% yield range, pulps of 60-67% yield are more sensitive to mechanical friction in processing (possibly related to lignin removal from secondary wall and hemicellulose distribution in outer layer), and as a result subject to large variations in drainage and properties. Despite their very good strength levels, they are more expensive to produce and less desirable from an operating point of view.

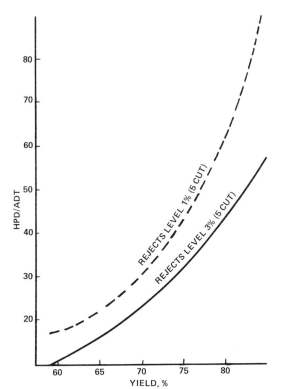

Fig. 139. Variation of refining energy with pulp yield at a fixed rejects level.

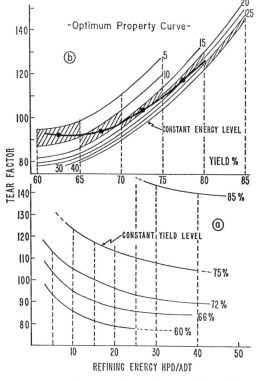

Fig. 140. Variation of tear factor with refining energy and yield.

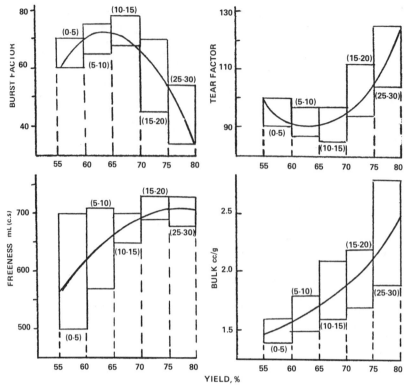

Fig. 141. Variation of pulp physical characteristics with yield [Numbers between brackets indicate applied refining energy levels (HPD/ADT)].

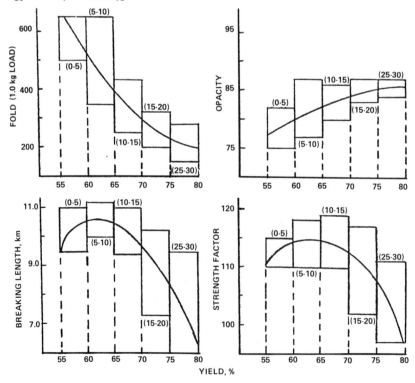

Fig. 142. Variation of pulp physical characteristics with yield [Numbers between brackets indicate applied refining energy levels (HPD/ADT)].

6. Comparison between HY pulps resulting from Na, Mg and NH₄ based cooking liquors

The recovery system for Mg and NH_4 based liquors are better established, cheaper to install and easier to operate than for the Na base. Before selecting any base for a specific mill, pulps resulting from these bases must be compared by production, processing and quality (as related to end application) standpoints.

To this end, Mg, Na and NH_4 base HY bisulfite pulps were prepared under similar cooking and refining conditions and were evaluated as previously described [11]. Plots of pertinent properties against yield are shown in Figs. 143, 144 and 145. It can be seen that, while a parallelism exists between certain pulp characteristics, the much lower brightness (10+ points) of the unbleached NH_4 base pulps, their higher lignin content and heterogeneity at higher yields (i.e. higher Kj No., and rejects) and their lower tensile strengths made them least attractive for use in newsprint furnishes. These results were also confirmed by other research findings [4, 18, 33]. For this reason, much of the industry's efforts were devoted to comparing the two competitive Na and Mg based processes. Typical data from pilot cooks comparing Na and Mg

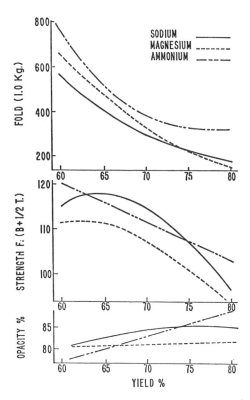

Fig. 144. Variation of fold, strength factor and opacity with yield.

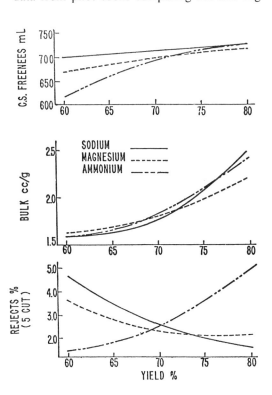

Fig. 143. Variation of freeness, bulk and rejects with yield.

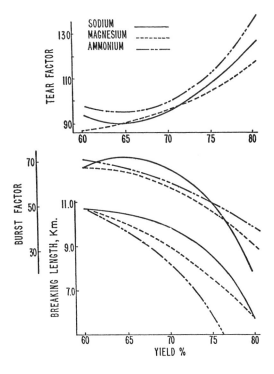

Fig. 145. Variation of tear factor, burst factor and breaking length with yield.

based pulps at 60-75% yield are illustrated in Table 23; Fig. 146 represents depletion curves of SO_2 during cooking for both bases.

Extensive mill scale trials to compare production and operation with Na and Mg based bisulfite pulps in newsprint manufacturing were conducted in a number of installations throughout the world. The following is an illustration of one mill's experience [11].

Port Alfred, a Division of Consolidated-Bathurst Inc., in Quebec, selected the Na base after experimenting extensively with both bases, by converting its HY pulping operation from Na to Mg and back to Na base bisulfite cooking. The pulps were used as the chemical furnish with groundwood for newsprint production. During these periods, variations in wood (spruce & balsam) density and moisture, stone ground-

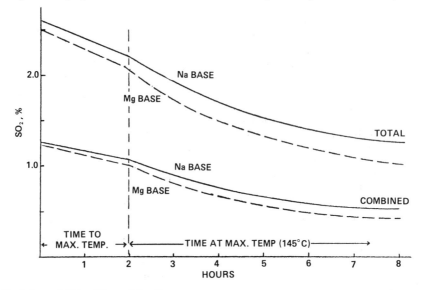

Fig. 146. Variation of SO₂ with cooking time.

Table 23. Comparison between sodium and magnesium bisulphite cooking data.

Base			Na	Mg	Na	Mg	Na	Mg	Na	Mg
Total Yield, %			[64]	64.0	62.5	64.0	68.5	67.0	74.1	72.5	73.1
Total SO₂, % [on O.D. Wood]			12.5	11.3	12.2	10.9	12.6	12.4	12.5	9.9	12.3
Liquor/Wood Ratio			5/1	4.5/1	5/1	4.5/1	5/1	5/1	5/1	4.5/1	5/1
Liquor Conc. SO₂, %	Total	Start*	2.40	2.74	2.44	2.76	2.51	2.52	2.60	2.48	2.60
		End	1.22	1.08	0.52	0.78	1.32	1.10	1.46	1.14	1.36
	Free	Start*	1.20	1.34	1.22	1.38	1.18	1.26	1.32	1.24	1.32
		End	0.70	0.66	0.56	0.52	0.74	0.64	0.78	0.64	0.72
	Comb.	Start*	1.20	1.40	1.22	1.58	1.14	1.26	1.28	1.24	1.28
		End	0.52	0.42	0.36	0.26	0.58	0.46	0.68	0.50	0.64
pH		Start*	4.1	4.0	4.3	4.4	4.1	4.2	4.1	4.3	4.4
		End	3.5	3.4	3.3	3.3	3.4	3.3	3.4	3.3	3.3
Time at Max. Temp. h			6.0	3.0	6.0	3.0	5.0	5.0	4.0	4.5	4.0
Max. Temp. °C			145	155	145	155	145	145	145	145	145
Kj Nos.			85	80	74	76	95	85	105	107	99
Drained chip consistency %			22.7	18.8	20.0	17.5	24.8	23.9	25.8	26.9	25.7

*NOTE: The cooking liquor was first added to the presteamed wood in the digester, then liquor circulated with indirect heating. When the temperature reached 100°C, the liquor was sampled and analyzed after cooling.

wood and HY pulp production and quality, as well as paper machine operating parameters and newsprint properties, were thoroughly monitored. Highlights of findings [22] are:

1. Liquor preparation with Na base was much easier, requiring less attention for controlling the concentration, pH and minimizing SO_2 losses through the absorption tower. With Mg base, it was found necessary to operate the absorption tower at pH below 5 to avoid the risk of serious problems with $MgSO_3$ precipitation, as confirmed by others [24].

2. Under comparable pulping conditions (liq. conc. 3% T. SO_2, pH 4.2, time 9.0 vs 8.8 h at max. temp. for Na and Mg), the Mg base liquors cook faster and produce an average pulp yield of 2.3% lower than with Na base (e.g. 70.2 vs 72.5%). At equivalent refining energy consumption of 1.6 GJ/t (20 HPD/ADT), they produced higher screening rejects (6.8 vs 4.5%), lower production of prime pulp, and more dirt at the decker. This was also recently confirmed by another mill [23].

3. The physical characteristics of both Mg and Na base pulps were generally competitive except that freeness, tear and coarse fibre fraction + 14 of the Mg base pulp were somewhat lower (Table 24).

4. Screening and cleaning operations with the Mg base pulp were sensitive and required much greater attention. This pulp behaved similarly to HY Ca-base acid sulfite pulp, i.e. it gave higher fibre stiffness. Similar findings were also reported by others [23].

5. Paper machine (Fourdrinier) operations at speeds of 1500 fpm (457 m/min) and higher were sensitive to Mg base pulp, requiring frequent adjustment of paper machine variables, particularly the draw. The sheet was softer; more breaks and lower paper machine efficiency were seen.

6. Indications were that variations in groundwood quality would have caused less trouble on the paper machine if Na base pulp had been used.

7. News sheets containing the Mg base pulp showed generally comparable strengths at equal percentages of HY pulp furnish, except for some 1-2 pts lower C.D. tear (Table 25).

8. At paper machine speed of 1600+ fpm (488 m/min), it was necessary to add 1 to 2% more Mg base pulp in the news furnish than with the Na base.

Considering the above findings, it was found more economical at that time, and for that specific mill, to adopt Na rather than Mg base.

B. HIGH YIELD SULFITE/BISULFITE PULPS FROM HARDWOOD

Hardwood species, such as white and yellow birch and poplar, can be pulped by bisulfite/sulfite liquors in the pH range of 3.5-6.0. Relatively light colored pulps are produced suitable for newsprint production, when blended in small proportion with HY softwood bisulfite pulps. Maple can also be used but, because of the lower brightness and strength of the resulting pulp, application is generally limited to corrugating medium.

Similar to low yield kraft, sulfite pulping shows significant differences in the cooking rates between and within softwood and hardwood species due to differences in chemical composition, density and morphological structures.

Table 24. Characteristics of HY bisulphite pulps. Average results of "regulator" samples.

Base	Sodium*	Magnesium Trial	Sodium**
Freeness, C.S. (mL)	704	682	699
Rejects (5 cut), %	0.07	0.16	0.13
Bulk cc/g	2.08	1.98	2.07
Burst Factor (B)	49.6	50.1	50.4
Tear Factor (T)	106	95	100
Strength Factor (B + 1/2T)	102.6	97.6	100.4
Breaking Length, (km)	8.10	7.95	8.30
Stretch, %	2.05	1.93	2.00
Fold (1.0 kg) MIT	322	326	336
Opacity (B & L) %	86.8	87.6	87.1
Fibre Classification			
+ 14	49.1	46.1	51.7
+ 30	28.7	28.9	25.6
+ 50	11.6	12.1	11.0
+100	5.7	5.8	5.6
−100	4.9	7.1	6.1

* Sodium; week preceding magnesium trial.
** Sodium; week following magnesium trial.

Refer to the previous section for discussions of liquor preparation and pulping softwood.

The following illustrates the production and properties of HY poplar and white birch pulps produced under simulated mill conditions [11]:

Cooking conditions (Pilot digester equipped with forced circulation and indirect heating used)
$NaHSO_3$ = 16-20% (O.D. wood basis)
Liquor to wood ratio = 4.4-4.8
Cooking temperature = 155 and 145°C
Time to temperature = 2 h
Time at temperature = 1 to 4 h

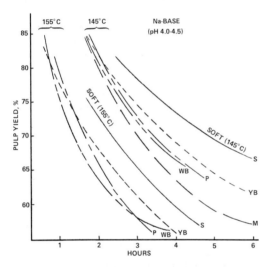

Fig. 147. Variation of pulp yield with cooking time for various wood species. W. Birch (WB); Poplar (P); Maple (M); Y. Birch (YB); Softwood (S).

After blowing, the chips were thoroughly washed then refined at different levels and the resulting pulps evaluated for their properties. Highlights of findings follow.

1. Pulp yield and chemical consumption

The cooking rates of hardwood species are not identical. They are dependent, for constant pulping conditions, on the wood density and lignin content.

Figure 147 is a comparative plot illustrating the variations of pulp yield with the cooking time at 145 and 155°C maximum cooking temperatures for yellow and white birch, poplar and maple. Also shown is a similar correlation for softwood (spruce + balsam).

For the same amounts of applied bisulfite, softwoods consume more chemicals at a given yield than hardwoods (birch and poplar consume practically the same amount) but the difference decreases with increasing pulp yield as illustrated below.

Average Sodium Bisulfite Consumption (on O.D. Wood)

Yield based on O.D. Wood %	Softwood (spruce & balsam) % A	Hardwood (poplar or birch) % B	Difference $\dfrac{A-B}{B} \times 100$
55	14.0	11.7	19.7
60	11.8	10.1	16.8
65	10.0	8.7	14.9
70	8.2	7.4	10.8
75	6.8	6.3	7.9
80	5.5	5.2	5.8

Table 25. Physical characteristics of newsprint sheets containing Na and Mg Base for high yield bisulphite pulps.

Paper machine Speed, fpm		1500			1550			1600	
Base		Na*	Mg.	Na**	Na*	Mg.	Na**	Mg.	Na**
Bisulphite pulp content, %		18.8	19.0	19.7	18.8	20.7	19.7	21.5	20.7
Basis Weight (lb/ream)		32.6	32.8	32.1	32.2	32.1	33.0	31.6	32.3
Caliper (1/1000")		3.44	3.42	3.30	3.54	3.43	3.41	3.38	3.44
Bulk (cc/g)		1.80	1.79	1.76	1.88	1.83	1.77	1.83	1.82
Burst Factor		10.9	12.8	12.4	11.6	10.9	12.4	12.3	11.2
Tear Factor	M.D.	36.2	38.1	37.7	37.9	36.0	35.8	36.8	38.3
	C.D.	49.4	53.1	54.5	52.0	49.0	52.0	52.4	53.3
Breaking Length,	M.D.	4490	4890	4700	4550	4190	4860	4345	4583
m	C.D.	1705	1960	1863	1760	1640	2000	1630	1865
Stretch %	M.D.	1.03	0.92	1.06	1.18	0.91	1.01	0.95	0.92
	C.D.	2.08	1.95	1.72	2.34	1.91	1.66	1.81	1.83
Fold (0.5 kg) MIT	M.D.	142	375	288	201	222	242	132	269
	C.D.	29	51	28	27	17	35	22	35
Opacity (Bausch & Lomb) %		92.3	92.4	92.8	92.5	92.1	92.1	91.7	92.0
Porosity (Gurley SPS)		39	69	38	56	43	44	40	37
Smoothness (Gurley SPS)		77	80	89	104	97	63	78	50

*Week preceding Mg base trial.
**Week following Mg base trial.

2. Physical characteristics

The variations of some properties of bisulfite pulps prepared from white, yellow birch, poplar and maple at various yield levels and different applied refining energies (i.e. 0.4, 0.8, 1.2, 1.4 & 2.4 GJ/O.D.T. at corresponding 60, 65, 70, 75 & 80% yields) are shown in Figs. 148 and 149. These properties are compared to those of softwood (spruce and balsam) bisulfite pulps.

It can be seen that the softwood pulps in the 60-80% yield range exhibit the highest strengths at the highest freeness level while maple pulp shows the lowest freeness and strengths with highest bulk in the 65-75% yield range.

It is worth mentioning that pure HY poplar bisulfite pulps become increasingly yellowish when exposed to air. The discoloration increases in presence of metal ions and also when in contact with bronze wire. This discoloration is more intense with higher yield pulps. The brightness of white birch is comparable to that of softwood pulps. For example, at 65% yield white birch unbleached pulp brightness can reach the 60-62% level.

3. Brite-chem type pulps

Cooking liquors with a pH lower than 5.5 are preferred for the production of HY softwood bisulfite pulps. Hardwood pulping at higher pH, up to 6.5, using steam phase technique was found effective and successful in mill application. Such type of pulps, called "Brite Chem" [90], are generally produced in the 75-80% yield range. They are used for displacement of low yield softwood pulps in newsprint furnish or other applications [11].

Table 26 illustrates some typical cooking conditions for Eastern Canadian hardwood species and a wood mix as well as the corresponding pulping results. Pertinent properties are shown on Figs. 150, 151 and 152.

Results show that:

1. Maple and beech pulp exhibit comparable physical properties. Their bulk and opacity are the highest but their strengths are the lowest as compared to the other pulps.

2. White and yellow birch and poplar produced strong pulps. Maple and beech in the wood mix reduce the strengths but improve the opacity and bulk.

3. In all cases, increased refining energy application resulted in decreasing pulp freeness, higher bursting, tensile, folding strengths, and lower bulk, but little effect on opacity.

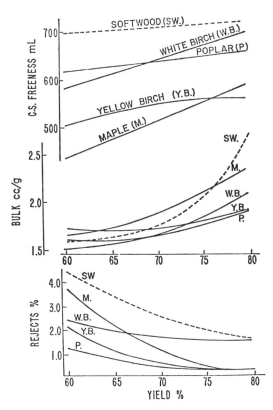

Fig. 148. Variation of freeness, bulk and rejects with yield.

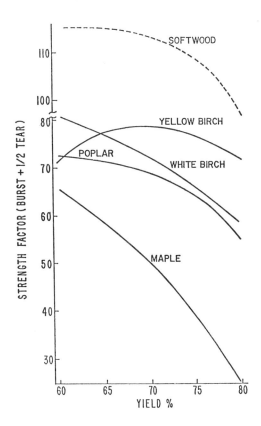

Fig. 149. Variation of strength factor with yield.

C. PRESENT AND POTENTIAL AREAS FOR APPLICATION OF HIGH YIELD SULFITE PULPS [11]

In general, major efforts were devoted to the use of HY softwood sulfite pulps, in the newsprint field. The use of hardwood HY pulp was restricted to a smaller proportion in the furnish. However, there are other areas of application which can be considered based on the properties of these pulps. A review of some of the potential areas and results of some pertinent mill trials will be highlighted.

Table 26. Laboratory pulping conditions for producing "Brite-Chem" type pulps.

RAW MATERIAL (Consolidated wood)		White Birch	Yellow Birch	Poplar	Maple	Beech	Mixture*
IMPREGNATION				MILL CHIPS			
Steaming time (a. pressure) min		10	10	10	10	10	10
Compression ratio		4.01:1	5.01:1	6.01:1	3.4:1	4.2:1	4.4:1
Compressing force (on chips) kg/cm²		140	140	140	140	140	140
Expansion time (a. pressure) min		10	10	10	10	10	10
Pressure impregnation time min		5	5	5	5	5	5
IMPREGNATION LIQUOR							
Chemicals as Na₂O g/L		62.3	61.4	62.3	61.5	61.5	61.4
pH of impregnation liquor		6.5	6.5	6.5	6.5	6.5	6.5
COOKING							
Steam pressure (absolute)	kg/cm²	8.0	8.0	8.0	8.0	8.0	8.0
Steam temperature	°C	170	170	170	170	170	170
Cooking time	min	30	30	30	30	30	30
Defibration time	min	0.5	0.5	0.5	0.5	0.5	0.5
pH after defibration		5.2	5.2	5.8	5.1	5.1	5.2
Yield after defibration	%	84.2	82.3	84.8	80.0	81.3	80.6
ABSORPTION OF CHEMICALS ON BONE DRY RAW MATERIAL (as Na₂O)	%	5.6	6.0	7.7	5.0	5.8	5.9
ABSORPTION OF CHEMICALS ON BONE DRY PULP (as Na₂O)	%	6.7	7.3	9.1	6.3	7.1	7.3

*(20% White birch, 10% yellow birch, 40% maple, 15% poplar, 15% beech)

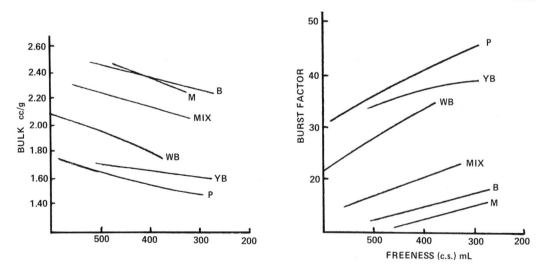

Fig. 150. Variation of hardwood pulp properties with freeness. W. Birch (WB); Y. Birch (YB); Poplar (P); Maple (M); Beech (B); Hwd. mixture (Mix).

1. Use of high yield hardwood pulps in various newsprint grades

Controlled trials were carried out to displace softwood LY chemical by HY birch sulfite pulps in the manufacture of standard, offset and superstandard sheets. Results showed that HY birch pulp can displace all the softwood LY sulfite pulp from a newsprint furnish when run on paper machine with speed not exceeding 305 m/min (1000 fpm). On the higher speed machines only partial replacement of the sulfite or kraft pulp can take place. In this case, the approximate displacement ratio found was up to two parts of HY birch to one part of chemical pulp. The minimum amount of softwood pulp required appeared to depend on the type of pulp (i.e. sulfite or kraft); paper machine (i.e. speed, open draw or pick up, double or single wires); and end application. Some typical results from mill trials are given below:

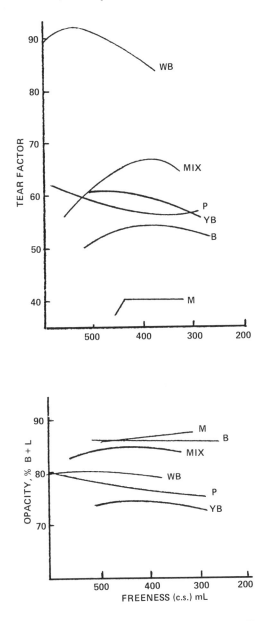

Fig. 151. Variation of hardwood pulp properties with freeness. W. Birch (WB); Y. Birch (YB); Poplar (P); Maple (M); Beech (B); Hwd. Mixture (Mix).

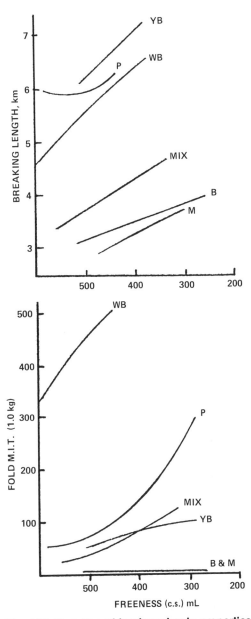

Fig. 152. Variation of hardwood pulp properties with freeness. W. Birch (WB); Y. Birch (YB); Poplar (P); Maple (M); Beech (B); Hwd. Mixture (Mix).

Type of newsprint	STANDARD			
Fourdrinier machine				
Speed (fpm)	845		2350	
	control	trial	control	trial
Groundwood, %	92	92	79	75
Swd. Sulfite, %	8	0	21	15
Swd. Semi-bl. Kraft, %	0	0	0	0
H.Y. Birch, Pulp, %	0	8	0	10
Trial duration, h		7		5

Type of newsprint	OFFSET			
Fourdrinier machine				
Speed (fpm)	845		1450	
	control	trial	control	trial
Groundwood, %	85	85	85	73-82
Swd. Sulfite, %	15	0	0	0
Swd. Semi-bl. Kraft, %	0	0	15	15-6
HY Birch, Pulp, %	0	15	0	12
Trial duration, h		7		12

Type of newsprint	SUPER STANDARD	
Fourdrinier machine		
Speed (fpm)	1295	
	control	trial
Groundwood, %	80	75
Swd. Sulfite, %	20	10
Swd. Semi-bl. Kraft, %	0	0
HY Birch, Pulp, %	0	15
Trial duration, h		5

In all of the above trials, no detectable differences were found between the printing characteristics and pressroom performance of both the control and trial sheets.

2. Use of high yield hardwood pulp in board manufacturing

High yield birch sulfite/bisulfite type pulps also offer potential applications in board manufacture. For example: a mill trial was conducted

for 4 h on a board machine running at 39 m/min (128 fpm), and producing 4 ton/h of non-folding board made from the following furnishes:

PULP FURNISH	CONTROL	TRIAL
Groundwood, %	42	42
Tab cards, %	14	14
Softwood sulfite pulp, %	44	22
HY birch pulp, %	0	22

The birch pulp was used in the top and bottom liner. No significant differences in machine operations were observed. The trial "0.040 caliper" non-folding board was found competitive in sheet curl, strength, stiffness, scoring and glueing properties to the control board.

3. Use of high yield hardwood pulp in kraft papers

A number of mill trials were conducted to assess the feasibility of displacing softwood kraft pulp by HY birch pulp in a variety of kraft paper grades. Conditions of these trials and furnish composition are illustrated below:

PAPER GRADE	MULTIWALL SACK		TOWELLING	
Paper Machine	F O U R D R I N I E R			
Speed (fpm)	1300		1500	
Trial duration, h	3		3	
Furnish components	control	trial	control	trial
Swd. unb. kraft	100	85	—	—
Swd. semi-bl. kraft	—	—	75	60
HY birch pulp, %	0	15	—	15
Groundwood, %	—	—	25	25

The resulting papers were found comparable in characteristics at similar paper machine efficiencies.

Since the use of HY hardwood pulps in sack papers offers great potential in savings, further successful mill trials were conducted with an available wood mix rather than with pure birch. The wood mix composition and pulping conditions were as follows: pulping of "50% birch, 25% poplar and 25% maple" mix was conducted for 0.5 h at 760 kPa (110 psi) in a steam phase continuous digester with a cooking liquor at a pH of 6.2 and 6.5 Na_2O (applied on wood).

The resulting HY hardwood pulp (after refining and processing) was used to displace various amounts of kraft softwood pulp in the manufacture of multiwall sack paper. Different runs on a commercial scale were made with the following trial papers.

40 lb kraft sheet containing 20% HY sulfite HWD pulp

50 lb kraft sheet containing 20% HY sulfite HWD pulp

40 lb kraft sheet containing 30% HY sulfite
  HWD pulp
50 lb kraft sheet containing 30% HY sulfite
  HWD pulp

Test results showed that, in the manufacture of multiwall bag paper, partial (20 and 30%) replacement of softwood kraft pulp by high yield hardwood pulp caused a reduction of the tearing and, to a lesser degree, folding strength of the sheets. However, the burst, tensile, stretch, and particularly the tensile energy absorption (TEA), indicative of the performance of sack paper, remained practically comparable to those of a pure kraft sheet. The hardwood containing sheet also showed superior formation. With these encouraging test results, conversion of the hardwood pulp containing sheet into 87½ lb cement bags for field trials (one ply of 50# internally coated with P.E. and 3 plies of 40#) was carried out, and results compared with regular all softwood kraft bags.

A field trial showed that no significant differences in performance, i.e. percentage of breakage, was observed. It was concluded that the superior formation imparted by the hardwood pulp accounted for the good performance, even though the paper strength was not improved. Thus at least 20% of HY hardwood pulp can be substituted for the softwood unbleached kraft pulp in the manufacture of multiwall bag paper with resulting equal or superior performance.

4. Use of high yield hardwood in absorbent grades

Trials were conducted on a mill scale to determine the feasibility of using hardwood HY sulfite pulps in towelling, tissue, and napkins. Findings from commercial evaluations showed that displacement of some 25 + % of kraft pulp by the corresponding amount of HY hardwood pulp is possible.

In all the above applications, the final justifications of the use of HY hardwood pulp produced on site or purchased, e.g. flash dried form, will have to be based on the economics specific to each mill location.

5. Use of high yield softwood pulps in linerboard

A mill trial was conducted on a linerboard machine running at 283 m/min (930 fpm) and producing 600 t/d of 42# liner to confirm laboratory findings. A base stock furnish containing up to 50% of softwood HYS and 50% kraft pulp was used. Linerboard with competitive strength properties to a pure kraft sheet was

produced. In the control (pure kraft stock) and trial (50:50 blend), the top liner component was the same but brightness of the base sheet containing HY sulfite pulp was 4 points higher. With the HYS containing furnish, no significant differences in stock behaviour, properties and paper machine operation were observed. The performance of the boxes, made out of both liners, was also found competitive. These findings indicate that replacement of kraft by HYS pulps for the manufacture of board has great potential if it can be economically justified.

D. VERY HIGH YIELD SULFITE/ BISULFITE PULPS (80% - 92% YIELD)

1. Introduction

As mentioned earlier, development and successful mill application of 60-75% yield sulfite/bisulfite pulps from soluble base cooking liquors started late in the 50's and received acceptance in the early 60's. At that time, HY pulp quality was found satisfactory to meet mill requirements and product performance. In the early 70's with increasingly restrictive Government pollution regulations, concerning particularly the BOD loadings of sulfite mill effluents, mills operating on the sodium base faced a number of alternatives (assuming there was economic justification to continue their operation):

1. Discontinuation of sulfite pulping and replacement of the pulp by a stronger mechanical pulp such as TMP and/or by the more expensive semi-bleached kraft pulp.
2. Installation of an expensive recovery system to operate on heat-deficient sodium base HY spent liquors.
3. Reduction of BOD loading by other methods, such as the two-stage biodegradation technique, to a level conforming to Government specifications.
4. Modifications in the HY pulping technique to increase the yield to such a level that the generated spent liquors do not contain more of the BOD loadings than regulated.

From the above, alternative No. 4 was chosen by some sulfite mills to minimize capital expenditures and operating costs, extend wood supply, and reduce the BOD loading to meet Government regulations.

The production of the ultra high yield (approximately 80 - 90%) pulp requires particular attention in cooking, refining and processing. As with all pulping processes, thorough penetration of the cooking chemicals prior to reaching maximum temperature should be obtained. This is particularly important in ultra high yield pulping, due to the very short cooking times generally used in either liquid or steam phase techniques. For this reason, one of the approaches adopted is to use shredded chips. Refining of the partially cooked chips will have to be carried out at high consistencies (20 + %). Sufficient energy must be applied to effectively separate the fibres and develop enough fractures and fibrillation of the secondary wall to improve flexibility and bonding ability. Particular attention will also have to be directed to proper pulp screening and cleaning.

One of the most important factors in developing fiber bonding in very high yield pulps is to insure a certain degree of lignin sulfonation. This renders the fibres more hydrophilic and permanently softens the lignin [34, 35]. When the level of sulfonate groups reaches 1.2% on O.D. wood, softening of the middle lamellae occurs leading to a complete fibre separation after refining. On the other hand, when more sulfonate groups are introduced to a level between 1.2-2.0%, softening of the fibre wall takes place and the long stiff fibres become more flexible, leading to increased fibre bonding.

Refining conditions also play an important role in achieving interfibre bonding. Refining, and pressing to some extent, creates slip planes and compression failures making the stiff fibres more conformable to each other. The interaction of energy, water, and surface hemicelluloses in the refiner increases the bonded area and bonding strength. See Volume 7 on Refining and Beating for further information on refining mechanisms and effects; and Volume 2 for information on mechanical pulping refiners.

Only small amounts of chemicals are added when mild sulfonation is required such as for modified TMP type pulps. Where stronger and longer fibred pulps are required, larger amounts of chemicals and stricter treatment conditions are necessary. The following section will not cover very mild treatments which will be studied with mechanical pulp.

Most of the softwood treatments are conducted with bisulfite/sulfite liquors in either the liquid or vapor phases. Short cooking times (30-45 min), pH 4-9, and temperatures varying from 140 to 160°C, produce VHY pulps in 80-92% yield range.

In general, sulfonation is carried out either as a pre-treatment cooking of whole or subdivided chips [32, 36, 41], as an interstage treatment of coarse pulp or screen and cleaner rejects, or as a post treatment of latent pulps at relatively high consistencies [42-44].

In order to tie together laboratory data and mill trial results, specific processes will be discussed to illustrate the effect of pulping variables on sheet properties. Only the following recently developed processes, planned or implemented in Eastern Canadian mills for newsprint operation, will be covered:
1) VHY Sulfite Pulping "VHYSP" Processes
2) Sulfonated Chemimechanical Pulping "SCMP" Processes
3) Thermochemimechanical Pulping "TCMP" Processes
4) "OPCO" Process

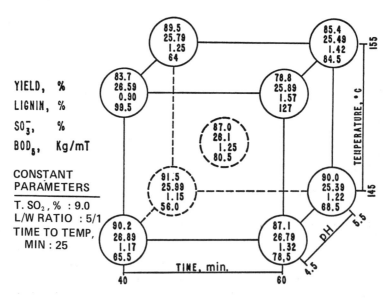

Fig. 153. Effect of time, temperature, and pH on pulp properties.

PFI REFINING

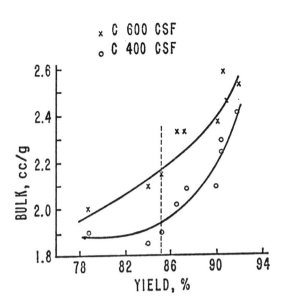
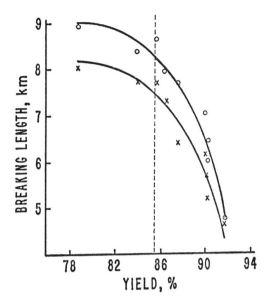

Fig. 154. Effect of yield on pulp properties.

2. Very high yield sulfite pulping process [22, 40, 46, 47] (Consolidated Bathurst, Inc.)

This process is based on maximizing pulp yield to 85%; strong clean and bright fibres are produced at half the TMP refining energy requirements. The resulting pulp can be used in newsprint furnishes with stone groundwood in equivalent or slightly higher amounts to low yield or high yield sulfite pulps. Comparable stock performance on the paper machine and sheet quality satisfactory to pressrooms are maintained.

Production of VHY pulp is also accompanied by the reduction of BOD loadings of the sulfite plant. This process would also require minimum capital outlays when considering conversion of existing HYS operation.

(a) Process and operating parameters

A statistical approach was adopted to obtain pulps at 80-90% yield from softwood (60:40 spruce/balsam) allowing definition of the interactions between cooking parameters for VHY pulping. Cooking variables (pH, time, temperature and SO_2 concentration) were selected within practical limits of existing HY pulping operations and experiences.

Figure 153 shows typical responses for yield, lignin, sulfonate content and BOD_5. Any response inside the cube can be calculated by

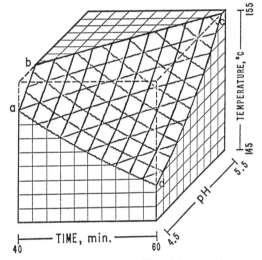

Fig. 155. Effect of time, pH and temperature on pulp yield.

regression analysis equations.

To determine optimum yield, two pertinent physical properties were measured at two freeness levels (600 & 400 CSF); results are shown on Fig. 154. It can be seen that pulp properties are significantly improved at lower freeness and drastic changes in bulk and breaking length curves occur at yields above 85%. The large variation of pulp properties seen for a modest increase in yield will result in difficulties in controlling pulp quality. This is why it was considered that a yield of 85% would be optimum, combining wood savings with pulp quality acceptable for news operations.

Another important factor in the production of very high yield pulps on a mill scale is the interchangeability between cooking parameters (i.e. pH, time and temperature) without affecting the pulp yield. An illustration of this is shown on Fig. 155. The cut "abcd" surface on the three-dimensional plot of the cooking parameters (i.e. pH, time, temp.) shows the relatively wide range of changes which could be made to produce a fixed yield pulp.

Properties of the 85% laboratory produced sulfite pulp are shown in Table 27. The following findings are of practical significance for producing VHY pulps.

Effects of SO_2 consumed

Figure 156 shows that, when more SO_2 was consumed, more lignin was dissolved while the sulfonate content of the pulp increased to approximately 1.4% where a levelling off occurred. In all cases, the amount of lignin removed from the fibres remained relatively

Table 27. Effects of cooking variables on 85% yield pulp characteristics.

| | COOKING VARIABLES | | | | PULP CHARACTERISTICS | | | | |
pH	TIME @ Max.Temp. min	TEMP. Max. °C	LIGNIN Content %	$SO_3=$ Content %	SO_2 Consumed %	DEBRIS (6 cut) %	BULK cc/g	BREAKING Length km	TEAR mN·m²/g
4.5	40	153	26.65	0.95	3.84	0.94	2.12	7.5	88.6
4.5	46	151	26.57	1.15	3.89	0.90	2.16	7.1	89.0
4.5	58	148	26.54	1.38	4.02	0.94	2.19	7.0	89.5
4.6	42	153	26.51	1.03	3.90	0.89	2.13	7.4	88.2
4.6	48	150	26.41	1.28	4.01	0.87	2.18	7.0	88.6
4.6	60	148	26.41	1.38	4.09	0.92	2.20	7.0	89.2
4.7	54	150	26.28	1.33	4.07	0.84	2.18	7.0	88.2
4.7	58	149	26.26	1.38	4.11	0.86	2.20	7.0	88.5
4.8	42	155	26.29	1.05	3.99	0.82	2.11	7.5	87.5
4.8	44	154	26.26	1.15	4.00	0.81	2.13	7.4	87.3
4.9	52	152	26.06	1.30	4.13	0.76	2.17	7.2	86.6
5.0	52	153	25.95	1.30	4.17	0.73	2.16	7.3	85.8
5.0	58	151	25.89	1.38	4.23	0.75	2.19	7.2	86.4
5.1	50	155	25.88	1.30	4.17	0.68	2.13	7.5	85.0
5.1	52	154	25.86	1.33	4.19	0.69	2.14	7.4	85.0
5.3	58	154	25.62	1.40	4.32	0.64	2.13	7.6	82.6

Constant: T·SO_2 % = 9 (O.D. Wood basis)   Time to Max. Temp. min = 25
Liquor/wood ratio = 5/1

85 % YIELD

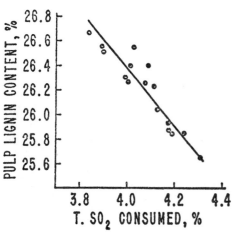
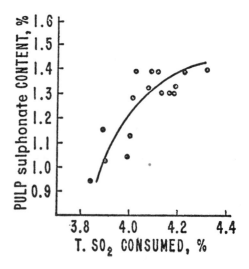

Fig. 156. Effects of SO_2 consumed on pulp properties.

small.

Effects of pH

Within the limits of time and temperature variations used in this study, a modest trend towards linear delignification appeared when the pH was raised from 4.5 to 5.5 as shown on Fig. 157. This was also reflected by variations in the amounts of rejects.

Effects of time-temperature

Figure 158 shows that, at a fixed temperature, the sulfonate content of the pulp increases with decreasing pulp yield (longer cooking time). At 85% yield the highest sulfonate content would be obtained at 145°C rather than 155°C. Despite this difference, no significant changes in pulp properties were noticed (refer to Table 23). Similar findings were recently reported [48].

Effects of chip dimension

To reduce chip dimensions, shredding (using a double disc refiner equipped with spike tooth plates) and pressing (using an Impressafiner, Model 560) techniques were used on a pilot scale. The three types of chips (i.e. original, shredded and pressed) were digested in a rapid cycle digester. The resulting VHY chips were refined in a double disc refiner. Results are shown in Fig. 159.

At any freeness level between 400-700 mL the original full size chips consumed substantially

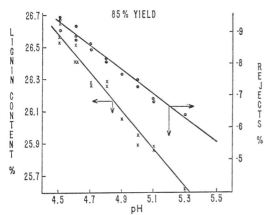

Fig. 157. Effects of pH on pulp properties.

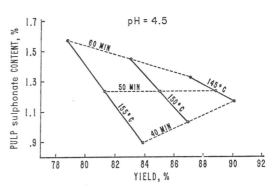

Fig. 158. Effect of time and temperature on pulp properties.

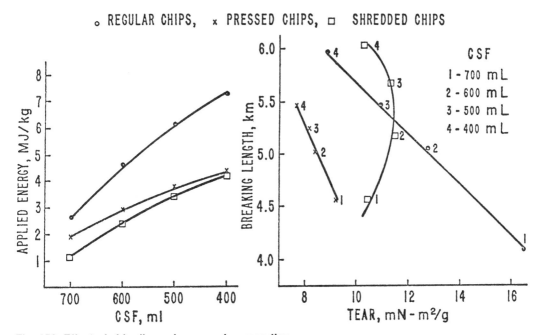

Fig. 159. Effect of chip dimension on pulp properties.

more refining energy than the other type of chips. The difference in energy consumption between the pressed and shredded chips was small particularly in the low freeness range. This can be explained by the fact that pressed or shredded chips are more uniformly cooked than regular chips, since short retention times do not always allow for thorough penetration of larger chips. It is also known that uniform chip size improves refiner stability, reducing refining energy, and resulting in improved product quality.

Figure 159 also shows that the original chips produced a better quality pulp than that obtained from pressed chips. On the other hand, if superior pulp is sought, chip shredding appears advantageous since pulps with higher tensile and tearing resistance are obtained at below 550 CSF.

(b) Newsprint furnish characteristics

Synthetic blends of mill-produced groundwood, low yield and high yield bisulfite pulps and a pilot plant produced VHY bisulfite pulp were prepared and resulting hand sheet properties compared to determine the replacement value of the VHY pulp in newsprint furnish. Results are given in Table 28.

Findings show that, by increasing the pulp yield, a significant reduction in pulp freeness (through more intense refining) would be necessary to produce acceptable strength pulp. Despite this drop in freeness no ill effect on pulp drainage is noticeable. This means that such high yield pulp would be well suited for fast speed paper machines.

Although the characteristics (bulk and strength) of the various yield sulfite pulps differ to a certain degree, their mixtures with groundwood pulp in the ratio of 20-30: 80-70 (sulfite: gwd) would produce newsprint sheets with competitive properties.

(c) VHY pulp fibre morphology

An analysis of the fibre morphology was undertaken using a scanning electron microscope. The following slides show differences between low yield bisulfite fiber (Fig. 160), and VHY bisulfite fiber (Fig. 161) pulps. In spite of the VHY fibers being bulkier, thicker walled, stiffer, less fibrillated and less collapsible (higher lignin content than the lower yield fibers), it was found that their bonding ability was sufficiently high for the production of acceptable shive free papers.

(d) Comparison between high yield and very high yield bisulfite pulps: mill scale

Mill trials were carried out in an existing HY sulfite plant by temporary conversion into VHY pulping to confirm laboratory findings. Conditions of the trial, pulp properties and characteristics of the resulting newsprint are shown in Tables 29 and 30.

Mill-produced VHY pulp was bulkier and showed lower burst and tensile strengths but higher tear and opacity than the HY pulp. These differences in quality did not result in observable detrimental effects on paper machine operations or on the quality of the newsprint sheet and its pressroom performance. The findings led to Consolidated-Bathurst's decision to convert its Laurentide Division (Grand'Mère, Québec) from regular LY to VHY bisulfite pulping. This VHYSP installation, which started operation early in 1982, is the first one of its kind.

Table 28. Effects of VHY pulp blends on newsprint furnish characteristics.

| | | SULFITE | | | MIXTURE : SULFITE / GWD | | | | | | | | |
| | | 50% | 70% | 85% | | 20/80 | | | 25/75 | | | 30/70 | |
	GWD	LYS	HYS	VHYS	LYS	HYS	VHYS	LYS	HYS	VHYS	LYS	HYS	VHYS
CSF, mL	140	648	558	513	223	178	178	244	191	200	252	186	198
Drainage time, s	14.7	4.4	4.4	4.3	9.5	12.5	10.4	9.1	10.9	9.5	8.6	12.4	8.6
Bulk, cc/g	2.60	1.68	1.86	2.13	2.58	2.67	2.77	2.61	2.66	2.68	2.47	2.62	2.60
Burst, kPa·m^2/g	1.7	4.5	5.6	4.7	2.2	2.1	2.2	2.3	2.2	2.3	2.5	2.3	2.6
Tear, mN·m^2/g	5.4	8.4	9.3	10.4	7.3	7.4	7.2	7.1	6.9	7.9	7.2	7.1	8.2
Breaking Length, km	3.76	6.88	8.60	7.31	4.38	4.28	4.32	4.49	4.31	4.35	4.75	4.62	4.76
Stretch, %	1.64	1.98	1.99	2.49	1.70	1.77	1.82	1.73	1.70	1.70	1.75	1.77	1.90
Smoothness, mL/min	243	191	374	420	324	322	352	296	291	289	258	321	328

CLASSIFICATION									
+ 14	5.0	49.6	49.8	47.9					
+ 28	17.3	18.3	10.3	14.5			TYPICAL VALUES		
+ 48	20.4	12.3	15.5	18.7	PULP		OPACITY (%)		BRIGHTNESS (%)
+ 100	20.6	10.4	8.4	11.3	GWD		97.0		58.0
+ 200	10.0	1.6	1.9	2.6	LYS		84.7		53.0
− 200	25.8	7.8	14.1	5.0	HYS		86.0		50.6
					VHYS		89.0		55.0

(e) Description of the Laurentide mill installation

The plant was designed for the production of 210 t/d of softwood pulp (spruce & balsam) at 85% yield and a freeness of 500 CS. The installed refining power is 5.5 GJ/t including rejects refining. The process flowsheet is shown in Fig. 162.

The chips are washed, shredded and blown into a small steamed bin. From the bin, two metering screws equipped with sodium sulfite spray nozzles feed the chips to two Bauer M & D continuous digesters. The sodium sulfite reduces the corrosiveness of the atmosphere in the gaseous phase of the digesters. Each digester has a capacity of 75% of the total required production.

After cooking and blowing, the stock is drained and conveyed to the 1st stage atmospheric double disc refiners. The resulting semi-refined stock is then press-washed to remove spent liquor prior to second-stage

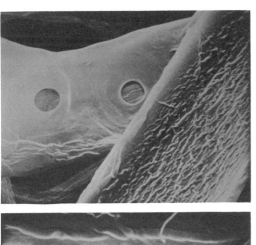
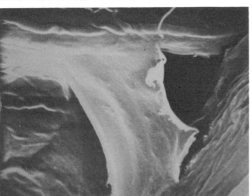
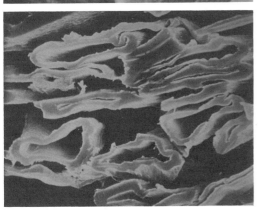

Fig. 160. Scanning electron micrographs of low yield bisulfite pulp fibre. Cross-section of "pulp" sheet.

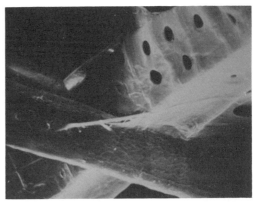
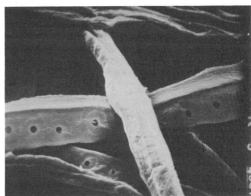
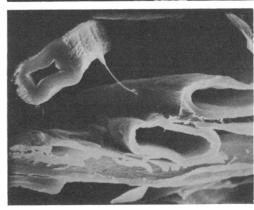

Fig. 161. Scanning electron micrographs of very high yield bisulfite pulp fibres. Cross-section of "pulp" sheet.

refining. After latency removal, the stock is conventionally screened (2 stages) and then centricleaned (4 stages) prior to storage.

Typical operating conditions for the production of 85% yield sulfite pulp are given below:

Yield range	: 82-87
T.SO$_2$ − Conc.	: 3-4%
− Applied on wood	: 8-10%

Table 29. Mill trial operating conditions and pulp properties.

COOKING & REFINING	CONTROL	TRIAL
Liquor conc. T. SO$_2$, %	3.0	3.0
pH	4.5	4.5
Max. Temp. °C	147	147
Cooking time, h	6	4
Yield, %	75	85
Refining energy, GJ/T	2.12	3.62
PULP PROPERTIES (DECKER)		
C.S. Freeness, mL	616	599
Bulk, cc/g	1.98	2.36
Burst, kPa·m^2/g	5.79	4.31
Tear, mN·m^2/g	9.22	10.0
Breaking length, km	8.99	6.95
Stretch, %	2.12	1.93
Smoothness, mL/min	388	538
Opacity, (TAPPI), %	83.4	88.2
CLASSIFICATION + 14	55.2	43.3
+ 28	21.8	23.3
+ 48	7.2	8.2
+ 100	7.7	8.8
+ 200	2.3	3.1
− 200	5.8	13.3

pH	: 4-6
Maximum temperature	: 150-160°C
Total Retention time	: 60-90 min
Refining Energy	: 1000-1400 kWh/t
Refining Consistency	: 20-25% 1st stage
	: 18-25% 2nd stage

The flexibility of the process permits wider operating conditions regarding liquor concentration, pH and temperature.

Properties of mill-produced VHY pulps resulting from regular and shredded chips are shown in Table 31. Overall gains in bulk, burst, tensile, brightness and a slight loss in tearing strength and fibre length classification were obtained with shredded chips with a 15% reduction in refining energy.

The competitive properties of supercalendered roto sheets containing LY and VHY sulfite pulps are shown in Table 32.

Based on this mill experience, the pulp characteristics are satisfactory to operate the paper machines [2 Fourdriniers and one 6 m (240″) Dynaformer running at 850-855 m/min (2800-2900 fmp) and producing 50% of the mill news production] efficiently with the complete displacement of the LY by the VHY pulp in the furnish.

It is estimated that a reduction of BOD loadings from 390 to 85 kg/O.D. ton of pulp and dissolved solids from 980 to 305 kg/O.D.T. would result from this conversion.

Table 30. Newsprint properties.

		TWIN WIRE		FOURDRINIER	
		CONTROL	TRIAL	CONTROL	TRIAL
Basis Weight, g/m^2		49.8	47.3	49.5	49.3
Caliper, μm		79	81	79	83
Burst, kPa		58.6	54.5	77.9	62.7
Tear, mN	MD	152	160	142	142
	CMD	230	248	228	219
Tensile, kN/m	MD	2.24	2.31	2.81	2.13
	CMD	0.85	0.82	0.98	0.91
Stretch, %	MD	0.73	0.88	1.07	0.85
	CMD	1.88	1.86	1.74	1.99
Smoothness, mL/min	TOP	133	159	85	124
	WIRE	142	135	122	127
Porosity, mL/min		384	874	175	252
Opacity, %	PRINTING	91.9	90.1	92.5	92.8
Brightness, %		57.6	57.9	58.1	58.0
Larocque Index	TOP	67.3	66.5	67.6	67.8
	WIRE	69.1	66.7	66.5	66.9
Dust		1	1	3	3
Lint		45	45	50	50
Furnish					
GWD %		40	40	75	75
TMP %		35	35	0	0
Sulfite %		25	25	25	25

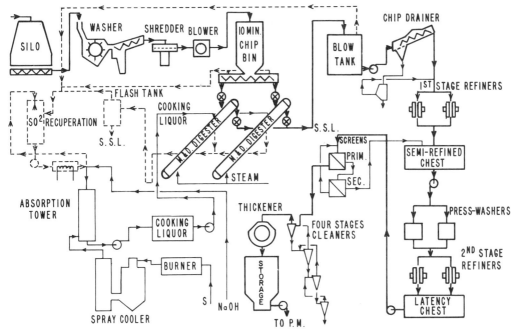

Fig. 162. Flowsheet of VHY sulfite plant (Laurentide Division, CBI).

Table 31. Effect of chip shredding on VHY pulp quality.

Chips		Unshredded	Shredded
Refining Energy, HPD/T		71.5	61.6
CSF, mL		408	415
Bulk, cc/g		2.54	2.40
Burst, kPa·m²/g		2.89	3.29
Tear, mN·m²/g		11.2	10.0
Breaking Length, km		4.84	5.45
Stretch, %		2.29	2.26
Porosity, mL/min		2298	2080
Smoothness, mL/min		520	360
Opacity, %		89.8	88.6
Brightness, %		52.4	53.7
Classification	+14	32.8	30.7
	+28	12.6	11.4
	+48	15.1	16.3
	+100	16.1	18.8
	+200	3.3	3.7
	−200	20.1	19.1

Table 32. Supercalendered paper properties.

Furnish			LYS/ GWD	VHYS/ GWD
Basis Weight, g/m²			49.4	47.4
Bulk, cc/g			1.49	1.32
Burst, kPa			58.7	50.0
Tear, mN	MD		152	225
	CMD		230	280
Tensile, kN/m	MD		2.33	2.17
	CMD		0.97	0.97
Stretch, %	MD		0.97	0.90
	CMD		2.02	1.85
Printing Opacity, %			95.4	95.5
Porosity, mL/min			170	92
Parker Print ⎱ Surf	Top		3.37	3.44
Smoothness ⎰	Wire		3.18	3.28
Formation Index			5.5	5.5

(f) The vapor phase — very high yield sulfite pulping process

A description of the Gaspesia mill at Chandler, Quebec [92]

A vapor phase variant of the above VHY sulfite pulping process has been developed recently. The first industrial plant began to operate in 1984 at the Abitibi-Price newsprint mill, Chandler, Quebec. The objective to be achieved with this installation are energy savings due to vapor phase cooking while maintaining newsprint quality.

The structure of the modified, inclined M&D digester developed for vapor phase pulping is shown in Fig. 102 and the principle of its operation is described in Chapter V, B, 5. The unit is 2.54 m (100 in) in diameter and 24.4 m (80 ft) long. It is designed for cooking 230 t/d of o.d. wood at 160°C (320°F) and 517 kPa (75 psig).

In this first installation shown in Fig. 163, washed and shredded softwood pinchips pass a presteaming vessel and enter the digester through a triple screw feeder with anti-blowback device which removes air and condensate by compression. On entering the digester, the compressed chips expand in re-cycled cooking liquor whose temperature is well below the 170°C of the vapor phase. This assures thorough impregnation of the wood during the 12 min of residence in the liquor. A yield of 90% is obtained in the vapor phase cook of 60 min duration.

A liquor circulation system with heat exchangers for fresh and spent cooking liquor permits exact control of impregnation and cooking temperature by indirect heating. It also minimizes dilution of cooking liquor by direct steam condensate, and thereby allows for good control of chemical concentration in the digester.

Based on experience gained in the previous liquid phase installation, the pH of the sodium base bisulfite/sulfite cooking liquor is held at 6 in order to prevent corrosion of stainless steel in the vapor phase region of the digester. In a true bisulfite cook at pH 4 some free SO_2 is present in the vapor phase which can cause formation of sulfuric acid in wall condensate via SO_3. Cladding with Inconel, good insulation of the vessel, and raising the liquor pH to bind free SO_2 have been found to provide the necessary protection.

Full automatic and/or manual electronic process control with flexible programming is provided.

3. Sulfonated chemimechanical pulping process (SCMP) [49-51]

The "SCMP" process was developed by CIP Research Ltd. as a partial or full replacement of the chemical pulp in newsprint furnishes. It emphasizes the need for a high degree of lignin sulfonation by treatment of the chips with strong sulfite solutions under protective cooking conditions to avoid substantial losses in wood components and maintain pulp yield at about 90%.

(a) Process Operating Parameters

Cooking temp., °C   = 130-160°C
Cooking time, min   = 20-60
pH   = 7.0-8.5
Conc. of Na_2SO_3, %   = 10-14

Wood chips are washed, pre-steamed and then cooked with 12% sodium sulfite solutions for 30-40 min at 140°C. The cooked chips are separated from the bulk of the spent liquor through a strainer, then pressed to about 50% solids prior to a two-stage atmospheric refining to 300-400 CSF. The pressate is combined with the rest of the weak liquor, fortified with sodium hydroxide, and sent to the absorption tower where sulfur dioxide is added to raise the liquor concentration to the desired level of 12% Na_2SO_3. This strong liquor, at pH 7.5-8.0, is

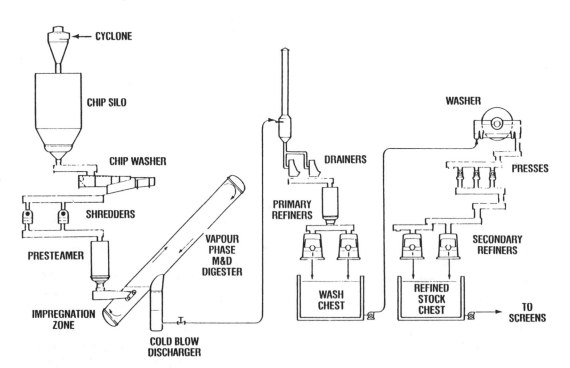

Fig. 163. The Gaspesia vapor phase BCMP system.

returned to the digester for cooking. There is a build-up of dissolved organics in the liquor system which reaches an equilibrium, in continuous operations. The dissolved organics appear to have no detrimental effect on pulp quality.

The effects of liquor concentration on softwood (spruce/balsam) pulp properties were studied extensively. Table 33 highlights some of the responses when the concentration of Na_2SO_3 varied from 50 to 180 g/L at 140°C and 30 min cooking time. There is no significant advantage in exceeding a concentration of 12% Na_2SO_3 with a corresponding 0.6% chemically bonded sulfur in the pulp. The importance of sulfonation on the

stretch properties of very high yield pulps was also confirmed by other researchers [52-55].

(b) Characteristics of SCMP and newsprint

Typical properties of pilot plant produced SCMP from a blend of spruce and balsam at different freeness levels are given in Table 34. A comparison of SCMP at the desired freeness level of 350 with other commonly used newsprint furnish components is shown in Table 35. It can be seen that the SCMP is very similar to mechanical pulps in yield and optical properties while its strength, fiber length and drainage characteristics are much closer to chemical-type

Table 33. Effects of liquor concentration on SCMP properties.

Liquor strength, as Na_2SO_3, g/L	50	70	90	120	150	180
Breaking length, km	3.52	3.87	3.91	4.57	4.48	4.59
Burst index, MN/kg	1.11	1.31	1.40	1.87	1.78	1.94
Tear index, N·m²/kg	7.0	6.9	7.6	7.7	7.6	7.8
Apparent density, kg/m³	365	390	390	405	405	415
Wet web tensile at 20% solids, N/m	15.4	17.2	19.5	23.0	18.8	18.5
Combined sulphur, %	0.45	0.52	0.56	0.60	0.62	0.66

Cooked at 140°C for 30 min, refined to 300 CSF.

Table 34. Properties of pilot plant SCMP.

CSF, mL	200	300	400	500	600
Breaking length, km	6.3	6.2	5.7	5.2	4.7
Burst index, MN/kg	3.9	3.5	3.1	2.8	2.4
Tear index, N·m²/kg	7.8	8.4	9.2	10.0	10.9
Apparent density, kg/m²	515	485	455	430	410
Wet web tensile at 20% solids, N/m	50	45	40	34	29
Opacity, %	91	92	91	90	88
Long fiber, % (> 48 mesh, Bauer McNett)	68	70	73	77	81
Fines, % (< 100 mesh, Bauer McNett)	21	20	18	16	13

Varied cooking conditions, but mainly 12% Na_2SO_3, 140°C, 30 min.

Table 35. Typical properties of newsprint pulps.

	SGW	RMP	TMP	SCMP	SBK	HYBS
Yield on wood, %	95	95	94	92	43	65
CSF, mL	90	100	150	350	550	550
Breaking length, km	3.2	3.5	4.0	6.0	8.0	9.5
Burst index, MN/kg	1.3	1.5	1.7	3.4	6.0	6.0
Tear index, N·m²/kg	5.0	6.0	8.0	8.8	14.0	7.0
Apparent density, kg/m²	400	335	400	470	660	625
Wet web tensile at 20% solids, N/m	25	25	35	42	55	50
Brightness (Elrepho), %	59	58	56	56	70	50
Opacity, %	97	97	96	91	75	82
Long fiber, % (> 48 mesh, Bauer McNett)	40	45	55	72	80	80
Fines, % (< 100 mesh, Bauer McNett)	50	38	35	19	8	20

pulps. On this basis, it was found that the SCMP is most suitable for replacement of part or all of the chemical pulp content in newsprint.

It was also found that the SCMP behaved like a mechanical pulp in its response to surface brightening agents such as hydrosulfite or peroxide. The brightness gain is dependent on the cooking parameters as well as on the metal ion content of the pulp as illustrated in Table 36.

Paper machine trials were carried out with pilot plant produced SCMP and with other chemical pulp furnishes on Fourdrinier as well as on twin wire machines for the manufacture of standard and roto news grades. Highlights of results are given in Tables 37 and 38. These trials demonstrate the feasibility of producing competitive quality newsprint from SCMP and groundwood with little or no added chemical (sulfite or kraft) pulps.

(c) Wood species

While spruce and balsam were used in most pulping studies, other softwood and hardwood

Table 36. Brightening SCMP.

Brightness obtained with:	Softwood SCMP 77 ppm Fe	60 SW/40 HW SCMP 185 ppm Fe	50 SW/50 HW SCMP
Unbleached pulp	57	53	55
1% Na hydrosulphite	64	61	64
2% H$_2$O$_2$ (alkaline) + 1% hydrosulphite	77	70	72

Table 37. Paper machine trials (see Table 16 for results).

Grade		Machine	Speed m/min	Furnish %	
(a)	Gatineau			SBK	SCMP
	News	Open draw	570	24-11	4-21
	News	Open draw	570	2	25
	News	Open draw	570	5	19
	News	Twin wire	610	3	25
	Lightweight	Pickup	670	4	27
(b)	Trois-Rivières			LYS	SCMP
	Roto	Open draw	400	1	34
	News	Open draw	640	1	34
	Roto	Open draw	400	1	34

Table 38. SCMP mill trials.

Mill Grade Furnish		Gatineau Standard news 25%		Trois-Rivières Standard news 34%		Trois-Rivières Newsroto 34%	
		0% SCMP 19% SBK	SCMP 2% SBK	0% SMCP 20% LYS	SCMP 1% LYS	0% SCMP 21% LYS	SCMP 1% LYS
		48.3	48.3	47.4	47.0	50.1	48.8
Basis weight, g/m²		86	86	89	84	66	69
Caliper, μm		0.73	0.76	1.19	1.35	1.69	1.54
Burst index, MN/kg	MD	3.70	2.90	3.37	3.30	2.51	2.79
Tear index, N·m²/kg	CD	5.79	4.81	4.82	5.15	3.72	3.96
	MD	1.80	1.98	2.00	2.08	2.35	2.26
Tensile strength, kN/m	CD	0.68	0.72	0.88	0.93	0.93	0.96
	MD	1.0	1.0	1.2	1.3	1.2	1.2
Stretch, %	CD	2.0	1.8	1.9	1.8	2.0	1.9
Roughness (Sheffield), cm³/min	Top	103	98	131	117	26	32
	Bottom	104	100	144	121	27	33
Brightness (Elrepho), %		59.8	57.5	59.4	57.2	57.4	59.9
Opacity, %		93.2	93.2	91.5	91.8	94.4	92.9
Larocque No., %	Top	79.8	79.6	66.8	67.7	72.6	72.4
	Bottom	81.4	81.5	69.7	69.6	75.8	74.8
Print through, %	Top	6.2	6.6	10.5	11.1	9.8	11.4
	Bottom	7.4	6.5	9.8	10.8	8.5	10.2

species were tried with encouraging results, as shown in Tables 39 and 40.

An advantage of the SCMP process is that, due to its very high yield level and recycling of spent liquors, strongly reduced BOD loadings (Table 41) are obtained. For example, it was calculated that a newsprint mill using 35% SCMP and 65% groundwood will have half the BOD load of a mill using 25% bisulfite at 65% yield and only one third the BOD load of a mill using 25% low yield sulfite with groundwood.

It was also suggested that, if regulations become more stringent, a recovery system based on the evaporation of the spent liquor and crystallization of Na_2SO_3 for reuse could be applied for the SCMP process (Fig. 164). This system would result in the reduction of Na_2SO_3 make-up from 150 to 100 kg/t of SCMP.

(d) Description of SCMP installation [52]

There are three installations in Eastern Canada operating with SCMP. The first (380 t/d liquid phase) is located at the CIP Gatineau mill (start-up July 1978); the second of 160 t/d liquid phase at Abitibi-Price, Thunder Bay (start-up in 1981); and the third of 380 t/d vapor phase at the CIP Dalhousie mill which started operations in 1982.

CIP Gatineau SCMP mill

The first liquid phase SCMP installation, which started operation in July 1978 at the CIP Gatineau mill in Quebec, is illustrated in Fig. 165.

Table 39. SCMP from various wood species.

	Black spruce	Balsam fir	Jack pine	Hemlock	Poplar	Maple
Breaking length, km	6.1	4.8	4.2	2.8	4.8	2.9
Burst index, MN/kg	2.9	1.9	1.6	0.8	1.5	0.8
Tear index, N·m²/kg	9.5	7.1	10.5	7.8	7.1	3.0
Apparent density, kg/m²	425	470	445	320	490	345
Wet web tensile at 20% solids, N/m	27.7	26.0	20.5	14.6	12.1	11.9
Brightness (Elrepho), %	57	56	45	57	66	61
Opacity, %	90	88	NA	86	82	90
Long fiber, %	75	74	60	73	58	39
Fines, %	20	16	27	16	21	28
Yield on wood, %	93	92	91	92	90	89

Cooked with 12% Na_2SO_3 solution at 140°C for 30 min; refined to 350 CSF.

Table 40. SCMP from hardwood-softwood blends.

Hardwood in blend, %	0	20	40	60	80
Breaking length, km	4.9	5.4	4.8	4.2	3.9
Burst index, MN/kg	2.3	2.4	2.2	1.8	1.6
Tear index, N·m²/kg	8.3	8.7	8.9	8.4	7.3
Apparent density, kg/m²	460	425	390	375	390
Brightness (Elrepho), %	58	57	57	56	55
Opacity, %	89	91	92	93	93
Long fiber, %	63	55	58	49	44
Fines, %	20	26	24	26	38
Yield on wood, %	93	91	90	88	87
Combined sulphur, %	0.61	0.58	0.56	0.56	0.38

Softwood: 62% spruce, 25% balsam, 14% jack pine.
Hardwood: 62% maple, 13% elm, 13% poplar, 8% beech, 4% others.
Cooked with 12% Na_2SO_3 solution at 140°C for 30 min; refined to 350 CSF.

Table 41. BOD₅ of newsprint furnish effluents.

Type of pulp	RMP	TMP	SCMP	LYS	HYS
Yield on wood, %	96	96	92	50	65
BOD₅, kg/t	23-28	20-25*	35-45	250	150

*Does not include steam condensate

Briefly, the chips are blown from the wood-room to a large live bottom chip silo, then washed and blown to the presteamer. Cooking is in the 500 t/d continuous M & D Bauer digester. The blown cooked chips pass through French Oil

presses (Model JL-88). The recovered liquor and that extracted from the digester through the flash tank are combined and recirculated. The pressed chips are blown via the surge bin to three atmospheric primary refiners [6700 kW (9000

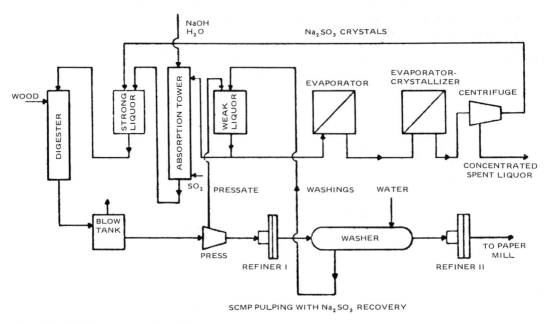

Fig. 164. SCMP process with sulfite recovery.

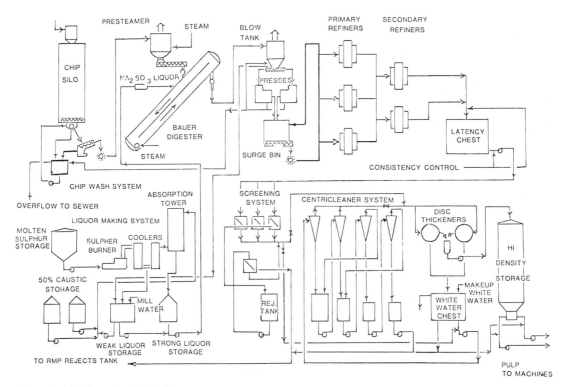

Fig. 165. Gatineau SCMP mill.

HP) Defibrator-RL54-5] followed by two in secondary stage. The refined pulp is treated for latency removal, screened (cleaners, by-passed), thickened and stored in the high density storage tank.

The resulting SCMP is used as partial replacement of the more expensive semi-bleached kraft (reduction from 18 to less than 7%) in the production of newsprint. Table 42 summarizes mill operating conditions, power usage and pulp quality over a three-year period. Some changes of the original design, machinery and process parameters were brought about for smoother operation and optimization of the process [52]. This led to increasing the original capacity from 380 t/d to 490 t/d of SCMP by early 1982, with 2⅓ operators per shift.

Starting in 1981, poplar was used in varying amounts, according to its availability, without loss in paper quality.

Thunder Bay SCMP installation [93]

The second SCMP plant, also liquid phase, commenced operation in the Fort William mill of Abitibi-Price in March 1981 under a license agreement with CIP, following the original design with only a few exceptions. The SCMP option was chosen because it provided news reinforcement pulp at 92% yield, compared with 63% produced in the existing HYS plant, at good runnability and a 55% reduction of BOD in the effluent.

Figure 166 shows the pulping flowsheet. Mill-made or purchased chips are screened, washed and stored in a silo from where they are added to a presteamer. They are then metered at the required production rate to the M&D continuous digester with a capacity of 135 o.d. t/d. The cooking liquor contains 12% Na_2SO_3 at pH 7.8 and 3:1 liquor-to-wood ratio. Cooking conditions are 155°C at 450 kPa (65 psig) for 45 min. The cooked chips are blown into a cyclone bin and pressed to 58% solids in a French oil press. Spent liquor withdrawn from the digester is mixed with pressate and recycled through the acid plant for reinforcement.

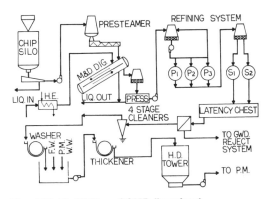

Fig. 166. Ft. William SCMP flowsheet.

Table 42. Gatineau SCMP mill — 3 years operation.

	1979		1980	1981
	Jan.-Aug.	Sept.-Dec.		(7 months)
PRODUCTION				
t/d	335	351	385	431
ENERGY CONSUMPTION				
Power, GJ/t	6.7	6.0	5.7	4.9
Steam, GJ/t	1.8	2.1	2.4	2.4
CHEMICAL CONSUMPTION				
Caustic, kg/t	98	107	107	94
Sulphur, kg/t	35	43	42	37
QUALITY				
Freeness (mL)	326	371	322	328
Tear Index (N·m²/kg)	9.7	9.2	9.3	10.2
Burst Index (MN/kg)	2.7	3.4	3.5	3.4
Density (kg/m³)	493	431	480	463
Breaking Length (km)	5.4	6.3	5.8	6.1
Brightness (%)	55.3	50.1	50.1	49.0
Yield (%)	89.4	89.0	88.8	88.9
OPERATING CONDITIONS				
Digester Pressure (kPa)	414	552	552	565
Digester Temperature (°C)	153	157	159	161
Retention Time (min)	45	45	40	40
Na_2SO_3 (% w/v)	13	13	13	13

Total installed power for the three primary and two secondary refiners (Sunds RSA 1000) is 13 000 kW (17 500 HP). After latency removal the stock is cleaned in 4 stages and thickened. To reduce dissolved solids, it is diluted again with fresh water from the economizer and excess white water, and thickened again for HD storage. High fines content of the white water makes it necessary to refine to 400-425 CSF from the second stage to achieve 325 CSF of the final pulp.

The dry strength properties of the SCMP pulp (Table 42) are found to lie midway between those of TMP and HYS pulp, typical for chemical pulp at very high yield. The wet strength of the SCMP pulp is superior to that of the HYS pulp when refined to a somewhat lower freeness. At 457 μm Elrepho SCMP and HYS pulps have about equal brightness. Brightening response to hydrosulfite is good provided that the pulp is well washed to remove dissolved solids. Physical tests made on newsprint sheets do not reveal any significant differences between HYS and SCMP furnish.

Dalhousie SCMP installation [94, 95]

This installation is the successful implementation of vapor phase SCMP pulping at the CIP Dalhousie newsprint mill. The adaptation of the original liquid phase process to vapor phase operation was supported by considerable R&D work, particularly regarding sulfonation of the lignin which was known to be of primary importance for developing strong pulps of very high yield. It was found that proper sulfonation can be achieved also in the vapor phase provided that the chips are thoroughly impregnated with sufficient chemical prior to cooking.

It was possible to meet this requirement by means of pressure impregnation of the steamed chips. The necessary equipment has become available in recent years in the form of the pressure screw plug feeder, which forces a tight plug of compressed chips into a vessel under pressure. Compression removes the air and part of the water in the chips and when the chips expand under cooking liquor inside the vessel like a sponge thorough impregnation follows. Thus, with a liquor-to-wood-ratio of 1.2-1.0 and a sodium sulfite liquor concentration of 12%, the chemical charge for producing vapor phase pulp at 92% yield would amount to 136 kg/t (300 lb/t) which matches the chemical usage of the liquid phase process.

An industrial demonstration of the SCMP process would demonstrate possible advantages over liquid phase operation, such as a simpler and less expensive unit, elimination of the cooking liquor cycle, and avoidance of solids deposition in the system.

The problem of build-up of dissolved solids, identified in the liquid phase operations, could also be corrected by employing two stages of press washing, one prior to the first stage and the second between refining stages.

Figure 167 shows the flow sheet of the VP-SCMP plant. Chips are fed from storage via a washer, a dewatering screen and the presteamer to the pressure screw plug unit which discharges into the bottom of the impregnation vessel filled with sulfite liquor at pH 8. Internal conveyors move the chips up to the top, providing 3-4 min retention time at 90°C, and over into the vapor phase digester. Usual cooking conditions are 30 min retention at 150°C.

The cooked chips discharge via a bottom screw conveyor and the discharge refiner into the blow chest where they are diluted to 3-4% consistency and fed to two primary presses, and on the three primary refiners. After a second dilution in the primary refiner stock chest, the stock is fed to the secondary presses and two secondary refiners. All refiners are Sprout Waldron, 6700 kW (9000 HP) units in the same arrangement as at Gatineau and Thunder Bay. Refiner feed consistencies are 30% and 25%. The residence time for latency removal is 45 min. This is followed by one-stage screening, four-stage centricleaning, thickening and HD storage.

Note that the two-vessel presteamer-VP digester plus discharge refiner combines the same purpose and principle of function as the original one-body Defibrator digester. The reason for separating impregnation vessel and digester in this case is the high production rate which would have made a single-body unit too large. Note also that the so-called preheater ahead of the primary refiner in CTMP installations is actually a rapid vapor phase cooking and sulfonation stage for the impregnated chips.

Table 43 shows that at a rather high freeness level good strength properties are obtained with VP SCMP, equal or approaching those of Gatineau liquid phase SCMP. Proper attention to impregnation together with extended cooking time assure a high and uniform distribution of sulfonation in the chip and provide a very high yield sulfite pulp capable of replacing the costly chemical pulp component in newsprint furnish.

4. Thermochemimechanical pulping process "TCMP" [56-58]

(a) Introduction

"Thermo" or "T" in this title does not stand for pressurized refining as in TMP and CTMP. It

DALHOUSIE SCMP PLANT

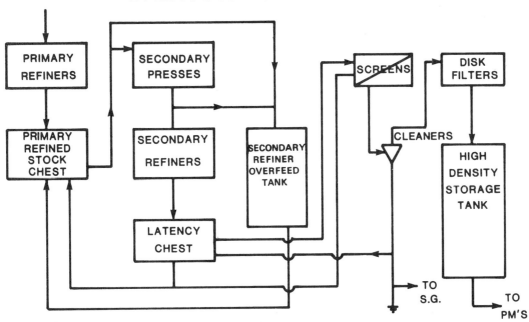

Fig. 167. Dalhousie SCMP plant.

indicates a sulfite impregnation treatment at 90°C, followed by atmospheric refining.

The TCMP process developed by Abitibi-Price as a result of extensive studies in the refiner pulping areas is a significant improvement over the thermomechanical pulping (TMP) process. The main emphasis was to produce pulp at highest yield satisfactory for inclusion with

mechanical pulp in the production of a good running newsprint sheet with the least linting tendency. To do so, efforts were concentrated on the study of operating parameters in chip pretreatment, refining and screening. It was found that:

1. Pretreatment of chips at atmospheric pressure close to 100°C with neutral sulfite solution assists in effective fiber separation.
2. At a given specific energy input, two-stage refining gives substantially better linting resistance than does single-stage refining.
3. Linting was primarily controlled by the specific energy input and by the refiner plate age and conditions.
4. Screen rejects rates had a very limited effect, since higher rejects rates caused only a modest drop in the linting resistance (Table 44).
5. Combined screen and cleaner rejects which have initially very low linting resistance can be readily refined to give essentially no linting.

Table 43. Comparison of Dalhousie and Gatineau SCMP Data.

	Dalhousie Vapor Phase	Gatineau Liquid Phase
Freeness	425	325
Tear	8.3	9.0
Burst	3.4	3.4
Breaking Length	6.0	6.1
Density	430	400
Wet Web	29	25
% Sulfur	0.55-0.60	0.60-0.65
Specific Energy (GJ/t)	7.0	5.5
Brightness	55	50

Table 44. Rejects rate and lignin.

Rejects, % of Production	Primary Refiner		Rejects Refiner		Deckered Pulp	
	kW	MJ/kg	kW	MJ/kg	MJ/kg	Linting Resistance
27	1343	6.25	560	9.88	8.92	4.8
62	1101	5.12	746	5.61	8.60	4.3

(b) Mill trials; pulp and newsprint properties

Tables 45 to 48 highlight findings from mill trials where a pilot-produced TCMP (Table 45) was used to displace HY sulfite pulp in the standard news (Table 46) and directory paper (Tables 47 & 48). Results show that the TCMP can effectively replace the HY chemical furnish.

(c) Description of the Kenogami "TCMP" process

The first commercial "TCMP" installation went on stream in 1980 at the Abitibi-Price Kenogami Mill in Quebec. It has a production capacity of 400 t/d: 2 lines of 2 stage refining (Sprout Waldron twin 50 C-4 refiners, motor = 9000 kW (12 000 HP), 1800 RPM, power factor 1.15. A simplified flowsheet is given in Fig. 168.

Purchased chips (mainly spruce and balsam) are blown from an outside storage bin equipped with a Wenburg conveyor, then passed through a shredder where sodium sulfite solution (15-20%; pH 10-12) is added at a rate of 4-6% on o.d. wood basis. The impregnated shredded chips progress to a preheater (where direct steam is added to raise the temperature to 90°C at atmospheric pressure), then into a reaction bin for a 20-30 minute retention time.

The treated chips are then blown through a cyclone to the two twin 50 C Sprout Waldron primary refiners. Rotor lifts feed the two secondary refiners. After the latency chest (20-30 min retention) the pulp is screened in four pressurized screens and cleaned (5 stages) before thickening on two disc filters. Rejects from the screens are thickened on two presses to 30-40% and refined on three refiners. The resulting stock is sent to the latency chest for processing with the incoming pulp.

There is a heat recovery system to produce hot water for the chip washer to heat the white water and the building.

Based on early mill experience, the total refining energy consumption is 125 HPD/ton o.d. basis (9.9 GJ/t); 60 and 50 HPD/T (4.8 and 4.0 GJ/t) in the 1st and 2nd stage respectively, and 70 HPD/ton (5.6 GJ/t) in rejects refining. This is based on an assumed 15-20% rejects at the screens. The total amounts of rejects from the 5th stage cleaner varied between 3-5 t/d. Main pulp characteristics, as produced at the mill, are listed below:

TCMP (after disc filter)

Freeness, CS	70-80
Burst, kPa·m²/g	2.55-2.75
Tear, mN·m²/g	7.85-8.82
Debris (6 cm) %	0.03-0.05
Brightness	62
Scattering coefficient cm²/g	580

The "TCMP" is used with stone groundwood and semi-bleached kraft: 50% in rotogravure, 50% in directory, and 60% in high brightness offset. All show good paper machine and pressroom performances.

(d) The Perlen very high yield continuous pulping process [96]

The Perlen paper mill near Lucerne, Switzerland specializes in the production of a variety of

Table 45. Comparison of mechanical pulp types. Properties at 100 CSF.

Pulp Type	SGP	RMP	TMP	TCMP
Burst Index, kPa, m²/g	1.37	2.37	2.86	3.35
Tear Index, mN	4.44	6.91	7.49	7.30
Bulk, cm³/g	2.30	2.65	2.40	2.40
Wet Breaking Length, m	170	270	290	360
Scattering Coeff., m²/kg	65.0	68.5	64.0	59.0
Debris Content, %	0.50	0.30	0.20	0.04
Specific Energy, MJ/kg	5.33	7.39	8.81	8.74

Table 46. Comparison: newsprint from standard and 100% TCMP furnishes.

	75% SGP 25% HYS	100% TCMP 1st trial	3rd trial
Basis weight g/m²	48.6	48.9	49.1
Burst, kPa	110	75	125
Tear, mN, M.D.	142	181	166
C.D.	202	236	279
TAPPI, Brightness, %	56.6	55.4	59.2
Sheffield Roughness, T	77	88	86
W	80	95	91

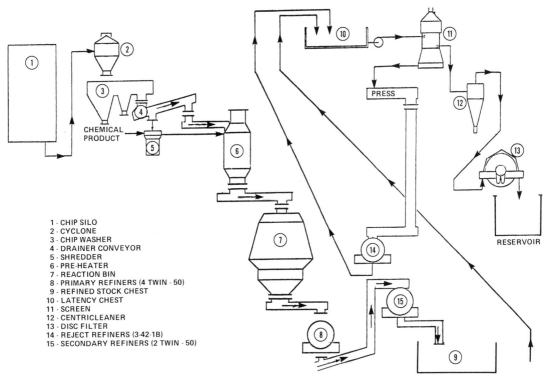

1 - CHIP SILO
2 - CYCLONE
3 - CHIP WASHER
4 - DRAINER CONVEYOR
5 - SHREDDER
6 - PRE-HEATER
7 - REACTION BIN
8 - PRIMARY REFINERS (4 TWIN - 50)
9 - REFINED STOCK CHEST
10 - LATENCY CHEST
11 - SCREEN
12 - CENTRICLEANER
13 - DISC FILTER
14 - REJECT REFINERS (3-42-1B)
15 - SECONDARY REFINERS (2 TWIN - 50)

Fig. 168. Kenogami "TCMP" process.

Table 47. Pulp properties of telephone directory paper furnishes.

	33% HYS 61% SGP 6% Clay	100% TCMP	94% TCMP 6% Clay
Freeness	167	103	100
Burst Index, kPa · m^2/g	3.06	2.96	2.94
Tear Index, mN	7.35	8.62	8.24
Bulk, cm^3/g	2.22	2.42	2.45
Wet Breaking Length, m	294	312	277
Debris, % on 0.15 mm screen	0.29	0.01	—
Brightness, TAPPI, %	59.6	60.5	60.3
Printing Opacity, %	95.8	95.5	95.9

Table 48. Properties of telephone directory papers.

Furnish	33% HYS 61% SGP 6% Clay	100% TCMP	94% TCMP 6% Clay
Basis Weight, g/m^2	36.1	37.0	36.6
Thickness, μm	56.4	62.5	59.2
Bulk, cm^3/g	1.61	1.69	1.62
Burst Index, kPa · m^2/g	2.33	2.40	2.16
Tear Index, mN, M.D	3.12	2.94	2.68
X.D	4.32	4.51	4.47
Breaking Length, km, M.D	8.43	9.01	8.85
X.D	3.26	3.19	3.10
Stretch, % M.D.	1.12	1.10	1.09
X.D	1.82	1.93	1.96
TAPPI Brightness, %	61.5	61.2	62.1
Printing Opacity, %, at 3.6 g/m^2	84.0	82.0	83.7

unbleached coated and uncoated papers, converted packaging materials, and newsprint with a total annual output of 100 000 t. The basis for this operation is the chemimechanical sulfite pulping plant (Fig. 169) producing 50-60 t/d of ultra high yield pulp at 85-90% yield in the first Kamyr continuous bisulfite digester (90 t/d) built in 1963. The second high yield bisulfite Kamyr digester was installed at Kapuskasing, Canada in 1964. Kamyr installations of that period were stand-alone units with small, horizontal presteaming vessels. A large vertical low pressure presteaming and impregnation vessel with extended residence time was added ahead of the high pressure digester system in order to provide optimum conditions for impregnation of the chips with cooking liquor.

The original intention was to make bleachable high yield bisulfite pulp at a yield between 55 and 60%. A number of local conditions, particularly stream pollution and wood supply required development in the direction towards highest possible yield of paper making pulps of a quality suitable for the grades produced at Perlen.

Increasingly higher pulp yields were attained by shorter cooking time and mechanical refining. Even a short cook softens the lignin bonds sufficiently for mechanical refining under lenient conditions. In 1975, the pulp produced had a yield of 80% and very good strength due to the high content of long fiber. With a pulp press and a Frotopulper added to the fiberizing stages, the trend to increase the yield continued to the present 90%.

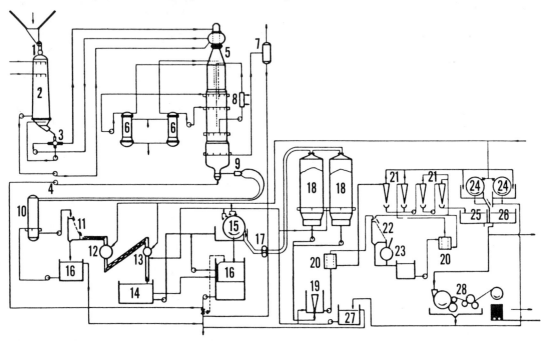

Fig. 169. Diagram of the Perlen semichemical pulp mill.

Cooking:
1. Low pressure feeder
2. Impregnation Vessel
3. High pressure feeder
4. High pressure pumps
5. Digester
6. Heat exchanger
7. Cyclone
8. Cooler
9. Blow valve
10. Blow tank

Washing/Defibration
11. Dewatering/inclined screen
12. 750 kW (1000 HP) Sprout Waldron refiner
13. 375 kW (500 HP) Sprout Waldron refiner
14. Storage tank

15. Wash filter
16. Spent liquor tank
17. High density stock pump
18. Pulp storage tanks

Screening-Cleaning
19. Collector-blending tank
20. Pressure screen
21. Centricleaners
22. Inclined screen
23. Beater
24. Thickener
25. Reject tank
26. Screened pulp tank
27. Recycle water tank

Dewatering
28. Voith band pressure filter

The wood supply consists mainly of waste lumber such as slabs, shreddings or sawmill chips at a ratio of 85-90% spruce and 10-15% beech wood. The wood is cooked under mild conditions for 5-6 h at 147°C/1175 kPa with sodium base bisulfite/sulfite liquor at pH 6. Cooking of mixed woods has been successful. Spent liquor is separated from the cooked chip charge on an inclined drainer and the chips are processed through two-stage refining, washing, pressure-screening, cleaning, and post-refining.

Physical properties of beaten and unbeaten chemimechanical production pulp and stone groundwood are listed in Table 49. When beaten in a Jokro mill, breaking length, elongation and burst increase, and tear decreases somewhat. Such strength levels are often exceeded in mill operation.

Optical properties are somewhat less favorable. The high lignin content results in a more yellowish-brown color; the whiteness of the fresh pulp is between 52 to 55% Elrepho. Brightening with sodium hydrosulfite increases pulp brightness to 60% Elrepho. At this brightness level the pulp is suitable for some paper grades. With a wood furnish containing 10 to 15% hardwoods (mainly beechwood), the opacity is 90%. This is high in comparison with sulfite or kraft slush pulp. Cooking of wood blends consisting of spruce and beech resulted in a softer sheet and a smoother paper surface. However, brightness reversion at this high lignin content is about the same as for groundwood. Bleaching investigations with peroxide have shown that a brightness of 65-70 Elrepho is attainable.

Production of 50-60 t/d of VHYS-CMP makes the mill quite independent of chemical pulp. For example, 45 g/m^2 newsprint is produced with a composition of 18% VHYS and 82% GWP, to which 5% filler is added. 40 g/m^2 light weight newsprint is produced from 12% VHYS, 12% kraft pulp, and 76% GWP at a machine speed of 722 m/min (2370 ft/min).

Calculated per tonne of dry pulp, the specific energy consumption is on the average 850 kWh, steam requirements are 1.2 t, and make up is 50 kg sulfur and 100 kg soda. In 1981, the mill opened its own small power plant utilizing a 2.7 m drop in the Reuss river.

This mill has come a long way towards eliminating environmental pollution of a typical high yield sodium base sulfite operation. Due to a number of internal measures it was possible to reduce effluent flow from 300 to 30 m^3 per tonne of pulp. It was also confirmed in laboratory studies that the biodegradable characteristics of organic compounds improve with increasing yield.

Consequently, it was decided first to prepare a closed cycle system for all effluents, and then to combine residual effluents of the paper mill and the CMP mill for biological treatment. Conventional recovery of chemicals by evaporation and burning in a recovery furnace was not practical owing to the high yield and comparatively small size of the pulp production.

The conventional calcium acid sulfite mill which operated up to 1970, produced 230 kg BOD$_5$ per ton of pulp, which was freely discharged. However, the BOD$_5$ load at the projected pulp yield of 85% amounts to only 70 kg BOD$_5$/ton of dry pulp. Biological effluent treatment is capable of lowering BOD$_5$ to 10 kg/ton of pulp, which corresponds to a 85% reduction in BOD$_5$.

5. "OPCO" pulping process [42,43,44,61]

(a) Introduction

Ontario Paper Co. Research studied extensively [61] the identification and optimization of properties responsible for paper machine efficiency. As a result of these studies, the OPCO process was developed for the production of pulps in the ultra high yield levels with a high

Table 49. Properties of Perlen pulps for newsprint.

| | Groundwood | Semi-chemical furnish | |
		Unbeaten	Beaten
Pulp Yield, %	95	80-90	80-90
Specific Energy consumption			
GJ/t o.d. pulp	6.1	—	3.6
Freeness, CSF, mL	90-75	670	425-360
Breaking Length, m	2500-3000	6000	6500-7500
Stretch, %	—	2.0	2.2
Burst, kg/cm^2	1.2-1.3	2.0	1.8-2.5
Tear, cm·g/cm			
Brecht-Imset, rel.	90-105	180-220	160-180
Brightness, % Elrepho	61-64	52-55	52-55
Opacity, %	92-95	—	88-90
	(50 g/m^2)		(70 g/m^2)

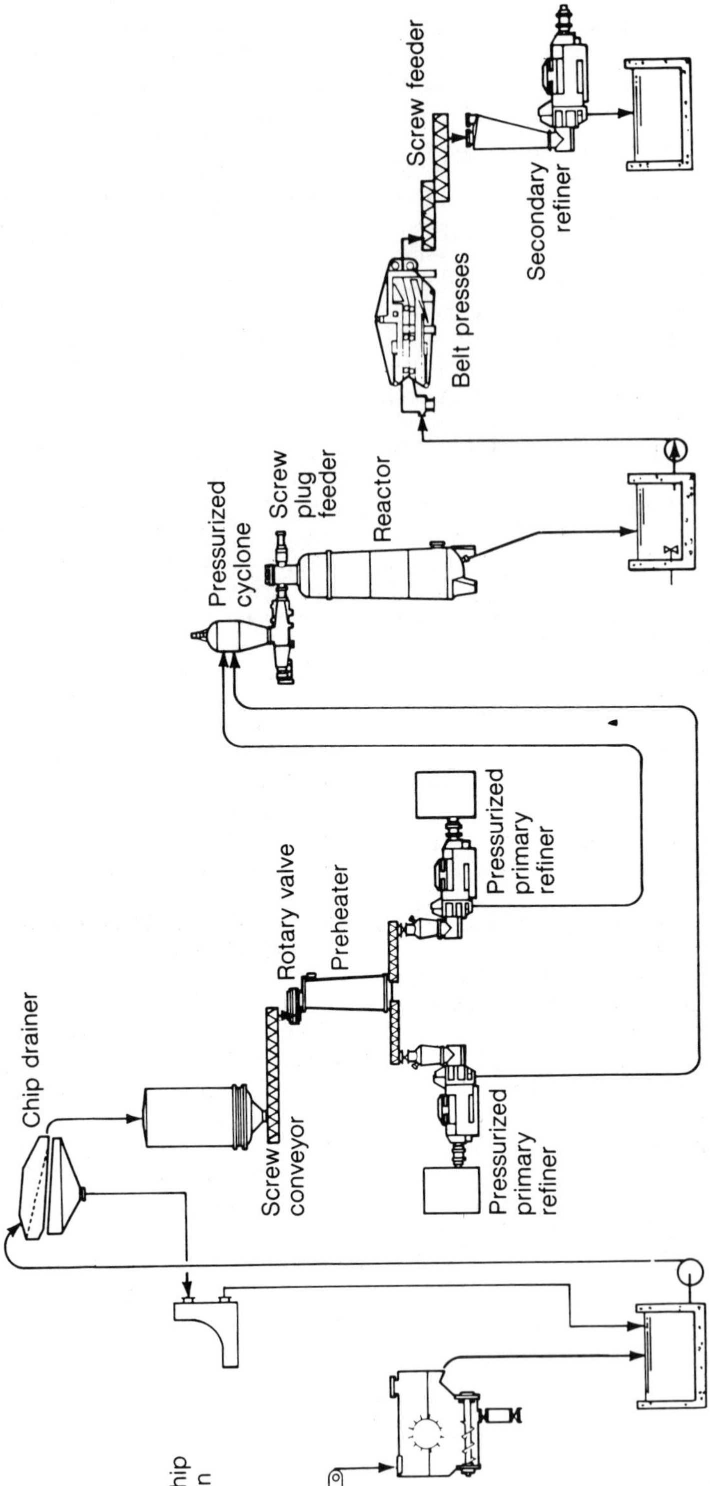

Fig. 170. Layout of the OPCO VHYS system.

degree of extensibility (i.e. stretch).

It was theorized and then confirmed that improvement of paper machine runnability is related to the degree of wet web stretch. "When the web is carried through the machine, stretch is traded off to supply the required tension at each point. Traded stretch is not regained when tension is relaxed but, when the web runs out of stretch, it breaks. Unused stretch provides the safety margin to absorb perturbation, and to carry and equalize strains around defects."

(b) Process operating parameters and main features

The 275 tpd process (Fig. 170), which started up in 1984 at Q.N.S. Paper Co. Ltd. [97], consists of a chemical treatment of a mechanical pulp such as TMP with sodium sulfite which maximizes wet stretch by stabilizing and permanently setting the curl in the pulp fiber.

The cooking reaction is normally carried out with a pulp consistency of over 10%, using 7-10% sodium sulfite (based on O.D. wood), at a temperature between 130-180°C and a time varying between 15 and 120 min to obtain a yield of 90%. Such treatment may be applied either in between the refining stages or as a post-treatment, with each approach offering advantages and disadvantages [42].

It was found that the product quality was not highly sensitive to reaction time and that the chemical treatment produces a permanent curl in the fiber not removed by a standard delatency treatment. It was shown [91] that the major contributor to the stability of the curl was the heat involved during refining.

The variations of wet tensile vs stretch to rupture of the original TMP (A,C, before and after latency removal respectively) and after treatment with Na_2SO_3 for 1 h at 130°C (B,D before and after latency removal) are illustrated in Fig. 171. At the same stretch to rupture, say 11%, the OPCO treatment significantly improved the wet tensile of the pulp (compare D^l vs C^l).

Variations of wet tensile at a fixed wet stretch to rupture (say at 5%) with a refining energy for untreated and for various types of OPCO treatments are shown in Fig. 172. It can be seen that the interstage treatment at 160°C produces a pulp having equivalent wet stress-strain properties at higher freeness but with 40% less energy consumption than the post treatment.

Figures 173A and B illustrate the variation of bulk and breaking length of some typical OPCO pulps with wet tensile. The treatment at 160°C was also found to offer superior pulp.

The following main features are claimed for the OPCO process:

1. A pulp yield in the order of 90% with a BOD leading of 56 kg/t of pulp.
2. Superior brightenss (2-4 pts) than TMP, with a good brightness stability (one point loss after three months of ageing) but with 15% less scattering coefficient)
3. Good levels of mechanical pulp properties with well fibrillated fibers of superior flexibility and extensibility, well suited for newsprint manufacturing and particularly helpful for improved paper machine performance and pressroom runnability.
4. Refining energy requirements lower than those required for TMP and comparable to stone groundwood (i.e. 5.5 - 5.9 GJ/ODTm) when interstage treatment is practiced.

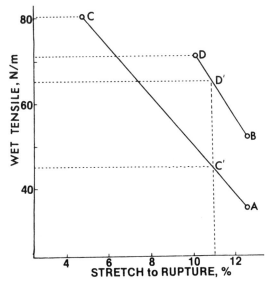

Fig. 171. Comparison of pulps at different states of latency.

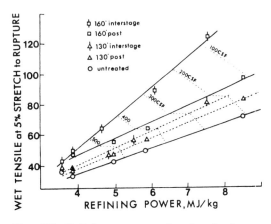

Fig. 172. Refining power and wet web stress-strain properties.

5. Process adaptable to continuous operation and automation.

(c) Mill trials and pulp and newsprint characteristics

Mill trials under OPCO conditions were conducted with softwood, and highlights of results are given in Table 50. Those relative to the paper machine operations (i.e. furnish composition, paper machine parameters and resulting paper characteristics) are shown in Table 51.

It was concluded that the OPCO pulp can displace the lower yield chemical pulp in the manufacturing of newsprint in view of the paper machine performance and pressroom runnability.

E. THE CTMP CHEMIMECHANICAL OR CHEMIREFINER PULPING PROCESS (92 + % YIELD)

Although the above title, when taken literally, would include all high yield and ultra high yield pulping processes, it is historically connected with specific developments [98,100] originating from purely mechanical processes. Since the process bridges the final gap between chemical and mechanical processes, refer to the Volume on Mechanical Pulping for additional background.

The first such process, stone groundwood pulping, was introduced about 140 years ago and marks the beginning of modern pulping. The technique consists of pressing a log against the coarse surface of a rotating mill stone, thus tearing fiber fragments out from the surface of the wood. Nowadays, wood grinding has evolved and uses sophisticated stones, steam-pressurized

housings, and process controls to produce a major portion of our present supply of filler pulp for newspaper and publication grades. The yield from wood is about 93%.

During the 1950's, the process was developed for making pulp by passing wood chips through a disc refiner (RMP). This refiner pulp has about the same yield from wood as stone groundwood, but its long fiber content and paper making quality are improved by about 15%. The introduction of the disc refiner marks the beginning of efforts to produce reinforcement fiber for newsprint and publication grade papers from mechanical pulp only. The original incentive was to convert "most of the tree" into useful fiber, that is to leave as much lignin as possible in the fiber, and this approach also served as a temporary complaisance towards effluent load regulations.

This line of pulping development brought about much activity in the design and manufacture of machinery for the mechanical fiberization of wood chips (see Vol. 2). However, attempts made in a number of installations have not been successful to achieve this goal.

Next, temperature was employed to help in the process of refiner fiberization of chips, as a means for softening the lignin to some degree, thereby facilitating fiber liberation. Chips are preheated to temperatures between 110-135°C and then fiberized in steam-pressurized refiner. The first such installation started up in Sweden in 1968. The process was named thermomechanical pulping, TMP, a broad term which does not give information on the actual process, and was promoted around the world for producing mechanical pulp for newsprint, paperboard,

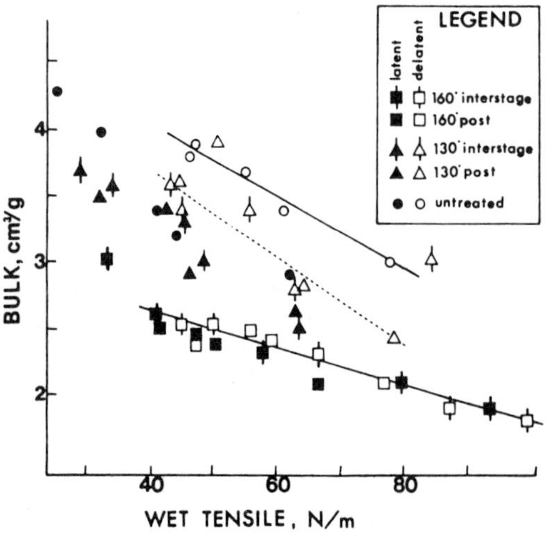

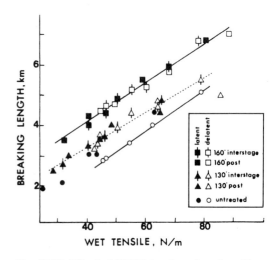

Fig. 173A. Effect of OPCO treatment on bulk.

Fig. 173B. Effect of OPCO treatment on breaking length.

Table 50. OPCO pulp — mill trial.

	Untreated		Treated	
	Latent	Delatent	Latent	Delatent
Freeness, CSF	493	265	186	178
Drainage, s.	0.59	0.74	1.19	1.31
Wet web properties[1]				
Solids — %	30.8	30.8	31.1	31.1
Tensile (N/m)	30	62	71	81
Stretch (%)	11.3	4.4	8.7	6.5
Caliper (mm)	0.450	0.387	0.335	0.318
Dry properties				
Bulk	3.55	3.20	2.17	2.02
Burst index	0.8	1.7	2.9	3.3
Breaking length	1300	3000	4600	5300
Stretch (%)	1.7	1.6	2.2	2.4
Tear index	8.3	11.5	9.0	8.0
Fibre classification				
R-14	—	21.0	—	16.0
R-28	—	32.8	—	33.0
R-48	—	13.6	—	14.4
R-100	—	6.8	—	7.2
R-200	—	1.0	—	6.0
P-200	—	24.8	—	28.8

(1) Basis weight of sheet — 52 g/m^2

Table 51. Paper machine trial.

		Before	During	After
Furnish				
Groundwood	(%)	57	50	58
TMP	(%)	25	22	26
Semibleached kraft	(%)	9	0	8
Unbleached sulphite	(%)	9	0	8
OPCO pulp	(%)	0	28	0
P.M. conditions				
Wire speed	(m/min)	693	694	695
Reel speed	(m/min)	724	724	724
Headbox pressure	(kPa)	63.2	63.8	63.3
Slice opening	(mm)	10	10	10
Supercalender speed	(m/min)	700	700	700
Linear pressure	(K·N/m)	190	190	190
Paper characteristics				
Basic weight	(g/m^2)	48.1	49.8	48.7
Bulk	(cc/g)	1.56	1.57	1.56
Tensile index M.D.	(Nm/g)	44.5	41.9	41.2
Stretch M.D.	(%)	1.05	1.13	1.11
Tear C.D.	(mN·m^2/g)	222	196	203
Burst index	(kPa·m^2/g)	1.43	1.31	1.32
Roughness 1 kg T.S.	(m/min)	86	101	87
W.S.		97	104	92
Brightness I.S.O.	(%)	60.9	59.2	61.2
Opacity I.S.O.	(%)	93.9	93.6	93.6
Porosity	(m/min)	224	213	216
Caliper uncalendered	μm	120	127	123
calendered	μm	75	78	76

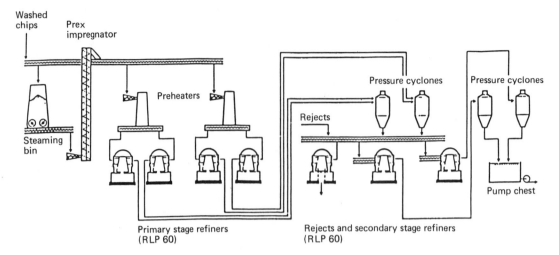

Fig. 174. Schematic diagram of CTMP plant at Bathurst.

molded products, and tissue. The yield from softwood is about 94%. Compared with groundwood, long fiber content and wet strength are increased by 40%, tensile by 30%, and tear by 60%. Although this trend has led to the installation of many plants worldwide, it is short lived as quality awareness continues to rise.

This was predictable because chemical treatment, sulfonating the lignin and rendering it hydrophilic, is more effective for modifying the physical properties of lignin than heat and water alone so that mechanical liberation of intact fibers from the wood structure is facilitated. Considering the nearness of mechanical wood fibers to the hard and brittle wood substance, the range of possible fiber quality improvements which could be produced by modifications of mechanical or thermomechanical technique is rather limited. A major step-up of fiber quality and step-down in power consumption occurs when the wood is softened by chemical treatment before mechanical treatment.

In 1978, also in Sweden, the first TMP plant was equipped with chemical pretreatment (CTMP[1]), and in 1984 the records of the Sunds-Defibrator Co. list 12 softwood and 3 hardwood CTMP plants, and 9 CMP plants (RMP plants equipped with chemical pretreatment). (Refer to Chap. V,B,7).

The treatment schedule of a recently installed

CTMP plant of Consolidated-Bathurst Inc. producing news furnish from a spruce-balsam blend (Fig. 174) consists of 30 min atmospheric presteaming, washing, draining, grit removal, a second 3 min atmospheric steaming, and adding the chips through a pressure screw plug feeder into the vertical impregnator, also with 3 min retention time. Sodium sulfite solution with DTPA (diethylene-triamine-penta-acetic acid) are added at the bottom of the up-flow impregnator as desired. The impregnated chips are then introduced via pressure screw plug feeders into pressurized preheaters where they are retained at 120-135°C for 2-5 min before distribution to the primary pressurized refiners operating at 350 kPa (51 psig). The pulp from the primary stage is blown to pressure cyclones (300 kPa, 44 psig) and conveyed to the secondary pressurized refiners (200 kPa, 29 psig) which discharge the refined pulp via pressure cyclones (150 kPa, 22 psig) and a high density pump into a common latency chest. Two-stage screening, followed by multistage cleaning, completes the treatment.

In order to reduce the relatively high energy demand of a CTMP operation, heat recovery systems have been developed for returning clean low pressure steam to other departments of the mill and/or heat to heat exchangers for heating mill process water, or glycol for space heating purposes.

Figure 175 shows a compact CTMP line for producing bleached fluff pulp. Note that the functions of pressure screw plug feeding, impregnation with chemical and preheating are combined in a single vessel.

The yield of CTMP pulp from softwood is 91-95%, from hardwood 88-95%. Compared with

---

1. Note: In the context of this volume on sulfite pulping the letter "C" in term abbreviations stands for chemical treatment, implying any of the sulfite compounds, bisulfite and/or sulfite, and alkaline sulfite. Non-sulfur alkaline chemimechanical processes, using caustic soda and/or soda for producing high yield pulps, are not included.

other mechanical pulps, CTMP pulp has a higher proportion of long fiber and greater tensile strength, lower shives and resin content, and higher brightness and bulk. In board manufacture the use of CTMP avoids the off-taste problem in milk containers made from composite liquid board using TMP. The low resin content renders this very high yield pulp useful for absorbents manufacture and its high content of long fiber helps to achieve satisfactory strength.

F. SUMMARY DISCUSSIONS OF VERY HIGH YIELD SULFITE PROCESSES

Reflecting on the present very high yield sulfite pulping processes contained in this section it becomes apparent that the conception, development and optimization in recent years originated from two distinctly different bases and followed two distinctly different lines. On the one hand, VHYS pulps were achieved from the basis of low yield pulping technology by using less and less severe cooking conditions while paying attention to good sulfonation and least dissolution of lignin as well as raising end use quality of the fiber by applying appropriate amounts of refining energy. On the other hand, VHYS pulps were achieved from the basis of very high yield mechanical pulping technology by reducing severity and energy demand of the mechanical treatment by means of marginal chemical treatment and in the shortest possible time, the attention being placed on providing fiber of satisfactory end-use quality at high productivity.

Although both lines of approach produce useful chemimechanical sulfite pulps at very high yield there is a distinct difference in the type of equipment developed for achieving this goal. On the chemical side, the emphasis has been on the adaptation of cooking equipment while refining equipment remained conventional and of secondary importance. On the mechanical side, new plants were built containing rapid impregnators and preheaters, and refiner departments for operation above atmospheric pressure. Systems for the recovery of steam heat are provided to reduce the relatively high energy demand.

Of the two, the chemical approach has the advantage that the option of improving pulp quality by lowering pulp yield remains open whereas with the mechanical approach taken pulp yield can be lowered only by installing additional cooking equipment which, at the same time, would weaken the need for pressurized refining.

G. COMPARISON OF OPERATING COSTS OF COMPETITIVE NEWSPRINT FURNISHES [46]

A comparison of newsprint furnish costs, containing mechanical pulp with sulfite pulps at different yield levels as well as other competitive ultra high yields pulps, is given in Tables 52 and 53. The costs of TMP & CTMP were made, for simplicity, without accounting for savings due to steam recovery. The highest cost, as can be seen, is that of low yield sulfite and groundwood, while one of the lowest is that containing very high yield pulp. It is expected that, in the foreseeable future, a constant increase in wood and energy costs will add weight to further expansion and wider applications of the ultra high yield pulps, not only in newsprint but also with some modifications in fine paper grades.

A more extensive comparison of manufacturing costs including capital and energy consumption were made [62]. While the results, as shown on Figs. 176 and 177 respectively, are not updated, they remain reasonably meaningful.

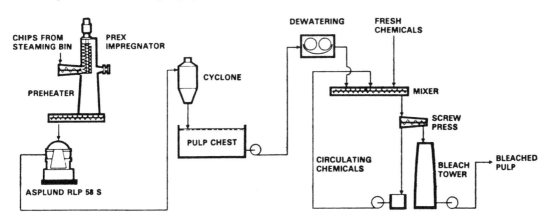

Fig. 175. Skoghall's bleached CTMP line for fluff pulp.

Table 52. Operating costs of various pulps used in newsprint production.

| Pulping Process | Wood @ C$100/T | | Energy @ 0.03$/kWh | | | | Chemicals | | | Total |
| | Yield % | $/T Pulp | Refining | | Steam | | S@$145/T; NaOH@$312/T | | | Cost |
			GJ/T	$/T	GJ/T	$/T	kgS/T	kgNaOH/T	$/T	$/T
LYS	50	200.00	—	—	6.4	53.3	143	154	43.1	296.4
HYS	72	138.89	2.0	16.7	3.7	30.8	60	75	32.1	218.5
VHYS	85	117.65	4.9	40.5	2.7	22.5	53	66	28.3	208.9
SCMP	89	112.36	5.7	47.0	2.4	20.0	37	94	34.7	214.0
OPCO	91	109.89	7.0	58.3	1.0	8.3	27	70	25.7	214.3
CTMP	92	108.70	8.5	70.8	—	—	11	27	13.4	192.9
TMP	94	106.38	7.8	65.0	—	—	—	—	—	171.4
GWD	95	105.26	5.5	45.8	—	—	—	—	—	151.1

LYS, HYS, VHYS: Low, high & very high yield sulfite (CBI)
SCMP: Sulfonated chemimechanical pulp (CIP) (1)
OPCO: Sulfite interstage treatment (including 1% EDTA) (1)
CTMP: Chemithermomechanical pulp (2)
TMP, GWD: Thermomechanical, groundwood pulps
Ref: (1) CPPA 1982 Ann. Meet. Prep.; (2) CPPA Mech. Subcom. Mtg.

Table 53. Costs of typical newsprint furnishes.

Furnish	Pulp Ratio	News Furnishes[1] Yield, %	Cost $/T
GWD/LYS	78/22	85	183.05
GWD/HYS	75/25	89	167.94
GWD/VHYS	70/30	92	168.44
GWD/SCMP	65/35	93	173.12
GWD/TMP/OPCO	45/25/30	94	175.13
GWD/CTMP	50/50	94	172.00
CTMP	100	92	192.90
TMP	100	94	171.38

[1]Based on initial wood

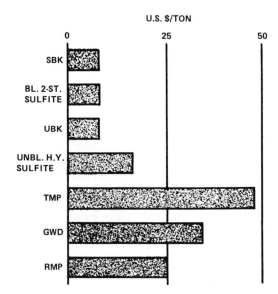

Fig. 177. External energy consumption of various pulping processes (locally adjusted, 1977 prices).

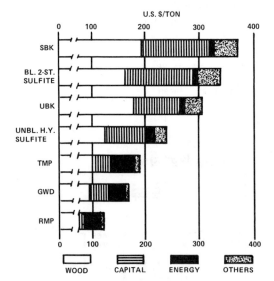

Fig. 176. Total manufacturing costs of various pulps (locally adjusted, 1977 prices).

A recent pulp cost comparison [63] done for a specific mill has indicated that the minimum (cooking and refining) costs for HY sodium bisulfite pulp would correspond to a yield of about 72%. Variations in wood costs (valued at the cost of 1500 kWh/t o.d. wood), pulp quality and recovery have some effect on the optimum pulp yield. (Figs. 178-180).

H. BLEACHING/BRIGHTENING OF HIGH AND VERY HIGH YIELD PULPS

Bleaching of low yield chemical pulp and surface bleaching of mechanical pulps received much attention and achieved a wide degree of mill-scale applications. Few efforts were devoted to improving brightness levels of HY and VHY sulfite pulps without significant losses of yield.

Conventional oxidative (peroxide) and reductive (hydrosulfite) methods and their combinations were tried as brightening methods for HY sulfite pulps [64-68]. Conflicting results were obtained as to their effectiveness. Some of the discrepancies could be explained by the fact that sulfonated lignins are sensitive to oxidation which results in the development of pink and grey color substances of stilbene quinone type chromophores [69]. Also, the brightness of unbleached bisulfite pulp is not directly proportional to the yield level (Fig. 181). The discoloration reaction of the pulp is also catalyzed particularly by the presence of metal ions Mn, Cu or Fe derived from the original wood material, refiner plates and water.

In a recent detailed bleaching study [66] on 55-82% yield spruce bisulfite pulps, brightness gains of 6 to 9 points were obtained in single-stage bleaching; a 7 to 14 point increase was seen, depending on pulp yield, in two-stage bleaching where peroxide was followed by hydrosulfite, as shown in Table 54 and Fig. 182.

The following brightening conditions (chemical charges based on O.D. pulp) were used:

	Peroxide H_2O_2	Hydrosulfite
Chemicals H_2O_2	: 1%	$Na_2S_2O_4$: 1%
NaOH	: 2%	–
Na_2SiO_3	: 2%	–
$MgSO_4$	: 0.05%	–

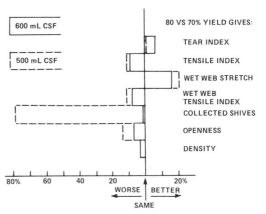

Fig. 180. Paper properties at 80 vs 70% yield.

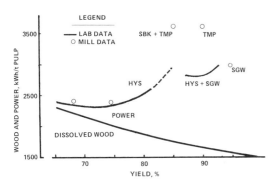

Fig. 178. Wood and power costs vs yield.

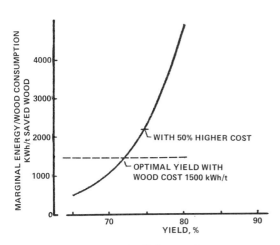

Fig. 179. Cooking and refining costs.

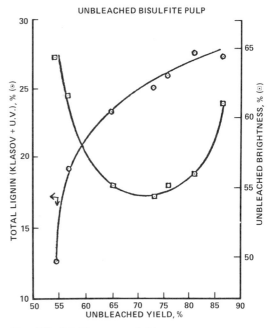

Fig. 181. Brightness vs yield.

	Peroxide H$_2$O$_2$	Hydrosulfite
Operating conditions pH	: 10.5	: 6
Consistency	: 10%	: 5%
Temp.	: 60°C	: 60°C
Time	: 2 h	: 1 h

The level of brightness gain for a 70% yield bisulfite pulp is dependent on the amount of chemicals applied as shown in Fig. 183. It can be seen that the brightness levels off with 1% hydrosulfite application while there is an increase in brightness with at least up to 1.5% peroxide addition. Significantly higher brightness gains are achieved when two-stage bleaching is applied (i.e. peroxide followed by hydrosulfite).

In general, equipment and techniques used in this area are the same as those for surface bleaching of mechanical pulps.

More innovative approaches to bleaching/brightening of HY sulfite pulps such as the use of oxygene/ozone [70-71], peroxygen compounds, e.g. cume peroxide [72] and peracetic acid [67, 72-75] as well as sodium borohydride [64, 65, 67] and thiourea dioxide [76-77] were investigated. While some results were encouraging, no mill implementation is considered due to the high costs of these chemicals.

I. TREATMENT OF SPENT SULFITE LIQUORS

Appendix 2 presents characteristics of spent sulfite liquors. A discussion of treatment methods and strategies will be presented in future volumes. See references 101-122 for additional information.

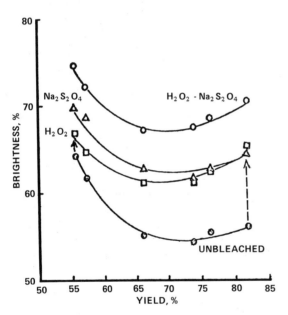

Fig. 182. Brightness of bleached spruce bisulfite pulp.

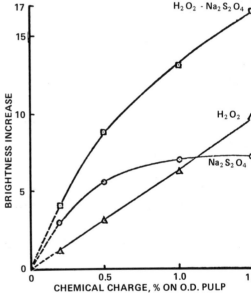

Fig. 183. Brightness increase on a high yield (70%) bisulfite pulp.

Table 54. The effect of peroxide hydrosulphite, and combination of P-Ha.

Unbleached Yield, %	Brightness %			
	Unbleached	H$_2$O$_2$	Na$_2$S$_2$O$_4$	H$_2$O$_2$-Na$_2$S$_2$O$_2$
81.8	56.1	65.3	65.0	70.7
76.6	55.2	62.7	62.7	68.6
74.0	54.4	61.3	61.6	67.7
65.9	55.1	61.2	62.9	67.3
57.2	61.6	64.8	68.8	72.2
55.5	64.3	66.9	70.1	74.6

REFERENCES

1. Bolviken, A., and Giertz, H.W.,: *Norsk Skogindustri* 10; 344 (1956); 12,235 (1958); Giertz, H.W.,: *Norsk Skogindustri* 11; 223 (1957).
2. Husband, R.M., *Tappi*, 36, 529 (1953); 38,577 (1955).
3. Hart, J.S., Strapp, R.K., and Ross, J.H., *Pulp Paper Mag. Can.* 55, 10, 14 (1954); Hart, J.S., Woods, J.M., *Pulp Paper Mag. Can.* 56, 9, 95 (1955); Kerr, W.D., Harding, S.A., *Tappi* 39, 308 (1956).
4. Dorland, R.M., Leask, R.A., McKinny, J.W., *Pulp Paper Mag. Can.* Tech. Section Proceedings: 121-31, Jan. 1958.
5. Tomlinson, G.H., and Tomlinson, G.H., 11, Bryce, J.R., and Tuck, N.G.N., *Pulp Paper Mag. Can.* Tech. Section Proceedings p. 132 (44th Annual Meeting Jan. 1958).
6. Rydholm, S.A., *Pulp Paper Mag. Can.* 68 (1); T2-22 (1967).
7. Private Communication. Literature Reviews "Bisulfite Pulping" Tech. Services Dept. Allied Chemicals Ltd. (Aug. 1965).
8. VanEykson, H.K., International Sulfite Pulping Conference, June 15-19 (1964); *Pulp Paper Mag. Can.* 65, 7, 86-91, 102 (1964).
9. Evans, J.C., and Dyer, H., *Paper Trade J.* 148, 26, 30-27 (1964).
10. Jensen, W., Palenius, I., and Makkonen, H., *Tappi*, 51 (8) 52A-55A, Aug. 1968.
11. Ayroud, A.M., Internal reports covering some 20 years of unpublished studies. Research Centre, Consolidated-Bathurst Inc., Grand'Mère, Quebec, Canada.
12. Hägglund, E., "The Chemistry of Wood", Academie, New York, N.Y., p. 425-31 (1951).
13. Stockman, L., *Svensk Papperstidn.* 54, No. 18, p. 621 (1951).
14. Schöön, N.H., *Svensk Papperstidn.* 65, No. 19, p. 729 (1962).
15. Rydholm, S.A., Pulping Processes, p. 417-420 and references therein: Interscience Publishes, N.Y. (1965).
16. Casey, J.P., Pulp and Paper Chemistry and Chemical Technology, 3rd Ed., John Wiley and Sons, New York. (1980).
17. Stockman, L., "Fortschritte in der schwedischen Sulfitzellstoff-Forschung" *Das Papier* 1960, No. 3, S 85 (1960).
18. Maloney, T.W., Gibbs, V., and Andrews, D.H., "Ammonium Base Bisulfite Pulps and their Use in Newsprint Blends". *Pulp Paper Mag. Can.* 74 (11):108 (Nov. 1973).
19. Nolan, W.J., *Tappi* 53 (7):1309-1315 (July 1970).
20. Jayne, V.G., *Das Papier* (Dec. 1960).
21. "Current Practice in H.Y. Sulfite Pulping" — A panel discussion moderated by Wilson, G., *Pulp Pap. Can.* 79 (8) T254 (Aug. 78).
22. Ayroud, A.M., In preprint "H.Y. Pulping Seminar" Tokyo, Japan, Oct. 2-6/78. Sponsored jointly by Japanese TAPPI & CPPA.
23. Luzzi, K., "Excessive Rejects with Mg Cured by Going to Na-base", *Pulp Pap. Can.*, Vol. 79, No. 8, p. T259 (Aug. 78).
24. Stevens, T., "Converting from Mg to Na Benefits Acid Making", *Pulp Pap. Can.*, Vol. 79, No. 8, p. T258.
25. Refer to papers delivered at a Panel Presentation on "H.Y. Bisulfite Pulping" sponsored by the Sulfite Committee during the CPPA Annual Meeting.
    a) Meunier, G.A., re. Consolidated-Bathurst Inc. — Port Alfred Mill, Quebec;
    b) Mosher, R.D., re. Ontario-Minnesota, P. & P. Co. Ltd. Kenora Mill, Ont;
    c) Bieweg, Ward Allan, re. Domtar, Trois-Rivières Mill, Quebec;
    d) Laberge, H.G., re. The Price Company, Alma, Quebec; Convention issue: *Pulp Paper Mag. Can.* (1971).
26. "Nova Scotia Forest Industries, New High Yield Sulfite Pulp Mill", reported in *Paper Trade J.*, p. 22-26, July 23/73.
27. Hassinen, Ilpo — "Continuous Digester for Bisulfite Pulp is Success at Finnish Mill" — *Pulp & Paper*, Vol. 49, No. 12, pp. 92-94 (Nov. 1975).
28. Fowler, F., "More Production from Part Conversion of Low to High Yield Mg" *Pulp Pap. Can.*, 79 (8); T260 (Aug. 78).
29. Keef, R.C., "Mg Bisulfite Recovery Start-up" *Tappi* 54 (4); 564 (April 71).
30. Binotta, A., "Effect of Chip Quality on High Yield Bisulfite Pulp" Data presented at 1980 CPPA Sulfite Committee.
31. Hartler, L., Lind, L., Stockman, L., "Decomposition of Sulfite Cooking Liquors" *Svensk Papperstidn,* S. 336-340 (May 15, 1964).
32. Atack, D., Heitner, C., and Karnis, A., "Ultra H.Y. Pulping of Eastern Black Spruce. *Pulp Pap. Can.* 82 (C), T103-T110 (1981).
33. Giertz, H.W., "Pulping of Chemi-mechanical Bisulfite Pulps with Ammonium Base Liquor". Also Andrews, D.H., "The Effect of Ammonium Base Sulfite Pulp on the Brightness of Newsprint Blends". Internal reports. Courtesy CIL, Private Communication.
34. Heitner, C., Beatson, R.P., and Atack, D., "Factors Affecting Sulfonation of Eastern Black Spruce Wood Chips" Preprints 19th Ekmann-Days 1981, Symposium entitled: International Symposium on Wood and Pulping Chemistry, Stockholm, Sweden, June 19-1981, Vol. 11, pp. 145-149.

35. Scallan, A.M., and Grignon, J., "The Effects of Cations on Pulp and Paper Properties. *Svensk Papperstidn.* 82, No. 2, 40-47 (Feb. 10, 1979).

36. Frederiksson, B., and Hoglund, H., "Chemi-Thermomechanical Pulps in Different Paper Grades" *Appita* 31, No. 5-365-370 (Mar. 78).

37. Ford, M.J., and Gardner, P.E., "Methods of Producing H.Y. Chemi-Mechanical Pulps" US Patent No. 4, 116, 758 (Sept. 26, 78).

38. Atack, D., Heitner, C., Jackson, M., and Karnis, A., "Sulfite Chemimechanical Refiner Pulp. Another option for newsprint". *Pulp Paper* 54, No. 6, 70-72 (June 80).

39. Beath, L.R., and Mihelich, W.C., *Tappi* 50, 12:77 (1977).

40. Ayroud, A.M., Goel, K., and Lemay, Y., *Trans. Tech. Sect., CPPA*, 6 (3) TR 65-72 (Sept. 1980).

41. Heitner, C., Atack D., and Karnis, A., "Sequential Peroxide-Ozone Treatment of Refiner Mechanical Pulp Rejects", *Paperi ja Puu* 63, No. 2, 53-61 (1981).

42. Barnet, A.J., Shaw, A.C., and Logan, C.D., "Drainage and Wet Stretch Improvement in Mechanical Pulps". Can. Patent No. 1071805 (Feb. 19, 1980).

43. Barnet, A.J., Leask, R.A. and Shaw, A.C., "The Opco Process, Parts 1 & 11, *Pulp Pap. Can.* 81, No. 10, T255-257, T258-T260 (Oct. 1980).

44. Shaw, A.C., "The Opco Process Applied to TMP Screen Rejects". *Pulp Paper Can.*, 84, 32-35 (May 1983).

45. Characklis, W.G., Bush, A.W., "Industrial Waste Water Treatment" — Chem. Eng. Deskbook issue (May 8, 1972).

46. Ayroud, A.M., and Lemay, Y., "Very High Yield Sulfite Pulping" paper presented at ESPRA Spring meeting (May 31-June 4, 1982).

47. Ayroud, A.M., and Lemay, Y., "Chemical & Physical Factors Affecting Productions and Characteristics of V.H.Y. Sulfite Pulps" — Paper presented at the 28th IUPAC Congress, Vancouver (Aug. 16-21, 1981).

48. Deshaye, R., Mihelich, W.G., "Cuisson de Différentes Espèces pour une Pâte de Sulfite à Très Haut Rendement", presented at Conf. Technologique Estivale. CPPA, Montreal, Quebec, April 1981.

49. Ford, M.J. and Gardner, P.E., "Method of Producing High Yield Chemimechanical Pulps" US Patent No. 4,116,758 — Sept. 26, 1978; Canadian Patent No. 1,051,681 April 3, 1979.

50. "The Story of Sulfonated Chemimechanical Pulp" reported in *Pulp Pap. Can.*, 82, No. 9, pp. 44, 47-48 (1981).

51. Mutton, D.B., Tombler, G., Gardner, P.E., and Ford, M.J., "The Sulfonated Chemimechanical Pulping Process", *Pulp Pap. Can.*, 83, No. 6, T 186-194 (1982).

52. Cayouette, D.J., Dines, R.E., Morey, L.P., and Ford, M.J., "Gatineau SCMP — Three Years Later" presented at the CPPA 68th Annual Meeting, Jan. 1982.

53. Hoglund, H., Bodin, O., "Modified Thermomechanical Pulp" *Svensk Papperstidn.*, 79, No. 11, pp. 343-353 (1976).

54. Atack, D., Heitner, C., Stationwala, M.I., & all — "Ultra High Yield Pulping of Eastern Black Spruce", *Svensk Papperstidn.* 81, No. 5, pp. 164-176 (1978).

55. Atack, D., Heitner, C., and Karnis, A., "Ultra High Yield Pulping of Eastern Black Spruce" — Part 11, *Svensk Papperstidn.*, 83, No. 5, pp. 133-141 (1980).

56. Beath, L.R., Mihelich, W.G., "Refiner Mechanical Pulping of Chemically Pretreated Wood", *Tappi* 50, (12):77 (1977).

57. Bertrand, G., "TCMP in Kenogami", paper presented at the Mechanical Pulping Committee Meeting in Montreal, March 1980.

58. Mihelich, W.G., "TCMP, A Strong Mechanical Pulp", presented at the ESPRA meeting (May 31-June 4, 1982) Trois-Rivières, Quebec, Canada.

59. Goel, K., Paquin, R., Consolidated-Bathurst Inc., Research Centre, Internal reports unpublished data.

60. Gilliespie, W.J., Marshall, D.W., and Springer, A.M., "A Pilot-Scale Evaluation of Rotating Biological Surface Treatment of Pulp and Paper Mill Wastes", *Tappi*, 57 (9):112-117 (Sept. 1974).

61. Leask, R.A., MacDonald, J.R., and Shaw, A.C., "High Yield Bisulfite — A Refining Pretreatment" *Pulp Paper Can.* 75, 59-63 (Oct. 1974).

62. Poyruy, J., *Norsk Skogindustri*, 31 (12):332, 1977.

63. Bernhard, R., Lawson, L., "Optimal Yield for H.Y. Sodium Bisulfite Cooking", CPPA Annual Meeting (Montreal) Preprints 66B: 79-82, Jan. 29-30, 1980.

64. Christensen, P.K., *Pulp Paper Mag. Can.* 69, No. 2,64-8 (Jan. 19, 1968).

65. Liner, P., Ferguson, D.W., *Tappi*, 47, No. 4: 205:10 (April 1964).

66. Liebergott, N., Joachimdes, T., "Bleaching of H.Y. Sulfite Pulps", A literature survey presented at the Intern. Sulfite Conf., Montreal 1978.

67. Yokoo, H., *J. Jap. Tappi* 20, No. 12: 641-6 (Dec. 1966).

68. Palenius, I., et al., *Pulp Paper Mag. Can.* 72, No. 21, 63-70 (Nov. 6, 70).

69. Adler, E., and Haggroth, S., *Svensk Papperstidn.* 53:321 (1950).

70. Liebergott, N., Bleaching Seminar PPRIC (Feb. 20, 1970).

71. Soteland, N., Loras, V., *Norsk Skogindustri* 28, No. 6: 165 (1974).

72. Bailey, C.W., Dence, C.W., *Tappi*, 49, No. 1: 9-15 (Jan. 1966).

73. Taniguchi, Euchi, *Japan Tappi*, 21, No.

74. Rapson, W.H. et al., *Pulp Paper Mag. Can.* (May 1965).
75. Damian, J.C. et al., *La Papeterie*, No. 6:558-62 (1969).
76. Turner, J.H.W. et al., US Patent 3,481,828 (Dec. 2, 1969), Can. Patent 769,632 (Oct. 17, 1967).
77. Kindron, R.R., and Houg, G.W.; Can. Patent 811,632 (April 29, 1969).
78. Wong, A., "Characteristics of Spent Sulfite Liquors — A CPPA/TS Sulfite Committee Survey" *Pulp Pap. Can.* 82 (6):T 184 (June 1981).
79. Refer also to Hartler, N., Stockman, L., and Sundberg "Effect of pH in Sulfite Pulping", *Svensk Papperstidn*, 64 (2):33 (Jan. 1961).
80. Meindl, N., Schimdt, J. and Schwarzl, "Continuous Vapor-phase Magnefite Pulping", *EUCEPA* 1978.
81. Kleinert and Marraccini, L.M., "Distribution of Chemicals in Commercial Wood Chips" *Tappi* (3):165 (1965).
82. Ginaven, M.E., and Adams, A.H., "Impregnation of Chips" US Patent No. 2,975,096 (March 14, 1961).
83. Alm, A., and Stockman, L., "Chip Impregnation in Sulfite Cooking". *Svensk Papperstidn*. (1):10 (1958).
84. Fogelberg, B.C., and Fugleberg, S., "Evaluation of Various Factors in the Evaluation of Chips" *Paperi ja Puu*, 47 (3):97-107 (1965).
85. Maass, O., "Principles of Penetration", *Pulp Paper Mag. Can.* 54(8) S.98 (1953).
86. Ogawa, E., and Gorbatsevion, S.N., "Vapor Phase Magnefite Pulping of Spruce", *Tappi*, 51 (4):171 (1968).
87. Jayme, G., Broschinski, L., and Matzke, W., "H.Y. Pulps from Spruce in Rapid Magnesium Bisulfite Process" *Das Papier* 18 (7):308 (1964).
88. Wenzl, H.F.J., "Progress in Continuous Sulfite Pulping", *Paper Trade J.* 57 (May 1975).
89. Annergren, G., and Backlund, A., "Continuous Sulfite Pulping" *Pulp Pap. Mag. Can.* T220 (April 1966).
90. "Brite Chem. Pulp?" Can. Patent #637,002, Feb. 20/62.
91. Page, D.H., Barbe, M.C., Seth, R.S., Jordan, B.D., *J. Pulp Paper Sci.* Vol. 10 (3):J74 (May 1984).
92. Garrie, M., and Doershuk, D.C., Gaspesia's vapor phase M&D digester. *Pulp Pap. Can.*, 85, 19-20 (Sept. 1984).
93. MacEwen, H.D., Richardson, R.T., and Rooney, R.J., Conversion from HYS to SCMP at the Ft. William Division of Abitibi-Price Co. *Pulp Pap. Can.* 85, T162 (June 1984).
94. Anon. NBIP's Dalhousie expansion saves dollars, increases production, improves sheet quality. *Pulp Pap. Can.* 85, 17-24 (Jan. 1984).
95. Dines, R.E., and A. Tyminski, The vapor phase SCMP process *J. Pulp Paper Sci.* Vol. 10 (5):J17 (Sept. 1984).
96. Herzog, P., Chemical mechanical pulp, *Wochenblatt Papierfabrikation* 103, 802-7 (No. 21, 1975).
97. Evans, J.C.W., First OPCO process in operation at QNS Paper, Baie Comeau, Que. *Pulp Paper*, 90-99 (Jan. 1985).
98. Evans, J.C.W., Con Bath's Canadian Plant produces CTMP for its British newsprint mill. *Pulp Paper*, 80-83 (Oct. 1984).
99. Smith O., Mason, J., Engstrom, B., and Jackson, M., Heat recovery at the Bathurst CTMP mill. *Pulp Paper Can.* 85, T275-282 (Nov. 1984).
100. Alsholm, O., and Swan, B., Sweden leads in CTMP development. *Pulp Paper Intern.*, 48-50 (Nov. 1984).

LITERATURE ON TREATMENT OF SPENT SULFITE LIQUOR

101. LeBlanc, P.J., "A review of the BOD test", *Pulp Paper Mag. Can.* 75 (3):79 (March 1974).
102. Jank, B.E., "Water Pollution Abatment Technology in the Pulp & Paper Industry", Environment Canada — Seminar notes (May 1975).
103. Duback, M., "Method and Apparatus for Sewage Disposal According to Activated Sludge Process" US Patent #3,390,076 (June 25, 1968) See also "Effluent Water Purification Process" US Patent #3,354,082 (Nov. 21, 1967).
104. "ARANOVA — System Attisholz" Report from Cellulose Attisholz A.G. Dept. AbWasser, Luterbach, Switzerland.
105. "Zurn-Attisholz System is chosen by Wausau for Pulp Mill Effluent" *Paper Trade J.* (Jan. 27, 1975). See also MacLeod, M. "Deinking mill gets outstanding results with pioneer treatment" *Pulp and Paper*, p. 87 (April '74).
106. Lo, S.N., Garceau, J.J., Pineault, G.; *ATIP* 2, 1975, Vol. 29 (1975).
107. Love, L.S., and Gukllaume, F., *Pulp Pap. Can.*, Vol. 76, No. 7, 62-66, July 1975.
108. Love, L.S. Associated Ltd; CPAR Project report 195-1, July 1974.
109. VanSoest, R. and Cheek, R.P., CPPA Tech. Sect. Papers Discus., No. 1, D17-22, July 1974.
110. VanSoest, R., *Pulp Paper Can.*, Vol. 76, No. 1, 1975; 70-71.
111. Tricart, G., Lo, S.N., Garceau, J., and Cholette, A., "Epuration Biologique de la Liqueur dans un Réacteur à Deux Stages" *Pulp Pap. Can.* 79 (2):T65 (Feb. 78).
112. Lo, S.N., Garceau, J., Wusunger, I., and Pineault, G., "Biological Oxidations of Concentrated High Yield Bisulfite Spent

10:538-44 (1967).

Liquors"; *Pulp Pap. Can.* 78 (1); T19 (Jan. 1977).

113. Goel, K., Paquin, R., Mehta, Y., and Lemay, Y., "Biological Treatment of Spent Sulfite Liquors from H.Y. Bisulfite Pulping Operation", *Trans. Tech. Sect. CPPA* 3 (1):6-11 (1977).

114. VanSoest, R., Scott, J., "Selection of Effluent Treatment Systems for Wausau Paper Mills Co." Paper delivered at Annual Meeting CPPA, Preprints A35-39 (1975).

115. Zajic, J.E., Hill, M.A., Manchester, D.F., and Muzika, K., "A foam activated Sludge Process for the treatment of Spent Sulfite Liquor" *J. Water Pollution Control*, Fed. 50 (5):884-895 (May 1978).

116. Cadotte, A. and Laughlin, B., "New Development in Waste Disposal", *Pulp Paper Can.* 80 (10):105-108 (Oct. 1979).

117. Peterson, G., Klein, J., and Martinsen, D., "A Pure Oxygen Activated Sludge System: Operating Experience", *Tappi* 61 (10):45-50 (Oct. 1978).

118. Othmer, D.F., "Oxydation of Aqueous Wastes: The Prost System". *Chemical Eng.,* p. 117-120 (June 20, 1977).

119. "Waste Water Processing Advances Underground", *Chemical Eng.* p. 46 (Aug. 4, 75).

120. Forss, K., Passinen, K., "Utilization of the Spent Sulfite Liquor Components in the Pekilo Protein Process"; *Paperi ja Puu*, No. 9:608-618 (1976).

121. Forss, K., Kokkonen, R., Sirelius, H., Sagfors, P.E., "How to Improve Spent Sulfite Liquor Use" *Pulp Pap. Can.* 80 (12): T411-T415 (Dec. 1979).

122. Andersen, R.F., "Production of Food Yeast from Spent Sulfite Liquors" Preprints CPPA, Annual Meeting.

VIII

DISSOLVING PULP MANUFACTURE

J.F. HINCK,   R.L. CASEBIER

& J.K. HAMILTON

Mr. Hinck, a graduate of the College of Forestry, Syracuse, New York and holding degrees in wood chemistry and forestry, has been involved for twenty-three years in research, technical marketing, and manufacturing of dissolving and specialty woodpulps for all end uses. His present position is Technical Director of Western Pulp Limited Partnership in Vancouver, British Columbia.

Dr. Casebier is Senior Technical Advisor reporting to the ITT Rayonier Corporate Technical and Quality offices. He holds a B.S. from Washington State University and a Ph.D. in organic chemistry from the University of Minnesota.

Dr. Hamilton is Director of the ITT Rayonier Research Center in Shelton, WA. He holds a B.A. and M.A. from the University of British Columbia and a Ph.D. in Agricultural Biochemistry from the University of Minnesota.

MANUSCRIPT REVIEWED BY

Alec Bialski, Technical Director Tembec, Inc., Temiscaming, Quebec

George Nelson, Process Engineering Supt. International Paper Co. Natchez, Mississippi

Wood derived celluloses made by the prehydrolyzed kraft and acid sulfite processes utilized for the manufacture of viscose rayon, cellulose esters (acetates, proprionates, butyrates, nitrates), and cellulose ethers (carboxymethyl, ethyl, methyl) are called dissolving pulps. Other terms synonymous with dissolving pulps are chemical cellulose and special high alpha grades. Since cellulose is virtually insoluble in ordinary solvents, the purpose of derivatization is to form cellulose compounds soluble in common solvents (e.g., aqueous caustic, acetone). The solubility then allows the formation of fibers, films, plastics, or water soluble derivatives by the appropriate technology. A tabulation of end uses for specific derivatives is shown in Table 55.

The degree of conversion or completeness of reaction to the derivative is important, since any impurities which do not react completely in the conversion process will likely remain insoluble and may therefore block the orifices of spinning jets (a typical rayon spinnerette contains 1200-30 000 holes each with diameters of 0.03-0.05 mm); or will show up as defects in the molded articles. The degree of conversion is estimated by the filtration value of the cellulose derivative. As an example, it should be noted that a good filtering viscose will contain no more than 0.04% insolubles by weight while a totally unsatisfactory viscose will contain approximately 0.1%. Dissolving pulps contaminated by as little as 0.2% paper pulp will reduce acetate filterability by 40%. Dissolving pulps are generally of high cellulose content (90-98%), low hemicellulose content, very little residual lignin, low extractive and mineral content, as well as having high brightness and very uniform molecular weight (chain length) distribution. Typical analytical properties are shown in Table 56 with molecular weight distribution curves found in Fig. 184. The specific influence of each of the properties to viscose and acetate processing will be found in the following sections.

In this chapter, we will discuss: (1) market overview, (2) description of the viscose and acetate processes with the effect of pulp analytical properties discussed, (3) the manipulation of sulfite, prehydrolyzed kraft pulping, and bleaching processes to achieve these requirements, and (4) future developments.

A. MARKET OVERVIEW

Worldwide dissolving markets require 4.6 to 5.0 million tons of chemical cellulose of which approximately 1.4 million tons are produced in North America. Major North American woodpulp producers include: ITT Rayonier, International Paper, Buckeye Cellulose, Tembec,

Table 55. Dissolving pulp end-use applications.

DERIVATIVE	END-USE APPLICATIONS	MM lbs Amount Produced[a]
Viscose Rayon		
Tire Cord	Tire and belting reinforcement	650
High Wet Modulus Staple	Apparel, furnishing	300
Regular Staple	Apparel	4600
Cellophane	Packaging	900
Continuous Filament	Apparel	1000
Miscellaneous	Sponges, sausage casing	
ESTERS		
Acetate		
Filament	Apparel, furnishing	650
Tow	Filter cigarettes	770
Plastics	Film, sheet, extruded articles	350
Mixed Esters		
Plastics	Sheet and extruded articles	200
Nitrates		
Lacquers		400
Film		
Explosives		
Esthers		
Carboxymethyl Cellulose	Detergents, cosmetics, food textile and paper sizes, well drilling muds	300
Hydroxyethyl Cellulose	Latex paints, emulsion polymerization, oil well drilling muds	
Methyl Cellulose	Food, paints, pharmaceuticals	
Ethyl Cellulose	Coating and inks	
Hydroxypropyl Cellulose	Foods and pharmaceuticals	
Carboxymethyl Hydroxyethyl Cellulose	Liquid detergents	

(a) Worldwide

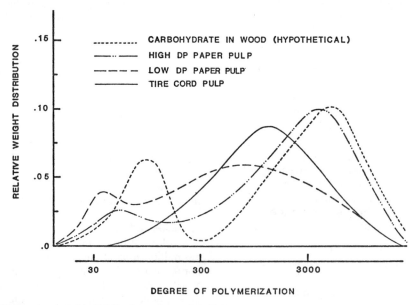

Fig. 184. Molecular weight distribution of wood cellulose.

Table 56. Typical composition of chemical cellulose from wood and cotton linters.

	Paper	Cellophane	Nitration	Plastic Filler	Textile Rayon	Tire Cord	Acetate	Acetate	Acetate Plastics
Source	Western hemlock	Western hemlock	Western hemlock	Western hemlock	Southern pine	Southern pine	Western hemlock	Southern hardwoods	Southern pine
Process	Sulfite	Sulfite	Sulfite	Sulfite	Prehydrolyzed kraft	Prehydrolyzed kraft	Sulfite	Prehydrolyzed kraft	Sulfite
Analyses[a]									
R_{10}, %	87.1	89.7	91.8	86.7	95.2	98.2	95.3	97.7	97.0
S_{10}-S_{18}, %	1.8	6.8	2.4	2.1	1.8	0.7	1.8	1.2	1.4
S_{18}, %	11.1	4.5	5.8	11.1	3.0	1.1	2.9	1.1	1.6
Xylan,[b] %	2.1	1.1	1.5	2.1	2.0	0.6	0.6	0.6	0.8
Mannan,[b] %	6.7	1.5	2.3	6.7	1.1	0.7	0.8	0.8	1.0
Extractives Soluble in Ethyl Ether, %	0.10	0.16	0.13	0.11	0.01	0.01	0.04	0.04	0.03
Ash, %	0.16	0.19	0.15	0.16	0.09	0.08	0.09	0.05	0.01
SiO_2, %	0.003	0.002	0.002	0.002	0.003	0.003	0.002	0.004	0.002
Fe, mg/kg	1.6	2.0	4.0	2.0	3.0	5.0	2.0	4.0	3.0
Cu, mg/kg	0.1	0.1	0.3	0.1	1.0	1.0	0.1	1.0	0.5
Mn, mg/kg	0.05	0.01	0.10	0.10	0.20	0.30	0.04	0.10	0.08
Elrepho Brightness, %	95.7	94.4	91.6	95.7	91.9	85.9	95.2	94.4	95.1
Cuene Intrinsic[c] Viscosity, dL/g	9.5	4.3	7.5	9.5	5.7	6.1	9.0	7.0	8.8

a) R_{10} = residue after treatment with cold 10% aqueous sodium hydroxide solution.
   S_{10} = soluble in cold 10% aqueous sodium hydroxide solution.
   S_{18} = soluble in cold 18% aqueous sodium hydroxide solution.
b) Xylose or mannose present in pulp hydrolysate calculated as xylan and mannan, determined by the paper chromatographic method.
c) This method is very similar, but not equivalent to ASTM D1795-62, other methods which are similar are TAPPI T 230 and SCAN C15-62.

Weyerhaeuser, Alapul, Western Pulp Limited Partnership and Louisiana Pacific.

Of the 5 million tons, approximately 90-95% is wood derived, with the balance originating from cotton linters. Although extensive work has been reported in the literature with regard to other sources of cellulose, e.g. sugar cane bagasse or bamboo, little is actually used in dissolving applications in North America. When used, such products are supplemented with wood based pulps in order to achieve satisfactory performance in production facilities. Operational problems arise because of the inclusion of high amounts of particulates and silica in non-woody plants. This situation does not normally occur in wood based pulps.

Wood derived pulps were first used in the 1900's to manufacture cellulose nitrate, primarily for gun powder applications. With the termination of World War I, demand for nitrocellulose dropped except for lacquer end uses which still remain as important markets today. Wood pulp manufacturers turned their attention to the viscose process of Cross and Bevan and later to cellulose acetate and ethers markets. These markets exclusively used cotton linters. Cotton linters, then as now, varied in price widely from one year to the next due to crop fluctuations. Wood pulps were introduced as being less costly and of uniform quality as well as having more stable price. One of the most significant factors in reducing rayon manufacturing costs was the introduction of wood based rayon pulp [1]. Since the introduction of wood based pulps, significant improvement (Fig. 185) in quality as measured by the R_{10}, % ($\sim$ alpha cellulose) has been attained. Not all products require the highest quality wood pulp but improvements in wood pulp purity have led to additional displacement of cotton linters. Cot-

ton linters are used only for the most demanding end uses (sausage casing, acetate plastics, and high viscosity cellulose ethers) and in those geographical areas where linters production is the by-product of a large agricultural base (e.g., China, Brazil, Argentina, India). The improvements in wood pulp quality are in part due to the following:

(1) 1930-45 – Improvements in hot alkaline refining and use of soluble bases in sulfite cooking.
(2) 1940-55 – Development of prehydrolyzed kraft pulps for viscose process.
(3) 1948-54 – Chlorine dioxide bleaching for production of high brightness pulps.
(4) 1955-63 – Cold alkaline extraction of kraft pulps for viscose markets.
(5) 1960 – Cold alkaline extraction of sulfite pulps for acetate markets.
(6) 1968 – Use of prehydrolyzed kraft cooking coupled with cold caustic extraction for acetate grades.

B. VISCOSE RAYON MANUFACTURE
1. Process description

The preparation of viscose starts with the steeping of pulp in NaOH at a concentration of 17-19% at temperatures ranging from 20-55°C. The steeping can take place either in sheet form (Fig. 186) or in a slurry (Fig. 187). The soaking of the pulp in NaOH greatly swells the cellulose and extracts hemicelluloses. This allows for further purification as well as for good diffusion of the reactants. Following soaking, the excess NaOH is removed by pressing. The pressing of the sheets or the slurry is very critical, since nonuniformity of pressing leads to poorly reacting alkali cellulose. The pressed alkali cellulose, which contains 32-35% cellulose and 14-15% NaOH is then shredded either continuously or batchwise. The purpose of shredding is to increase the surface area of the alkali cellulose, thus ensuring rapid aging and uniform xanthation.

The shredded pulp is "aged" or depolymerized in a controlled manner by reacting with atmospheric O_2. In this step, the degree of polymerization (DP) is reduced from about 700-1100 in the pulp to the DP desired for the specific end product. In the case of tire cord rayon (for which strength is important), a DP of about 500 is required; while for regular rayon and cellophane, DPs lower than 350 are required. Aging

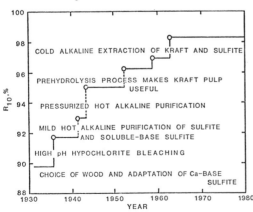

Fig. 185. Progressive change in purity of viscose-grade chemical cellulose and principal processing changes.

improves the molecular weight distribution of cellulose. Following aging, the shredded alkali cellulose is reacted with about 25-38% CS_2 by weight to form the sodium cellulose xanthate. At this point, the product color becomes yellow-orange which is caused by by-products formed in the reaction between CS_2 and NaOH. Cold dilute NaOH is then added and mixed well to dissolve the derivative. This is called *viscose*. The solution contains 6-9.5% cellulose and 5.0-7.0% NaOH by

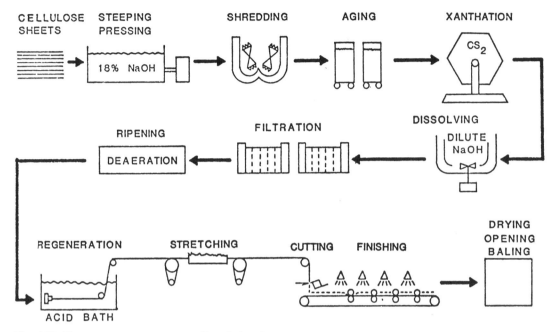

Fig. 186. Viscose process — conventional steeping.

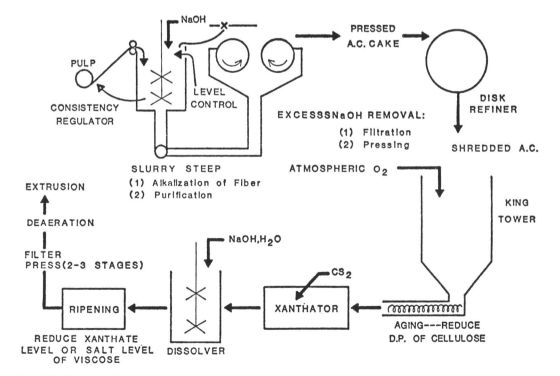

Fig. 187. Viscose slurry process.

weight. The lower DP celluloses for the less demanding end uses give viscoses having the highest concentration of cellulose and lowest concentration of NaOH. High cellulose maximizes plant throughput, and low NaOH will minimize usage of H_2SO_4.

The viscose is ripened to give the optimum degree of xanthate substitution for spinning and then is filtered to remove unreacted particles and partially reacted fibers (gels). Finally, the viscose dispersion is deaerated and then spun into a diluted H_2SO_4 bath to regenerate cellulose as rayon. The fiber or film (in the case of cellophane) is washed and dried and sent to converters. The chemistry of this process is shown below. More complete information is found in referenced publications [2, 3].

Viscose Process Chemistry
ALKALIZATION:
$R_{cellulose}OH + NaOH \cdot \chi H_2O \rightleftarrows$
$R_{cellulose}OH \cdot NaOH\chi H_2O$

DEPOLYMERIZATION (Aging):
$\cdot R + O_2 \rightarrow ROO\cdot$
$ROO\cdot + RH \rightarrow ROOH + \cdot R$
$ROOH + H_2O \rightarrow ROH + H_2O_2$
$ROH^* + OH^- \rightarrow$ chain scission

*oxidized cellulose, e.g. (HO-C-OH)

(A) Free Radical Mechanisms of Alkali Cellulose Degradation

(B) Depolymerization by β-alkoxyl Elimination

XANTHATION:

$R_{cellulose}OH \cdot NaOH\chi HO + CS_2 \rightarrow$

$$R_{cellulose}O\text{-}\overset{\displaystyle S}{\overset{\displaystyle \|}{C}}\text{-}S^{-\,+}NA$$

2. Influence of pulp properties

Pulp factors will be discussed in relation to the

end use properties which they causally affect. These are summarized in Table 57.

As noted earlier, dissolving pulps are generally of higher cellulose content than paper grade pulps. The analytical procedures for determining the cellulose content include the S_{10} and S_{18} methods [4] (i.e., the solubility of pulp in 10% and 18% NaOH, respectively). Ten percent NaOH at 20°C will result in optimum swelling of a pulp. This results in the dissolution of not only the degraded cellulose fraction but the hemicellulose fraction as well. With somewhat reduced swelling power, 18% NaOH solubilizes essentially the shorter chemical residual hemicellulose fraction. These, along with mineral and organic solvent extract analyses, characterize the purity of the cellulose. Generally, the S_{10} fraction contains those polymer molecules with less than 150 DP while the S_{18} fraction contains polymers less than 50 DP. Figure 188 (based upon gel permeation chromatography) shows this diagrammatically where S_{10}-S_{18} represents the degraded short chain cellulosic fraction in pulp. The R_{10} or $(100-S_{10})$ corresponds to the old alpha measurement and is a measure of the long chain cellulose content. For viscose pulps, 0.88 $(100-S_{18})$ is proportional to viscose process yield. Other estimates of viscose yield include the $R_{21.5}$, i.e., $(100-S_{21.5})$ determination.

The range of R_{10}, S_{10}-S_{18}, and S_{18} for the various rayon grades is shown in Table 55. Those viscose end products requiring the highest strength have the highest R_{10} lowest S_{10}-S_{18} and S_{18}. The contribution [5] of the lower molecular weight short chain material (S_{10}-S_{18}) to lower product strength is clearly shown in Fig. 189.

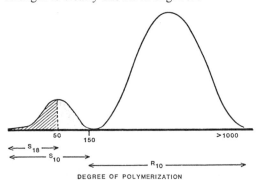

DEGREE OF POLYMERIZATION

S_{10} – MAXIMUM SOLUBILITY, DISSOLVES BOTH DEGRADED CELLULOSE AND HEMICELLULOSE

R_{10}(or 100-S_{10}) IS A MEASURE OF UNDAMAGED LONG CHAIN CELLULOSE

S_{18} – REPRESENTS HEMICELLULOSES

S_{10}-S_{18} – REPRESENTS DEGRADED CELLULOSE

Fig. 188. Idealized DP distribution of cellulose pulps.

Table 57. Influence of pulp analytical properties on end-use properties.

	R₁₀ S₁₀⁻/S₁₈, %	S₁₈, %	I.V., dL/g	Limit I.V., dL/g	E.E., %	ELB, %	Xylan, %	Mannan, %	Ash, %	SiO₂, %	Ca, %	Fe, %	Co, %	Mn, %	Sheet % O.D.	B.W.	Density	Dirt Count
Viscose																		
Steeping	S				L		L	L			M				M	M	S	
Slurry Drainage	S																	
Pressing Rate	S																	
Aging Rate					L				S			S	S	S				
Filterability		M			S				S	S	M	L						
Spinnability		S			S					S	M	L			M		S	L
Yarn Strength		S	S		M					M	M	L						L
Yarn Whiteness						S						L			M			
Yield	S						S								M			
Acetate																		
Shredding		S					S								S	S	S	
Acetylation Rate			M	S				M							S	S		
Haze		S			L		S		M		M							
Color, Solution Plastics						M			S			M	M	M	S			
Filterability					S		S	S	S	S		M			S			S
False Viscosity					M		S	S			S							
Spinnability		M	M		L					L					S			
Yarn Strength		M	M		L					L								
Yield	S																	

S) Strongly influences   M) Moderately influences   L) Slight influence   N) No influence

KEY DEFINITIONS:

Acetylation Rate — Measured in terms of clearing time or point of maximum temperature exotherm.

Aging Rate — Rate of loss of anhydroglucose units during alkaline oxidation.

Color — Solution — appearance of cellulose ester dispersion when viewed with transmitted light. Plastics — the development of color after heating the cellulose ester at elevated temperature.

False Body Effect — The increase in concentrated solution viscosity of wood pulp based cellulose acetates compared to cotton linters based acetates of the same intrinsic viscosity.

Haze — The degree of light scattering of ester dispersions.

Pressing Rate — The rate at which NaOH can be expressed from alkali swollen sheets during the press cycle of conventional steeping.

Steeping — The ease of alkali absorption, air removal, etc.

Slurry Drainage — The rate of which NaOH is filtered from an alkali swollen slurry.

Spinnability — The relative ease of extrusion or the number of yarn defects per length of fiber spun.

Yarn Strength — The intrinsic yarn physical properties: tenacity, elongation, modulus, tensile energy absorption, loop strength, knot strength, etc.

Yield — The efficiency of conversion of cellulose into the specific derivative.

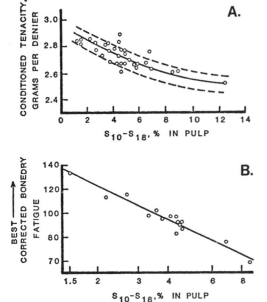

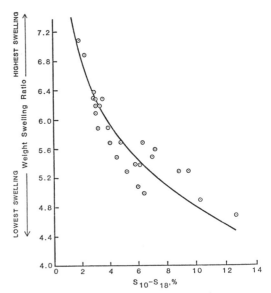

Fig. 189. a) Relationship of S_{10}-S_{18} to tire cord strength; b) relationship of S_{10}-S_{18} to cord fatigue resistance — prehydrolyzed kraft pulp from pine.

Fig. 190. Plot of weight swelling ratio vs. S_{10}-S_{18}, %.

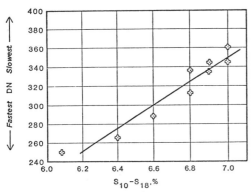

Fig. 191. Effect of S_{10}-S_{18}, % upon drainage number (DN).

The S_{10}-S_{18} will also influence both degree of swelling (Fig. 190), and rate of NaOH expressed from an alkali cellulose slurry (Fig. 191). The first will influence the completeness of reaction (hence the filterability); the latter relates to throughput of a slurry press. Fiber morphological factors as well as pulping process also affect drainage of NaOH from the slurry.

Pulp yield in the viscose process is proportional to R_{18}. Hemicellulose levels (S_{18}) affect the steeping characteristics by increasing the viscosity of the NaOH and the rate of NaOH drainage [6]. They slow the penetration of NaOH into the fiber and reduce fiber swelling. High hemicellu-

lose levels also adversely affect the viscose filterability by affecting the steeping as above, and also the xanthation by preferentially reacting with carbon disulfide. High hemicellulose levels adversely affect strength in the end product [7].

The specification for R_{10}, S_{10}-S_{18}, and S_{18} imply a much more uniform molecular weight distribution of dissolving pulps in comparison to paper pulps leading to a viscose of less polydispersity. The elimination of a large proportion of S_{10}-S_{18} and S_{18} fractions by the pulping/bleaching process is clearly seen by comparing plots in Figs. 192 and 193 [7a].

To minimize the time of aging, viscose processors require dissolving pulp at close to the desired final viscosity. The aging step is used to fine tune the DP as well as to narrow the molecular weight distribution of the final product. A comparison of starting pulp DP is shown in Table 58 compared to the final end use DP's.

As shown, for products in which strength is not critical, pulps of low DP can be used. It has been mentioned previously that this allows for a high concentration of cellulose in the viscose at low NaOH levels thereby maximizing plant throughput as well as minimizing usage of NaOH and H_2SO_4. Demanding fiber properties for certain end uses require the highest DP pulps. For all end products, the uniformity of DP within grade is extremely important since a small deviation, if carried into the viscose, will cause large changes in the viscose viscosity. Large changes in viscose viscosity will cause disruption in the spinning of the viscose.

Pulp resin content, i.e., extractives, can affect the viscose process both positively and negatively. For those pulps which are high in resin content (e.g., hardwood or softwood sulfites), more complete diffusion of CS_2 [8] into the alkali cellulose is achieved due to the lowering of surface tension. This gives more efficient xanthation and a viscose which does not contain excessive amounts of gel particles originating from partially reacted fibers which block filters (Fig. 194).

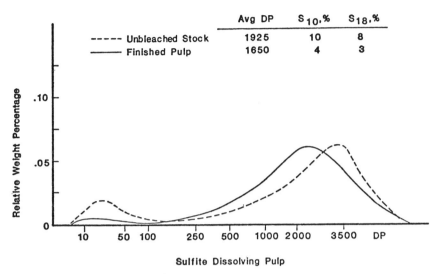

Fig. 192. DP distribution dissolving pulp.

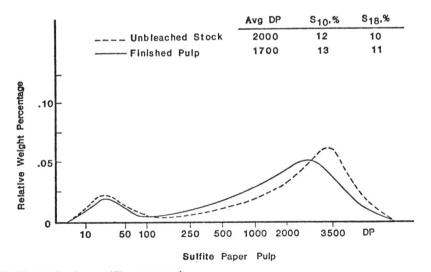

Fig. 193. DP distribution, sulfite paper pulp.

Table 58. Relationship of starting pulp degree of polymerization to end product quality factors.

End Product	DP Pulp	DP	Tensile Conditioned, g/dL	Wet Tensile, g/dL
Polynosic Fiber	1925	580	4.4	3.2
Tire Cord	1545	475	5.3	4.2
High Wet Modulus Staple	1475	400	4.0	2.6
Regular Staple	1000-1440	325	2.8	1.8
Cellophane	980	300	—	—

Above about 0.20-0.30% ether extractables, however, the viscose becomes more turbid due to the insolubility of the resin in the viscose. In the spinning orifice, itself, the high shear forces are enough to deposit the resin, causing blocked or partially blocked holes. In particular, it has been found that hydrocarbons and waxes deposit preferentially in the jet orifices [9]. Rayon tire cord made from pulp of high resin content also loses strength more rapidly when heated (simulating a hot road condition) than similar pulps of lower resin content (Fig. 195) [9]. The mechanism for this is the liberation of and subsequent

attack by HCl of the cellulose chains. The HCl originates from chlorinated resin components. To eliminate the effects of HCl, tire cord pulps require less than 0.05% ether extractables.

The ash and mineral content can be a measure of particulates which can block filters or, if small enough, can cause a deterioration in fiber strength. The reduction of viscose filterability by solid particulates is shown in Fig. 196. Very small particles (e.g., < 3 nm) will cause strength defects by passing through the filters and becoming a discontinuity in the fiber. The fiber will break at the discontinuity when a stress is applied. Very small gel particles in the viscose can also give this effect [9a].

Some metals catalyze the rate of the viscose aging reaction markedly. Cobalt, manganese, and iron act as accelerants while copper acts as a retardant. Aging rates involving some of these metals are shown in Table 59. Pulp manufacturers practice strict quality control procedures to ensure that the composition of reactive metals does not vary from specified limits from shipment to shipment. Some customers with capacity constraints in the aging step have the pulp manufacturer add controlled amounts of metal ion (usually Mn ^{2}) to the pulp during manufacture while others require the lowest possible amount to avoid process upsets.

The brightness of pulp (at the same resin content) correlates generally with rayon yarn

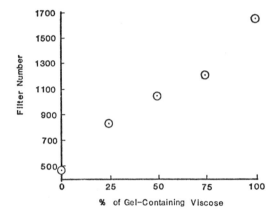

Fig. 194. Effect of gel-containing viscose on filter number.

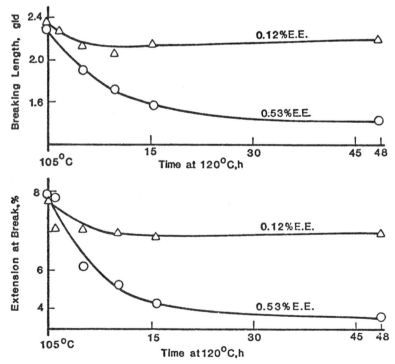

Fig. 195. Heat stability of high and low resin containing pulp.

color with pulp brightness levels of at least 93% ELB (Elrepho brightness) being specified. Normally, rayon tire cord pulps are at 88-90% ELB since yarn color is not important.

Visible speck count or dirt count must be very low to ensure good filterability. For instance, pulp shives (i.e., fiber bundles of unbleached pulp) can cause significant reduction in viscose plant filterabilities. Power boiler fly ash as well as other contaminating particulates can contribute similar negative influences.

The physical properties of the pulp sheet are extremely important for the sheet steeping process but of lesser importance for the slurry process. The former requires: (1) uniform absorption of the steeping caustic and good sheet formation, (2) optimal and uniform rate of NaOH absorption (the sheet must be easily wetted), (3) effective air displacement (the sheets should be of high sheet density) and such that sheets do not float when caustic is added to the steeping tank, and (4) the sheets should have sufficient mechanical strength to be handled adequately. There is a specific rate of NaOH flow into the steeping tank for each pulp grade. The rate depends upon, among other factors, wood species, R_{10}, S_{18}, sheet density, and basis weight. Not optimizing steeping caustic flow will yield poor filtering viscoses. For sheet steeping, the sheet density should be $> 0.80 \, g/cm^3$, but for the

slurry process a density of < 0.7 is optimum. This will assist in sheet break-up.

C. CELLULOSE ACETATE MANUFACTURE
1. Process description
During the acetate process, in contrast to the viscose process, a stable derivative is formed which is soluble in common organic solvents. Usually the acetate ester is formed but at times, for different end use reasons, the butyrate, propionate, or mixed esters can be made from the corresponding acids and anhydrides. Also, in contrast to the viscose process, there are almost as many process variants as there are manufacturers. Generally speaking however, the acetate process falls into four categories: (1) low catalyst, (2) high catalyst, (3) methylene chloride (MeCl₂), and (4) fibrous. For the first three processes the derivative gradually dissolves in the reaction medium (either acetic acid [HOAc] or a mixture of HOAc and MeCl₂). The solubilization is due to formation of the ester and reduction of degree of polymerization. This chain shortening can either be carried out concurrently with the esterification of the cellulose or prior to the actual esterification. In the fibrous acetylation, hydrolysis and derivatization is carried out in a diluent and will not be be discussed here.

All three of the aforementioned processes generally start with dry defibering. Exceptions to the rule are the newer continuous processes [10, 11] which depend upon defibering the pulp sheet in a water or H₂O/HOAc slurry, then the countercurrent removal of the dispersant with HOAc. For other manufacturing processes, dry defibering is a critical step. Defibering is carried out in hammermills, disk refiners, or pin pickers. The acetate process is a permutoid reaction such that each of the individual fibers need be exposed to the acetylating reagents for complete reaction.

Following defibering, the shreds are allowed to tumble with 0.35 to 2.5 parts HOAc by weight.

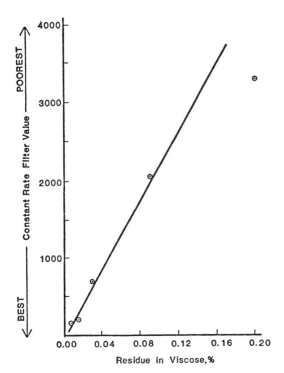

Fig. 196. Effect of insolubles found in viscose upon filtration value.

Table 59. Percent increase in rate of aging upon control pulp.

METAL ION	%Increase
MANGANESE[1]	
13 ppm mg/kg	34%
20 ppm mg/kg	43%
26 ppm mg/kg	54%
COBALT[1]	
33 ppm mg/kg	46%
43 ppm mg/kg	72%

[1] Amount of catalyst added to pulp. (as manganous or cobaltous ion)

This swelling step is called the pretreatment. In the low catalyst procedure, ~ 2-4% H_2SO_4 on the weight of cellulose along with additional HOAc is then added to depolymerize the cellulose at temperatures of 30-50°C. This is the activation step of the low catalyst procedure. Acetic anhydride is then added to cause esterification. Both the high catalyst and $MeCl_2$ processes differ from the low catalyst procedure in that the cellulose hydrolysis and esterification occurs simultaneously. All three procedures are stopped by addition of water to react with the excess acetic anhydride when the desired solution viscosity is reached. At this point the degree of substitution is about 2.9 with some of the remaining hydroxyl groups substituted with sulfate esters. The high catalyst procedure uses about 14% H_2SO_4 on the weight of the cellulose as catalyst, while the $MeCl_2$ process uses < 1.0% H_2SO_4 or $HClO_4$. The unique advantage of the $MeCl_2$ process is that $MeCl_2$ is a much better solvent than HOAc for cellulose triacetate. This allows for the formation of a cellulose acetate(CA) relatively free from undissolved particles. Recovery of the $MeCl_2$ from the reaction dispersion, however, complicates the process.

Since the 2.9 degree of substitution (DS) product is not soluble in acetone, the cellulose acetate while still in solution form is hydrolyzed back to an acetone soluble derivative having a DS of about 2.4. (This means that a greater proportion of the monomers are substituted with two acetyl groups rather than three.) After this hydrolysis, the cellulose acetate (CA) is precipitated in dilute HOAc, washed and dried, then converted into fiber, film, or plastics. The 2.9 substituted derivative (the so-called triacetate) has desirable mechanical and clarity advantages over the 2.4 DS product (referred to as diacetate). If the triacetate is desired, a short hydrolysis prior to precipitation is carried out to remove the combined sulfate groups. The requirement of a mixed solvent system to dissolve triacetate for spinning restricts the greater production of this derivative.

If the diacetate CA is used for filament or film manufacture, the CA is dissolved at 25% concentration in acetone, filtered, deaerated, then spun by the dry spinning procedure; or cast into film by the casting wheel technique. Plastics manufacture requires compounding of the dried acetate flake with plasticizers and stabilizers prior to plastics extrusion. Commonly used plasticizers include diethyl phthalate and dimethyl phthalate. Stabilizers [12] for this process include 2,6-di-t-butyl substituted phenols, alkali methyl oxalates, and anionic stabilizing salts (e.g., cyclohexane carboxylic acids). Plastics acetate flake is sometimes filtered in the hydrolyzed acid gum phase. Process conditions are shown in Table 60 with chemistry [13, 14] of the sulfuric acid catalyzed acetate process shown below:

ACETYLATION MECHANISM
Cellulose OH + H_2SO_4 → Cellulose OSO_2OH
Cellulose OSO_2OH + Ac_2O → Cellulose OAc + $AcOSO_2OH$
Cellulose OH + $AcOSO_2OH$ → Cellulose OAc + H_2SO_4

2. Relationship of pulp properties to acetate quality

In contrast to the viscose process in which purification of the pulp takes place in pulp steeping, no purification takes place in the acetate process. All cellulose and impurities put in the reactor take part in the reaction. Impurities in the pulp cause low filterability, yellowness in yarn, poor solution clarity, as well as a lack of thermal stability. Up to about 1940 dissolving wood pulp purity was insufficient to allow penetration into a cotton linters based acetate market. The first acetate pulp derived from wood which gained market acceptance was Rayonier's Rayaceta. This was achieved by a combination of sulfite cooking changes, as well as the use of pressure hot caustic extraction in bleaching in order to drop the S_{18} to 3.0% or lower. Conventional kraft cooking to comparable S_{18} levels led to unsatisfactory cellulose acetate product quality.

Even today, the use of pulps with S_{18} higher than 3.0% leads to cellulose acetate of low filterability, high false viscosity, and high turbidity of acetate dispersions unless the acetylation process is specially modified to accommodate pulps of lower purity. These data are presented in graphical form (Fig. 197) [15]. In acetate processing, false viscosity body effect is defined as the increase in concentrated solution viscosity of a wood pulp derived cellulose acetate in comparison to a cotton linters derived acetate, both at the same I.V. (intrinsic viscosity) and concentration. This factor, in turn, relates to the ease of pumping concentrated dispersions. Haze and filterability relate to the completeness of reaction and are indirect measures of the amount of insoluble particles in solution. The effects of pulp processing on haze deterioration, false viscosity effect, and reduction in filterability as well as color development have been studied in detail.

Wilson and Tabke[16] showed the influence of pulp hemicellulosic components upon CA properties (the types of hemicellulose added were those found in finished pulps). In short, they found that glucomannans adversely affected

Table 60. Acetylation process conditions[a].

Stage	Low Catalyst	High Catalyst	Methylene Chloride
Sheet Disintegration Pretreatment	Mechanical, dry defibering HOAc (2.4:1)	Dry defibering HOAc (0.35-0.5:1) 25-35°C, 1 h	Dry defibering HOAc (0.35-0.5:1) 25-35°C, 1 h
Cellulose Depolymerization	HOAc H_2SO_4 (0.02-0.04:1) 35-45°C	Occurs simultaneously. HOAc, acetic anhydride H_2SO_4 (0.14:1)	Occurs simultaneously. HOAc, acetic anhydride, methylene chloride H_2SO_4 (0.01:1).
Cellulose Esterification	HOAc, acetic anhydride + additional H_2SO_4 0.04:1. Reaction proceeds adiabatically.	Reaction proceeds adiabatically.	Reaction controlled by distillation of methylene chloride.
Hydrolysis of Acetyl Groups	With or without some catalyst neutralization, amount of water controlled to give optimum distribution of acetyl groups.		
Precipitation	Dilute HOAc	Dilute HOAc	Methylene chloride distilled off, then precipitated in dilute HOAc.
Washing	Similar	Similar	Similar
Drying	Similar	Similar	Similar

a) All process reagents to O.D. wood pulp.

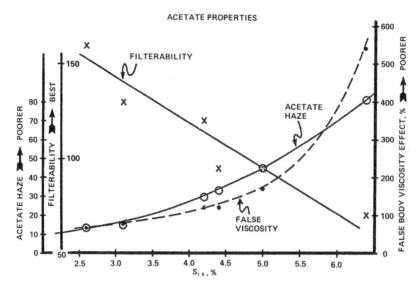

Fig. 197. Effect of pulp S_{18} upon acetate properties (in conventional acetylation).

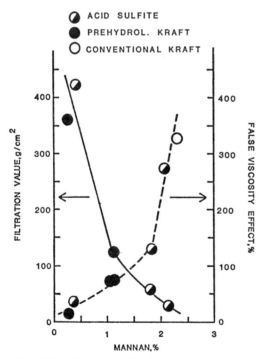

Fig. 198. Effect of pulp mannan content on diacetate filterability and false viscosity.

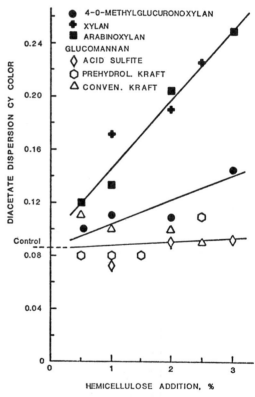

Fig. 199. Effect of hemicellulose addition on diacetate dispersion color.

acetate filterability (Fig. 198), solution haze and acetate false body effect. On the other hand, 4-0-methylglucuronoxylan is a minor factor in these acetate properties but severely affects dispersion color and causes thermal instability. Xylan and arabinoxylan originating from prehydrolzyed or kraft cooks give rise to haze and color problems as well as being a factor in acetate

heat instability (see Figs. 199, 200) [17].

The reason for the adverse effects of the hemicelluloses upon acetate end use solution properties is due in large part to their solubility in comparison to pure cellulose acetate. In terms of relative degree of swelling (not solubility), the glucomannan acetates have the highest degree, the xylan acetate the lowest, with the 4-0-methylglucoronoxylan intermediate. The large gel-like mannan acetates help plug filters and cause abnormal viscosity behavior as well as being present as haze. The smaller, less soluble

xylan acetate particles from wood pulp give acetates their characteristic yellow color by preferentially scattering in the blue end of the visible spectrum. The inability of conventional kraft pulp to perform as raw material for the manufacture of CA lies, in part, in the poor solubility characteristics of the higher DP arabinoxylan acetate as well as morphological factors.

The pulp S_{18} component, however, is not wholly responsible for the adverse CA properties noted. High carboxyl levels in pulp, brought about by an improper bleaching sequence along with high pulp calcium contents, give rise to higher false body viscosity effect levels.

Chlorinated wood resin components generate HCl during the elevated temperatures of injection molding (Fig. 201). The HCl attacks the cellulose leading to poor internal stability as evidenced by a discolored and low I.V. product. It is also suspected that wood resin components, by virtue of their poor stability, tend to deposit in regions of high shear (e.g. in in a spinning jet) which may lead to yarn of variable cross-section and may affect dyeability and strength.

As with viscose rayon, particulates present in the pulp lead to poor filtering acetates. Good filterability is of utmost practical importance, since a reduction of filter throughput of about 25% causes about a 50% increase in acetate spinning production costs, as well leading to product of lower strength. The reduction of filterability is shown in Fig. 202.

High and variable cations in pulp can neutralize a portion of the sulfuric acid catalyst leading to incompletely acetylated acetates

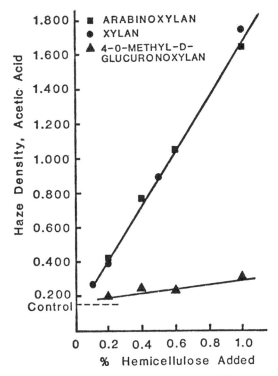

Fig. 200. Effect of hemicellulose addition on triacetate haze density in acetic acid.

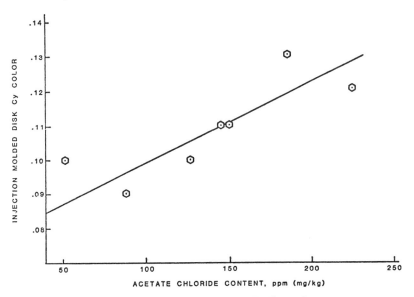

Fig. 201. Effect of acetate chloride content upon acetate plastics color.

which will not filter properly. This is especially apparent with the methylene chloride process that requires less than 1% H_2SO_4 catalyst.

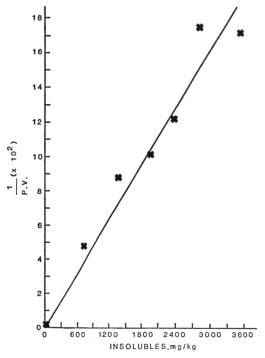

Fig. 202. 1/Plugging value ($\times$ 10²) vs insolubles in acetate dope.

The nature of the pulp sheet prior to dry defibering can affect profoundly the quality of the resulting cellulose acetate. For that reason, acetate pulp sheets are supplied at low sheet densities ($<$ 0.55 g/cc) with basis weights ranging from 650 to 830 g/m². Too high a density of basis weight sheet will lead to incompletely defibered fluff which will not acetylate uniformly.

In addition, pulp which is insufficiently dried will consume an inordinate amount of acetic anhydride, thus raising the cost of the process markedly. Conversely, reducing the pulp moisture to less than 5% causes marked reduction in the filterability due to collapse of fibers and subsequent incomplete penetration of acetylation reagents. This is shown in Fig. 203.

Both the finished pulp I.V. and limit I.V. as well as wood species morphological properties will affect acetylation rate. Of the three mentioned the latter two are most important. Limit I.V. or level-off DP represents the chain length of the crystalline material in the cellulose. A plot (Fig. 204) of limit I.V. versus esterification time shows that those pulps of the largest crystalline size show the slowest acetylation rate.

Pulp brightness, which can be considered an estimation of residual lignin content, relates to yarn colors with the higher brightness pulps generally giving the best yarn and plastics color.

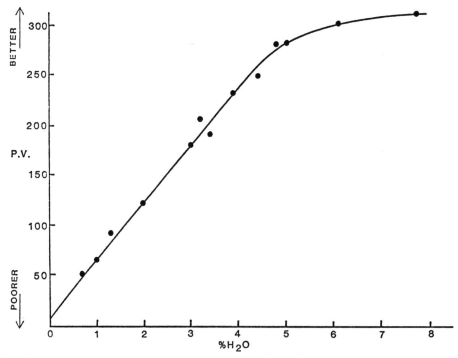

Fig. 203. Effect of pulp moisture content upon acetate filterability.

Lastly, yield in acetate process is dictated by the quantity of lower molecular weight cellulose acetate which is soluble in dilute HOAc. The R_{10} is generally used as a guide to potential yield in the process.

D. PULP MILL OPERATIONS

The previous sections illustrated process descriptions and the influence of pulp properties on rayon and acetate end-uses. In this section the manipulation of pulping, bleaching, sheet forming and drying variables to achieve the desired end-use properties will be discussed. Both sulfite and prehydrolyzed kraft pulping have the same objective in the preparation of unbleached pulps for bleaching: (1) reduction of lignin levels in pulp before excessive cellulose degradation occurs, (2) depolymerization of hemicellulose facilitating their removal either in cooking (prehydrolyzed kraft) or in subsequent bleaching operations (sulfite) and (3) controlled depolymerization of the cellulose to viscosity levels required for smooth operation in the bleach plant as well as to achieve levels required for various end-uses.

1. Sulfite pulping conditions

(a) Conventional

Sulfite dissolving pulp processing conditions (Table 61) differ from paper pulp production

conditions considerably. In the latter, hemicellulose is retained for strength development, while in the former the hemicelluloses are either removed or degraded substantially to ease in removal during pressure hot caustic extraction (HCE) in bleaching.

The main differences between dissolving and paper pulp cooking conditions include faster temperature rise times, higher maximum temperatures, higher acidity and lower combined SO_2. Examination of the energies of activation for delignification ($\approx$ 22 Kcal mol^{-1}) [18], hemicellulose hydrolysis (28 Kcal mol^{-1}) [19], and cellulose hydrolysis ($\approx$ 28-44 Kcal mol^{-1}) [20], explain the lower temperatures used to produce a stronger paper pulp; while higher temperatures are required to remove both lignin and hemicelluloses in order to produce a dissolving grade pulp.

In addition to higher temperatures, higher acidity is also required. Hot pH profiles (Fig. 205a, b) during cooks which have followed the above schedules show clearly the increased cook severity required for dissolving pulp cooking as compared to paper type cooking. (Refer to Chap. I,A and D,1.)

During the initial phases of the sulfite cook, the temperature rise causes the apparent dissociation for SO_2 (Eq. 1) to decrease and causes the equilibrium to shift to the left. The increasing

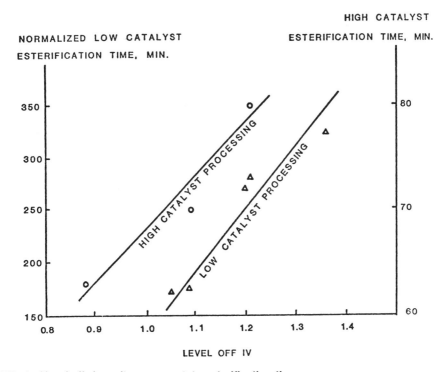

Fig. 204. Effect of level off viscosity upon acetate esterification time.

Table 61. Typical acid sulfite cooking condition.

	Dissolving Celluloses	Paper
Wood		
Liquor to Wood	4.0-5.5	4.0-5.5 (depends on wood species)
Cooking Acid		
Base	Na, NH_3, Mg, Ca	Na, NH_3, Mg, Ca
Combined SO_2, %	0.60-1.15	0.9-1.2
Free SO_2, %	> 6.5	> 6.5
Temperature rise, H:M	3:30-4:00	4:00-5:00
Total cook time	4:30-7:00	5:00-9:00
Maximum temperature, °C	135-150	125-150
Pressure, psi	95	95
Unbleached Pulp		
Yield, %	44-46	46-48
Tailings, %	0.5	0.5
K Number, mL	8-20	7-15
I.V., dL/g	7-11	10-12
S_{10}, %	9-10	12-13
S_{18}, %	8-9	11-12
Xylan, %	1.5-3.5	2-4
Mannan, %	5-6	6-8

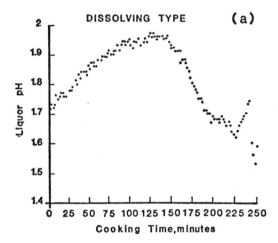

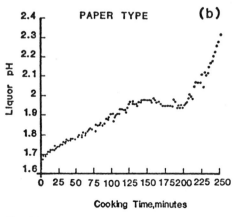

Fig. 205. In-process pH of (a) dissolving and (b) paper type cook.

temperature physically drives the SO_2 from the digester through the relief system, thus increasing pH:

$$H_2O + SO_2 \rightleftharpoons \text{``}H_2SO_3\text{''} \rightleftharpoons H^+ + HSO_3^- \qquad (1)$$

Further temperature increases force the reaction and consumption of bisulfite ions causing an increase in acidity due to the formation of very highly acidic lignosulfonic acids along with the sugar derived carboxy and sulfonic acids. The extent of base present apparently buffers the digester contents. As shown in Fig. 205b, the higher amount of base (combined SO_2) of the paper cook prohibits the very low pH which is required in the dissolving cooks. It is this difference which promotes the ease of hydrolysis of hemicelluloses in dissolving pulps and, in contrast, retention of hemicelluloses in paper pulps.

The reduction in combined SO_2 levels to achieve the desired acidity, however, is very subtle as shown in the data [21] depicted in Fig. 206. As shown, increased combined SO_2 helps preserve hemicellulose, with low combined SO_2 levels giving lower hemicellulose levels. Note, however, that while low combined SO_2 levels result in low hemi levels, continued pulping under these conditions causes the formation of significant quantities of degraded cellulose (S_{10}-S_{18}). Therefore, not only are the hemicelluloses being hydrolyzed to low levels but, at high levels of acidity and prolonged conditions, cellulose is also being attacked leading to unacceptably high levels of degraded cellulose. This, of course, is carried into the bleached pulp and affects the end-use properties as discussed earlier. Further evidence for the additional cellulose hydrolysis is the reduction of the intrinsic viscosity for a given pulp. Coincident with these high levels of acidity are increased lignin condensation reactions as shown by the dramatic increase in pulp permanganate number. Therefore, combined SO_2 levels are adjusted to the required degree of hydrolysis at an easily accommodated permanganate or K No. (note that there are differences between permanganate and Kappa numbers).

Cooking temperatures (Fig. 207) are usually in the range of 135-150°C. The proper temperature is selected on the basis of obtaining the required degree of carbohydrate hydrolysis to obtain satisfactory dissolving pulp end use properties while not generating excessive quantities of S_{10}-S_{18} material or raising the K No. too high.

Free SO_2 levels are maintained at levels high enough to insure adequate delignification. This is usually ~ 7.0% free SO_2. Free SO_2 within the range from 5-10% free SO_2 does not affect the cellulose hydrolysis rate.

In a sulfite dissolving mill, time is manipulated to achieve the desired pulping results at given starting conditions. Since sulfonation occurs early in the cook while hydrolysis occurs later,

stopping the cook early leads to pulp having a high K Number without the desired amount of hemicellulose or cellulose hydrolysis. The proper amount of hemicellulose hydrolysis is gauged by the intrinsic viscosity of the unbleached stock. As the hydrolysis reactions are accelerating at the end of the cooks (in contrast to kraft cooks), due to the combination of increased acidity and high maximum temperatures, considerable attention is paid to procedures used to terminate the pulping reactions. Prediction of when to end the cook are made by (1) examining the viscosity of the previous cooks, (2) determining the color and rate of color development, (3) calculating the degree of cooking with a modified "H" or "S" factor or (4) pH measurement. All have limitations. Full details of cooking models are found in Chapter III, C, 1-4.

The action of acidic sulfite cooking conditions on hemicelluloses in dissolving pulps corresponds to the hydrolysis reactions of cellulose. These include the acidic hydrolysis of glycosidic bonds, causing a loss of DP leading to their ultimate solubility in the cooking liquor and degradation to sugars. The hemicellulose hydrolysis precedes cellulose hydrolysis. In the analysis of spent sulfite liquor during the course of a softwood cook, it has been shown that arabinose and galactose are the first sugars liberated [23]. Usually this occurs at a temperature of about 100°C. Xylose appears next followed by mannose and small quantities of glucose at a cooking temperature of 130°C. This is substantiated by analysis of the sugars in the side relief [23]. Arabinose and xylose make up more than 60% of the total sugars found in the stream. In

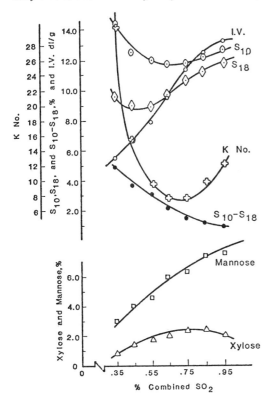

Fig. 206. Effect of combined SO_2 upon acetate end use properties.

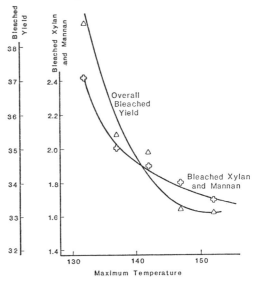

Fig. 207. Effect of maximum temperature on acetate properties.

softwoods, the arabinose originates from the acid labile arabinofuran side unit on a 4-0-methylglucuroarabinoxylan. Similarly, galactose originates from the galactoglucomannan. In contrast, in the spent liquor, arabinose makes up only 15-25% of the total sugars found [23]. The polymer nature and reaction mechanisms have been detailed by Hamilton [24]. Furanosidic bonds are cleaved much faster than pyranosidic bonds. This accounts for the rapid dissolution of arabinose in the sulfite cook. However, morphological aspects outweigh chemical considerations, in that the rate of hydrolysis of glucomannan and xylan is about 15-30 times greater than cellulose, which is about 4-5 times that predicted by studies with model compounds. This fact is taken advantage of in the design of the cooking curve for dissolving cellulose as noted earlier.

In addition to the glycosidic hydrolysis reactions, acetyl groups on the hemicellulose are hydrolyzed yielding acetic acid.

Other sugar reactions taking place in the dissolved liquors include conversion of aldoses into alpha hydroxysulfonic acids as well as aldonic acids. The alpha hydroxysulfonic acids represent a large proportion of the loosely combined SO_2 in spent liquors. Under sulfite pulping conditions xylose can be dehydrated to furfural while glucose is decomposed to levulinic and formic acid from hydroxymethylfurfural. Dehydration reactions are favored by low pH high temperature such as is found in dissolving-type cooking. Full description of the decomposition reactions of the spent liquor is found elsewhere [25].

(b) With congruous prime parameters

It is evident from the above that the three major acid bisulfite cooking parameters, combined SO_2, total SO_2 and temperature exert a most significant effect on the quality of the dissolving pulp produced. Quality considerations include cellulose purity and reactivity, dope filtration behavior, ease of removal of low DP fractions in steeping, and color of the regenerated fibers. While pulp quality is of importance in all dissolving processes, the demands are particularly stringent for acetylation grade pulps.

It was of interest, therefore, to examine the above three parameters with regard to optimum configurations for producing dissolving pulp of very high quality under the most efficient conditions of cooking.

This was achieved for Northeastern spruce wood and calcium base within the practical ranges of 1.2-0.4% combined SO_2, 5-15% total SO_2 and 147-123°C maximum temperature [25c]. The results of the evaluation of simulta-

neous interactions within the three parameter ranges with regard to total cooking time to a viscosity of 20 mPa • s (Tappi T 230, 0.5% CED), and celluose acetate acetone scatter are shown in Figs. 208A and B. Acetone turbidity and acetone color follow a similar trend of major improvement with decreasing combined SO_2, when supported by increasing total SO_2 and decreasing cooking temperature.

The requirement of the greatest possible degree of purification of the cellulose during the cooking state is met by the optimum interaction of the three cooking parameters. The acidity of the liquor is raised substantially by means of drastically reducing the base content and simultaneously greatly increasing the SO_2 content at reduced temperature so that the equilibrium in Equation 1 can supply large amounts of hydrogen ion as well as bisulfite ion. On the one hand, the very low pH level meets the purpose of attacking hemicellulose and existing cross linkages. On the other hand, together with the ample supply of bisulfite ions, it accelerates the completion of the cook while preventing lignin

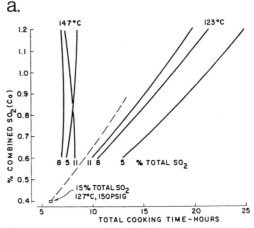

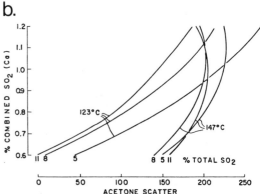

Fig. 208. Effect of combined SO_2, total SO_2, and cooking temperature on A) cooking time and B) acetone scatter.

re-condensation and liquor breakdown (sulfate formation), the major reason for uncontrolled cellulose degradation.

Table 61A summarizes the results of optimizing the three parameters for three levels of combined SO_2 in the cooking liquor. Improved acetylation pulp quality is achieved with decreasing concentration of any of the four bases (Ca^{++}, Mg^{++}, NH_4^+, and Na^+) commonly employed in acid bisulfite pulping. There is evidence, however, that the bivalent alkaline earth cations cause a greater improvement in quality than the monovalent alkali cations. Apart from acetylation grade pulp, beneficial effects on the properties of other types of dissolving pulp are also found.

2. Prehydrolysis

Prior to 1950, dissolving grade pulps were produced almost exclusively by the sulfite process [26]. As noted earlier, the strongly acidic pulping stage coupled with a subsequent hot alkaline extraction serves to remove the bulk of the lignin and hemicellulose fractions, resulting in a pulp with high alpha cellulose content. Since the conventional kraft process stabilizes residual hemicelluloses against further alkaline attack, it is not possible to obtain acceptable quality dissolving pulps through subsequent treatment in the bleach plant. In order to prepare dissolving type pulps by the kraft process, it is necessary to give the chips an acidic pretreatment before the alkaline pulping stage. A significant amount of material, on the order of 10% of the original wood substance, is solubilized in this acidic phase. Under the prehydrolysis conditions, the cellulose is largely resistant to attack, but the residual hemicelluloses have been degraded to a much shorter chain length (about 30% of the original DP) [27] and can therefore be removed to a large extent in the subsequent kraft cook by means of the well-known peeling reaction and other hemicellulose hydrolysis reactions. Primary delignification also occurs during the kraft cook. However, because of lignin condensation reactions during the acid stage, residual lignin in unbleached pulp from a prehydrolyzed kraft process is more difficult to remove during bleaching compared to conventional kraft pulp.

The value of prehydrolysis is evidenced in a study on structural changes occuring in wood during this step. The prehydrolysis stage normally involves treatment of wood at elevated temperature (150-180°C) with dilute mineral acid (sulfuric or aqueous sulfur dioxide) or with water alone requiring times up to 2 h at the lower temperature [28]. In the latter case, liberated acetic acid from certain of the naturally occurring polysaccharides (predominantly the mannans in softwoods and the xylan in hardwoods) lowers the pH to a range of 3 to 4 [9]. This acidic treatment hydrolyzes a substantial portion of the hemicelluloses to short chain polymers and, to a lesser extent, to monomer form as a result of chemical treatment [30]. Whereas successive treatments of alkali and acid partially convert hemicelluloses to a resistant fraction remaining in the cell wall, the reverse sequence (acid-alkali) results in extensive hemicellulose fragmentation to readily solubilized components. Extensive studies on the chemistry and mechanism of prehydrolysis for both softwood [31a] and hardwood [32] have been carried out. The quantities and fates of the extracted constituents were determined as reaction conditions were varied. It is thus possible, depending on the severity of prehydrolysis, to produce pulps with varying alpha cellulose contents. A minimum level of purity must nevertheless be acheived for suitability in dissolving pulp applications.

Recently, the catalytic effect of anthraquinone in the alkaline phase of a prehydrolyzed soda dissolving pulp production has been presented [31b]. No adverse effects on end-use quality were noted for pulps prepared by the procedure.

3. Bleaching

Typical bleach sequences for dissolving pulps are shown in Table 62, and are discussed in detail in subsequent sections.

Following the bleaching, most pulp mills universally treat the bleached stock at 2.5-3.0 pH for approximately 5 min at low consistency to

Table 61A. Effect of combined SO_2 on acetate solution purity [25c].

Cooking conditions					Acetylation test			
% Comb. SO_2	% Total SO_2	Max. temp.	Press. PSIG	Time Hours	Turb.	Col.	90° Scatter	Visc. *
1.15	5.5	147	85	8:15	12.4	24.1	118	84
0.65	7.0	130	95	10:15	8.0	15.1	60.6	100
0.4	15.0	127	150	5:50	6.0	12.8	31	100

* After constant acetylation time.

remove cations associated with carboxyl groups. SO_2 is normally used to adjust this level of acidity, although HCl or H_2SO_4 could equally be used. Following acid treatment, only soft water is used on the post SO_2 washer and in the machine white water system. This treatment removes those cations which affect both acetate and viscose processing.

The pH of the machine white water system is between 5-7. Solid particulates of density different from cellulose are removed from the bleached stock using centricleaners.

4. Extractive control

It was previously shown that regular rayon dissolving pulps require from 0.1 to 0.2% ether extractives to achieve satisfactory filtration levels while acetate yarn and plastic grades as well as rayon high wet modulus and tire cord grades require essentially extractive free pulps to achieve satisfactory performance levels. Extractive controls lie in the following:
1. Wood and chip aging
2. Parenchyma cell removal

Table 62. Bleaching sequences for dissolving pulps.

End Use	Cook Type	Bleach Sequence
Plastics Filler	Sulfite	CEDED, CEDPD
Cellophane	Sulfite	CE⁰H, CE⁰CH
	Prehydrolyzed kraft	CEHDED
Nitration	Sulfite	CE⁰CH, CE⁰CHD
	Prehydrolyzed kraft	
Textile Rayon	Sulfite	CE⁰CHD, CE⁰HD, CE⁰CH
	Prehydrolyzed kraft	CEHDED, CHEDED
Regular Acetate	Sulfite	CE⁰CHD, CE⁰DED, E⁰CEHD
	Prehydrolyzed kraft	XDEDH, CEHXDED, CHEDXD, CHDXD, XCEHDED
Tire Cord	Prehydrolyzed kraft	CEHXD, CEHDX, XDEDH, CHEDX, XCEHDED
Plastics Acetate	Sulfite	E⁰CEHXD
Where	C = Chlorination	
	E⁰ = Pressure hot caustic extraction (HCE)	
	E = Mild (less than atmospheric HCE)	
	D = Chlorine dioxide	
	P = Peroxide	
	X = Cold caustic extraction	

3. Substitution of chlorine dioxide for chlorine in bleaching
4. Hot caustic stage conditions
5. Surfactant usage

(a) Aging

In Scandinavia, wood is stored in the chip form for three to four months to achieve about 50% resin reduction. The same amount of reduction is achieved in the Southeastern United States in 10-14 days. The chip piles are normally exothermic but the reactions will proceed as long as the initial temperature is greater than 3°C. The first reaction in outdoor chip aging is the enzymatic cleavage of fatty acid esters of the triglycerides and fats. The liberated fatty acids undergo further decomposition reactions which are usually oxidative. The negative aspect of chip aging lies in its excessive wood loss. It has been shown that, for every month of storage, wood losses of approximately 1% per month [33] will be seen. This has prompted pulp manufacturers to investigate other procedures.

(b) Cooking

With prehydrolyzed kraft cooking of conifers, fats and waxes are usually saponified. These liberated fatty acids and resin acids form a stable emulsion system for the nonsaponfiables. Thus, low pulp extractives levels (approximately 0.01%) are found with these pulps.

Hardwoods, due to the fact they contain virtually no resin acids but a higher level of fats and nonpolar nonsaponfiables than conifers, require surfactants during the kraft phase of the cook. In addition, the same techniques as used for the highly resinous sulfite pulps are used.

With sulfite cooks of both softwood and hardwood, the formation of soluble soaps does not occur. The only reaction which does take place is ester hydrolysis and to a lesser extent some sulfonation. In spite of this, about 50-60% of the resin (mostly the fatty acid component) is removed. The high level of resin remaining after sulfite cooking of unaged chips requires specialized techniques as discussed below.

(c) Fiber fractionation

Since approximately 50-70% of the resin in hemlock, spruce or southern pine, as well as all hardwoods, is found in the ray parenchyma cells, fiber fractionation offers a means of resin reduction. Removal of the ray cells by passage of a low consistency stock ($\sim$ 1-2%) slowly over an inclined screen achieves substantial resin reduction. One to four percent removal of fiber in softwoods and 4-6% in hardwoods causes a reduction of unbleached pulp resin levels from

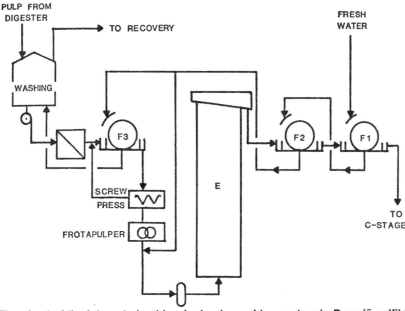

Fig. 209. Flowsheet of the integrated unbleached pulp washing system in Domsjö sulfite mill.

about 0.7% to 0.3%. However, the use of inclined screens also causes substantial fiber loss and again alternative means of extractive reduction are favored. The size of the screens, number of inclined panels and consistency are varied to achieve the desired level of fractionation.

(d) Chlorination

Chlorine, which enhances lignin removal by selective substitution and oxidative fragmentation, also reacts deleteriously with the unsaturated resin components. Both chlorine substitution and oxidation occur with the double bonds of fatty acids, resin acids and unsaponfiables. Molecular weight increases have been noted. This means that even polymerization reactions may occur. The chlorine substitution reactions on the resin components causes the resins to become more hydrophobic which complicates their removal. Preferential oxidation of the double bonds by chlorine dioxide does not allow the chlorination reactions to proceed. In fact, during chlorination about 50% removal of the resin can be achieved with substitution of about 30% of the active chlorine by chlorine dioxide. Chlorine dioxide substitution in the first stage of bleaching allowed the CIP Hawkesbury sulfite mill to manufacture satisfactory dissolving pulp from hardwood [34, 35].

(e) Hot caustic extraction

Along with the inclined screen, the hot caustic stage remains the major step in resin reduction

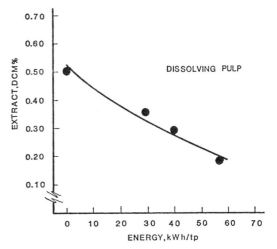

Fig. 210. Content of extractives in fully bleached dissolving pulps as function of energy input in a Frotapulper.

for sulfite pulps. Three reactions take place: 1) saponification of fats, waxes and other esters, 2) solubilization of the fatty acids liberated by formation of soaps, and 3) micelle formation allowing the solubilization of the nonsaponifiables. The micelle formation is enhanced by hardwoods and conifers by the addition of surfactants. The most effective surfactant is a nonyl phenol with attached polyoxyethylene chain. The hydrophilic/hydrophobic balance of the polyethoxylated nonyl phenols solubilizes the non-saponifiables. Phosphoric acid addition, as trisodium phosphate (TSP), allows more

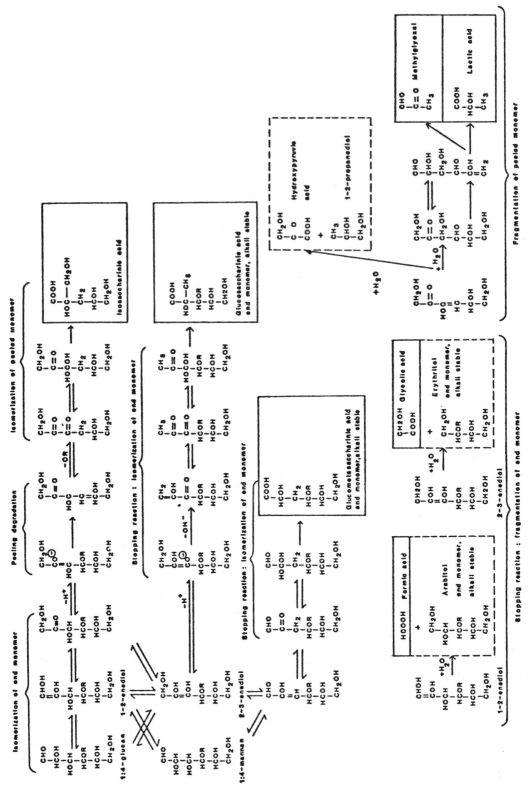

Fig. 211. Reactions of 1:4 linked hexosans in hot alkali.

effective surfactant usage. TSP helps build and stabilize micelles at low surfactant levels.

Agitation, as well as increased pulp consistency in the hot caustic stage, greatly enhances resin removal. This probably arises from the increased physical forces allowing removal of the encapsulated resin from the ray cells. In Fig. 209 a new process developed by MoDo, now marketed by KMW, involves the use of a Frotapulper along with added caustic for the deresination of sulfite pulps. Use of the Frotapulper with added caustic allows approximately a 75% removal (Fig. 210) of the resin prior to bleaching. Frotapulping can replace chip aging and to a limited extent fines removal. The distinct advantage of Frotapulping is that wood losses associated with chip aging and fiber losses associated with sidehilling operation are eliminated. Installations are currently in production use in Scandinavia.

5. Hot caustic extraction (HCE)

The purpose of the hot caustic extraction with acid sulfite cooked pulps is the removal of chlorinated lignin components and the reduction of pulp S_{18} levels to the desired end-use target. With prehydrolyzed kraft pulps the function of the hot caustic extraction stage is to mainly remove chlorinated lignin components. Cold caustic extraction, which is used to achieve the lowest S_{18} levels with the highest quality end-uses, will be discussed later.

The temperatures used in HCE stages for sulfite pulps range from ~ 95 to 145°C with NaOH additions ranging from about 4 to 12%. The lower extremes will give about a 6.5% S_{18} pulp while the higher extremes will give a 2.7-2.8 S_{18} pulp.

The most rapid reaction in the HCE stage is the neutralization of acids carried over from primary chlorination. The solubilization of chlorinated and oxidized lignins also occurs relatively rapidly. In dissolving cellulose manufacture, the major caustic consumption is the result of chemical refining. The cellulose and hemicelluloses undergo a stepwise peeling [37] reaction of the reducing glucose end groups. This stepwise degradation proceeds by way of beta alkoxy elimination reaction to yield isosaccharinic acids. After 40-50 end groups have been removed this stepwise reaction is terminated through the formation of a terminal glucometasaccharinic acid which is stable to further alkaline attack (Fig. 211).

Stepwise degradation has a small effect on the DP of the cellulose. However, the hemicelluloses which start at a much lower DP are readily solubilized during this stage. In actual practice, an I.V. increase is shown across the stage. This reflects the hemicellulose removal and a shift to a less bimodal molecular weight distribution. This stage is normally done at the highest possible consistency to save on energy as well as to effect mechanical softening of the fibers, and resin removal.

Temperature and time are adjusted to achieve a minimal caustic residual for the caustic added. Caustic consumed relates to cellulose yield and the ultimate pulp S_{18} level (Fig. 212). The yield loss from hot caustic extraction in achieving an S_{18} level of approximately 3% is on the order of 25%, whereas for a viscose grade of 4.5 S_{18}, the loss is 18%.

Other factors influence the caustic consumption. Wood quality is a major factor with wood furnishes containing a high rot content, high chip fines or damaged chips giving rise to high caustic consumption. Pulps with high unbleached viscosity or those made with high ratio of combined SO_2 to wood, all things being equal, cause increases in caustic usage. These latter factors are, of course, associated with the higher DP levels of these hemicelluloses.

The hot caustic extraction stage and cooking determine the overall yield for sulfite pulps. The

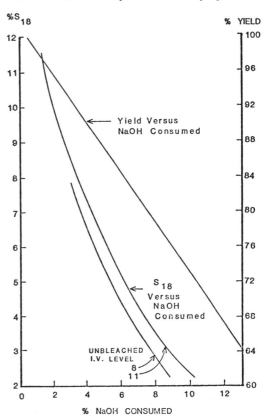

Fig. 212. **HCE yield pulp S_{18} vs HCE NaOH consumption.**

hot caustic extract is easily treated both in primary and secondary treatment systems. The potential for the recovery of specialty chemicals from the hot caustic extraction effluent, however, is great [38]. Lastly, a process innovation [39] permits the placement of the hot caustic extraction stage ahead of the chlorination stage. This allows evaporation of a chloride-free concentrate satisfactory for reuse as soda in a kraft mill, or recycling the extract along with the SSL in a soda-base recovery system.

6. Cold caustic extraction

The swelling of cellulose in the presence of sodium hydroxide is enhanced by increasingly lower temperatures. This allows the removal of hemicelluloses which are soluble in sodium hydroxide. The extraction is especially effective if it is preceded by hot caustic extraction in the case of sulfite pulps or prehydrolyzed kraft cooking (PHK). This allows preparation of wood pulp [40] grades for the highest quality end-uses approaching the market areas dominated by cotton linters (i.e., cellulose acetate plastics and rayon high wet modulus end-uses). In softwoods, prehydrolyzed kraft cooking coupled with cold caustic extraction will produce pulps with mannan levels of 1%. Similar treatment of hardwoods will give pulps of about a 0.5% mannan and 1% xylan. The extraction of mannan can be enhanced through the formation of a soluble complex with the addition of borates

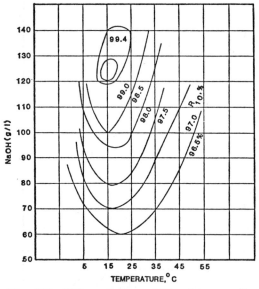

Fig. 213. Alpha-cellulose obtained by caustic soda extraction of refined sulfite wood pulp. The lines represent temperature and caustic concentration conditions for obtaining the levels of alpha-cellulose represented by the numbers on each line.

[41a, b]. However, use of borate complicates the recovery of sodium hydroxide from hemicaustic effluent.

The conditions of extraction include the dispersion of cellulose in 9-10% sodium hydroxide for 3-15 min at 25-40°C. The lower temperatures cause a high degree of swelling and give sodium hydroxide of high viscosity. This factor is a capacity constraint on the washers. At temperatures approaching 50°C, less than optimum extraction occurs with the cellulose beginning to enter into alkaline peeling reactions. Normally a 1.2-1.8% loss in yield is seen for a 1% gain in R_{10}. The range of conditions available [42] for efficient extraction is shown in Fig. 213.

At temperatures approximating 25°C with 10% caustic a shift in the crystalline structure of cellulose is seen. This is mercerization. These conditions cause a change in the cellulose crystalline lattice structure which, when subjected to drying, alters the fiber structure and subsequent reactivity. For example, dried mercerized cellulose does not react under the normal processing conditions used during acetylation. These structures, however, are usually reactivated in the viscose process; hence mercerization conditions are used to produce rayon pulps while conditions slightly less than mercerizing are used for cellulose acetate. Schemes have been devised [43] to use dried mercerized pulp for cellulose acetate. These prevent the collapse of the highly swollen crystal structure upon drying.

The large excesses of sodium hydroxide used to effect cold caustic extraction require elaborate washer control as well as various reuse plans for the hemi containing caustic. Countercurrent extraction utilizing 6-7 washers are normal with 3-4 used in the addition of caustic and three for removal. Full details of the material balances around cold caustic extraction is found in TAPPI Bleaching Monograph No. 27 [42b].

Cold caustic extraction [44a, b] can also be done with filtered white liquor. Depending upon the consistency of the extraction stage, the white liquor may need to be concentrated. Additionally, the white liquor is required to be well clarified to prevent retention of particles by fibers. Sodium sulfide does not assist in more efficient extraction [44c] in the stage, but does increase the conversion of cellulose I to cellulose II unless the sulfidity levels are typical of normal kraft liquor. The presence of cellulose II will prevent the facile conversion to cellulose acetate if the pulp is machine dried.

The position in the bleach sequence is also open to debate. Cold caustic extraction may be placed in the beginning of the bleach sequence, where less capital is required since the brown stock washing system can be used to wash in the caustic. The cold caustic stage may be last where the extraction is more selective. Both schemes are used.

7. Hypochlorite bleaching

Hypochlorite, a non-specific bleaching agent, is used in the manufacture of dissolving cellulose to reduce residual lignin material (therefore brightening the pulp) as well as to control the pulp viscosity to the degree of polymerization required for the specific end-use. With both prehydrolyzed kraft and sulfite pulps, the hypochlorite stage usually is placed early in the sequence. Reactions conditions dictate 1-2 h retention time at 45-70°C with 0.5-1.0% active chlorine having an initial pH of about 11.0. Pulp consistencies higher than 13% are used to force the reaction. The initial pH of 11 is required to eliminate the possibility of bleaching with hypochlorous acid. As shown in Fig. 214,

hypochlorous acid is not present above a pH of 9.5. Bleaching of dissolving pulps should only be done in the presence of hypochlorite ion. Figure 215 shows that bleaching near pH 7 leads to excessive and uncontrollable viscosity drops. At the same time, pH 7 bleaching with hypochlorous acid leads to glycosidic linkages which are especially labile to alkali as well as giving pulp of lower brightness and brightness stability. These factors are explained by the formation of neighboring carbonyls in the anhydroglucose units, which increase peeling reactions in subsequent alkali stages. Bleaching at pH 7 also leads to alkali celluloses which show a faster rate of aging in the viscose process. Because of those factors, dissolving pulp hypochlorite bleaching is done at high enough initial pH to yield a final pH of 10.0. Figure 216A shows the effect of hypochlorite consumption at varying temperatures versus time while Fig. 216B shows the rate of viscosity loss.

Both curves have been combined in Fig. 217 to

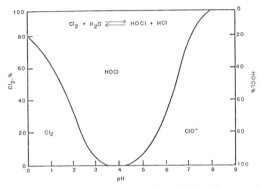

Fig. 214. The composition of a 0.01 N solution of chlorine in water at various pH's.

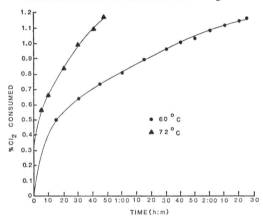

Fig. 216A. Correlation of chlorine consumed with time in the hypochlorite stage.

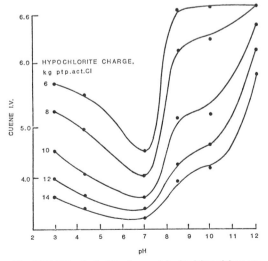

Fig. 215. Effect of pH in hypochlorite bleaching on the resulting pulp viscosity.

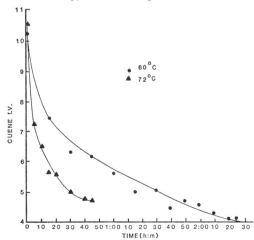

Fig. 216B. Correlation of I.V. drop with time in the hypochlorite stage.

show that hypochlorite consumption relates directly to the amount of viscosity loss. Temperature is the driving force. In actual mill practice, the amount of hypochlorite added (tower or batch cells) and time (batch cells) are the main control variables. As the chlorine consumption increases (hence, lower viscosity), increases in pulp S_{10}-S_{18} are obtained (Fig. 218). These affect end-use properties as noted earlier.

The amount of hemicellulose entering the hypochlorite stage affects the amount of chlorine consumption markedly. Hence, a lower loss in viscosity is shown at a given chlorine consumption as pulp S_{18} entering the stage increases. A 1.5 unit addition in entering S_{18}, requires 60% more active chlorine to achieve the same I.V. (Fig. 219). Hemicelluloses consume a disproportionate amount of hypochlorite relative to the cellulose viscosity drop.

Hypochlorite is also an effective shive bleaching chemical. Improvements to both acetate and viscose filterability are seen with bleach sequences containing hypochlorite.

8. Chlorine dioxide and hydrogen peroxide

Both these bleaching agents are used to make the highest quality dissolving pulps. The processing conditions used with them to achieve the highest brightness and brightness stability with minimum viscosity loss for dissolving pulps are usually those specified for paper pulp. With the exception of the already mentioned use of early chlorine dioxide to achieve low pulp resin levels, both hydrogen peroxide and chlorine dioxide (ClO_2) are normally placed after CE portions of the bleach sequence when the partially bleached pulp contains less than 1% residual lignin. The effective use of ClO_2 in promoting shive bleaching should be noted.

9. Pulp drying and finishing

The technical aspects of the manufacture of dissolving pulp does not end in the bleach plant. Both the sheet analytical properties of the dried sheet and the methods by which sheet formation and drying are carried out play a pronounced role in the acceptance of a dissolving pulp in a customer's operation.

Acetate pulps, as noted in Table 60, normally are shredded dry prior to pretreatment. Shredding is done by hammermills, pin pickers, or disk refiners. This opening facilitates the diffusion of acetylation reagents allowing a smooth reaction to be carried out. Because of this, acetate pulps are made at low sheet densities (0.45-0.51 g/cm) with basis weights ranging from 650 to 800 g/m².

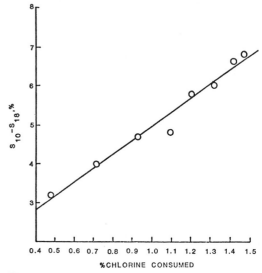

Fig. 218. The rate of increase of S_{10}-S_{18} as the severity of hypochlorite bleaching increases.

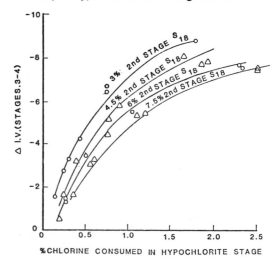

Fig. 219. The influence of incoming pulp S_{18} upon viscosity drop in the hypochlorite stage.

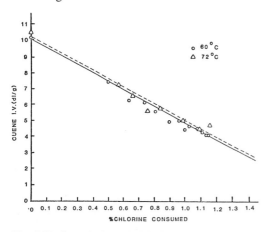

Fig. 217. Correlation of chlorine consumed with cuene I.V. in the hypochlorite stage.

The target percent O.D. for acetate pulps is 93.5-94.0%. Too high an O.D. will give an acetate of poor reactivity while too low of an O.D. will cause excessive anhydride usage. Viscose pulps for conventional steeping require sheet densities greater than 0.80 g/cm^2 while viscose slurry steeping pulps typically have a density of $\sim$ 0.68 g/cm^2, although a wider range can be tolerated. Basis weights are adjusted for optimum caustic uptake, and lack of floating during conventional steeping, and for ease of sheet breakup in slurry steeping. Overdying causes fiber hornification which does not allow efficient sheet rewetting.

Fourdrinier variables are adjusted to give uniform cross machine sheet physical properties. Maximum water removal without sheet crushing is carried out on wire and wet presses. It is desirable to avoid exposure of the pulp sheet to excessive temperatures which causes fiber inactivity. Vacuum assisted (Minton type) dryers and air supported dryers (Flakt type) are used to minimize contact of the pulp sheet with hot cylinder rolls. The sheet density of viscose pulps is achieved by a smoothing press located between the second and third dryer stacks. Such pressing is done when the pulp sheet is at 60-75% O.D. Sheet coolers (pull through dryer) are almost universally installed at the end of the pulp machine. This is done to minimize brightness reversion of the pulp when stored in the jumbo roll.

Alternative means of drying dissolving pulps include flash drying. While consuming less energy than sheet drying, flash drying generally gives pulp with fish eyes or nits which are extremely difficult to wet in both acetate or viscose slurry operations.

Recutting of sheeted pulp into sizes compatible with a customer's processing equipment is called finishing. Pulp is cut into either rolls or bales. On bales cut for viscose customers, additional sheet density up to 0.9 g/cm^3 is attained in the presses following cutting. Two jumbo rolls are cut simultaneously when baled pulp for viscose customers is required.

E. FUTURE DIRECTION

The dissolving pulp market can be considered overall to be in the mature product phase of its life cycle. Associated with the mature market phase, dissolving pulp manufacturers are tapping new markets, tailoring product improvements to market requirements, as well as optimizing manufacturing conditions.

Improved rayons with the cover and hand along with the durability and launderability of cotton have been introduced [45, 46]. Simplifications of the rayon process with the direct solution of wood pulp in appropriate solvent systems have been proposed [47-51]. However, although no large commercial plants have been recently constructed, the future for derivatives in this area looks bright. The application of membrane technology to recover chemicals from the viscose process and simultaneously resolve some long-standing environmental problems is of considerable interest. In addition, a new derivatization process to produce filament or staple from cellulose has been described [52].

Wood pulps approaching the viscosity of cotton linters have been described [53, 54]. Preparation of wood pulps by these and other methods can allow displacement of cotton linters as the starting materials for high viscosity cellulose ethers.

The use of liquid ammonia to enhance pulping reactions has been recently reviewed [55]. Adaption of this swelling agent could be promising. Reduction of bleaching costs through the use of oxygen in the manufacture of dissolving pulps can be beneficial as will cooking changes to promote delignification. Biochemical reactions specific for hemicellulose degradation reactions and lignin removal could well be the key to future dissolving pulp preparation methods so that pulps could be more specifically tailored to their intended end-use.

ACKNOWLEDGEMENTS

We acknowledge the contributions of George Nelson of International Paper and Alec Bialski of Tem Fibre Inc. for their assistance in reviewing this manuscript for content as well as for suggesting additional material.

REFERENCES

1. Hollander, S., The Sources of Increased Efficiency: A Study of DuPont Plants, 171-172 (1965) MIT Press.
2. A. Entwistle, D., Cole, E.H. and Wooding, N.S. *Textile Research J., 19*, No. 9, 527-546 (Sept. 1949).
   B. Entwistle, D., Cole, E.H., and Wooding, N.S., *Textile Research J., 19*, No. 10, 609-624 (Oct. 1949).
3. Ott, E., Spurlin, H., and Grafflin, M., Vol. V. Part II 673-1026 (1954).
4. TAPPI Standard Method, T 235 (1976).
5. Bachlott, D., Miller, I., and White, W., *Tappi, 38*, No. 8 (1955).
6. Ringstrom, E., *Svensk Papperstidn. 58*, 145-153 (1955).
7. Siclari, F., *Pure and Applied Chemistry, 14*, 423-433 (1967).
7a. W.J. Alexander, and T.E. Muller, *J. Poly. Sci. C*, Symposium No. 36, 87-101 (1971).
8. A. Schleicher, H., Jacopian, V., and Schmiedeknecht, H., *Zellstoff und Papier, 17*, 53 (1968).
   B. Schmiedeknecht, H., *Chemiefasern, 15*, 788-792, 879-884 (1965).
9. Croon, I., Jonsen, H., Olofsson, H., *Pure and Applied Chemistry, 14*, 3-4, 307-316 (1967).
   A. Gray, K.L. More, J.W., and Yorke, R.W., *Tappi 46*(12), 1963).
10. Genevray, H., and Robin, J., *Pure and Applied Chemistry, 14*, 3-4, 489-506 (1967).
11. A. Seymour, G.W., White, B.B., and Plungian, M., U.S. Patent 2,603,634 (July 15, 1952).
    B. Horne, C. Jr., and Howell, C.J. Jr., U.S. Patent 3,631,023 (Dec. 28, 1971).
    C. Campbell, K.C., et al., U.S. Patent 3,755,297 (Aug. 28, 1973).
12. A. Williams, R.F. Jr., and Lowe, C.S., U.S. Patent 2,899,315 (Aug. 11, 1959).
    B. Rouse, B.P., and Hill, R.D. Jr., U.S. Patent 2,899,316 (Aug. 11, 1959).
13. Steinmann, H.W. (to be published).
14. A. Malm, C.J., Tanghe, L.J., and Schmitt, J.T., *Ind. Eng. Chem., 53*, 363 (1961).
    B. Lamanen, L., and Sihtola, H., *Paperi ja Puu, 46*(4A), 159 (1964).
    C. Akim, E.L., *Pure and Applied Chem., 14*, 475 (1967).
    D. Rosenthal, A.J., *Pure and Applied Chem., 14*, 535 (1967).
15. Tedesco, F.M., Conca, R.J., Thelman, J.P., and Auerbach, A.B., TAPPI International Dissolving Pulps Conference, Chicago Papers: 84-87 (Sept. 28-30, 1977).
16. Wilson, J.D., and Tabke, R.S., *Tappi, 57*, 8/87, 77-80.
17. Conca, R.J., Kircher, H.W., and Hamilton, J.K., *Tappi, 45*(11), 644-648 (Nov. 1963).
18. A. Yorston, F.H., Studies in Sulfite Pulping, *Dominion Forest Service Bulletin* (Ottawa, Canada), 97, 20-29 (1942).
    B. Morud, B., Studies on the Chemistry of Sulfite Cooking, Thesis, Trondheim, N.T.H. (1958).
19. Konkin, A.A., and Shukalova, E.A. *Zhur. Priklad, Khim., 32*(5), 1076-80 (1959).
20. A. Rydholm. S.A., Pulping Processes, Interscience Publishers (1965), 126.
    B. Moelwyn-Hughes, E.A., *Kinetics of Reactions in Solutions*, 2nd Ed., Oxford (1947), 321.
    C. Wise, L.E., and Jahn, E.C., *Wood Chemistry*, 2nd Ed., Vol. 2, Reinhold (1952), 873.
    D. Stamm, A.J., Wood and Cellulose Science, Ronald Press, NY, 308.
    E. Saeman, J.F., *Ind. and Engineering Chem.*, Vol. 37(1), 43-52 (1945).
    F. Sarkanen, K.V., and Lai, Y-Z., *Cellulose Chemistry and Technology* I, 517-527 (1967).
    G. Vink, H., *Die Makromolekulare Chemie 94*, 1, (1966).
21. Gray, J.P., 185th ACS meeting, Seattle, Washington (Mar. 1983).
22. Eriksson, E., and Samuelson, O., *Svensk Papperstidn, 65*, 600-605 (1962).
23. Wenzl, H., *Chemical Technology of Wood*, Academic Press (1970), 399.
24. Hamilton, J.K., *Pure and Applied Chem., 5*, Nos. 1-2, 197-217 (1962).
25. A. Wenzl, H., IBID, 427-431.
    B. Rydholm S., IBID, 523-539.
    C. Ingruber, O.V. and coworkers, Wood pulping process. Can. Patent 817,837, July 15, 1969; also Process and product, US Pat. No. 3,525,667, Aug. 25, 1970.
26. Rydholm, S.A., IBID, 439.
27. Bernerdin, L.J., *Tappi, 41*, 491-499 (1958).
28. Wenzl, H.F.J., and Ingruber, O.V., *Paper Trade J., 150*(48), 51-56 (Nov. 28, 1966).
29. Brasch, D.J., and Free, K.W., *Tappi, 48*(4), 245-248 (1965).
30. Jacobian,V., and Casperson, S., *Faserforsch. Textiltech, 17*(16), 267 (1966).
31. A. Casebier, R.L., Hamilton, J.K., and Hergert, H.L., *Tappi, 52*, 2369 (1969).
    B. Production of Dissolving Pulps by Prehydrolysis Soda — Anthraquinone Cooking, T.J. Baker, J.K. Hamilton, L.G. Haruff and J.D. Wilson, presented at the Fifth International Dissolving Pulps Conference, Vienna, Austria (Oct. 1980).
32. R.L. Casebier, J.K. Hamilton, and H.L. Hergert, *Tappi, 52*, 135, 150 (1973).
33. A. Forssblad, L.H., *Das Papier*, 20(10a), 726-731 (1966).

B. Andersson, O., *Pulp Paper Intern.,* 8(13), 79, 82-83 (1966).

34. Rapson, W.H., *Pulp Paper Mag. Can., 57,* No. 10, 147 (1966).
35. Rapson, W.H., and Wayman, M., U.S. Patent 2,716,058 (1955).
36. Östman, H., and Lindqvist, B., Presentation at Suppliers' Technical Information Days, (May, 1981).
37. Rydholm, S.A., IBID, 596-609.
38. Reintjes M., and Cooper, G.K., 185th ACS meeting paper-presentation, Seattle, Washington (Mar. 1983).
39. Conca, R.J., Edge, D. Jr., Wick, H.J., Canadian Patent, 966,312 (Feb. 8, 1982).
40. A. Wayman, M., Sherk, D.L., and Kraska, J.V.A., U.S. Patent 2,882,965 (Apr. 21, 1959).
    B. Mitchell, R.L., Hamilton, J.K., and Smith, D.K., U.S. Patent 3,148,106 (Sept. 8, 1964).
41. A. Partlow, E.J., U.S. Patent 3,305,433 (Feb. 21, 1963).
    B. Nells, F.L., Schattner, W.C., and Ekwell, L.E., *Tappi,* 54(4), 525-529 (1971).
42. A. Wayman, A.M., The Bleaching of Pulp, 84 (1963), Mack Printing Co.
    B. Wayman, A.M., IBID, 82-91.
43. K.D. Sears, J.F., Hinck, and C.G. Sewell, *J. Appl. Poly. Sci., 27,* 4599 (1982).
44. A. Brinkley, A.W. Jr., et al., U.S. Patent 3,294,623 (Dec. 27, 1966).
    B. Brinkley, A.W. Jr., and Kilborn, J.F., U.S. Patent 3,345,250 (Oct. 3, 1967).
    C. Dobrynin, N.A., and Zarudskaya, O.L., *Tsellyoloza Burnaga Karton* No. 17:5 1981, also ABIPC, 54(5), 304, Abstract No. 2764 (1983).
45. Daul, G.C., and Barch, F.P., *Modified Celluloses,* Academic Press (1978) 81-93.
46. Dyer, J., and Daul, G.C., I and EC Product Research and Development, *20* (June 1981), 222-230.
47. Turbak, A.F., et al., U.S. Patent 4,302,252 (Nov. 24, 1981).
48. Turbak, A.F., et al., U.S. Patent 4,352,770 (Oct. 5, 1982).
49. Turbak, A.F., et al., *ACS Symposium Series, No. 58,* American Chemical Soc. (1977), 12-24.
50. Hergert, H.L., Hammer, R.B., and Turbak, A.F., *Tappi,* 61(2) 63-67 (1978).
51. Hergert, H.L., *Das Papier, 33* (12) (1979).
52. Cooper, G.K., Hinck, J.F., and Sandberg, K.R., *J. Appl. Poly. Sci., 26* (11), 3827-3836 (1982).
53. Starr, L.D., and Casebier, R.L., U.S. Patent 3,513,068 (May 19, 1970).
54. Fleming, B.I., and Reid, A.R., Conference Notes, 1981, Pulping Conference, Denver, CO, 431-437.
55. Herrick, F.W., *J. Appl. Poly. Sci.,* Applied Polymer Symposium, 37, Part II, 993-1024 (1983).

IX

RECOVERY OF CHEMICALS AND HEAT

O.V. INGRUBER

Retired Senior Research Associate from CIP, now an independent consultant

MANUSCRIPT REVIEWED BY
David Johnston, Manager Pulping
Nova Scotia Forest Industries
Port Hawkesbury, Nova Scotia

Recovery of cooking chemicals and heat has been practiced on soda and kraft pulping from the beginning of their industrial application to render these processes economically feasible. In sulfite pulping, all four bases of industrial interest were examined in the beginning and calcium base was chosen because of its low cost as the basis for the stupendous growth of the sulfite industry during its first fifty years. Low cost of the base meant avoidance of recovery altogether. Meanwhile, the kraft process with recovery rose to provide 80% of all pulp produced, its position being challenged only in recent years by alkaline sulfite processes with recovery.

Recycling cooking chemicals and heat, and/or by product recovery, is becoming obligatory for all pulping processes, including the calcium base sulfite process; partly in response to the grave environmental pollution load of many large pulp production units, partly in response to the scarcity of fuel oil and therefore rapidly rising cost of purchased energy.

Quite apart from these urges imposed from external causes, there is a good intrinsic reason for adopting and perfecting recycle processes as an integrated part of pulping systems, thereby divorcing the problem from politics and legislature: sound engineering, sound technology. All recovery practices help to keep the chemicals in the mill where they belong and keep air and water clean.

Sulfite spent liquor recovery systems considered and those currently in use are collected in Table 63.

Characteristics of spent sulfite liquors are presented in Appendix II [28]. A discussion of effluent treatment will appear in later volumes.

A. SPENT LIQUOR RECOVERY AND PULP WASHING

At the end of the cooking stage, the digester contains solid lignocellulosic material at various degrees of hardness depending on the pulp grade produced, and cooking liquor containing the original inorganic chemicals as well as dissolved or colloidal lignin fragments, carbohydrate material and resinous substances. The next treatment stage required consists of separation of solid fibrous material and solution and also some preliminary mechanical fiberization of the cooked material.

A large portion of the used cooking liquor is recovered by drainage from the chips and/or pulp, either directly from the digester or from blow pits, blow tanks, continuous filters, or presses after blowing the digester contents.

Initial defibration occurs when digester charges are blown under pressure into blow pits or tanks against stainless steel targets, by the rotating discharge ploughs at the bottom outlet of continuous digesters; and for hard or high yield cook by disk refiners located within or without the pressure and temperature cone of the digester.

Further recovery is accomplished by washing methods for displacing used pulping liquor from the pulp by water. These have reached high levels of efficiency in the last two decades.

Although disposal of spent cooking liquor into natural waters is still practiced by some smaller pulp mills, considerations of environmental impact as well as of economic management of energy and chemicals have produced sufficient incentive to discontinue such practices; and to encourage complete spent liquor recovery in the near future. Even calcium base, the only "cheap"

Table 63. Sulfite recovery systems considered and currently used (in boxes). (Adapted from Arthur D. Little, 1972).

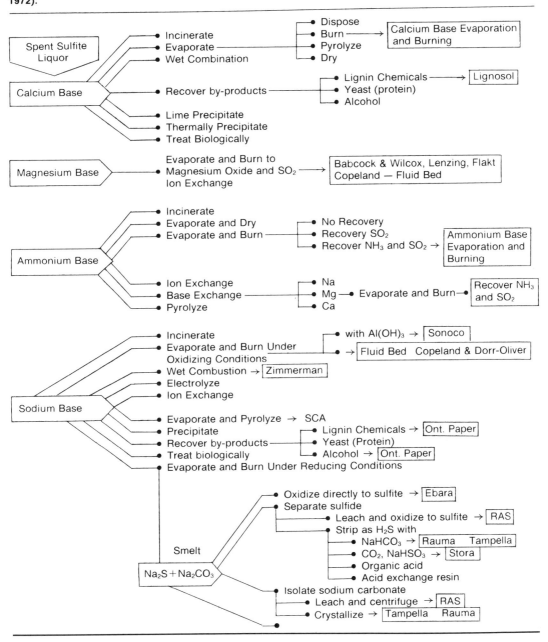

and "non-recoverable" base, is now increasingly committed to small specialty pulp mills with valuable by-product operations, and appears on the market as calcium ligno-sulfonate or gypsum ash. Sodium base, on the other hand is becoming increasingly important in sulfite pulping, and its cost makes recovery from all product units necessary as has been the case for kraft pulping from the beginning.

The two most important factors in the collection of spent pulping liquor are the amount of dry substance recovered, and the dry solids concentration of the liquor. The first determines the efficiency of the recovery of inorganic and organic substance from the cook, and the second the amount of steam energy required for evaporating the collected liquor to a concentration of at least 55% for self-sustained burning in incinerators or recovery furnace-boilers.

The original reason for pulp washing was to obtain a pulp free of soluble impurities. However, the further developments of washing methods was mainly influenced by obtaining highest efficiency of solids recovery at minimum dilution, or even with a gain in concentration of these solids. The efficiency of solids recovery in pulp washing is expressed as the weight percent inorganic solids reclaimed, as compared with the total inorganic solids applied to the cook. The efficiency of reclaiming liquor at high concentration from pulp washing can be expressed in different ways, of which the dilution factor and displacement ratio, as a concept for measuring the effect of multi-stage washing on filters, are common.

The dilution factor is the weight ratio of wash water added to the recovered spent liquor (= shower water less water leaving with the pulp) over dry pulp processed. The displacement ratio is the weight ratio of shower water (that is the weight of displaced liquor plus excess shower water to recovery) over the water leaving with the washed pulp. Ideally the liquor in the pulp should be completely displaced by the wash water without dilution but this is not the case because some of the liquor solids remain physically or chemically held in the pulp and some wash water passes into the displaced liquor.

The displacement ratio is defined in the following way:

$$DR = \frac{C_i - C_o}{C_i - C_w}$$

$$= \frac{\text{actual reduction in solids}}{\text{maximum reduction in solids}}$$

where

C = concentration of dissolved solids or salt

cake in the liquid phase

i = pulp entering the displacement
o = pulp leaving the displacement
w = wash liquor

When water is used as shower, $C_w = O$, and

$$DR = \frac{C_i - C_o}{C_i}$$

The dilution factor can be explained by example:

	lb/ton pulp
Total wash water added	15 663
Total water leaving wet washed pulp	9 425
Dilution	6 238

Dilution Factor = 6238/2000 lbs pulp = 3.12

1. Blow pit washing

Traditionally sulfite pulps are washed in the blow pit into which the digester contents are blown (Fig. 220). The forces of the blow fiberize the pulp from soft cooks which are consequently easier to wash. Blow pits are constructed of tiled concrete or wood and have a perforated false bottom made of stainless steel through which spent liquor and wash water can drain. The blown pulp is washed by displacement of the liquor by water flooded onto the top of the pulp from large moveable water jets. In the case of mills with recovery requirement, the solids content of the displaced liquor can be improved by washing first with weak liquor followed by fresh water and by keeping the temperature of the weak liquor wash high. About 70% of the spent liquor is usually recovered in blow pit washing, but a maximum of 93% has been reported.

The present trend is to convert open blow pits into blow tanks or to install new blow tanks with gas scrubbers which, in acid sulfite operations, permit control over SO_2 emissions. Presently, spent liquor reclaim is more and more transferred from blow pits or blow tanks to counter-current multistage washing on drum or belt filters and in diffusers. This conversion, in favor of waste reduction and environment protection, has led to spent liquor recovery efficiencies between 96 and 99%. The use of hot wash water at 65-70°C has improved the washing efficiency and reduced steam consumption for evaporation.

2. Vacuum washers

Drum filters have first been applied in pulp bleaching and, after 1930, first for the recovery of

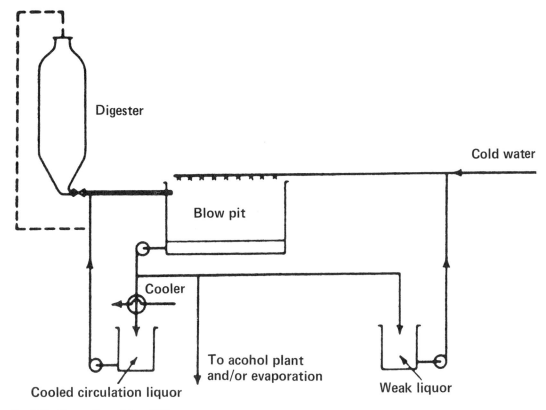

Fig. 220. Blow pit washing [39].

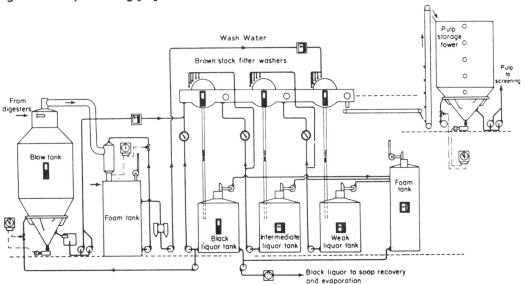

Fig. 221. Three-stage brown stock drum filter washing [39].

spent cooking liquor from kraft pulping, and much more recently from sulfite pulping. A large drum covered with a fine mesh screen rotates partly immersed in a vat into which the pulp and liquor slurry is pumped. The inside of the drum is connected to a vacuum leg which ends in a seal tank one or two floors below (Fig. 221). As the drum rotates a pulp mat of about 25 mm thickness is picked up from the slurry by the vacuum effect, passes one or more shower banks and is scraped off on the opposite side as the washed product. A set of washers contains 2-4 units, the washing efficiency increasing with the number. The consistency of the pulp leaving a rotary or drum washer is usually 12-15%.

With indirect heating of the digester and blow pit washing, about 85% of the cooking chemicals can be recovered. If a three-stage brown stock washing plant is added, the recovery yield rises to 96% and if, in addition, liquor displacement in the digester is practiced, a recovery yield of 98% is reached [1].

Table 64 is an overview of factors affecting washing efficiency.

Rather than applying vacuum, the spent liquor can be removed from the pulp mat on a drum washer by air pressure. The pressure is supplied by a blower to the totally enclosed hood of the washer (Fig. 222) and contained by a seal roll on the side of pulp discharge. By a simple shower arrangement, up to three washing stages can be applied on a single drum. A major advantage of the closed pressurized system is the containment, ease of collection, and re-use of air pollutants from the cooking liquor, be it kraft or acid bisulfite. A disadvantage is the difficulty of cleaning or replacing the drum wire mesh. The pressure washer is patented and supplied by Rauma Repola Company of Finland or its licencees. The washer has been installed in acid bisulfite mills covering all four bases and also in bisulfite and neutral sulfite mills.

3. Horizontal belt washer

A very recent development in the field of multi-stage washing is the horizontal filter (Fig. 223). Rather than forming the pulp mat on the filter surface of a drum, an endless fabric screen is driven horizontally and fed by a headbox while liquid is withdrawn by vacuum boxes, similar to the principle of the paper machine. Vacuum drop legs and seal tanks are thereby eliminated. Wash liquid is added countercurrently at designated locations over the extension of the horizontal filter, the number of stations which can be accommodated depending on the flat length of the design. Five to six stages are common. There is no interstage dilution. Pulp mats formed on the wire may be 75-100 mm thick and "near-perfect" control of liquor displacement with little mixing can be achieved reaching 99% liquor recovery efficiency at 1.0 dilution factor in some cases. At equal capacity, less space is required for a horizontal washer installation than for conventional 2-3 stage drum washers. Fully enclosed models for air protection and heat savings and also for operation under pressure are available [2].

4. Diffusion washing

Continuous diffusion washers are an offshoot of continuous digesters. The majority of installations are in the kraft sector but there are several operating in the sulfite sector.

As shown in Figs. 224 and 225, the diffuser contains a series of concentric screen rings mounted on structural arms that radiate from the center of a tower. Each ring consists of two sets of screens through which the wash liquor is collected and removed from the unit. The pulp flows upward through the tower and the screen rings assembly is lifted upward at approximately the flow rate of the pulp. When it reaches the top of its 10 cm stroke, the extraction flow is stopped and the screen drops rapidly to its bottom limit

Table 64. Factors affecting washing displacement efficiency.

Fiber Characteristics	— pulping process — stock hardness — stock freeness — species
Shower Characteristics	— temperature — distribution — method of application
Sheet Formation and Thickness	— specific loading — vat consistency — drum rotational speed
Operating Factors	— dilution factor — stock temperature — air in stock (foam) — liquor solids level — fabric mesh — characteristics fabric fouling

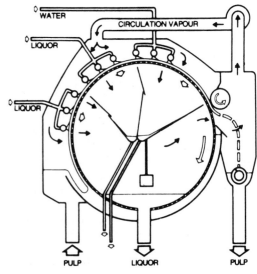

Fig. 222. Pressurized rotary washer (Rauma Repola Oy).

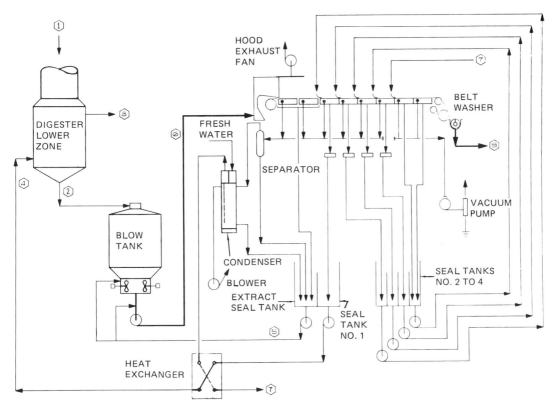

Fig. 223. Horizontal filter washer (Powesland).

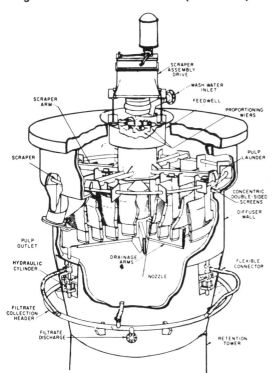

Fig. 224. Single-stage continuous diffusion washer (courtesy Kamyr Inc.).

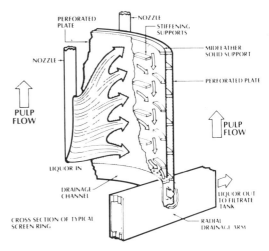

Fig. 225. Schematic of Kamyr diffusion washer screen assembly. The washing medium is introduced through rotating distribution nozzles. The displaced liquor is collected through the screens and flows into the drainage arms.

thereby clearing the screen openings of fiber. The wash liquid is added to the annular space between the screen rings through rotating nozzles connected to the top scraper assembly. During the rotation the nozzles pass between the screen rings and leave a path of washing medium in the pulp. As the presence of large particles in the stock interferes with wash-liquor flow, diffusers are more efficient if knots are removed prior to washing.

The advantages over rotary filter systems are much longer residence time of the pulp in the diffusion zone, allowing more time for residual chemicals to diffuse out of the fiber wall, and uniform stock flow together with excellent distribution of wash liquor. There is no exposure to air so that defoamer is seldom needed. The system is very simple to operate and can be readily controlled from a remote location. It is possible to accommodate more than one diffuser in a single tower and consequently the overall space requirement for a diffuser system is relatively small compared to that of a filter system of equal washing efficiency. Diffusers use less pumping energy because no dilution is required between stages. The enclosed design allows diffusers to operate with minimized vapor discharge, and the release of odorous substances to the atmosphere is much lower than that encountered in vacuum-filter operation.

5. In-digester diffusion washing

Extraction of free used cooking liquor prior to discharge of the cooked material is a feature of both vertical Kamyr and the Ingersoll Rand-ESCO continuous digester systems. In-digester washing of the cooked pulp is, however, exclusive to the Kamyr system (Fig. 226) except for early models. In the countercurrent washing stage which may be designed for up to 4 h duration, the pulp is simultaneously cooled in order to minimize fiber damage by the rotating discharge mechanism. In the ESCO digester the charge is also diluted and cooled to prevent damage in discharge, but the pulp is washed after leaving the digester. In the M&D system a vertical wash reactor can be installed at the outlet of the inclined cooking tube, and the high

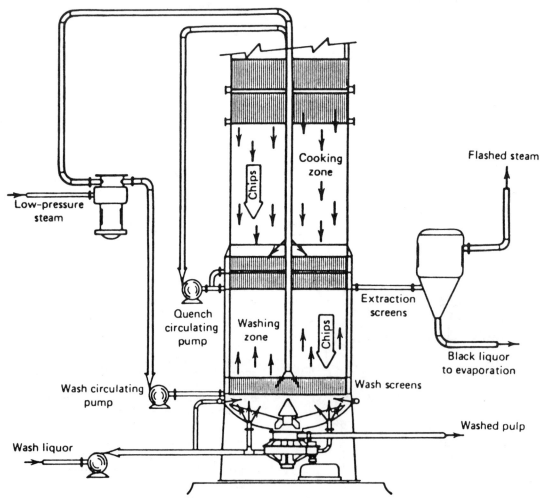

Fig. 226. Typical internal wash zone in a Kamyr digester.

pressure rotary valve can be removed from the tube and added to the bottom discharge of the wash reactor, if high pressure washing is wanted. One advantage of in-digester or high-heat washing is the great increase of the rate of diffusion. The diffusion coefficients of sodium compounds in kraft-cooked chips rises from $1.1 \cdot 10^{-5}$ cm^2/s at 40°C, to $2.1 \cdot 10^{-5}$ at 70°C (filter washing) to about $5 \cdot 7 \cdot 10^{-5}$ at cooking temperature. However, the material in the washing zone of the digester is mainly undisintegrated chips, and the diffusion rate of dissolved solids from chips depends on the smallest chip dimension, the thickness. Also, the rate decreases with the square of the thickness. This is the reason that the time allowed for in-digester washing is much longer than for filter washing of pulp, and that one or two external washing stages are always added in order to bring the efficiency of solids recovery to the required level of 97% or higher. In some cases the bottom section of the Kamyr digester provided for washing has been converted into an extension of the cooking section, but without circulation in the case of the hydraulic digester, in order to increase productivity and/or uniformity of the cook. As in all cases, the optimum configuration can be determined from quality demands plus economic calculations.

6. Screw presses

The use of presses for liquor recovery has increased greatly in recent years. Earlier pressing techniques used by the sulfite industry were based on the drum washer principle. One model, the twin roll nip press (Fig. 227), uses two filter drums rotating in opposite sense immersed in the vat. The pulp mat forms and is pressed in the close space between the rolls. The other model uses a press roll on the pulp discharge side of the rotary filter to increase consistency to 35-50% after displacement washing by showers on the filter. The consistency of the pulp mat leaving a drum washer is normally 12-15%. These presses were known to have high energy consumption and high maintenance cost.

Modern presses are of the screw type with adjustable compression ratio (Fig. 228). They are used on all pulp grades in combination with filters, or in high yield and ultra high yield pulping processes in combination with refiners.

7. Wash press

The development of the wash press started as early as 1970. Today this machine has proved to be a very efficient tool in order to both increase washing efficiencies and reduce the load on the evaporation plant.

The reason why the washing efficiency is so high is the combination of displacement and pressing to high outlet consistency, which means as high as 40%. The inlet consistency can be between 3 and 5%. (Fig. 229)

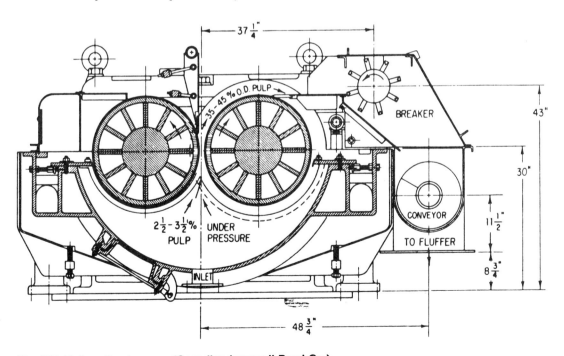

Fig. 227. Twin-roll pulp press (Canadian Ingersoll Rand Co.).

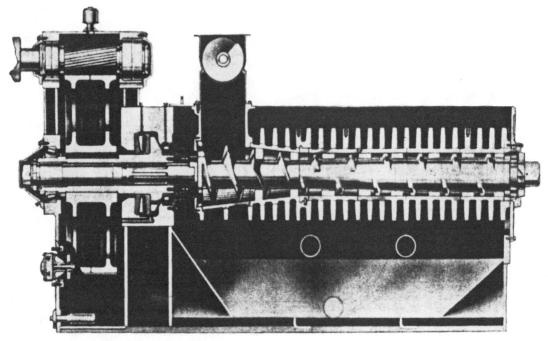

Fig. 228. Screw press (Bauer Bros. Co.).

8. High yield pulps

High yield pulps are obtained by fiberizing partly cooked chips in mechanical disk or cone refiners. If the mill does not yet practice chemical and heat recovery the chips can be lifted out of the blow tank by a conveyor-drainer and the spent liquor discharged. After refining, the high yield pulp is then washed clean in the screening stage.

If chemical recovery is practiced liquor recovery must be as complete as possible. In this case the blow liquor is collected and the drained chips are passed through one or more pressing stages to 35-40% consistency with some intermittent dilution to reach 80-90% liquor recovery. This will be satisfactory for pulp cleanliness at this stage and to meet environmental standards. Pressing of the partly cooked high yield chips opens the solid structure; it has the beneficial effect of reduced energy consumption and fiber damage in the subsequent refining stage.

Chemical recovery by evaporation and burning is possible from pulping operations up to 75% yield and requires higher than 90% liquor recovery. This can be obtained with an additional filter washing stage. Above 75% yield (very high yield processes) the organic content of the spent liquor is too low for burning in a boiler or pyrolyzer and the inorganic cooking chemicals can be recovered for re-use by evaporation crystallization [3]. The small amount of remaining mother liquor can be disposed by incinera-

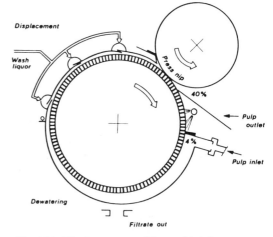

Fig. 229. Wash press working principle.

tion in a suitable way. Another way of utilizing chemical and heat values of liquors from ultra high yield cooks is to pre-evaporate the spent liquor and to pump or ship it to a large pulp mill for make-up (see Chapter VII).

B. EVAPORATION

The spent cooking liquor collected needs to be concentrated before it can be burned. The degree of concentration depends on the required heat (calorific) value of the dry solids and on the incineration process. In most cases the spent liquor is concentrated to the highest degree possible in order to maximize heat recovery. The

Table 65. Basic data of native wood components [4].

		Softwood Lignin	Hardwood Lignin	Carbohydrates
Elementary composition				
C %		64	60	46
H %		6	6	6
O %		30	34	48
Formula of basic unit		$C_{10}H_{11.2}O_{3.3}$	$C_{10}H_{12}O_{4.2}$	$C_5H_{8.3}O_{4.2}$
Mean molecular weight		187	200	135
Heat Value (BHV)	$\dfrac{MJ}{kg}$	26.4	24.7	17.6
Heat of formation	$\dfrac{MJ}{kg}$	-0.59	-0.71	-0.82

upper limit of concentration is determined by the severity of evaporation scaling, and/or problems associated with the handling of highly viscous liquor.

The combustion of concentrated pulping spent liquors is one of the central unit operations in all chemical recovery processes. This operation will often prove to be an economically feasible method of reducing the stream pollution of the pulp mill, because it makes it possible to regenerate the pulping chemicals and simultaneously to utilize the fuel value of the spent liquor [4]. Generally it can be stated that the higher the ratio of both active and inactive inorganic chemicals the lower the heat value of the liquor. Since in many of the chemical recovery processes the make-up chemicals are fed to the process before combustion, they have the effect of decreasing the heat value.

The chemical reactions during the cook will change the composition of the dissolved wood components as well as the inorganic compounds. Reactions which cause changes in the elementary composition of the spent liquor dry solids are of major significance to heat values. Such reactions are the splitting of water from, or absorption of water into, the liquor dry solids; and the formation of volatile compounds which will strip off from the liquor when the cook is degassed, during liquor expansion or in the evaporation plant.

For conventional fuels such as oil, coal, and wood there are international standard testing methods and definitions, but not for pulping spent liquors because of the considerable difference in their chemical composition, and of the combustion products. For these reasons data found in the literature may not be comparable. The heat value most commonly reported is that determined by igniting the dry spent liquor solids with oxygen in a bomb under isothermal conditions. This means complete oxidation of the organic and inorganic compounds (calorimetric or bomb heat value). Calorimetric heat values of the wood components are given in Table 65. In the actual recovery situation, complete oxidation is not the rule and the actual combustion products formed, and therefore the recoverable heat, may vary for a given spent liquor. The basis for the necessary corrections is the detailed knowledge of the elementary composition of the spent liquor solids and of the composition products formed, as well as the heat of formation values.

When the actual heat value of the spent liquor, including all necessary corrections, has been found, the net amount of heat recoverable in a furnace can be calculated. This also requires that a number of technical process variables, influencing the heat recovery, have to be known or assumed. These include the compositions, amounts and temperatures of the spent liquor, combustion process air, feed water, flue gas and inorganic solids output, as well as the heat losses through radiation and convection. Evaluations show, for example, that for a given spent liquor burnt in a reductive recovery boiler, the heat value which could be effectively utilized was only 68% of the corrected bomb heat value [4].

Table 66 contains uncorrected and corrected bomb heat values of a variety of sulfite spent liquor solids.

The calcium-base process has been popular since the beginning of sulfite pulping of wood because of the low cost of calcium obtainable from limestone. Provided there was a convenient stream or other water-course to absorb the spent liquor, it was economical to operate the mill without any chemical or heat recovery. However, since the beginning of the process, attempts have been made to recover the heat in the liquor, which was always known to be appreciable. A difficult problem was the scaling of the steel tube surfaces when evaporating calcium-base liquors, because of the low solubilities of calcium and, to make it worse, the inverse temperature effect on the solubility of calcium sulfite, sulfate, and carbonate. Soluble bases for cooking liquor seemed desirable, in an effort to greatly reduce these scaling problems which cannot be eliminated from any system, since calcium is introduced with the wood and water used. Also, magnesium and sodium base held promise for recovery of chemicals in addition to heat, because heating values alone were not attractive except in areas of very high fuel costs. It therefore followed that the mills which lead the field in the evaporation of calcium-base liquor were encouraged to do so for one of two reasons: (1) stream pollution was unacceptable and (2) the recovery and marketing of lignin chemicals; either liquid at about 50% concentration or in dry form, was possible and financially attractive.

Before the major changes from calcium to soluble base there were about forty mills employing the calcium process in Sweden. Most of the research and development work on evaporation and burning of calcium-base liquor was done there, and the Ramen and Rosenblad evaporator systems were developed to overcome this scaling difficulty. One of the problems was the corrosive action of sulfite liquors in the acid range on carbon steel, which had been used from the beginning for the evaporation equipment of kraft and soda liquors. Since stainless steel of the 316 or 317 type had to be used for the acid liquors, there was an incentive to reduce the number of stages of the multiple effect vacuum evaporator in order to lower the cost of the installation. This incentive has become much

less, with the increasing use of stainless steel also in alkaline evaporators. Instead, heat economy is taken into consideration (however, heat economy of a vapor compression-type plant is independent of the number of units).

The rate at which concentrated spent liquor is produced by a common heat-exchange evaporator depends on the concentration of dry liquor solids in the collected cooking liquor and on the evaporation (water removal) rate which, in turn, depends on the rate of heat transfer from steam to liquor.

In the equation $q = UA\Delta t$, q is the rate of heat transfer, A is the heating surface, Δt is the mean temperature difference between steam and boiling liquid, and U is the overall heat transfer coefficient. Under ideal conditions, the latter is the resultant of individual thermal resistances including steam film, metal wall, and water film

$$U = \frac{1}{\dfrac{1}{h_1} + \dfrac{1}{h_2} + \dfrac{1}{h_3}}$$

In practice, a fourth item, the scaling or fouling resistance must be attended to, since with increasing deposition of scale-forming material on the tube surface, evaporator efficiency and consequently the overall efficiency of chemical and heat recovery drops necessitating periodic cleaning.

As in the construction of alkaline digesters, the use of stainless steel has increased also in evaporator design for alkaline spent liquors to prolong life of the unit, and modern equipment can be expected to have much flexibility over the pH range available for sulfate pulping. For production units planning to make use of the inherent product flexibility of the sulfite system, it will be necessary to select proper corrosion protection over the pH range chosen.

The concentration of sulfite spent liquor solids from digesters with indirect heating is 11-17% depending on pulp grade and liquor-to-wood ratio. Free SO_2 present in acid bisulfite spent liquors is commonly stripped off and recovered so that the pH of the liquor entering evaporation is at or near the bisulfite levels. After regular pulp washing the concentration is 10-15%, the highest possible value being approached by recent installations of displacement washing with minimum dilution. In liquors from high yield cooks and in acid bisulfite operations coupled with by-product operations, the organic solids content will be reduced correspondingly. In all cases of low liquor solids, preconcentration by an auxiliary evaporator or by reverse osmosis will be required for reasons of energy economy

Table 66. Calorimetric ("bomb") heat values of spent sulfite mill liquor dry solids from various sources

Process	Pulping Yield %	Heat Value MJ/kg
Acid bisulfite		
Ca-base, spruce	45	18.3
Mg-base spruce/balsam	50	14.7
Mg-base spruce	54	14.4
NH₃-base, HW, dissolving	44	18.9
Na-base, softwood	45	18.0
Bisulfite		
Mg-base, spruce/balsam	49	16.4
Mg-base, spruce/balsam	54	14.5
Na-base	70	12.1
NSSC (NH₃) Birch	80	16.7
NSSC (Na) Gum	75	14.2
Poplar	70	13.7
Oak	60	12.2
Aspen, Maple	75	12.3
Spruce	78	13.0
NSSC, Sonoco Aspen	75	10.8
SCMP, CIP No. Softwood	90+	8.3
Kraft No. Softwood	46	14.3
AS-AQ (80-10-10)		
No. Softwood	51	11.5

Note — Net heat value of 55% dissolved solids is about 40% of dry solids.

before entering the major evaporation step, for achieving a firing concentration of 55% solids and above. The upper solids limit of indirect multiple effect evaporation for acid bisulfite and bisulfite spent liquors from softwood cooks to low lignin levels appears to be 55-58% for the metal bases.

Ammonium base liquors are difficult to evaporate to higher than 50% solids content, though 54% has been achieved industrially [8], because of a substantial rise in viscosity to butter-like consistency due to polycondensation of lignosulfonate at elevated temperature. Neutral sulfite spent liquors from high yield hardwood cooks (NSSC) can be easily concentrated in standard evaporators to above 50% solids but becomes extremely viscous at a concentration approaching 60% solids. Formation of calcium oxalate or calcium sulfite scale may occur.

Strongly alkaline sulfite spent liquors have been found to be less viscous than kraft liquors and can therefore be evaporated to higher levels of solids concentration. The behavior of moderately alkaline sulfite spent liquors is now under study.

In all cases, liquors can be raised to higher concentrations than quoted above by direct contact evaporation. This is common practice in kraft recovery where the spent liquor is evaporated to 50% and then raised to a firing concentration of 62-65% by direct contact with hot flue gas or by means of producing more steam from the hot flue gas and adding an indirect concentration unit to the evaporation plant. Application of such systems to alkaline sulfite liquor evaporation is most appropriate but has not been reported so far.

1. Scaling

The most common routine to alleviate scaling is to wash the evaporator tubes frequently with water and/or acid solution. This is effective since the solubility of $CaSO_4$ is 0.223 g per 100 mL of water. Of the original systems, the Rosenblad uses the channel-switching technique for scale control, whereby the liquor and steam sides of the heat exchanger are alternated, and the acid steam condensate removes the slight scale formed during the liquor period. In the Ramen system, one effect at a time is removed from production for de-scaling by acid condensate. In many cases extra evaporation capacity is installed so that one unit can be "on wash" at all times. In the horizontal spray film evaporator [5] the outside wall of the steam tubes can be easily washed for scale removal. This type of evaporator is widely used in North American sulfite mills

(see Chap. X,B,2).

Scale deposition on the heating surfaces of a multi-effect evaporator set increases greatly between the first effect, receiving weak, spent cooking liquor, and the last or product effect where the concentration of dissolved solids is greatest. Daily washing with acid condensate is adequate for removing caramelized liquor deposits and calcium sulfate scale. It is not effective, however, for removing more permanent scale which, in most cases, consists of 50 to 90% calcium oxalate, soluble only in strong mineral acids. Nitric acid has been in use for many years for this purpose, offering several advantages such as strong oxidizing power which aids in both organic scale removal and rendering stainless steel surfaces "passive" against acid attack. A recent serious accident at a Canadian sulfite recovery plant [6] has emphasized the need for extreme care to prevent any contact between concentrated nitric acid and hot, concentrated spent liquor containing 75% of its dry weight as organic substance derived from wood. When brought together, the exothermic reaction starts immediately causing a thermal explosion. Alternatives to nitric acid for evaporator cleaning are being examined. The most promising so far is the sodium salt of EDTA, a well-known chelating (complexing) agent. It is safe and permits evaporator cleaning during normal operation, rather than only during shutdown as with nitric acid. This increase in recovery efficiency (availability) could compensate for the higher cost of the substance [6,7].

The evaporation systems in use for concentrating sulfite spent liquors are briefly described below. Irrespective of the type of evaporation plant, the individual evaporation units are of similar design. Acid bisulfite mills with alcohol production usually integrate spent liquor evaporation with alcohol distillation.

2. Multiple effect vacuum evaporation

This technique has been in use for 135 years and is the most common throughout industry. The principle is explained in Fig. 230a. Each unit of the three-effect evaporator shown consists of a heat exchanger, an expansion vessel or separator and a liquor circulation pump. Fresh or primary steam is applied to the heat exchanger of the first effect. The steam given off by the heated liquor in the expansion vessel heats the liquor circulating through the heat exchanger of the second effect, and so on. Thus, the heat supplied by primary steam is utilized at a decreasing temperature and pressure level, to evaporate water from the liquor in successive effects until it is removed as warm water by the vacuum ejector or vacuum pump.

Steam ejectors are common in North America, while vacuum pumps are routine in Europe. The spent liquor passes continuously from one unit to the other in such a way that it reaches the first unit last, where the high temperature decreases

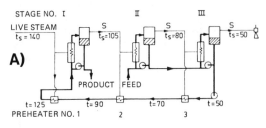

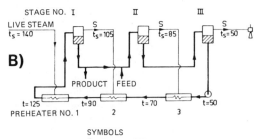

SYMBOLS
S   Steam and Vapour
t   Temperature °C
t_s Saturation temperature °C

Fig. 230. a) Conventional three-stage evaporator; b) Multi-flash three-stage evaporator.

the viscosity of the thickened liquor. Forced liquor circulation, as shown, is necessary for more viscous spent liquor; and for liquors with scaling tendency whereas for dilute liquors the falling film type evaporator effect can be used (Fig. 231). In this system, the liquor is introduced at the top of the vertical tubes and flows down, covering the walls while the vapor rises in the free annular space of the tube. Heat transfer coefficients are high at the relatively low film velocities, and losses of operating temperature are less than in the case of forced circulation. However, tubes which do not receive feed liquor may run dry, and show severe scaling, especially at higher solids concentration. Recirculation can

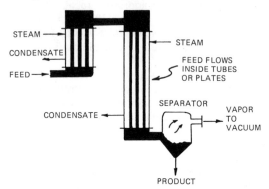

Fig. 231. Single effect falling film evaporator.

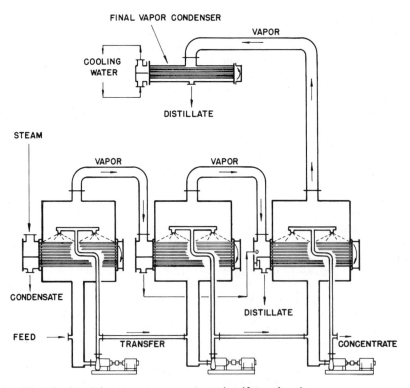

Fig. 232. Three-effect horizontal spray vacuum evaporator (Aquachem).

be used to overcome such problems.

A special horizontal tube falling (spray) film evaporator [5] with liquor circulation is shown in Fig. 232. The evaporator was developed for producing fresh water from sea water and successfully applied to sulfite spent liquor evaporation. Rather than passing the feed liquor through the tubes, the liquor is sprayed onto the steam-heated tube banks and trickles down while being evaporated. The evaporator has a heat transfer efficiency comparable with forced circulation and falling film evaporators, but at much reduced pump power and requiring less head room. The three-effect system shown can be extended to include more effects depending on the efficiency required and process details. By arranging tube banks in a number of zones which can be isolated from production scale, washing and dissolving can be accomplished during operation. This evaporator model is also used in vapor recompression systems with one or more concentration steps (Fig. 235) (see Chap. X,B,2).

A modification of the conventional vacuum evaporator with forced circulation is represented by the "multi-flash" system shown in principle in Fig. 230b. The preheater and individual circulation pump of the conventional effect are eliminated, and a single pump is used to circulate the liquor from the last effect back to the first. On the way, the liquor is heated in heat exchangers by the flash vapors from the expansion vessels; and finally by primary steam to flash effectively in the first effect where the concentrated product liquor is drawn. This scheme allows a multi-effect flash evaporator to be constructed as a single vertical unit as in the Lockman design [9] in Fig. 233. Experience with the Lockman evaporator in Canada and Scandinavia indicated that it is efficient as a pre-concentrator to about 35% liquor solids, particularly when secondary steam (low pressure) steam is used (see Chap. X,A,d). Evaporator applications for the utilization of stripper and flash steam from a continuous digester, and of relief steam from a batch digester plant are shown in Fig. 234 [10].

3. System heat efficiency

The number of effects designed into a vacuum evaporation system is based on steam consumption and capital cost. The efficiency of the system is expressed as the weight ratio of water removed to steam used. Ratios of 4-5:1 are common.

Heat extracted from the steam supplied to an evaporator must be sufficient to heat the feed solution to the boiling temperature and evaporate the water. Usually, the latter is by far the largest item. Thus, the vapor given off an effect is steam in an amount approximating the amount of steam condensed, only somewhat lower in temperature due to the temperature difference required to accomplish heat transfer. This steam given off, which is usually termed "vapor" to distinguish it from the prime steam, may be used to heat another unit boiling at lower temperature (multiple-effect operation), or may be recompressed to serve as part or all of the prime steam (vapor recompression operation). These

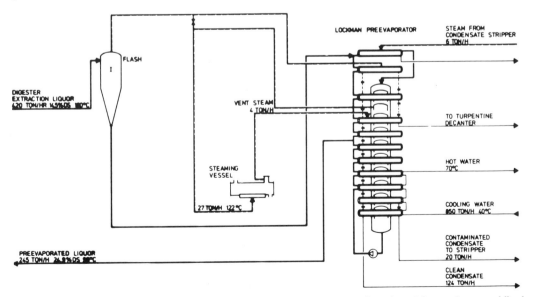

Fig. 233. Lockman pre-evaporator for 135 ton/h extra evaporation capacity using stripper steam and flash steam from a Kamyr digester for 1000 adt/day

means of conserving energy requires a more expensive evaporator than would otherwise be needed, but sometimes save enough in other investment costs to justify its use on the basis of capital cost alone. Multiple-effect operation reduces the amount of prime steam needed and hence the cost of the boiler. It also reduces the amount of cooling water needed to condense the final vapor and hence the cost of the condenser and water supply system. Vapor recompression operation eliminates the need for condenser

water entirely, and also either eliminates or greatly reduces the size of steam boiler needed (see below).

4. Vapor compression evaporation

In the vapor recompression operation, low-grade steam (vapors) leaving the evaporator effect is upgraded by a compressor to its original values of temperature and pressure to replace part of the prime steam required by the system. Recompression by high-pressure steam jets may

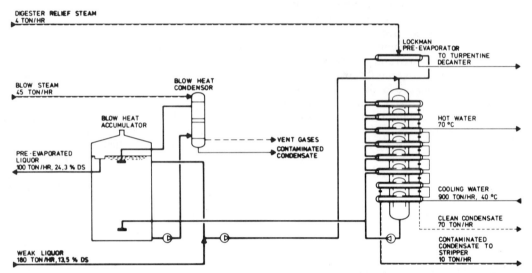

Fig. 234. Lockman blow heat evaporator for 80 ton aq/h from batch digesters of a 1000 adt/day plant.

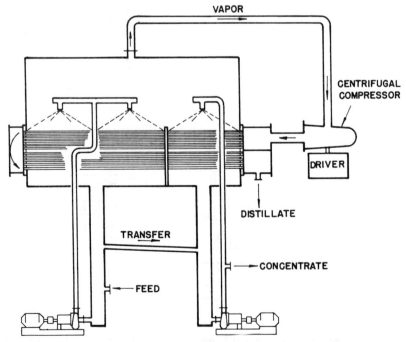

Fig. 235. Two-effect horizontal spray vapor recompession evaporator (Aquachem).

be used and requires less capital for equipment, but recovers less heat than mechanical recompression (Fig. 235). The compressor may be driven by a steam turbine or an electric motor, of which the latter is common. A single-stage compressor can increase the pressure of the vapor by about 50%; a two-stage compressor by about 100%. Under proper conditions, the mechanical (electrical) energy required by the compressor is low compared to the energy recoverable in the recompressed vapor. The method finds its widest use in areas where fuel is expensive and electric power is relatively inex-

pensive, i.e. where hydroelectric power is the major source, such as in Canada and in the Pacific Northwest of the United States. Generally, interest in vapor recompression has been mounting in recent times in parallel with the rise of cost of fuel oil. Limiting factors are (1) capital cost; (2) a large rise in boiling point of the concentrated liquor offsetting the temperature increase that can be obtained economically; (3) vapor corrosivity; and (4) liquor foaming.

The amount of net process steam available from different vacuum evaporator sets is compared in Table 67.

Table 67. Net process steam available using various vacuum evaporators.

Evaporator	Energy required to evaporate 1 kg of water, kJ	Total steam generated per tonne pulp, MJ	Total steam reqd. for evap. and atomization, MJ	Excess for process, MJ
Triple effect	1023	10 060	7290	2770
Quadruple effect	814	10 060	5850	4210
Sectuple effect	582	10 060	4245	5815
Vapor compressor	116	10 060	1035	9025

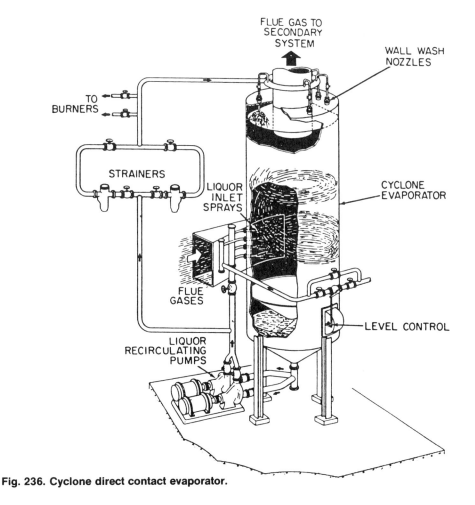

Fig. 236. Cyclone direct contact evaporator.

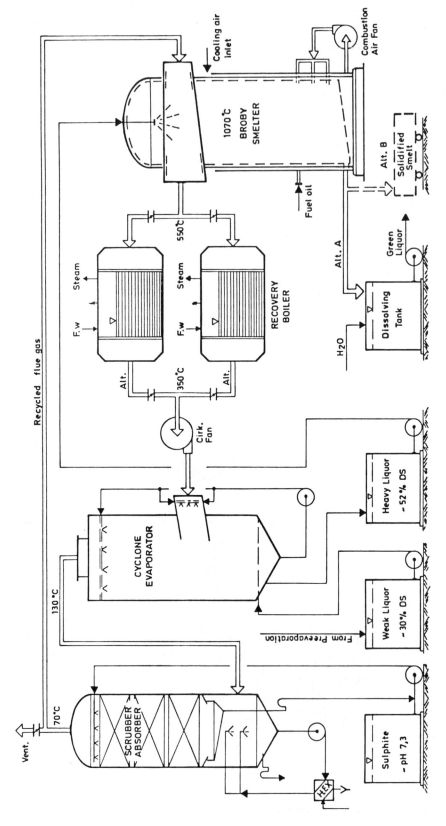

Fig. 237. Incinerator-smelter for NSSC spent liquor with DC evaporator, steam generator, and SO$_2$ recovery (Broby).

5. Direct contact evaporation

In direct contact (DC) evaporation hot flue gas from an incinerator or furnace is used to evaporate water from the spent cooking liquor. This technique is employed for raising the solids content or pre-concentrated liquor to the final level necessary for proper ignition and burning of the organic substances of the liquor. For improved heat economy, evaporation is often taken above 60-70% solids, in special cases even to dryness. This cannot be done with evaporators described in the previous sections because of viscosity (fluidity) limitations. A schematic of the common cyclone direct contact evaporator is shown in Fig. 236.

Direct contact evaporation can be designed into the body of the recovery furnace-boiler (see later) of large pulp mills, a common practice in North American kraft mills; but not in Europe where complete evaporation to firing concentration prior to the furnace is preferred. Direct contact evaporation of kraft liquor with flue gas containing CO_2 and SO_2 greatly increases the severity of the air pollution problem due to hydrogen sulfide emission, whereas with neutral and alkaline sulfite liquors no such problem exists and a choice between the two options is made on the basis of capital cost, energy balance and maintenance cost.

For smaller sulfite mills up to 200 tons of pulp per day, particularly those producing high yield pulp, inexpensive and efficient liquor concentration and burning units have been developed in recent years. Such a unit typically consists of a brick-lined incinerator with a large cyclone evaporator attached. Interposition of a steam generator is optional and depends on the amount of heat available from the liquor solids; that is, their unit heat value and amount produced.

Figure 237 shows an evaporation-incineration-smelter unit for NSSC spent liquor consisting of a Broby-type furnace-smelter, a cyclone-type DC evaporator and a steam generator in the first cooling step of the flue gas [11]. The cyclone DC evaporator raises the dry solids content to 52% for burning. Steam generated by the boiler supplies part of the required amount for pre-concentration of the spent NSSC liquor to 30% dry solids. Flue gas is recycled through a SO_2 absorber for preparing fresh sulfite liquor. Fuel oil is used to start up the furnace but is not required to sustain burning during regular operation. There are two alternatives for treating the sodium base smelt running off the inclined bottom of the furnace: a) addition to the dissolving tank to produce green liquor, consisting of an aqueous solution of carbonate and sulfide for conversion into fresh cooking liquor, or shipping to a user, such as a kraft mill for make-up, or b) preparation of dry flake of the salts for shipping to users or for storage.

The MEI unit shown in Fig. 238 features a venturi-type DC evaporator. In both cases the spent sulfite liquors are pre-evaporated to 12-14% solids and raised to 55% solids by the heat energy content of the flue gas. Pre-concentration can be accomplished with standard vacuum evaporators or, as shown in Fig. 239, from a very low level of solids content by a unique combination of air, vacuum, and DC evaporation in combination with a heat recovery column. The incinerator in Fig. 238 is a smelter with inclined bottom for sodium base smelt recovery; whereas the incinerator in Fig. 239 is of the burner type with an ash separator attached as used for calcium or magnesium type liquors. For ammonium base liquor burning, this latter unit can be used, or the ash separator can be omitted since there is no smelt or ash produced. The MEI burning and heat recovery system can be used with any of the sulfite spent liquors from acid to alkaline, including the four bases. It provides a simplified and very versatile approach to evaporation, incineration and recovery of inorganic liquor solids.

Direct contact evaporators of the cyclone type are also used in some magnesium base recovery units for raising the concentration of the red liquor from the limit of about 55% from the multiple effect evaporators. The concentration of concentrated liquor feed to the MgO recovery furnace boiler may vary from 55% to 70% solids, the minimum value determined by the liquor heating value required to provide a furnace

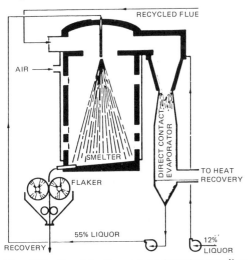

Fig. 238. Combination of incinerator-smelter, venturi DC evaporator and smelt flaker for sodium base green liquor chemical recovery (Marathon Engineering).

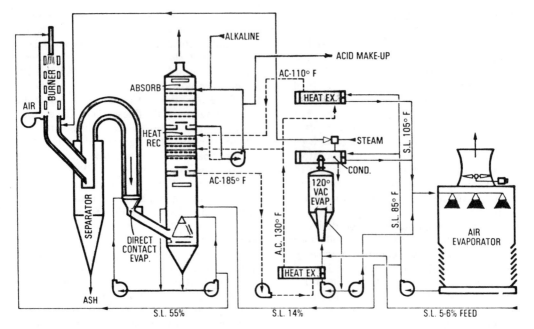

Fig. 239. Combination of ammonium, magnesium, or calcium base spent liquor burning with DC, vacuum and air evaporation for three-stage heat recovery (Marathon Engineering).

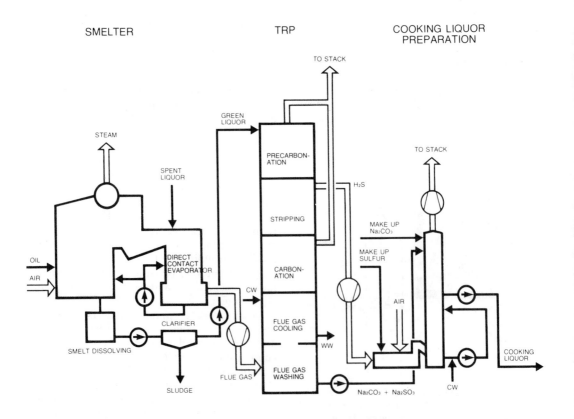

Fig. 240. Tampella recovery system for sodium sulfite pulping chemicals.

temperature for complete combustion. Liquor is evaporated in the direct contact evaporator at the expense of the high pressure steam output of the recovery boiler. Factors other than heat balance may influence the selection of direct contact evaporation; such as the trend to higher pulp yields, which demands a higher firing concentration of the liquor or the need for higher combustion temperature in the furnace. The reason for the latter can be the requirement for very clean ash for direct injection into the cook, or the higher heat requirement of a water-cooled furnace compared with a refractory wall furnace.

The most recent addition to this line of equipment is a compact smelter plant supplied by Tampella for sodium base sulfite recovery consisting of a furnace-boiler and a direct contact evaporator (Fig. 240). The furnace is made of water-cooled, gas-tight membrane walls and provides complete reduction of the smelt as in common large soda furnace-boilers. The heat is recovered as a steam, but the flue gas temperature is reduced to a level suitable for the DCE without using tube bank heat transfer surfaces susceptible to fouling and plugging. The heat remaining in the flue gas is utilized for the final evaporation-concentration of the spent liquor for burning. In this adaptation to smaller scale sodium recovery operations, both simplification of the equipment and better heat economy were achieved. The unit is intended to be used in combination with the Tampella multistage reaction column for chemical conversion (see Section D,1c of this Chapter).

6. Air evaporation

The principle of an air evaporator is explained in Fig. 241. A similar design is used for the common cooling towers of power plants. Cold

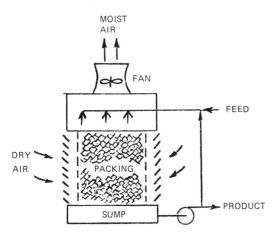

Fig. 241. A typical air evaporator.

dry air picks up moisture in contact with hot spent pulping liquor. Its simple construction makes it an inexpensive method of pre-evaporation. Air evaporation is more effective in winter when the inlet air is drier. Depending on the liquor composition, the moist air leaving the unit may carry volatile compounds which are not acceptable by environmental standards.

7. Reverse osmosis (ultrafiltration)

Another technique available for initial or pre-concentration of spent sulfite liquors is membrane filtration. Natural membrane filtration is called osmosis. It occurs when solutions of different solute concentration are separated by a suitable membrane and water passes through the membrane from the dilute solution into the concentrated solution. Reverse osmosis (also called pressure filtration, ultrafiltration, hyperfiltration) is produced artificially by applying pressure to a solution and thereby forcing water preferentially through a water-swollen membrane leaving a more concentrated solution behind. Cellulose acetate has been in use as membrane material, and new plastic membranes have been developed.

Some typical operating conditions for the concentration of SSL are given as follows:

Pressure	: 3.8-4.8 MPa
Temperature	: 0-30°C
pH	: 2-9
Average flux	: 16 $L/m^2 \cdot h$

The flux is inversely related to the concentration obtained. Electricity for pumping is the only form of energy required, the typical range of consumption being 6-20 Kwh/m^3 of filtrate [12].

Practical applications include the concentration of SSL to 12% DS or higher. Low molecular weight organics, e.g. methanol and furfural are generally not retained efficiently by these membranes. In particular, acetic acid in the SSL will be retained in the concentrate only if the SSL is not neutralized prior to membrane processing.

Membrane fouling is a particular problem of this type of process. Even with regular back washing and cleaning with special detergent, membrane life is still fairly short, although one manufacturer of membrane processing equipment is prepared to give a guarantee of one-year life on the membranes. The replacement of membranes could account for 30-50% of the operating cost of the process.

Figure 242 illustrates one version of reverse osmosis equipment. There are several units of this type operating at North American and Scandinavian sulfite mills.

C. COMBUSTION

1. Loddby burning with steam generation

A way for the efficient burning of calcium base SSL for steam generation was found in Sweden in 1949 [10,11]. Prior to this date spent liquor at 50-55% solids was burnt by feeding in the form of an atomized spray directly into the boiler. There were several difficulties with this method, notably in increasing the heat value of the liquor by increasing the solids content and the need to use much supporting fuel. Another problem was ash deposits on the boiler tubes when the liquor solids were burnt wet. The simple solution found was to let both liquid drying and combustion take place in a separate brick-lined chamber with tangential air-intake to form a cyclone-shaped flame where the movement of gases and particles could be controlled to give optimum conditions for drying and burning. An important feature of the flame cyclone effect is that the particles which resist combustion most are forced out towards the circumference where the concentration of oxygen is greatest. The particles must be dry enough not to stick when they strike the wall. Figures 243 and 244 show Loddby (the name of the mill where the cyclone incinerator was developed) combustion chambers attached to a horizontal style boiler of moderate capacity and a large upright combination furnace boiler for power generation.

This unit is equipped for auxiliary burning of coal or oil in order to maintain a given level of steam production even when the supply of SSL is low or interrupted.

The Loddby SSL combustion technique is used by a considerable number of small or medium sulfite mills. Application has expanded from the original calcium base to magnesium and ammonium base. With ammonium base liquors, the process is clean because very little ash is formed, but excess oxygen in the firing zone must be monitored and controlled to about 1% to prevent ammonium sulfate formation.

Loddby burning cannot be used for sodium base liquor recovery because of the low melting point of sodium salts. Even with the other bases, some stickiness or caking of the ash may occur when sufficient sodium is present, such as brought in with logs transported in seawater.

2. Recovery furnace-boilers
Sodium base

The most common equipment for spent liquor combustion and steam generation of chemical pulp mills of any size is the reductive soda furnace-boiler, with water-cooled walls and discharge of carbonate/sulfide smelt to make green liquor. The method of reductive burning applies only to sodium base spent liquors. It is the only liquor burning method available for producing, in one operation, the fully reduced sodium-sulfur compound (sulfide, Na_2S), which is used directly for cooking as in the kraft

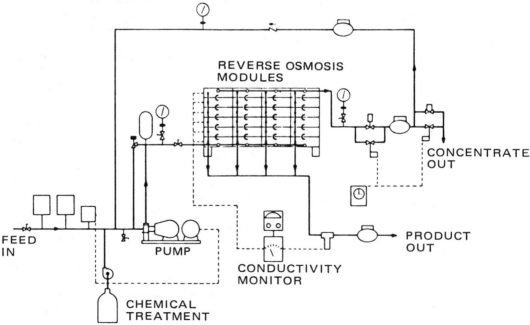

Fig. 242. Schematic diagram of pilot plant reverse osmosis system for sulfite liquor.

process; and which can be converted into bisulfite and sulfite for the different sulfite processes. Oxidative burning produces sodium sulfate which is inert in a cook.

The first efficient models were designed and introduced industrially in the thirties (G.H. Tomlinson Sr., Babcock & Wilcox Co., Combustion Engineering, Inc.) for the recovery of chemical and heat from kraft pulping operations; and have grown with the rapid increase of the production capacity of kraft pulp mills to enormous sizes, capable of processing spent liquor from pulping operations exceeding more than 1000 tons of pulp per day. A clear layout of the basic soda furnace-boiler and its major components is shown in Fig. 245. Design changes other than size have been few over the years, and those made are mainly concerned with reducing corrosion and increasing the safety of operations. These two items are of fundamental importance for operating this type of equipment, since any leakage of water from the tube walls or boiler tube banks can cause a smelt-water explosion. However, to the knowledge of the author, no instances of major explosions from sodium base sulfite recovery units have been reported so far. This could be due to far fewer installations compared to kraft, or to better control.

Addressing the corrosion and safety aspect became particularly important with the continuing need to increase steam pressure and temperature; consequently, the operating temperature of the furnace boiler to meet the demands of back pressure co-generation of electrical power by the mill. In the 1930's the first units operated at a saturated steam pressure of 3-4 MPa (234°C); in 1950's leading mills increased to 6.4 MPa (278°C); and in the 1960's to 8.5 MPa (300°C). At these higher levels, alarming corrosion damage of the tubular walls became apparent at the smelt level. Measures to make the steel tubes in the lower part of the furnace resistant to smelt corrosion included stainless steel overlay welding on the exposed part of the surface with some success (Fig. 246), but eventually culminated in the present practice to gradually replace all water-wall steel tubing in this area by compound stainless steel tubes (Fig. 247). The cross-section of four of the compound tubes forming the water-cooled wall in Fig. 248 shows the basic tube made of carbon steel (ASMESA-210 grade A1) and the tube links and tubes covered by a metallurgically bonded layer of 304 stainless steel (cf. Chap. X,C).

The floor tubes in these furnaces are not exposed but are covered with a special ceramic mixture which is resistant to the action of the hot, corrosive smelt. The heat from the cooled walls and floor of the furnace serves to pre-heat the boiler feed water prior to entering the lower boiler drum.

The basic furnace boiler design in Fig. 245 includes a direct contact evaporator for final liquor concentration. With the establishment of air purity standards in recent years, this practice which has been common in North American kraft manufacture has been put in doubt, and more and more kraft mills are eliminating the DC evaporator from their furnace boilers, either

STEAM GENERATER

CYCLONE BURNER

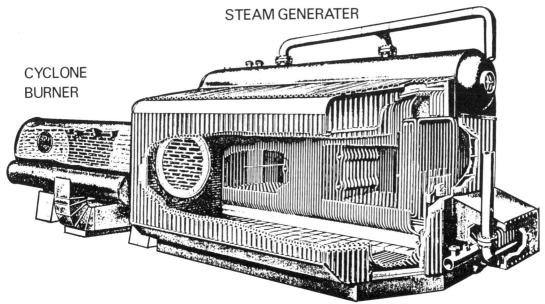

Fig. 243. Loddby cyclone burner attached to horizontal steam generator.

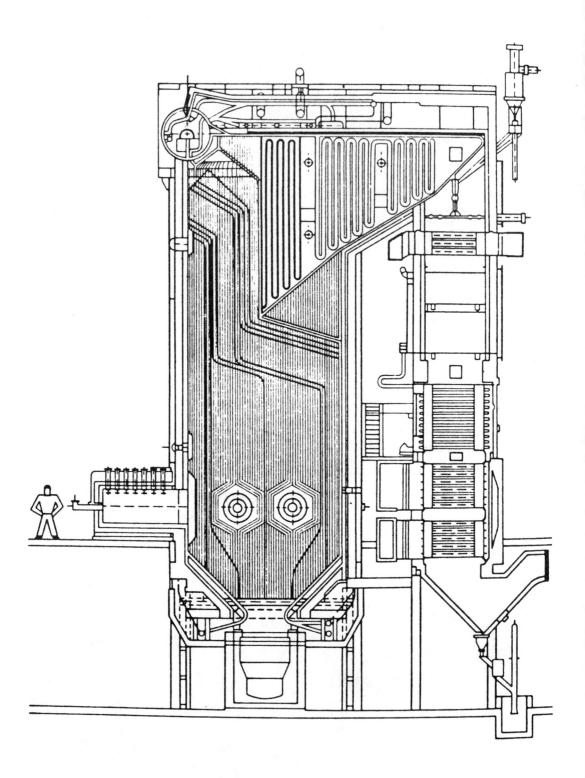

Fig. 244. Loddby burner attached to large upright furnace boiler.

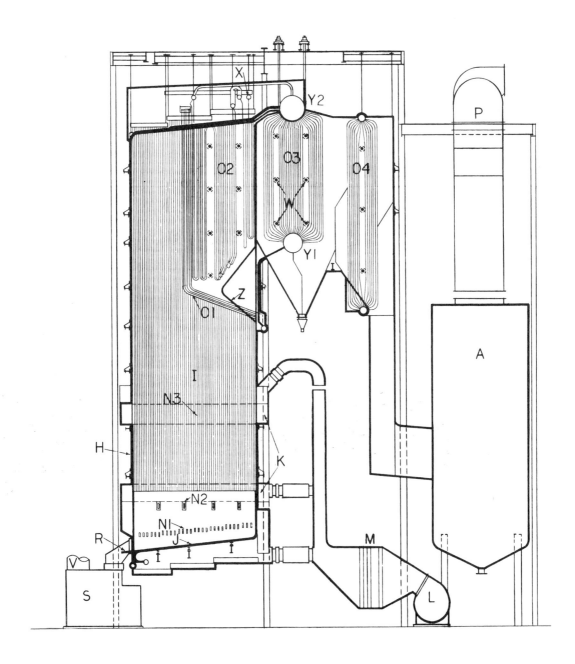

Fig. 245. Side elevation of Babcock & Wilcox Co. sodium base recovery unit with cyclone-type direct-contact evaporator. Key to components: A, direct-contact evaporator; B, venturi scrubber; C, cyclone separator; D, salt cake make-up mix tank; E, flow box; F, liquor heater; G, heavy black-liquor pump; H, black-liquor spray nozzles; I, furnace; J, hearth or floor; K, auxiliary fuel burners; L, forced-draft fan; M, steam air preheater; N, air ports; 01, screen; 02, superheater; 03, boiler bank; 04, economizer; 05, gas air heater; P, induced-draft fan; Q, economizer bypass damper; R, smelt spout; S, smelt-dissolving tank; T, smelt-dissolving tank vent; U, dissolving tank agitator; V, attemporator or spray-type desuperheater; W, soot blower location; X, superheated steam outlet; Y, boiler drum; Z, nose baffle.

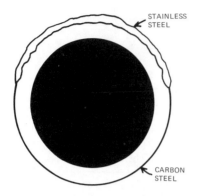

Fig. 246. Water wall furnace — boiler steel tube with overlay stainless steel (Tampella).

Fig. 248. Stainless steel clad compound tubes to form the water wall of the hearth of a sodium base furnace-boiler (Tampella).

Fig. 247. Water wall hearth section made of stainless steel compound tubing; carbon steel tubing higher up (Tampella).

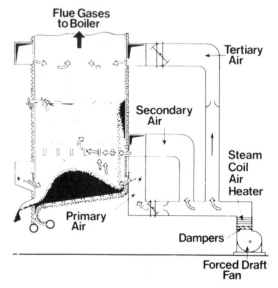

Fig. 249. Char bed and air distribution for B&W recovery furnace.

by modifying the existing unit or installing a new unit. This restriction does not apply to furnace boilers of sulfite mills, because the spent liquor does not contain foul smelling substances which could be liberated in DC evaporation.

DC evaporation reduces the amount of ash particles in the flue gas by a washing effect and decreases the load to the precipitators.

There are three principal manufacturers of water-walled smelting furnace-boilers for recovering sodium-base SSL: Babcock and Wilcox Co. (with Gotaverken in Europe), Combustion Engineering Inc., and Tampella Oy of Finland. The units are similar in design except for the mode of spraying the concentrated spent sulfite liquor into the furnace. B & W use their standard kraft-type feeding method, an oscillating splash-plate nozzle (Fig. 249) which directs the spent liquor against the opposite walls of the furnace where the liquor adheres, dries, and the solids fall to the char bed for subsequent reduction. C-E, on the other hand, uses a special method whereby coarse sprays from wall-mounted nozzles are directed out over the width of the furnace (Fig. 250) so that the liquor droplets in falling lose a significant part of their moisture before reaching the top surface of the char bed. Tampella uses an even coarser spray from one centrally mounted nozzle at a somewhat higher elevation to allow increased flight distance for drying.

In a recovery unit as in Fig. 245, about 45% of the combustion air is fed to the primary air inlets and directed to impinge on the char bed at an angle to form its conical shape. About 25% of the air may be fed to the secondary inlets, and the remainder 30% to the tertiary inlets below the boiler banks. There are three relatively well defined zones in a smelting furnace: a reducing zone at the bottom, a drying zone roughly in the vicinity of the liquor ports, and an oxidizing zone roughly in the vicinity of the secondary air ports. Combustion of the volatilized and pyrolyzed organic matter rising from the char bed takes place in stages as the ascending gas mixture reaches and passes the level of the secondary and tertiary air ports. It is assumed that combustion is complete before the hot gas passes over the screen tubes in the upper part of the furnace. Control of the amount of oxygen in the exiting flue gas is critical for the chemical composition and quality of the flue gas.

There is a fundamental difference in the chemical value of the flue gas from kraft liquor burning and from sulfite liquor burning. In the case of kraft operations the flue gas has no value besides its heat content and its sulfur content carries the liability of air pollution. In the case of sulfite operations, the flue gas from complete combustion of the organic substances carries a considerable proportion of SO_2 gas which varies with the sulfidity of the spent liquor from the particular sulfite process used. Sulfidity is expressed as the ratio of mol sulfur-bearing chemical to mol sodium bearing chemicals, or

mol S/Na_2. The lower the ratio, the more of the sulfur will be bound by sodium in the combustion process and end up in the furnace smelt; the higher the ratio, the more sulfur will leave the furnace boiler with the flue gas. Sulfidities of spent sodium base liquors range from low values such as 0.3 for kraft to 2.0 and higher for bisulfite and stripped acid bisulfite spent liquors. Intermediate values are about 0.45-0.50 for a strongly alkaline SSL and 0.8 for a NSSC or weakly alkaline SSL. Unlike in the kraft process, the sulfur dioxide leaving the recovery furnace-boiler is a sulfite cooking chemical. It is returned to liquor making via a full-size flue gas scrubber following the electrostatic precipitator (see Fig. 271). In this way the sulfite operation with chemical recovery provides a closed cycle of the gaseous sulfur component and avoids air pollution. Recycle values for SO_2 directly from the combustion stage are of the order of 30% for sodium bisulfite processes, and 10-15% for neutral and alkaline sulfite processes. The higher the SO_2 recycle ratio the lower is the heat loss in the furnace for reduction.

Values of recoverable SO_2 are, of course, much higher in the flue gases from magnesium or ammonium SSL burning, approaching complete and direct recovery from the furnace.

The partial oxidation of the surface of the bed is sufficient to supply the heat needed to reduce Na_2SO_4 formed near the surface to Na_2S by reaction with the carbon in the char. The bed temperature also has to be high enough to melt the sodium salts at all stages of the reduction process. The molten smelt which is composed mainly of sodium carbonate and sodium sulfide flows through spouts from the furnace floor into the dissolving tank containing water, or a weak liquor where it forms the green liquor.

The reaction between the dry solids of the liquor and the primary air in the char bed produces hot flue gas and smelt. The main part of the heat generated is carried away by the flue gas and utilized in the heat transfer surfaces of the boiler.

A quantitative impression of the performance of a recovery unit is given in Fig. 251. It can be seen that only 6%, by weight, of the total material flow introduced is discharged as smelt.

The quality of the smelt is characterized by the sulfidity %S and the reduction %R defined by the following formulas:

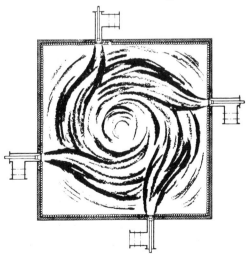

Fig. 250. Plan view of Combustion Engineering furnace showing tangential entry and mixing pattern for secondary air.

$$S\% = \frac{Na_2S + Na_2SO_4}{Na_2S + Na_2SO_4 + Na_2CO_3} \times 100;$$

$$R\% = \frac{Na_2S}{Na_2S + Na_2SO_4} \times 100$$

where the chemical symbols stand for molar proportions. The sulfidity gives the relationship between sulfur bound sodium and total sodium of the smelt. This parameter is chosen for the process to give the best cooking results. The reduction figure gives the recovery efficiency of the unit; usual values are between 95 and 98%.

The relationship of smelt composition and melting point is shown in Fig. 28. The higher the sodium sulfide content of the smelt the higher is the melting point. However, the temperature needed is considerably lower than indicated by the diagram. Smelts from bisulfite SSL require about 950°C (instead of 1100°C) in the char bed as compared with about 750°C for smelts from the kraft process. This is probably due to two factors, the removal of a considerable portion of the sulfur load of the system with the SO_2 in the flue gas, and the presence of sodium salts other than sulfide and carbonate in the smelt.

The hot gases from the combustion zone carrying some ash in the form of sodium salts (mainly Na_2SO_4) pass through the steam generation zone containing screen tubes, super heater, boiler and economizer. When seawater-rafted wood is used for pulping, the furnace dust also contains accumulations of sodium chloride which is volatile at combustion temperatures. The water-cooled walls of the furnace and the screen tubes reduce the temperature of the ash carryover to a level where it loses its stickiness and does not foul the tubes of the steam generator. Material adhering to the tubes of the screen and superheater is dislodged by automatic soot blowing and falls back into the furnace hearth. The amount of ash carried in the flue gas depends on the temperature in the combustion zone, the position of the spent liquor nozzles

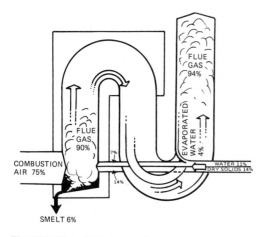

Fig. 251. Material balance, by weight, of a sodium base recovery furnace-boiler (Tampella).

relative to the secondary air ports, loading of the furnace, and gas velocity.

The heat exchange surface of the economizer is dimensioned to cool the flue gas to 120-150°C in units without DC evaporation and to 300°C with DC evaporation where further cooling takes place.

Design changes of the original sodium furnace-boiler became necessary for adapting the unit to magnesium and ammonium base sulfite operations.

3. Magnesium base

The magnesium oxide recovery furnace-boiler differs from the conventional model in that the chemicals to be recovered are not collected as a smelt but as MgO dust and SO_2, which are swept through the boiler section and are recovered from the flue gas as light MgO ash and SO_2 gas after leaving the boiler. Since these two chemicals form fresh cooking liquor when brought together in an absorption system, this process is unique as it provides the required cooking chemicals directly from the spent liquor combustion reaction. In distinction from the reductive sodium base recovery process, the magnesium base recovery process is characterized by controlled oxidative combustion.

The only remaining steps to reform the cooking liquor are cleaning of the MgO ash by washing, slaking in water to form $Mg(OH)_2$ (milk of magnesia) and absorption of SO_2 gas (see Section C, 8 of this Chapter).

The process was developed before 1940 jointly by Howard Smith Paper Mills and Babcock & Wilcox Co. in Canada, and independently by Weyerhaeuser Co. in the United States. In 1948 Weyerhaeuser's 275 t/d calcium base sulfite mill at Longview, WA. was the first in the world which had been converted to magnesium base operation. Meanwhile a considerable number of acid bisulfite and bisulfite mills have been converted to Mg base and new mills built. In Sweden and middle Europe the majority of existing sulfite mills have now been committed to this base.

Conversion of a mill from magnesium base to sodium base operation would require complete replacement of the recovery system.

The particular design of a typical furnace-boiler for Mg base SSL ("red liquor") [11] includes an air-cooled refractory-lined combustion zone needed for reaching combustion temperatures of 1250-1400°C (Fig. 252) to produce clean MgO ash without carbon residue at low excess oxygen. Too high temperature and long furnace residence time can lower the reactivity of the MgO by forming a crystalline

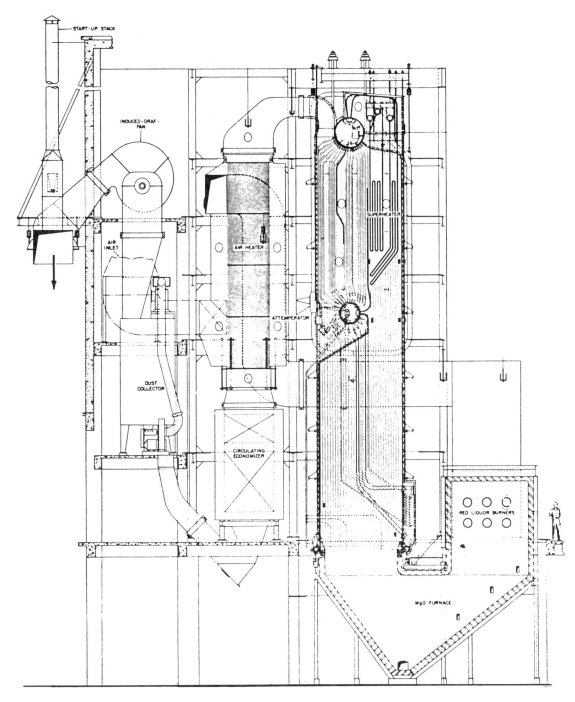

Fig. 252. MgO furnace-boiler with extended refractory lined combustion chamber (Babcock and Wilcox).

form of MgO, less soluble in H_2O. More recently MgO furnace-boilers have been built with water-tube walls and refractory lining (Fig. 253) so that the required high combustion temperature can be maintained and auxiliary fuel oil can be fired together with the spent liquor without overheating. Unlike the reductive soda furnace, the MgO furnace is designed to fully oxidize the base component of the SSL.

The concentrated spent liquor at 55% solids, or higher if DC evaporation is used, is fired into a furnace through circular burners similar to oil

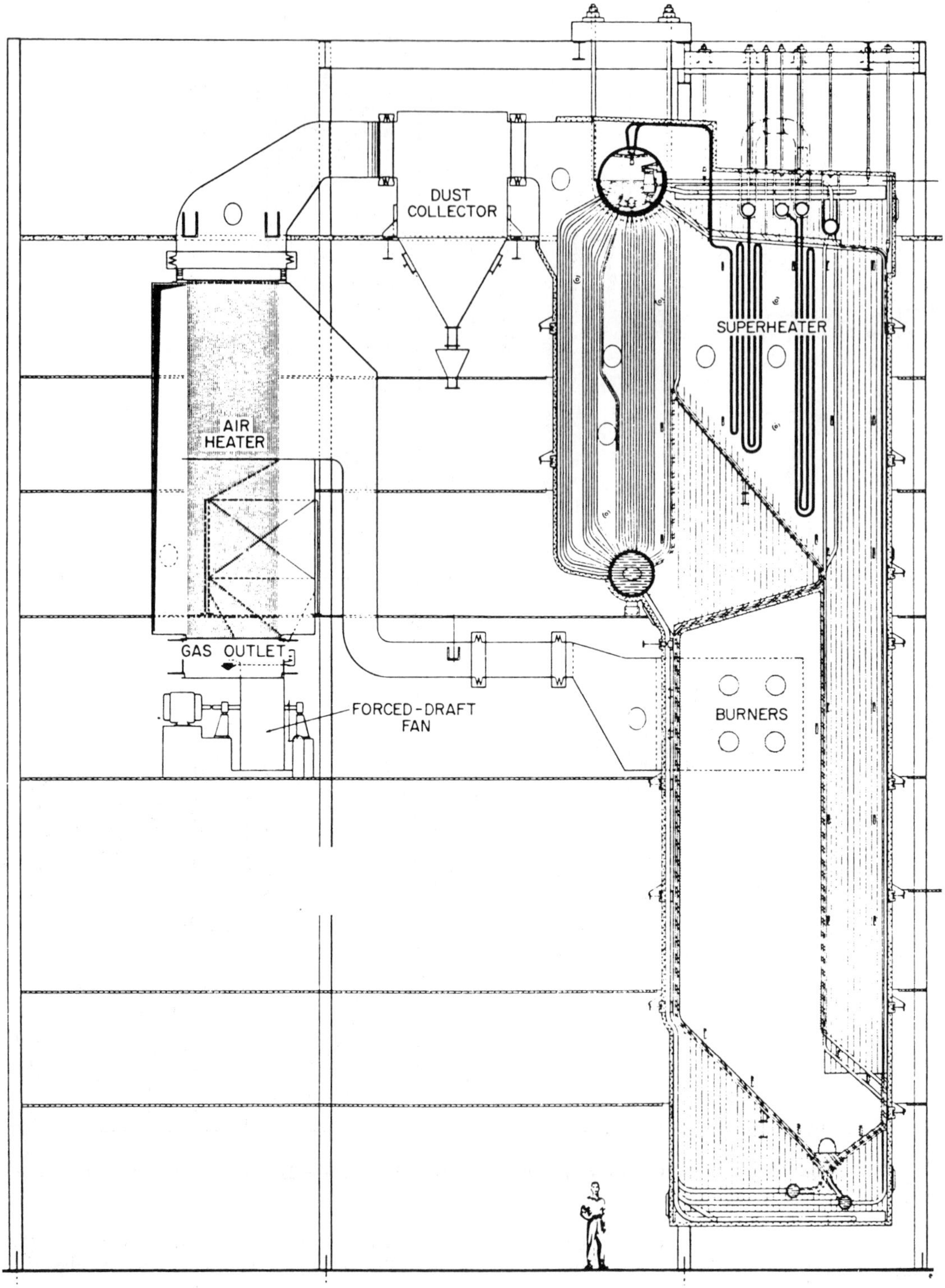

Fig. 253. MgO furnace-boiler with integrated water-cooled and refractory-lined combustion chamber (Babcock and Wilcox).

burners. The burners are located in opposed positions in the side walls of the furnace, thereby providing turbulence, and are equipped with sprayer plates, where, in combination with steam, the liquor is atomized to facilitate rapid and complete evaporation and combustion. The liquor burns in suspension in an oxidizing atmosphere with excess air as low as is compatible with complete combustion. Excess air is minimized to avoid the formation of SO_3 and of magnesium sulfate in the ash.

The air to support liquor combustion is introduced through the circular burner coincident with the liquor flow. To ensure rapid and complete combustion of the liquor, it is desirable to have the air at temperatures in the range of 315 to 370°C. These temperatures can be obtained with a tubular air heater located in the path of the combustion gases or with a direct-fired air heater located adjacent to the furnace. Furnace temperature is controlled by liquor concentration so that the flue gases leave the combustion chamber above 1300°C to assure a carbon-free MgO ash with 2-10% $MgSO_4$ and a minimum of SO_3 in the sulfur dioxide gas.

The gases from the furnace pass through radiant wall tube screens to reduce the gas temperature to a level where sticking of ash particles to super-heater and boiler tubes is prevented. The super-heater is used to raise boiler steam temperature to higher levels such as required for co-generation of power. Flue gases then pass through a two-drum boiler, which can be designed for pressures up to 10.4 MPa. Ash accumulations on the surfaces of the boiler and super-heater are removed by retractable soot blowers. The soot-blowing schedule is once per shift. Ash deposits on heating surfaces would seem impossible at first glance, with an ash having a melting point of 2500°C, but the deposits are due to contamination by sodium and potassium derived from the wood ash, water, and chemicals.

Primary recovery of the chemicals of the magnesium base sulfite process is thus accomplished in the furnace process. Separation of MgO from the flue gas and preparation of fresh cooking liquor (secondary recovery) is explained in Section C, 8.

An interesting modification of the MgO furnace-boiler was introduced commercially in 1965 [15, 16]. The Lenzing furnace-boiler was developed at about the same time as the Loddby cyclone burner, with the difference that the liquor combustion chamber was accommodated in a compartment of the furnace-boiler. At the same time the mode of spent liquor injection and of air addition was radically altered for provid-

ing countercurrent combustion conditions which are favorable for uniform and efficient burning of concentrated Ca and Mg base sulfite liquors. The principle is illustrated in Fig. 254 which shows the combustion chamber integrated with the furnace portion of the boiler; the location of the liquor nozzle which directs the steam atomized liquor spray upward; and the admission of combustion air (350°C) at the top of the combustion chamber. As a result, the rising liquor jet is ignited at the point where it meets with the air and the flame burns downwardly through the combustion chamber into the main furnace, drying the droplets of the liquor jet on its way. This flame design provides a sufficiently long combustion path for the particles of liquor solids to burn throughout. Thorough burning and MgO/SO_2 dissociation in the combustion stage take more time because of the presence of molten $MgSO_4$ (melting point 1127°C). For comparison, $CaSO_4$ has a melting point of 1297°C and would not melt at the furnace temperature of 1250°C used in this process.

Again, unlike burning of calcium base sulfite liquor for steam generation only, the magnesium base recovery process demands controlled oxidation conditions to minimize the formation of $MgSO_4$ and SO_3 and a clean MgO ash with maximum reactivity in contact with water. The

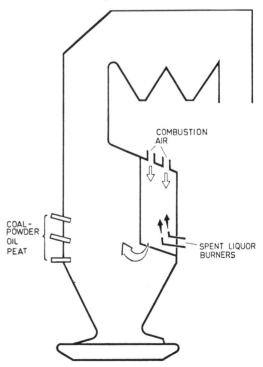

Fig. 254. Operating principle of the Lenzing MgO furnace-boiler (Lurgi-Lenzing-Steinmüller).

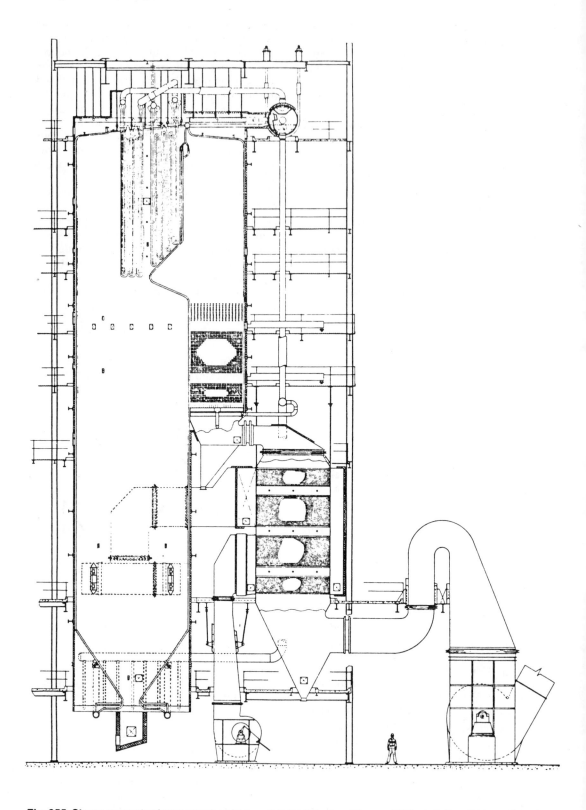

Fig. 255. Steam generator for ammonium base acid bisulfite spent liquor. 450t/d of dissolving and market pulp (Combustion Engineering-Superheater Ltd.).

first condition is met by maintaining a concentration of 1.5-2.0% excess oxygen in the flue gas. The second condition is met, firstly by producing ash of uniform quality by countercurrent burning, and secondly by a sufficiently high combustion temperature in the chamber due to predrying the spray and covering the tube walls with a 40 mm thick refractory lining containing chromium. The furnace incorporating these features (Lurgi-Lenzing-Steinmüller) is operated at 1250°C and produces a cake on the MgO filter (see later) of 65% brightness. A brightness of over 60% indicates that the MgO is almost free of carbon.

4. Ammonium base

The original water wall soda furnace-boiler was adapted for the recovery of SO_2 gas and heat from ammonium base SSL as shown in Fig. 255. The model shown was installed in 1977 in a mill producing 450 tons daily of acid bisulfite dissolving and bleached market pulp from NW Pacific softwoods [8]. The installation of the furnace-boiler was part of an investment scheme for changing from calcium base to an ammonium base operation and to drastically reduce water pollution, recover sulfur dioxide, and to generate electrical power with a back pressure steam turbine to help meet the electrical power demand of the mill.

Similar to Mg SSL burning, this process also uses combustion with controlled oxidation. As explained earlier, the heat value of the organic liquor solids is augmented by the heat value of the ammonia in the SSL. About 80% of the SO_2 in the spent liquor can be recovered from the flue gas. Note the replacement of the smelter floor by a conical collector and a little bin for collecting the small amount of ash formed.

The ammonium base SSL ("red liquor") is first steam-stripped of loosely bound SO_2 and then evaporated in a six-effect stainless steel evaporation unit. Multiple effect evaporation rather than vapor compression was chosen in Order to make maximum use of the steam discharge from the turbo-generator. The red liquor is evaporated to 54% solids rather than the 52% originally intended, resulting in a substantially higher steam production. The concentrated SSL is sprayed into the furnace from eight burners. Use of extra fuel oil to support burning was discontinued after increasing the temperature of the firing zone by 165°C by lining furnace wall sections above and below the liquor burners with refractory. As shown in Fig. 255, the heat of the flue gas is exchanged with incoming combustion air. The gases from the induced draft fan of the boiler exit at 260°C into a spray cooler (75°C), an absorber cooler (30°C), and final SO_2

absorption is accomplished in a tray-type absorber (see Chap. X, D, 4) with 20% ammonia solution as absorbing medium to produce ammonium bisulfite solution. The latter is adjusted to cooking strength in the liquor making system as described earlier.

5. Fluidized bed incineration

As an outgrowth of a method invented in 1942 for the catalytic cracking of crude oil, fluidized-bed techniques are widely used in the petroleum, chemical, and metallurgical industries. They were first introduced to the Pulp and Paper Industry in the mid-sixties for incinerating spent pulping liquors and reclaiming inorganic liquor components.

The principles of fluidized-bed operations are understood from Fig. 256. A bed of solid particles of controlled size can be set in fluid motion by passing a stream of gas up through the bed. Maintaining the bed in a fluidized state requires careful control of the flow rate of the fluidizing gas. In this condition, the fluidized bed resembles a boiling liquid and obeys many of the laws of thermodynamics.

The outstanding advantage of this technique is the ease with which the temperature of a reaction can be controlled and maintained due to turbulent agitation, high heat capacity and high heat transfer rate. In practice, this means that the large heat sink due to the solid charge permits large variations in feed liquor supply or even periods without liquor supply with little change in temperature. A typical MgO pellet bed 6 m in diameter and containing 75 tons of material at 925°C would lose about 38°C in a 5-day shutdown and could be started up again without auxiliary fuel.

Fluidized-bed incineration of concentrated SSL results in a completely oxidized product. Its application is limited to solids systems which would not contain or form a softened or molten component that could cause agglomerations or caking. This is not a problem with magnesium base because of the high melting point of the ash. Careful control of the bed temperature below an allowable level is necessary in the case of sodium compounds.

Two models of fluidized-bed incinerators or reactors have been used by the Pulp and Paper Industry. The Copeland reactor is discussed by example of Mg base SSL recovery [17] and the FluoSolids reactor by example of Na base SSL incineration [18]. Either system is capable of processing any SSL, and kraft liquor.

6. Copeland reactor — Mg base

The special characteristic of magnesium base

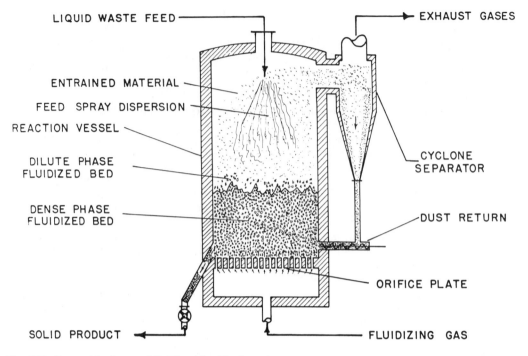

LIQUID WASTE FEED ——

——► EXHAUST GASES

ENTRAINED MATERIAL

FEED SPRAY DISPERSION

REACTION VESSEL

DILUTE PHASE
FLUIDIZED BED

DENSE PHASE
FLUIDIZED BED

CYCLONE
SEPARATOR

DUST RETURN

ORIFICE PLATE

SOLID PRODUCT ◄——

——— FLUIDIZING GAS

Fig. 256. General features of fluidized-bed incinerator.

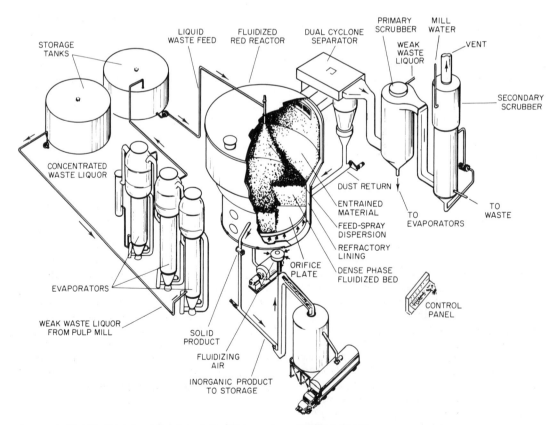

STORAGE
TANKS

LIQUID
WASTE FEED

FLUIDIZED
RED REACTOR

DUAL CYCLONE
SEPARATOR

PRIMARY   MILL
SCRUBBER WATER

WEAK
WASTE
LIQUOR

VENT

SECONDARY
SCRUBBER

CONCENTRATED
WASTE LIQUOR

DUST RETURN

ENTRAINED
MATERIAL

FEED-SPRAY
DISPERSION

REFRACTORY
LINING

DENSE PHASE
FLUIDIZED BED

ORIFICE
PLATE

TO
EVAPORATORS

TO
WASTE

CONTROL
PANEL

EVAPORATORS

WEAK WASTE LIQUOR
FROM PULP MILL

SOLID
PRODUCT

FLUIDIZING
AIR

INORGANIC PRODUCT
TO STORAGE

Fig. 257. Fluidized-bed magnesium base liquor recovery (Copeland).

SSL to yield the original cooking chemicals in the oxidizing combustion process has been dealt with. Fluidized bed incineration can be used, as well as burning in other incinerators, or in the recovery furnace-boiler. The advantages of fluidized-solids technology which have been mentioned before provide excellent conditions of treatment: the bed is isothermal, both vertically and horizontally; excellent air-solids contact results in efficient combustion; and close control over the composition of the exhaust gases is provided by proportioning spent liquor feed and air supply [17].

The closed cycle of the reconstituted Mg base bisulfite liquor is expressed chemically in simplified form:

$$Mg(HSO_3)_2 + O_2 + C \rightarrow MgO + SO_2 + H_2O + CO_2$$
$$\downarrow$$
$$+ H_2O$$
$$\downarrow$$
$$Mg(HSO_3)_2$$

A schematic diagram of the operation itself is given in Fig. 257.

Weak spent liquor at 11-12% solids is concentrated in a multiple-effect forced-circulation evaporator to 40-42% solids consistency, sufficient to maintain autogenous combustion in the fluidized bed. The inherent thermal efficiency of the fluidized bed requires less concentration of the spent liquor than the spray combustion methods. Experience has shown that the concentration for optimum overall economy, depends on the heat value of the spent-liquor solids and steam generation requirements. The stainless steel evaporators are of a special design, developed to minimize tube fouling and to ensure maximum efficiency. No switching effect is provided, but periodic boil-out with acid condensate is scheduled once weekly. No significant scaling problems are encountered.

The reactor shell is constructed of carbon steel and lined internally with cast and formed refractory material. A specially designed orifice plate assures positive distribution of the fluidizing air supplied by a multi-stage turbo compressor. An air heater and gas burners are provided for start-up and auxiliary use.

The concentrated spent liquor is introduced into the space above the fluidized bed (freeboard area) as a conical spray with air dispersion. Temperatures in the freeboard space will be in the 1023°C range, with the evaporation, and some of the combustion, taking place in this zone. A temperature of approximately 927°C is maintained in the fluidized-bed zone. At these temperatures, the organic matter in the spent liquor is converted to carbon dioxide and water vapor. The magnesium complexes formed during the pulping reaction are decomposed to form magnesia and gaseous sulfur dioxide. The MgO product is collected as cyclone dust and as comminuted bed particle material, cooled, slurried with water, washed to remove soluble impurities, and slaked to form magnesium hydroxide slurry at 10% solids. This slurry is used in subsequent absorption operations to regenerate cooking acid (cf. following section).

The "magnesia availability" of the reactor product, that is the magnesia present in reactive form available for preparation of cooking liquor, is 95% or higher, up to 99.5% having been obtained. Losses are due to small amounts of $MgSO_4$ in the MgO ash which are dissolved in washing, and of some additional sulfur with other soluble impurities.

Sulfur dioxide, the other product of the decomposition of the magnesium sulfur complexes, passes from the oxidation furnace into the exhaust gases. The mixture of gases passes through the cyclones, and to a waste heat boiler, at a temperature of 950-1000°C. By-product steam is generated at a rate of 3.6 t per ton pulp produced, and the gas temperature is lowered to about 230°C. Residual particulate material is removed in a venturi scrubber, and the gases are passed through a cooling stage (to about 60°C) and into a flooded-bed absorption tower. The $Mg(OH)_2$ slurry is fed countercurrent into the absorption tower, where it reacts with the SO_2 to form magnesium bisulfite cooking acid. Make-up SO_2 to compensate for losses in cooking and liquor processing is added from a small sulfur burner. Recovered liquor has a pH of 4.5-4.8 and is fortified for mill use with SO_2 to a pH of 3.2.

The quantity of steam generated in the waste heat boiler depends to a large extent on the organic content of the spent liquors. Steam can be fed to a turbine which powers the fluidizing air blower; it then exhausts at low pressure to supply the evaporators. Excess steam goes into the normal mill distribution system. If the power-fuel ratio does not favor this approach, the fluidizing air blower can alternatively be operated by an electric motor. Some twenty units of this reactor have been installed for SSL processing and recovery, of which some five burn magnesium base liquor, the remainder sodium base liquor from acid or neutral sulfite pulping operations.

A solution of the problem of ammonium base recovery has been found by combining the bisulfite pulping operation with fluidized bed magnesium base recovery. The ammonium base

SSL is neutralized with MgO to raise the pH to about 8.0. The resulting alkaline slurry is fed to an ammonia stripping and rectification system to release the ammonia. The residual liquor is substantially magnesium base spent liquor. The MgO is then recovered in a closed loop, thus releasing the SO_2 for recombination with the ammonia to form ammonium bisulfite cooking acid.

6. FluoSolids fluidized system (Fig. 258)

The FluoSolids fluidized bed system for the incineration of NSSC spent liquor is built on the same basic principles as the Copeland Mg base SSL reactor. It features some changes in design, however, which permit smooth processing of a chemical product having a substantially lower melting point and provide a different mode of heat utilization and recovery [18].

The incentives for installing the recovery system in the NSSC mill were given by the need for drastically reducing stream and air pollution, while producing a pelletized inorganic sodium ash that could be marketed, and providing low level heat recovery to lessen the steam demand of the mill. Unlike for Mg base, oxidative cumbustion of Na base SSL produces an intermediate recovery product consisting of sodium sulfate, Na_2SO_4, and sodium carbonate, Na_2CO_3, which is not suitable for re-use as sulfite liquor without further conversion. The readily soluble product is suitable for chemical make-up in a standard

kraft recovery system or may be more readily disposed of without detriment to the environment, because of the inertness of these products.

Although a chemical-recovery system for sodium base sulfite processes based on this technique has not yet been commercially developed, the process enjoys a substantial cost advantage over conventional combustion-type processes, even for simple disposal. The fluidized-bed reactors represent a much smaller investment than a conventional recovery furnace; no noxious gases are involved; and by virtue of the low spent-liquor concentration required, i.e. 35% solids vs 55 to 65% solids, lower evaporation costs are incurred. The latter is particularly important for NSSC liquors, which become troublesome at solids contents much above 50%.

To feed the reactor, the collection of spent liquor solids was increased to 70% without installing drum washers by diluting the digester content prior to blowing, recovering 60% of the liquor from the blown chips by a screw press, and combining it with the blow liquor from the digester. The combined spent liquor has about 13% dry solids and is evaporated to 15-20% solids in a conventional evaporator effect. There is virtually no evaporator scale formation with this type of liquor. The liquor is then evaporated to the reactor feed concentration of 40-45% in a venturi-scrubber-evaporator, utilizing the heat

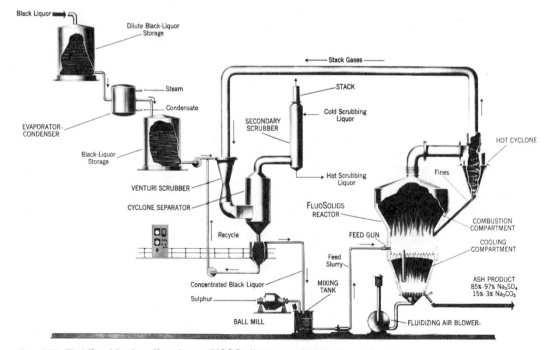

Fig. 258. Fluidized bed sodium base (NSSC) liquor recovery (Dorr Oliver).

of the combustion gases leaving the reactor at 730°C. Finally, the concentrated liquor is injected into the bottom of the fluidized bed through a number of guns arranged around the periphery of the reactor. A simple and reliable orifice feeding system has been developed for this purpose, requiring only one centrifugal pump with adjustable speed.

The reactor is again made of a steel shell with refractory lining. The design comprises a fluidized combustion bed maintained by the air passing through the refractory constriction plate. Above is an expanded combustion compartment (freeboard section); and below a fluidized compartment into which the bed product is continuously withdrawn by a transfer pipe for cooling and heat recovery. Decreased gas velocity in the expanded freeboard space allows entrained particles to fall back onto the bed, but there is always some carry-over of particles which are returned to the reactor by the scrubber.

Fluidized bed operation aided by the above design provides excellent temperature control of combustion. The variation between bed and freeboard temperature is only ±6°C which indicates that autogenous burning in the bed is complete. With an ash product having a melting point of only 36-66°C higher than the combustion bed it is obvious that uniform temperature control is essential for undisturbed operation. For smooth temperature control it is also necessary to maintain the solids concentration of the feed liquor at a desired level. This is accomplished by monitoring the concentrated liquor by a refractometer and adjusting the solids level by varying the liquor flow to the venturi evaporator.

Ash pellets form and grow continuously in the fluidized bed and would become increasingly coarse, leading eventually to defluidization of the bed. To prevent this, small seed particles must be provided by diverting a portion of the product to a roll crusher and returning it to the bed to serve as nuclei. The product is free-flowing pellets discharged from the cooling bed at about 190°C.

Pellets made from the mill liquor solids contain 65% Na_2SO_4 and 35% Na_2CO_3. Their market value as salt cake make-up for kraft mills is greatly increased by supplying SO_2 gas to the cooling bed for increasing the Na_2SO_4 content to 90-95%.

Eighty-five percent of the heat content of the reactor gases can be recovered as low level heat. With the fluidized-bed liquor processing system, the BOD discharge could be reduced by 75% and the SO_2 discharge brought to a negligible level.

Although economics would differ from case to case, the process has the advantage of minimal capital investment compared with other liquor burning processes, and would cover operating and fixed cost or even generate nominal profit where sufficient value can be realized from the chemical product and steam savings. The latter has been enhanced greatly by the rise in fuel oil price.

7. Pyrolysis of Na base SSL

Spent liquor pyrolysis is practical only with sodium base SSL. Unlike spent liquor combustion in reductive furnace-boilers, it avoids the formation of smelt and therefore the danger of smelt-water explosions. The latter was the chief incentive for developing pyrolysis processes. The first attempt was made at the Pulp and Paper Research Institute of Canada in the 1950's to develop the Atomized Suspension Technique (AST). It has not achieved industrial application because of the very high reactor temperature used, producing corrosive gas components which necessitate specialized and costly construction materials.

The second attempt made in Sweden in the 1960's by Svenska Cellulosa jointly with Billerud used a lower process temperature in a reactor with refractory lining, and did achieve successful application with eight systems operating at the time of writing. The only system installed in North America has been used for several years to recover NSSC cooking liquor.

The heart of the system shown in Fig. 259 [19] is the refractory-lined bottle-shaped pyrolysis reactor (1) with upward curved exit duct at the bottom. The top carries a special oil burner and ports for the admission of combustion air. Spent sulfite liquor evaporated to 55% solids concentration is injected under high pressure in a finely atomized spray through several nozzles arranged around the wall of the pyrolysis chamber.

In operation, heat for liquor pyrolysis is supplied from fuel oil burnt in the top section with a very slight excess of air. The hot gases from the oil burner travel downward at high velocity and contact the liquor sprays at high turbulence and rapid mixing. The unique feature of the process is this "shock pyrolysis" which converts all sodium in the liquor instantaneously into a solid ash powder composed mainly of carbonate; and the other constituents into a gas containing virtually all of the sulfur in the form of hydrogen sulfide (H_2S) mixed with carbon mono- and dioxide (CO, CO_2), hydrogen (H_2), methane (CH_4), nitrogen (N_2) and water vapor. Melting of sodium compounds is prevented by holding the pyrolysis temperature at 715-745°C.

As this is too low to gasify carbon, varying amounts will be present in the powdered soda ash depending on conditions. Wall deposits of sodium sulfate form in time; they are periodically melted down by raising the reactor temperature and removed through the bottom outlet.

The gas-duct suspension from the reactor is passed through a simple single pass waste heat boiler (2), and then the dust is separated out in a cyclone (3). The gas continues through a spray cooler (4) to remove water vapor by condensation and the cooled gas is burnt together with oil in a boiler (5) equipped with a superheater which also superheats the steam from the first boiler. Here the H_2S gas is converted into SO_2 gas while the other combustible gases are burnt to carbon dioxide and water. The SO_2 is then recovered in a flue gas scrubber (8) by absorption into recovered sodium carbonate solution. The partially sulfited carbonate solution is then fortified in a second absorption unit with make-up sulfur dioxide from a sulfur burner to form cooking liquor. The reactor dust from the wet scrubber is slurried with hot water in a tank (6) and the carbonate solution filtered from suspended carbon on drum washers (7).

The temperature of reaction conditions necessary for instantaneous H_2S release, and thus to maximize Na_2CO_3 formation in the reactor, coincides with the formation of considerable amounts of carbon. This reduces the quantity of heat available from the spent liquor and requires considerable amounts of supplemental fuel oil to run the pyrolysis process. Also, additional equipment is necessary for collecting and processing the carbon. The carbon is of low activity and of no immediate value for absorption and water purification purposes. It is returned to the process for improving the heat balance. Early models of the pyrolysis system used a cyclone burner attached to the gas boiler. More recent installations either co-burn the wet carbon in a bark boiler, which is an inefficient

way for recovering its energy, or recycle it to the SSL evaporators for gasification in the reactor.

The system is reliable in operation and achieves high chemical recovery efficiency for sodium (95%) and sulfur (85%).

8. Major processes for the preparation of Mg base sulfite liquor from furnace-boiler product

A historical note on the development of the magnesium-base sulfite pulping process with chemical recovery is found in the section on "Evaporation and Burning", where the design and operation of the special furnace-boiler is treated.

(a) B&W venturi Mg system

The original Babcock and Wilcox system built for recovering acid bisulfite liquor at low base content from the furnace-boiler MgO product used packed SO_2 absorption towers. Later when bisulfite pulping was introduced by Howard Smith Paper Mills in Canada (Magnefite Process) the cooking liquor required a larger base content and a higher pH, with the result of achieving a shorter cooking time at higher temperature and a higher pulp yield. The first special Magnefite mill was built in Sweden in 1962. In the process of producing magnefite solutions at high chemical concentration, the best results were obtained when supersaturated solutions of magnesium sulfite were used in the SO_2 absorption stage. To avoid precipitation and plugging, the absorption system should be open and moving, which led Babcock and Wilcox to design individual absorption units as venturi scrubber-absorbers. A further venturi scrubber was added for gas cooling. The resulting standard model [14] then included one venturi scrubber cooler and three venturis for SO_2 absorption, and has been the landmark of this type of operation since. The complete recovery

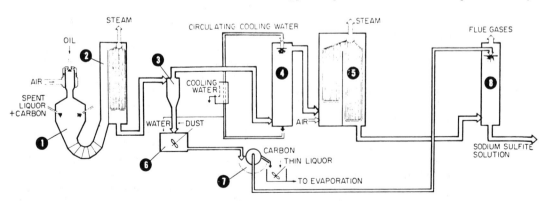

Fig. 259. Pyrolysis recovery plant for sodium base sulfite spent liquor (SCA).

unit is shown in Fig. 260.

Furnace ash collectors are usually of the cyclone type. Collection of 80-90% of the ash is sufficient since the remainder is taken up in the liquor making system. From the collector, the ash is flushed into a retention tank required to level out fluctuations in ash production and to dissolve salt impurities. The MgO slurry passes from the retention tank to a vacuum-drum or belt type ash washer, where calcium, potassium, and sodium in the form of sulfates are washed from the ash. A small quantity of the magnesium and sulfur is also lost at this ash washer in the form of magnesium sulfate.

The washed ash passes to the slaking system, where a slurry of recovered and make-up MgO is maintained at a temperature of about 85°C. In this slaking system about 95% of the magnesia slurry is converted to the highly reactive magnesium hydroxide for the rapid absorption of sulfur dioxide in the subsequent absorption system.

The SO_2 absorption system consists of the three venturi units arranged in series. Stainless steel is specified as construction material. Cooled flue gas containing about 1% SO_2 by volume passes through the system where the SO_2 is absorbed in $Mg(OH)_2$ slurry flowing counter-currently to the gas. Incremental amounts of $Mg(OH)_2$ slurry with pH control are used in each venturi stage to compensate for the decreasing concentration of SO_2 in the gas and the low solubility of the $MgSO_3$ formed. The absorption slurry is circulated at high rate in each unit to ensure adequate gas-liquid contact. Only about 0.003% SO_2 by volume leaves the last tower and is vented. The slurry leaving the last venturi unit is settled and filtered and fortified to cooking concentration with make-up SO_2 from a sulfur burner and digester vent gases. Liquors are passed through sand filters for final cleaning.

Quantities of chemical make-up from 10-20 kg MgO and 20-40 kg SO_2 depending on whether acid bisulfite or bisulfite liquor is used and on the type of equipment. Magnesium sulfate, $MgSO_4$, may be used for make-up since it does not interefere with combustion and forms MgO and SO_2 in the furnace.

(b) Lenzing Mg base system

The design of the Lenzing Mg base recovery system which started up in 1964 deviates in a number of significant ways from the classical B&W model. Apart from the special recovery furnace-boiler described in Section C, 3 of this Chapter, ash collection and absorption were modified to meet three principle requirements: (1) use of natural magnesite ($MgCO_3$) for make-up containing 13-14% impurities and including silicates; (2) collection of ash with minimum losses due to dusting; and (3) preparation of MgO ash of maximum reactivity in the hydration process.

A schematic diagram of the main components of the installation [20] is seen in Fig. 261. The flue gas/dust suspension from the boiler is blown into a wet separator where the ash is collected by a fine water spray in the form of a thin sludge. The flue gas is cooled from 200°C to 67°C for subsequent SO_2 absorption. Thus, ash separation and flue gas cooling are combined in one stage. Hydration of the MgO in the thin sludge takes place in the clarifier where a thick sludge separates from the supernatant clear solution. Make-up MgO carrying natural impurities is also added at this stage. In contrast to the conventional process, stages for filtering and slaking the MgO ash are eliminated. The sludge drawn from the clarifier is diluted and added to the top of the three-stage tray tower for countercurrent SO_2 absorption. The gas exhaust contains only 0.01-0.02% SO_2 by volume. Raw

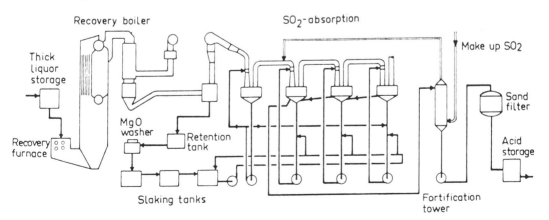

Fig. 260. Magnesium base sulfite liquor recovery system (Babcock and Wilcox).

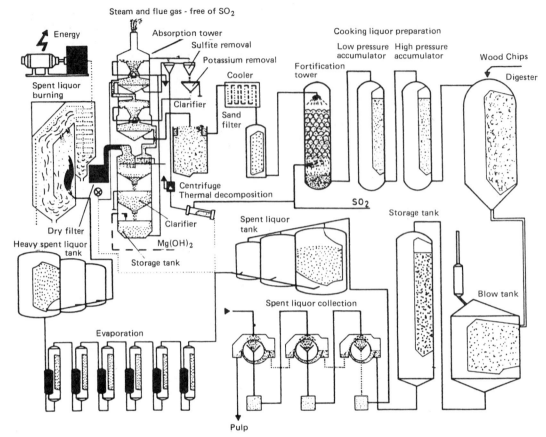

Fig. 261. Lenzing Mg liquor recovery system.

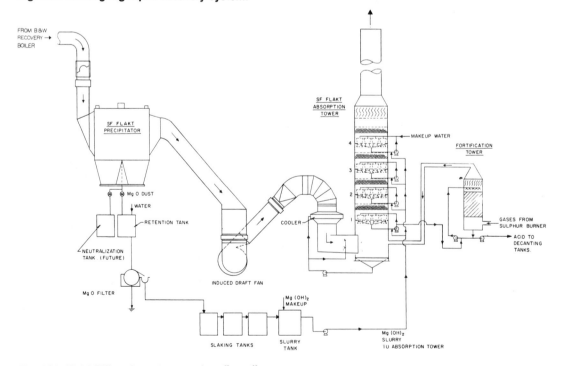

Fig. 262. Flakt-Billerud recovery system flow diagram.

liquor from the absorber is then purified of ballast material by clarification with pH control and addition of flocculation agents, and filtered through a gravel bed before final fortification to cooking liquor strength.

The process recovers 70% of the sulfur and 85% of the magnesium oxide applied to the cook. The composition of the acid bisulfite liquor produced is 6% total SO_2 and 1.2% combined SO_2 (0.75% MgO); it is used to manufacture dissolving pulp from beech wood.

(c) The Flakt Mg base system

The Flakt magnesium base liquor preparation

system was introduced industrially in Sweden in the mid-1970's [21] and has been chosen for the recent conversion of two older Canadian ammonium acid bisulfite plants to magnesium base operation (see Chap. X, B).

Shown in Fig. 262 [22], it deviates from conventional ones by using an electrostatic precipitator unit rather than the common multi-cyclone separators for reclaiming the MgO ash from the flue gas suspension. The premise for this choice was the argument that the MgO ash from Loddby furnace burning is very fine and its collection in multi-cyclones inefficient due to high pressure drop. Another drawback of the Loddby unit is the erosion of the equipment by MgO ash in dynamic systems. Electrostatic precipitators do not suffer from these problems. Yet, the quoted rate of ash recovery is 85%, the same as for the cyclone system.

Washing and slaking of the ash are conventional. The Flakt absorption system operates basically in the same manner as the conventional venturi scrubber system already described. Typically, however, the four stages are mounted vertically above each other in a tower and each stage performs a dual absorber function, venturi absorption and turbulent contact absorption as in the TCA unit discussed in Chapter IV, E, f. The design of the tower and its individual stages is shown in Fig. 263. Each stage consists of two levels, a set of eight venturis with slurry tray and circulation pump at the bottom and a compart-

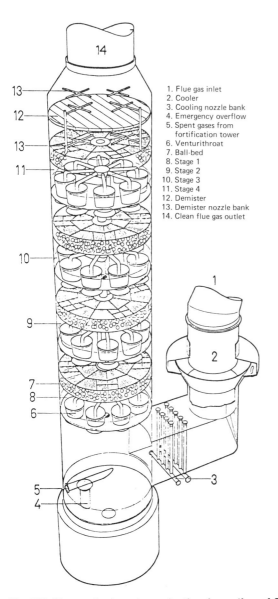

1. Flue gas inlet
2. Cooler
3. Cooling nozzle bank
4. Emergency overflow
5. Spent gases from fortification tower
6. Venturithroat
7. Ball-bed
8. Stage 1
9. Stage 2
10. Stage 3
11. Stage 4
12. Demister
13. Demister nozzle bank
14. Clean flue gas outlet

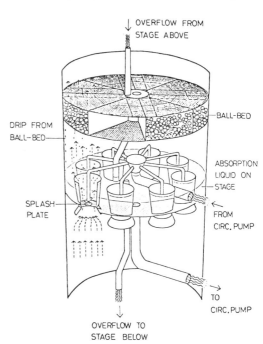

Fig. 263. Magnesia slurry tower for the absorption of SO_2 from flue gas (Flakt).

mented turbulent bed above, filled with polypropylene balls. The absorption slurry is sprayed by nozzles into the throats of the venturis against the rising gas. As in the conventional system, incremental amounts of slurry are metered into subsequent stages controlled by a pH-measuring system. The gas liquid mixture then rises into the ball bed where the liquid separates and flows back into the conical tray. Above the top absorption stage, a demister spray bank reduces the amount of droplets entrained in the stack gas.

The flakt system is completed by a fortification tower, also constructed of stainless steel. Liquor circulation and means for adjusting the flow rate of the slurry are provided. The gases exiting from the tower are piped back to the scrubber-absorber which operates under elevated pressure.

D. PROCESSES FOR CONVERTING FURNACE SMELT OR GREEN LIQUOR INTO SODIUM BASE SULFITE LIQUOR

1. Processes employing carbonation of green liquor

The product obtained from burning any sodium base spent liquor in the reductive furnace boiler is a smelt consisting predominantly of sodium sulfide (Na_2S) and sodium carbonate (Na_2CO_3). This smelt on flowing out of the furnace bottom can be solidified by cooling and claimed in the form of flake or powder with special machinery, or it can be quenched and dissolved in water to make a solution with a green color due to traces of iron. The latter processing step is used almost exclusively and provides the intermediate for 80% of the pulping liquor manufactured in the world. Most of it goes into making kraft liquor which uses the sulfide of the smelt unchanged, but requires conversion of the carbonate into caustic soda (hydroxide) by a causticizing step with burnt lime. For the causticization treatment it is not necessary to separate the sulfide from the carbonate.

By contrast, production of sodium base sulfite liquor from green liquor requires chemical conversion of the sulfide into sulfite while carbonate is used directly for liquor preparation. For the conversion of sulfide into sulfite it is necessary to reduce the pH of the green liquor to below 10.5 with carbon dioxide for releasing the sulfur content as H_2S gas, and to burn it to SO_2 gas which is then combined with the carbonate to make sulfite pulping liquor.

At the time of writing there are three mature processes for the recovery of sulfite liquor employing green liquor carbonation in commercial use, each with industrial installations in different countries [23]. The present status of these systems is the result of the continued efforts of the few companies active in this field to improve the original versions first implemented in the early fifties.

(a) The Stora System

This process was developed by Stora Kopparbergs Bergslags AB of Sweden in order to reduce the consumption of high-price imported coal during WWII. Maintaining the original concept, unit operations went through several modifications based on laboratory development and plant experience, the most significant improvements and simplifications having been made in the Canadian operation [24, 25]. This system supports a 600 000 t/y pulping operation producing two-stage spruce and hardwood sulfite paper grade pulp; and spruce high yield bisulfite news grade pulp.

The process of sulfite conversion is described in Fig. 264. Basically the plant consists of a reaction tower with self-cleaning Glytsch cap trays for stripping H_2S gas from green liquor and a set of two Claus reactors for converting a mixture of H_2S and SO_2 gas to elemental molten sulfur. The reason for taking this route for H_2S-to-SO_2 conversion is the low concentration of H_2S in the gas from the stripping reaction. At least 60% H_2S would be required for direct burning. Not shown are the sulfur burner with waste heat boiler and the SO_2 absorption tower.

The green liquor is first diluted to 80 g/L Na_2O on entering the reaction tower which contains 42 trays. In the upper $\frac{2}{3}$ of the tower the sodium sulfide is completely stripped of H_2S by countercurrent CO_2 gas at 85°C. The tower gas containing 16% H_2S and CO_2 is blown through steam and electric heaters into the two-stage Claus conversion plant where the H_2S is reacted at 400°C with pure SO_2 to form liquid sulfur and steam. The alumina catalyst lasts several years and costs less than 1¢ per ton of pulp to replace. Two Claus lines are needed for 1000 t/d pulp production. The CO_2 is returned to the reaction tower and the liquid sulfur is separated out and burnt to SO_2. For alkaline sulfite an equivalent amount of bicarbonate is decarbonized by heating to form carbonate. If higher alkalinity is required the necessary amount of carbonate is causticized. Both the residual steam and the liberated CO_2 gas from the Claus reactor operation are utilized in the main reaction column. In the bottom $\frac{1}{3}$ of the tower bisulfite solution is added for decarbonizing the oncom-

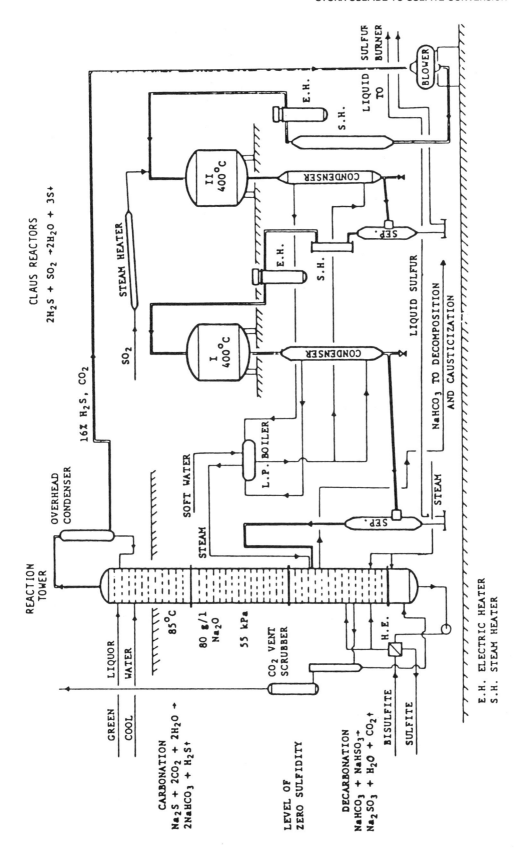

Fig. 264. Stora sulfite conversion from green liquor.

ing sodium bicarbonate solution, the liberated CO₂ rising again in the tower and the excess being vented to the atmosphere. The sulfite formed in the bottom'of the reaction tower returns to cooking liquor preparation where it can be used as such, or reacted with SO₂ gas in the absorber to form bisulfite, depending on the required pH level of the cook.

Sodium hydroxide for make-up of sodium losses can also be used for pH adjustments; sulfur losses are made up by burning additional liquid sulfur. The sulfur burner/boiler generates superheated steam at 480°C, 5 MPa. The whole conversion system operates at a pressure of 55 kPa. Its conversion efficiency for sodium is greater than 98%. A full size SO₂ scrubber for the recovery furnace-boiler is also provided.

Control of thiosulfate content in the recovered liquor is critical in acid bisulfite and bisulfite cooking. Formation of thiosulfate in the reaction tower is continuously monitored and maintained at low levels by adjusting operating conditions.

Rayonier, Inc. was the first to introduce sodium base sulfite recovery into North America by adopting the Stora process. The recovery plant was built in the late fifties as part of a program including increased production, conversion of the Grays Harbor mill from ammonium base to sodium base pulping, and depollution. Operation started in 1962 with the modification that the furnace smelt was cooled and dissolved at high concentration, and the dissolved Na₂S separated from the Na₂CO₃ crystals. The isolated sulfide with some remaining carbonate was then carbonated for releasing the H₂S gas. This modification was abandoned after a few years and regular green liquor carbonation used.

(b) The Rauma system

Both the Rauma system and the following Tampella system are technologically advanced offsprings of the original Sivola system which was installed by Combustion Engineering at the sulfite mill of Rauma Repola Oy in Finland in 1952 in the changeover from calcium base to two-stage sodium base pulping. A later Sivola version was installed by Lurgi A.O. at the Heinola mill of Tampella Oy in Finland in 1961. These systems characteristically use distinct processing stages for producing highly concentrated H₂S gas (95 + %) from green liquor which is then burnt directly to SO₂ gas. Figure 265 illustrates a simplified flow sheet of the Sivola-Lurgi plant.

At Rauma, the original Sivola installation was modified in 1962 when two-stage cooking was

changed to three-stage cooking. In 1975, when the start-up of a new high yield bisulfite line brought increased recovery capacity requirements, a major modification of the existing process was undertaken. The principle target was to increase the efficiency of recovering sulfur as hydrogen sulfide and to minimize the cost of steam required in the separation process. The new version of the system (Fig. 266) has been in operation since 1978 [26].

Cooling and precarbonation of green liquor from a pH of 13.5 to 10.5 is performed in a packed tower (PC) with scrubbed and precooled flue gas. The resulting solution is then fully carbonated with recycled concentrated CO₂ by feeding it tangentially into the ejector of a small, continuous mixing device (C) with forced circulation (Fig. 267). The operating pressure is kept higher than the sum of the partial pressures of CO₂, H₂S, and water so that no H₂S can separate. The heat of this reaction is also stored in the liquor stream for utilization in subsequent stripping. Bicarbonate crystals form in this stage due to lower solubility.

The carbonated solution is flashed in the adjacent H₂S stripping column (ST) equipped with valve trays and operated below atmospheric pressure. Heat required in this stage can be supplied by low-cost secondary steam from the liquor evaporation plant.

The introduction of this new method and equipment for pressure carbonation and strip-

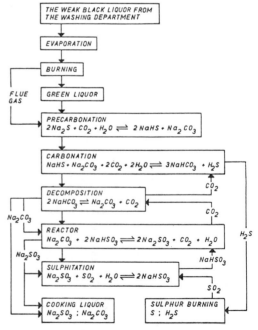

Fig. 265. Simplified flow sheet of the Sivola-Lurgi plant.

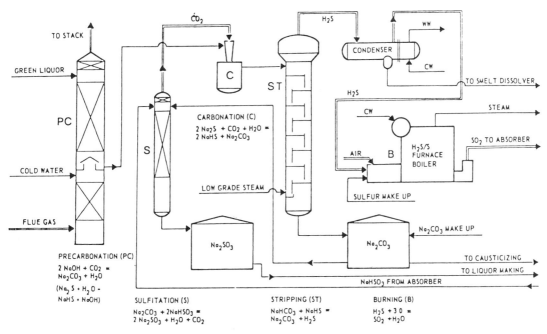

TO STACK

GREEN LIQUOR

PC

COLD WATER

FLUE GAS

PRECARBONATION (PC)
2 NaOH + CO$_2$ =
Na$_2$CO$_3$ + H$_2$O
(Na$_2$S + H$_2$O =
NaHS + NaOH)

CO_2

CARBONATION (C)
2 Na$_2$S + CO$_2$ + H$_2$O =
2 NaHS + Na$_2$CO$_3$

S

C

ST

H$_2$S

CONDENSER   WW

CW

TO SMELT DISSOLVER

STEAM

H$_2$S

CW

AIR   B   H$_2$S/S
FURNACE
BOILER

SO$_2$ TO ABSORBER

SULFUR MAKE UP

LOW GRADE STEAM

Na$_2$SO$_3$

Na$_2$CO$_3$

Na$_2$CO$_3$ MAKE UP

TO CAUSTICIZING

TO LIQUOR MAKING

NaHSO$_3$ FROM ABSORBER

SULFITATION (S)
Na$_2$CO$_3$ + 2NaHSO$_3$ =
2 Na$_2$SO$_3$ + H$_2$O + CO$_2$

STRIPPING (ST)
NaHCO$_3$ + NaHS =
Na$_2$CO$_3$ + H$_2$S

BURNING (B)
H$_2$S + 3 O =
SO$_2$ + H$_2$O

Fig. 266. Rauma sulfite conversion system from green liquor.

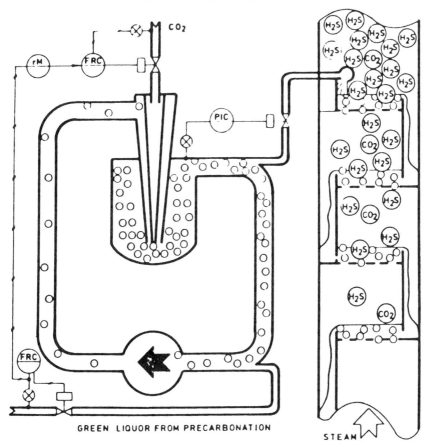

CO_2

rM

F RC

PIC

H$_2$S H$_2$S
H$_2$S H$_2$S H$_2$S
H$_2$S H$_2$S H$_2$S
H$_2$S CO$_2$ H$_2$S
H$_2$S H$_2$S
H$_2$S H$_2$S

H$_2$S

H$_2$S CO$_2$ H$_2$S

H$_2$S

H$_2$S CO$_2$

H$_2$S

H$_2$S

H$_2$S

CO$_2$

FRC

GREEN LIQUOR FROM PRECARBONATION

STEAM

Fig. 267. Pressure carbonation of precarbonated green liquor and H$_2$S stripping at reduced pressure (Rauma Repola Oy).

ping at reduced pressure has greatly increased the efficiency of H_2S liberation in concentrated form while reducing the size of the equipment.

The H_2S gas produced has a concentration of $95 + \%$ after condensation of the water vapor and is burnt at $\simeq 1500°C$ with air in a high-intensity burner together with make-up sulfur. The heat of combustion is recovered in a waste heat boiler generating about 140 kg of superheated steam at $350°C$ and 3.5 MPa per kmol of H_2S.

Carbon dioxide for the carbonation reactor is regenerated from carbonate solution leaving the stripping column by reaction with bisulfite solution in a packed-bed sulfitation reactor. This reactor operates under sufficiently high pressure to meet the requirements of the carbonation reactor. Bicarbonate crystallization is optional for cases where pure cooking liquor free from thiosulfate, sulfate, etc. is required.

(c) The Tampella system

Based on the experience gained from operating the Sivola system, Tampella developed the modified system at the Heinola NSSC mill and started the first industrial size plant in 1968 [27]. The immediate function of the plant was to balance chemical recovery in a dual purpose pulp mill producing kraft grade and bisulfite grade. Nine plants have been installed in Finland, France, Eastern Europe, Russia, and Japan, in mills producing a variety of sodium base sulfite grades. Three are stand-alone plants in sulfite mills. The installation in France combines both Stora and Tampella technology. The installation made at United Paper Mills, Valkeakoski, Finland has become widely known as a representative example of the process and its operation. As in the first installation, the Tampella system serves to balance the recovery of chemicals from green liquor for co-production of bisulfite and kraft pulp in the same mill. This is also the first mill known where moderately alkaline sulfite-anthraquinone pulp is produced since 1981, without modification of the sulfite recovery unit previously used for bisulfite liquor production (refer to Chap. X, C).

The flow sheet of the Tampella recovery process and its chemistry are seen in Figs. 268 and 269. The simple inorganic reactions involved are effectively controlled in the separate processing stages and full use is made of the carbon dioxide present in the flue gas while absorbing its sulfur dioxide content [28].

In the packed precarbonator the sodium sulfide of the green liquor is converted to sodium bisulfide without liberating H_2S by maintaining a sufficiently high pH in the upper part. Simultaneously, sodium carbonate returning from the stripper section is carbonated with flue gas to bicarbonate.

$NaHS$ and $NaHCO_3$ are then reacted together in a stripper unit with 5 to 15 specially designed trays under conditions of high concentration, steam dynamics, heat and vacuum to liberate H_2S gas at over 95% dry concentration. The resulting sodium carbonate is split into two streams, one for reforming bicarbonate used in stripping, the other for cooking liquor preparation by absorption of SO_2 gas from the H_2S burning process.

The H_2S gas is piped over a short distance to a surface condenser, the I.D. fan and the H_2S/sulfur burner with steam generator. Maintaining reduced pressure in stripping and piping has the triple advantage of counteracting bicarbonate decomposition by lowering the temperature, removing H_2S from the reaction, and preventing H_2S emissions in case of leakage.

If required, carbonate of 99.5% purity can be crystallized by raising the concentration of the effluent from the stripping stage and separated in a centrifuge. The quality of cooking liquor made from process carbonate is the same as made from purchased soda ash. With carbonate crystallization and separation, ballast chemicals formed, such as sodium thiosulfate and sodium carbonate, are left in the mother liquor which is returned to the combustion-reduction stage. This option is provided for acid bisulfite and bisulfite pulping operations using extended cooking schedules for the manufacture of dissolving grade and other pulp for special uses. It is the most positive method for preventing cooking liquor breakdown by keeping the thiosulfate content at a low level.

Recently, the unit operations of the Tampella recovery system were combined in a multi-unit tower which incorporates the flue gas scrubber at the bottom (Fig. 270). This arrangement does not change the chemistry or relative dimensions of the units of the system, but reduces investment and operating cost by requiring less piping and floor space and fewer pumps. By placing the flue gas scrubber at the bottom the heat economy of the operation is improved by using the latent heat. Since the Tampella concept aims at full integration with the recovery furnace-boiler the flue gas scrubber uses a highly efficient packed layer design and is part of the cooking liquor preparation unit.

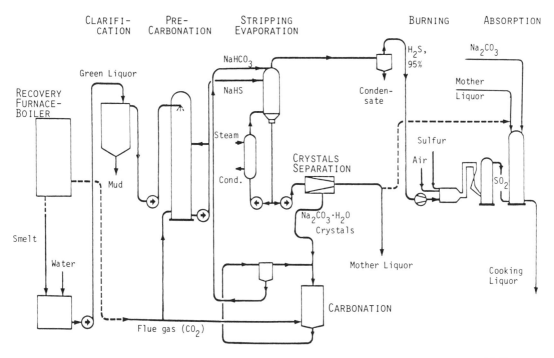

Fig. 268. Flow sheet of the conversion of green liquor into sulfite cooking liquor by treatment with flue gas (Tampella).

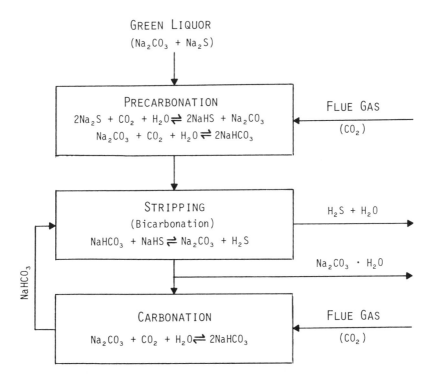

Fig. 269. Chemistry of H₂S gas stripping from green liquor by flue gas (Tampella).

These three leading processes have the following major distinguishing features:

Function	Stora	Rauma	Tampella
CO_2 supply	CO_2 recycle	Scrubbed flue gas and CO_2 recycle with storage	Scrubbed flue gas
H_2S liberation	CO_2, steam $NaHSO_3$	$NaHCO_3$ and CO_2, pressurized carbonation, steam, vacuum, $NaHSO_3$	$NaHCO_3$, steam, vacuum, evaporation
H_2S concentration	16%, CO_2	95 + %, CO_2	95 + %, CO_2
H_2S conversion	sulfur (Claus), heat recovery	—	—
Burning to SO_2	liquid sulfur (steam generator)	H_2S (steam generator)	H_2S (steam generator)
Unit operation	separate	separate	separate or multiple unit column

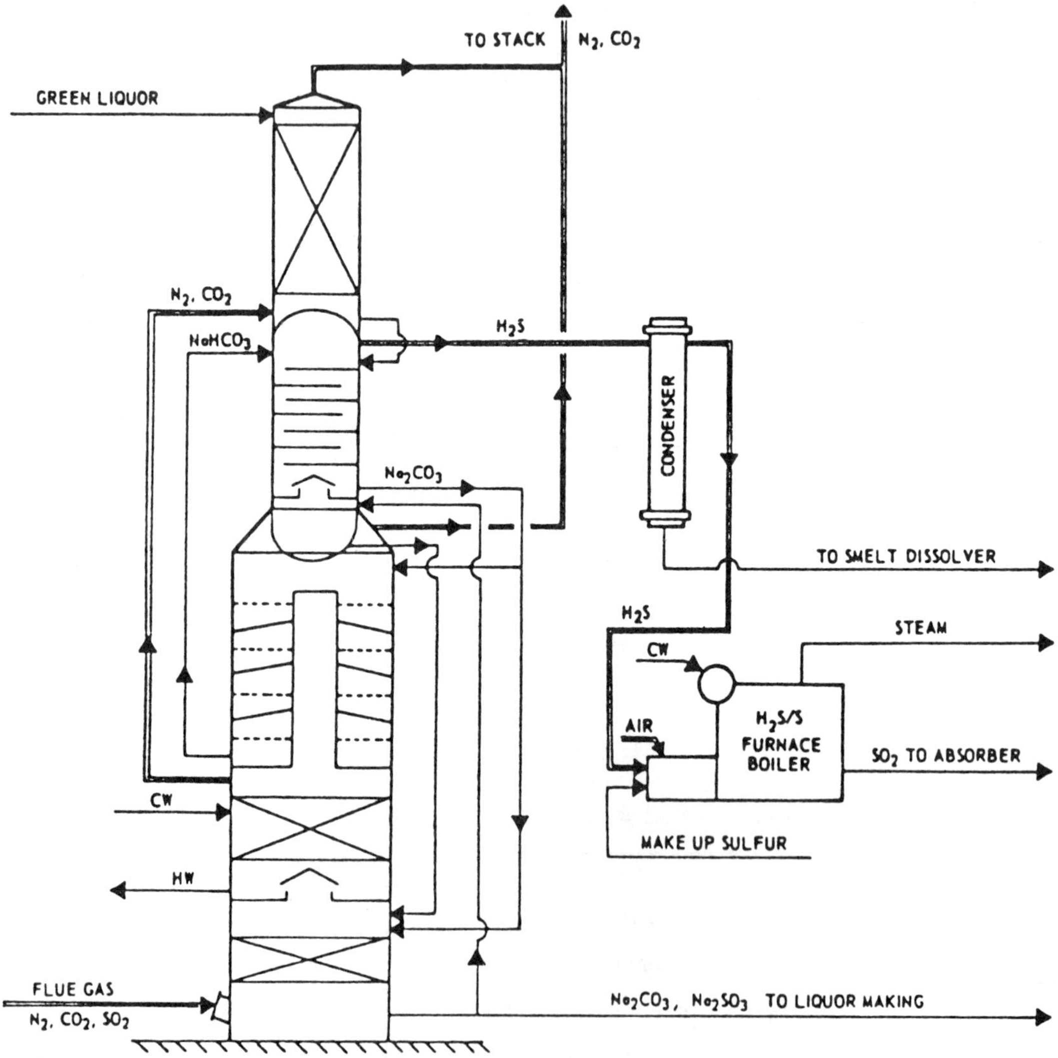

Fig. 270. Multiple-unit column for stripping of concentrated H_2S gas from green liquor with flue gas (Tampella).

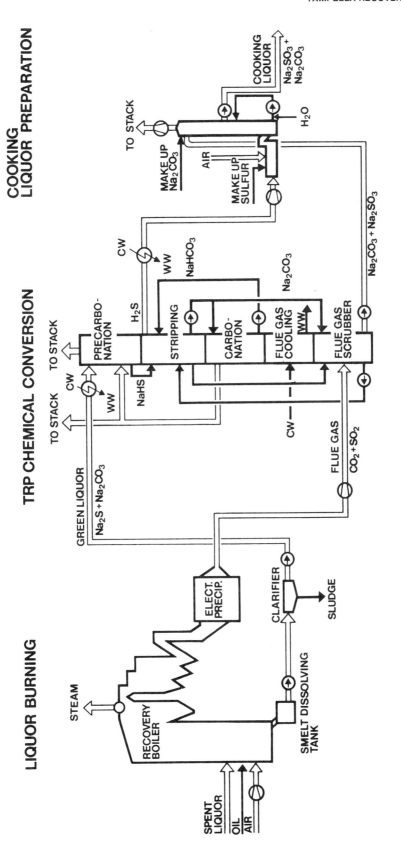

Fig. 271. Tampella recovery system for sodium base sulfite pulping.

The degree of integration with the liquor combustion process varies greatly. The Stora concept aims at considerable independence from the furnace-boiler operation. With complete CO_2 recycle it can function independently for periods which depend on the storage capacity of green liquor and some provision for carbonate make-up. The Rauma concept takes an intermediate position using flue gas only for the precarbonation of the green liquor, and recycling CO_2 for carbonation. The Tampella concept aims at full integration and has achieved this goal (Fig. 271) to a degree where the cycle of volatile sulfur compounds and of gas-borne particulates is closed, even under the most stringent air quality criteria.

In 1981 a cooperative agreement was reached between Rauma-Repola and Tampella whereby Tampella obtained the exclusive rights to the Rauma recovery process. This allows for even greater adaptability and efficiency of the available sodium base recovery equipment [29].

Most recently the special conditions provided by the H_2S stripping stage of Sivola type sodium base sulfite recovery have been made use of for developing a simpler and more efficient sodium chloride removal and recovery process from the liquor of closed cycle bleached pulp mills [30]. The basis for this process is the higher solubility of sodium chloride in carbonate solution as compared with caustic soda solution. In the Tampella-Rauma system, the heat content in the flue gas can be used effectively to evaporate the solution stripped of H_2S gas to achieve carbonate crystallization, as described. Carbonate recovery in crystal form is as high as 95% with all sodium chloride remaining in the small volume of mother liquor. This small volume is causticized without dilution and the resulting caustic soda solution evaporated further to precipitate the sodium chloride. Although the process can be applied to mills producing bleached kraft pulp, it is particularly suited to a closed cycle mill using the moderately alkaline sulfite-anthraquinone process. The process has been developed to the pilot plant stage.

2. Other processes

(a) The Mead process

This process, using carbonation of green liquors with flue gas and H_2S burning in the recovery furnace [31], was developed in the mid-fifties and operated for several years in two proprietary NSSC mills of the Mead Corp. The recovered cooking liquor typically contains 11% thiosulfite rendering it unsuitable for the acid sulfite processes. Modifications to reduce air

pollution from hydrogen sulfide added to the complexity of the system which has been abandoned.

(b) The Marathon Engineering (MEI) process

The MEI spent liquor evaporation and combustion process was discussed earlier in this text. Its simplicity and economy has been demonstrated and proven in industrial installations. The MEI sodium base spent liquor recovery system, on the other hand, has gone through a series of proposals and modifications culminating in a patent granted in 1976 [32]. The process design (Fig. 272) performs the required chemical reactions of green liquor carbonation and H_2S liberation, under modified conditions and in different equipment than described earlier under the Sivola-type processes. Liquid-liquid rather than countercurrent gas-liquid reactions are emphasized, as is the use of pure CO_2 from a pressure carbonation stage, with the aim of simplifying and reducing the size of the equipment by increasing reaction rate and efficiency.

So far the process has not been tested and evaluated on the pilot plant or industrial scale. In the opinion of the writer its merit under the present status lies in providing a possible alternative to the existing processes and adding to the options for designing efficient processes of this type.

(c) The IPC green liquor sulfitation process

This process was developed in the early 1950s by the Institute of Paper Chemistry for the use of its member companies. Green liquor is clarified and sulfited directly in a packed absorber by using 16 to 18% SO_2 gas from a conventional sulfur burner. Sodium sulfite is formed in a two-step reaction which can be combined for simplification in the following equation:

$$Na_2S + SO_2 + H_2O \rightarrow Na_2SO_3 + H_2S$$

Even with careful control, 10-20% quantities of thiosulfate are formed during the sulfitation, making the process unsuitable for acid bisulfite or bisulfite processes. Similarly, since no sodium carbonate is produced, it is not suitable for the more alkaline sulfite processes. Applications were thus restricted to NSSC mills. Losses of sulfur, as SO_2 from the furnace-boiler and as H_2S from the sulfitation stage, could have been corrected by adding a Claus reactor. The process was used for some time in two NSSC mills of Owen-Illinois and was abandoned in 1972.

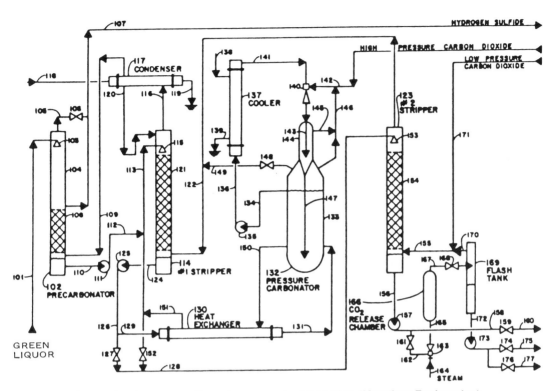

Fig. 272. H₂S stripping from green liquor by pressure carbonation (Marathon Engineering).

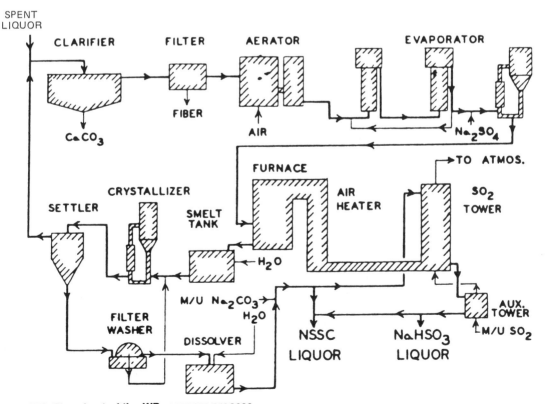

Fig. 273. Flowsheet of the WP recovery process.

(d) The Sulfide Recycle and the Western Precipitation (Bradley) processes

Developed by Consolidated Power and Paper Co. in the late fifties, these two processes attempted to avoid the green liquor carbonation - H_2S - SO_2 route by separating the Na_2S from the soda furnace smelt and to use the recovery furnace to produce the SO_2 gas from it. The net reaction can be expressed as:

$$Na_2S + CO_2 + 1\frac{1}{2} O_2 \rightarrow Na_2CO_3 + SO_2$$

This represents the simplest of all sodium base sulfite recovery models based on reductive furnace operation.

As a first stage a sodium carbonate-sodium sulfide smelt is obtained by the usual evaporation-recovery furnace route, with the production of steam for power generation and process use. Separation of the carbonate depends on its lower solubility relative to the sulfide. To effect this separation the smelt is dissolved in the minimum amount of water; then the green liquor is clarified and subjected to evaporative crystallization. As water is removed, the solubility of carbonate is exceeded and it crystallizes out as the monohydrate, $Na_2CO_3 \cdot H_2O$, leaving the sulfide in solution. The solid carbonate is separated from the mother liquor in a two-stage filter-washer and then sent to a dissolving tank. The mother liquor containing all the sulfide and some saturated sodium carbonate solution is added to the evaporated spent liquor and returned to the recovery furnace, for partial conversion to carbonate and sulfur dioxide. Presumably, hydrogen sulfide is formed as an intermediary, but is completely burned to sulfur dioxide in the furnace.

In the Western precipitation process [33] the sulfide mother liquor is first mixed with spent liquor and oxidized with air and then evaporated and burned (Fig. 273).

In either process the recovered sodium carbonate is used to absorb sulfur dioxide from the furnace gases, final liquor of the required composition being attained by the absorption of makeup sulfur dioxide from a sulfur burner in a secondary absorption system. Unlike in the direct sulfitation (IPC) process, thiosulfate formation is not a problem; therefore, sulfide recycle is applicable to all sodium-base sulfite-type processes. However, it is essential in the crystallization step to wash the solid sodium carbonate completely free of sulfide ions.

The processes were tested in an industrial size pilot plant at Consolidated and good results were reported. Soon after, the facility was shut down without explanation. Obvious weaknesses of the two processes are the inefficiency and lack of control of the production of recovery SO_2 gas; also its very low concentration in the flue gas and consequently large size of absorption equipment. Passing sodium compounds and sulfur back through the furnace would lower the heat energy available for steam production. Air oxidation and the necessity for controlling the proportion of recycled sulfide to spent liquor complicate the Western Precipitation version.

(e) The Mitsubishi RAS and MRC processes

The first of these two processes became known in 1974 and has been in operation in NSSC mills in Japan [34]. The flow diagram is seen in Fig. 274. It follows initially the same route as the above two processes, employing evaporative crystallization to separate Na_2S and Na_2CO_3 of the green liquor. The following step means an achievement, however, since it appears that the inventors have accomplished the long sought conversion of sulfide in the alkalinized mother liquor to sulfite by direct oxidation with air at $110°C$. Consequently, the sodium sulfite precipitates from the concentrated liquor due to its relatively low solubility and is centrifuged, washed, dissolved and used for liquor making. For technical reasons alone, this process deserves close consideration.

The Mitsubishi MRC process precedes the RAS process historically since several patents have been issued since 1967 and 1977. It was the subject of a paper [35] given at the Recovery Seminar in Helsinki, 1968.

What is attempted here is to treat solidified and powdered furnace smelt with CO_2 and steam at $100-300°C$ to drive off H_2S gas and leave a residue of sodium carbonate. Although the carbonation reaction is heterogeneous, reaction times of only 5 to 10 min were reported for achieving 90% conversion of the sulfide. Data given in the Helsinki paper show that most of the Na_2S is decomposed in this treatment, but what is formed is predominently Na_2SO_4, some $Na_2S_2O_3$ and very little Na_2SO_3. However, in June 1978 the Sulfite Committee of CPPA was informed in writing by Mitsubishi Canada Limited that the process was abandoned due to its complexity and high capital cost.

(f) The Ebara process

This process presents again a workable method for two-stage oxidation of the Na_2S in a concentrated slurry of soda furnace smelt and to obtain a high yield of Na_2SO_3 [33]. Figure 275 gives an impression of the flow sheet of the slurry oxidation stages.

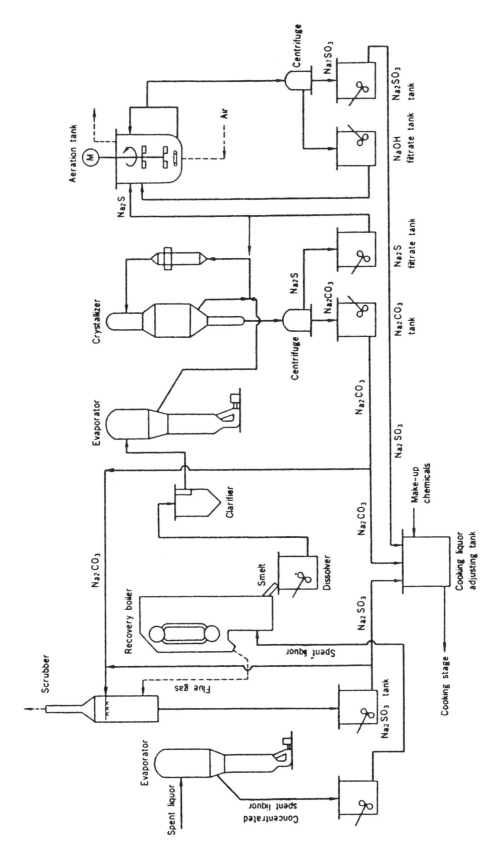

Fig. 274. Flow diagram of RAS recovery process.

Preliminary oxidation with air forming thio-sulfate involves a smelt hopper, a smelt screen, green liquor tank and circulating tank. Smelt stream from the recovery boiler is dispersed by a steam jet in the smelt hopper, producing droplets as small as possible to facilitate decomposition of the smelt which solidifies in the circulating tank. This small size promotes oxidation. The smelt screen feeds coarse smelt formed at the smelt spouts directly into the green liquor tank. The smelt droplets fall into the circulating tank and form the slurry solution as described below.

The temperature of the circulating slurry is maintained between 75 and 85°C by cooling coils installed in the tanks, and the concentration of the circulating slurry is maintained between 45 and 60% so that most of the sodium carbonate is present in crystal form and most of the sodium sulfide dissolves.

A portion of this slurry is continually recycled to the smelt hopper, and is used to wash the inner wall to prevent smelt from adhering to the wall. Air is also introduced into the smelt hopper to cool the smelt and oxidize part of the sodium sulfide in the smelt and slurry to sodium thiosulfate. The primary reaction in this first oxidation stage is:

$$2Na_2S + H_2O + 2O_2 \rightarrow Na_2S_2O_3 + 2NaOH.$$

About 65% of the heat of oxidation of Na_2S to Na_2SO_3 is removed in this stage.

The partly oxidized slurry is then prepared for feeding into the continuous main oxidizer. It is necessary to adjust the S/Na ratio to that of the original smelt and also to increase the consistency. A stable feeding operation is obtained by using a combination of centrifuge and flaker for making the necessary adjustments to the wet cake required.

The oxidation stage consists of the main oxidizer and a sub-oxidizer for providing more retention time for unreacted thiosulfate and sulfide. The main oxidizer is a kneader-type mixer with two axles; it stirs intensely hot powder consisting of sodium carbonate and sodium sulfite. Sodium carbonate crystal, sodium sulfide solution, and preliminary oxidized sodium thiosulfate are sprayed onto this hot powder. The solution to be oxidized sticks to the surface of the dry powder, and this enlarges the solution's contact area with air. By the reactions described above, sodium thiosulfate and sodium

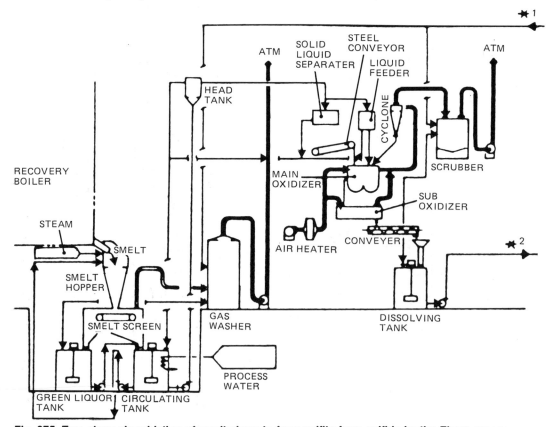

Fig. 275. Two-stage air oxidation of smelt slurry to form sulfite from sulfide by the Ebara process.

Table 68. Chemical analysis in each stage.

wt%	1st		Centrifuge (2nd)		Flake (2nd)	Flaker Over-flow (2nd)	Main oxidation product (3rd)	Sub-oxidation product (3rd)	Cooking liquor (4th)
	Smelt	Slurry	Wet cake	Filtrate	Flake	Over-flow			
Na_2S	43.8	42.1	18.0	50.0	44.2	41.9	2.1	0.9	0
Na_2CO_3	50.6	32.8	63.8	22.2	30.3	32.7	42.2	40.9	3.9
Na_2SO_3	0.5	—	—	—	—	—	44.5	47.6	82.6
$Na_2S_2O_3$	0.4	12.5	9.3	14.0	13.9	12.8	3.6	2.3	4.2
Na_2SO_4	3.0	4.6	6.7	3.9	4.4	4.6	6.2	6.9	7.9
S_{x-1}	—	5.1	1.0	6.5	4.3	5.1	—	—	—
NaCl	1.7	2.9	1.2	3.4	2.9	2.9	1.4	1.4	1.4
S/Na_2O	0.556	0.926	0.455	1.081	0.934	0.925	0.556	0.556	0.987

sulfide are oxidized directly to sodium sulfite by contact with oxygen:

$$Na_2S_2O_3 + 2NaOH + O_2 \rightarrow 2Na_2SO_3 + H_2O \qquad (1)$$

$$2Na_2S + 3O_2 \rightarrow 2Na_2SO_3 \qquad (2)$$

The reaction temperature used in the oxidizer varies between 150 and 250°C with little effect on sulfite yield. At lower temperature more thiosulfate tends to remain, and at higher temperature more sulfide and sulfate. The temperature is maintained by preheating the air and by the heat of reaction. The product is a dry powder consisting mainly of sodium carbonate and sodium sulfite.

The data in Table 68 help to understand the process and its stages. The resulting NSSC cooking liquor, after fortification with SO_2, has no sulfide left which is necessary for avoiding cooking odor. The thiosulfate content would be too high for acid bisulfite or bisulfite cooking. In the NSSC cook at the higher pH it only adds to the load of the ballast (dead load) chemicals, Na_2SO_4 and NaCl. Only about 1% of the smelt processed is lost in this conversion.

The process was installed in four NSSC mills in Japan between 1971 and 1975 to cope with tight pollution control regulations. To evaluate this part-solid phase process technically and economically relative to existing liquid-gas phase processes, many details concerning capital cost, availability, operating performance and maintenance of equipment, and energy consumption need to be known.

(g) The Sonoco process

Among the sodium base sulfite recovery options the process introduced in 1974 by Sonoco Products Inc. is unique in that it replaces the recovery furnace-boiler or incinerator by a calciner or roaster (for example, a lime or cement kiln or ore roasting furnace), and uses recycled alumina, Al_2O_3, as a medium for carrying the sodium component in the conversion of spent liquor chemicals to cooking chemicals.

The process arose out of a series of attempts to reduce the BOD of the effluent from a 200 t/d NSSC mill into a creek. The significant results from introducing this process were the complete elimination of liquid effluent, SO_2 control in the atmospheric discharge, heat recovery, and relatively simple equipment with low maintenance and easy control of the process [37].

The flow sheet is shown in Fig. 276. Effluent from the then existing acetic acid recovery stage was mixed with alumina cake from the ash washer at 35% consistency and the slurry was evaporated to 60% solids. The present system is based directly on spent neutral sulfite liquor at 10% solids, and evaporated to 60% solids. The concentrated spent liquor is mixed with wet aluminum hydroxide cake at 35% solids from the cooking liquor filter. The slurry is pumped to feed mixer where ground ash from the kiln (calciner) is added and solid pellets of 0.3-1.25 cm diameter are formed. Groundwood can be added at this stage to improve the heat distribution and economy in the combustion stage, and the efficiency of sodium sulfate reduction to sulfide in the chemical conversion process, by increasing the carbon content of the system. The pellet feed passes through stages of drying, combustion, chemical reactions and cooling while in the rotary kiln. The net effect is oxidative combustion with intermediate reduction, similar to the smelter operation, but with a solid sodium compound as product. No auxiliary fuel is required. An average temperature of the hot zones of the kiln of 850-900°C and a low level of oxygen in the combustion gas ($\simeq 2\%$), conducive to efficient reduction, have been found advantageous to reach a conversion yield of 98%. A conversion of 85% is typical for the system without addition of wood.

Some of the proposed reactions involved in the kiln process are given below:

$$2\, Al(OH)_3 \xrightarrow{T} Al_2O_3 + 3H_2O$$

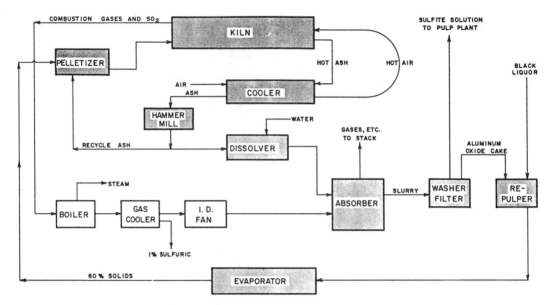

Fig. 276. Flowsheet of Sonoco NSSC Recovery Process. Concentrated spent liquor and ash recycled from kiln are mixed to form solid pellets from which organic pollutants are burned out in the rotary kiln.

2 lignin $SO_3Na \rightarrow Na_2SO_4 + SO_2$ plus water and carbon dioxide

$Na_2S + 2O_2 \rightarrow Na_2SO_4$

$Na_2S + H_2O + SO_2 \rightarrow Na_2SO_3 + H_2S$

$Na_2S + SO_2 + \frac{1}{2}O_2 \rightarrow Na_2S_2O_3$

$Na_2SO_4 + 2C \rightarrow Na_2S + 2CO_2$

Reactions with alumina:

$Na_2CO_3 + Al_2O_3 \rightarrow Na_2O \cdot Al_2O_3 + CO_2$

$Na_2S + Al_2O_3 + O_2 \rightarrow Na_2O \cdot Al_2O_3 + SO_2$

The ash temperature is lowered to 150°C in a rotary cooler and reduced to a coarse powder in a hammermill. Part of the ash is recycled to the pellet feed mixer and a measured part is made into an aqueous slurry in the dissolving tank and pumped to the top of the flue gas absorber for reforming cooking chemicals.

The combustion gases leave the kiln at 600°C and pass through a cyclone to remove particulates to the waste heat boiler where the temperature is reduced to 150°C so that no condensation of the water vapor occurs. They are further cooled to 75°C before entering the absorber at the bottom. In a modification of the system [38] the waste heat boiler is replaced by a direct contact evaporator to raise the concentration of the spent liquor.

The cooking chemicals, sulfite and carbonate, are reformed by the reaction of acidic SO_2 or CO_2 with the alkaline sodium component of the aluminate formed in the kiln:

$Na_2O \cdot Al_2O_3 + SO_2 + 3H_2O \rightarrow Na_2SO_3 + 2Al(OH)_3$

$Na_2O \cdot Al_2O_3 + CO_2 + 3H_2O \rightarrow Na_2CO_3 + 2Al(OH)_3$

$Na_2CO_3 + SO_2 \rightarrow Na_2SO_3 + CO_2$

Aluminum hydroxide precipitates in these reactions which are performed in stages to obtain the required particle size for good filterability of the final cooking liquor on the drum washer.

The amount of alumina in the cycle is stoichiometric or only slightly higher. Make-up is added prior to the pellet feed mixer. Sulfur make-up in the form of SO_2 from the sulfur burner is added directly before the waste heat boiler.

The system is run by one operator with minimum testing and modest instrumentation. No claims for the applicability to sulfite processes other than NSSC have been made.

E. BY-PRODUCTS

Chemical products of sulfite pulping will be covered, along with kraft byproducts, in Volume 6 on Pulp Mill Operations.

REFERENCES

1. Ahonen, A., and Männistö, H., General concepts of the recovery of chemicals in sulphite mills. CPPA Sulphite Committee: Chemical Recovery Workshop, Montreal 1979. Papers pp. 1-25.
2. Horizontal washer, Black Clawson, Powesland Engineering Ltd.
3. Mutton, D.B., Tombler, G., Gardner, P.E., and Ford, M.J. The sulphonated chemimechanical pulping process. *Pulp Paper Can.* 83, T189-194 (June 1982).
4. Gullichsen, J., Heat values of pulping spent liquors. Proceedings of the Recovery Symposium, Helsinki 1968, pp. 211-34.
5. Goeldner, R.W., and Leitner, G.F., A horizontal spray film evaporator for sulfite liquor concentration. *Tappi* 47, 185A-187A (November 1964).
6. Livingstone, L., Chemical cleaning of evaporator scale at Nova Scotia Forest Industries. CPPA — TS Atlantic Branch Report, November 19, 1982.
7. Basciano, C.R., and Livingstone, L., EDTR treatment of evaporator scale — a safe cost effective alternative to nitric acid cleaning. TAPPI 1983 Pulping Conference; Proceedings pp. 353-355.
8. Betts, G.J., Philips, F.E., and Serenius, R.S., Spent sulfite liquor (NH_3) burning and recovery. 1979 *TAPPI* Pulping Conference, Preprints, pp 123-7.
9. Jönsson, S.-E., The Lockman evaporator, a new liquor evaporation system. Proceedings of the Recovery Symposium, Helsinki, 1968, pp. 137-52.
10. Olauson, L., Heat economy improves with Lockman pre-evaporation using digester flash steam. *Pulp Paper Can.* 80, T65-8 (March 1979).
11. Lunden, B., Stora Broby recovery process for semi-chemical pulp mills. *Tappi* 53, 1726-31 (September 1970).
12. Claussen, P.H., Membrane filtration of SSL for by-product recovery and pollution control. *Pulp Paper Can.* 79, T81-5 (March 1978).
13. Simmons, T., The Loddby cyclone burner. *Svensk Papperstidn.* 56, 121-3 (4/1953).
14. Clement, J.L., Magnesium oxide recovery system. *Tappi* 49, 127A-134A (August 1966).
15. Kann, F., and Fuchsel, R., Comparison of basic characteristics in burning thickened cooking liquors obtained from magnesium bisulfite and calcium bisulfite pulping. *Das Papier* 21, 174-9 (April 1967) English.
16. Hulden, B., On the combustion of Ca-base sulphite waste liquor in Finland. Proceedings of the Recovery Symposium, Helsinki, 1968, pp. 387-405.
17. Arvold, W.V., Ray, R.K., and Hanway, J.E. Jr., Fluidized bed treatment of magnesia-base spent sulfite liquor. Proceedings of the Recovery Symposium, Helsinki, 1968, pp. 429-442.
18. Miles, H., Spent liquor disposal via FluoSolids Combustion at a NSSC mill. *Paper Trade J.* 154, 26-31 (No. 33, 1970).
19. Horntvedt, E., A sodium base sulfite recovery process based on pyrolysis. *Tappi* 53, 2147-52 (November 1970).
20. Hornke, R., Technological and economical consideration of processes for the elimination of pulping effluents and waste gases from sulfite mills. *Wochenblatt Papierfabr.* 98, 895-900 (23/24, 1972).
21. Ahman, S., and Arne, H., New sulfite chemical recovery system proven at Swedish mills. *Pulp Paper*, 80-82 (July 1975).
22. Herrlander, B., Experiences with the Flakt-Billerud recovery process. *Pulp Paper Can.* 83, T208-13 (July 1982).
23. Ingruber, O.V., Alkaline sulfite for the elimination of kraft odor. AIChE Symposium Series 200, Volume 76, 196-208 (1980).
24. Johnston, D., Calder, K., and Ho, L., Stora recovery process at NSFI. *Pulp Paper Can.* 80, T184-7 (June 1979).
25. Bernhard, R. and Backman, K., The Stora process for conversion of green liquors from sulfite pulping. TAPPI Pulping Conf., Proc.: 323-9 (Oct. 1981).
26. Reilama, I., The Rauma recovery process. CPPA Sulphite Committee: Chemical Recovery Workshop, Montreal 1979. Papers, pp. 186-206.
27. Anon. Kemi starts multi-purpose Tampella recovery process. *Pulp Paper Intl.*, 32-3 (August 1968).
28. Vuojolainen, T.J., and Romantschuk, H., The Tampella recovery process (TRP) for regeneration of sodium sulphite pulping chemicals. International Sulphite Conference, Montreal, 1978. Preprints I. 19-24.
29. Hauki, T. and Reilama, I., The chemical cycle and recovery of chemicals in NS-AQ pulping. *Pulp Paper Can.* 85, T120-3 (May 1984).
30. Davies, C.J., Bohmer, V.J., and Raubenheimer, S.J., Pulping and bleaching chemicals recovery in a closed (cycle) sulphite mill. *Tappi J.* 66, 47-50 (July 1983).
31. Markant, H.P., The Mead recovery process. *Tappi* 43, 699-702 (August 1960).
32. Farin, G.W., Methods for recovery and recycling of chemicals from sodium sulfite and sodium bisulfite pulping operations. U.S.

Patent No. 4 148 684, April 10, 1979.

33. Berry, L.R., and Larsen, A.D., The WP recovery process. *Tappi* 45, 887-9 (November 1962).

34. Ono, M., and Yoneda, M., Chemical recovery process for spent cooking liquors. Mitsubishi Heavy Industries, Ltd., Technical communication paper, March 1974.

35. Nishizawa, A., Fukui, S., and Ono, M., Chemical recovery process by direct carbonation of smelt. Proceedings of the Recovery Seminar, Helsinki, 1968, pp. 659-73.

36. Mizuguchi, S., and Naito, T., NSSC recovery process used direct oxidation. *Pulp Paper Can.,* T251-3 (August 1978).

37. Cook, W.R., and Parkinson, J.R., NSSC recovery process eliminates BOD_5 load and recovers chemicals. *Pulp Paper Intnl.* 45-47 (October 1976).

38. Cook, W.R., Wood addition to Sonoco recovery. 1981 Int'l. Conference on Recovery of Pulping Chemicals, Preprints, pp. 19-22.

X

DESCRIPTIONS OF MODERNIZED SULFITE PULPING OPERATIONS

O.V. INGRUBER

Retired Senior Research Associate from CIP, now an independent consultant

A.M. AYROUD

Director of Research, Consolidated-Bathurst Inc., Montreal, Quebec

In the previous chapters various technological steps and units employed in sulfite pulping and recovery systems were dealt with. In this final chapter, several examples are given of technologically advanced operations which will help to understand the interrelated function of the various types of equipment and unit operations, and the level of economical balance achieved regarding chemicals and energy. These cases will also illustrate the flexibility of sulfite pulping operations in product and by-product.

In recent years, a remarkable consolidation of wood pulping facilities producing chemical sulfite grades has taken place in North America. These mills do not intend to compete with kraft mills on the market but to provide various grades with paper making characteristics which cannot be produced by the kraft pulping process, while simultaneously reducing, where applicable, their dependence on kraft reinforcement pulp by upgrading the strength of their own sulfite pulp.

Mills discussed in the chapter include:

1. Nova Scotia Forest Industries mill, Port Hawkesbury, Canada, illustrating two-stage sodium base and bisulfite; and single-stage high yield bisulfite for newsprint with full recovery.
2. Edmundston and Atholville mills, Canada, illustrating magnesium bisulfite pulping for coated and uncoated groundwood paper, fine paper, and market pulp and full recovery.
3. Kemi and Valkeakoski mills, Finland, illustrating joint sulfite and kraft operation with full recovery, and AS-AQ pulp procedure.
4. Finch Pruyn, New York, illustrating ammonium base bisulfite pulping in a Kamyr digester with SO_2 and heat recovery.
5. Port Alfred mill, Canada, illustrating high yield sodium bisulfite pulping.
6. Great Northern Millinocket mill, illustrating high yield magnesium bisulfite pulping and full recovery.

Refer also to Chapter VII for additional mill descriptions of high yield mills.

A. TWO-STAGE SODIUM BASE ACID BISULFITE MARKET PULP; AND SINGLE-STAGE HIGH YIELD BISULFITE NEWS FURNISH

Nova Scotia Forest Industries at Port Hawkesbury, N.S., a subsidiary of Stora Kopparberg, Sweden, operate a mill complex which is rather unique in its concept. The two products, bleached sulfite market pulp and newsprint are closely integrated to take full advantage of chemical recovery from the sulfite operation and the high yield sulfite plant for newsprint furnish.

The original market pulp mill which started operating in 1961 [1] was designed for a capacity of 130 000 tons/year utilizing the two-stage sodium bisulphite "Stora" process for cooking as well as recovery of heat and chemicals.

The Stora system was developed in the 1950's by NSFI's parent company, Stora Kopparberg Bergslags AB of Sweden. Stora Kopparberg produces pulp and paper products from several locations in Sweden. NSFI is presently the company's only foreign operation. This was the first sulphite mill to be built in Canada in many years. It is also the first sodium-base sulphite mill for Nova Scotia. The mill's chemical recovery system is the most advanced design of its type to date [2].

1. General

The general lay-out of the mill followed modern practice regarding operational departments as shown in Fig. 277. Future expansion has been an important consideration. Precedents were established in pulp mill design, equipment and operation.

Concurrently with the installation and start-up of a newsprint complex of 160 000 tons/year in 1971, the capacity of the market pulp mill was increased to 150 000 tons/year, and a 42 000 tons/year high yield plant was added to supply chemical pulp to the newsprint operation [3].

The capacity of the entire recovery system had to be increased by approximately 60%. Many of the design features in the recovery system were modified during the expansion phase to embody the improvements called for by ten years of experience.

The concept of integrating chemical recovery with the existing market pulp not only results in savings of chemicals and heat but also makes significant contributions toward protection of the environment. The plant is essentially a closed system as far as chemicals are concerned, and little BOD is discharged to the receiving waters. The design of the plant is also such that only a small amount of suspended solids are discharged to the clarifiers, resulting not only in operating economy, but also in maintaining the high standards of effluent quality already achieved at the mill.

2. Wood supply

The mill uses mainly softwood. The spruce/balsam ratio used to be 50/50 but due to budworm problems is now about 30/70. Actually, a SW deficiency developed and the mill has been successful in accommodating hardwood in some of the market pulp grades. The daily use of barked roundwood is 1500 cords, including the demand of the groundwood plant of the

Plot plan showing the mill site and building layout.

1 scale house	6 warehouse
2 wood yard	7 dock
3 wood room	8 chemical recove-
4 pulping and	ry, steam and
bleaching	power
5 pulp drying	

9 chemical conversion
10 water treatment

11 maintenance shops
12 mill offices and control laboratory
13 administration

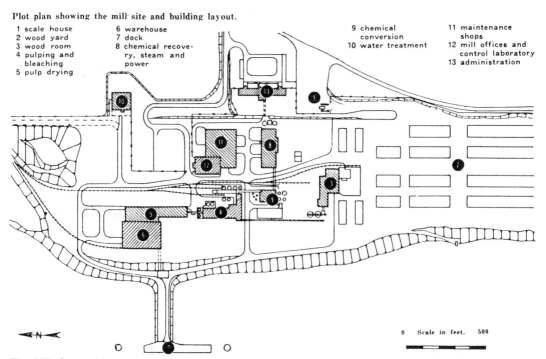

Scale in feet. 0   500

Fig. 277. General layout of the mill.

newsprint department. Chip storage capacity is 1200 cords.

The wood room is designed to handle eight foot logs of up to 36 cm diameter. Present equipment consists of 5 drum barkers, and two large Rauma chippers and a smaller unit. Addition of chip screening and recycling of poorly barked roundwood are under study. Both chip silos and outdoor storage are used. In order to prevent hang-up in winter, the chip silos are insulated and can be heated just sufficiently to prevent freezing.

3. Cooking Department
(a) Market pulp — two-stage pulping

Seven 280 m³ digesters produce two-stage sulfite market pulp at 51% yield; two 200 m³ digesters (since 1972) produce HY bisulfite pulp at 72% yield. All nine digesters are of identical design with stainless steel cladding, bottom strainers, one-man top closures, hydraulically operated bottom valves, forced circulation, indirect heating, and 120 psig (830 kPa) maximum pressure (cf. Chap. V, A, 2). The stainless steel linings have been very successful.

The two-stage market pulp process uses about 56 odt wood chips per digester, added by means of a built-in Svensson packer consuming 9752 kg of steam per cook. Provisions for presteaming exist but are not used routinely. The breach-type closure is operated by one lever and does away with the many nuts and bolts and associated labor still used on many conventional batch digesters. The first stage has the function to equalize chemical charge in the wood and stabilize some hemicellulose. Liquor for this stage contains about 20 g/L Na₂O (Na₂SO₃ + NaHSO₃) at pH 6.1-6.3 and is made from sulfite solution leaving the carbonation-stripping tower of the recovery system at 70-80 g/L, pH 7.4 by dilution with recycled and filtered high yield spent liquor from the blow tank. The remaining digester volume is filled with the drained liquor from previous first stages stored in a spherical tank and boosted with a pump to 830 kPa. The active chemical in this make-up liquor amounts to 10-12 g/L Na₂O, but is not accounted for. The full digester is then heated to 140°C in 1.5 h, held for 15 min and then a measured amount is pumped back to the sphere in 20 min, leaving enough liquor in the digester for maintaining second stage circulation. The second stage requires 1 h SO₂ charging (635 kg) at the bottom of the circulation line to a pressure of 830 kPa (≅7% total SO₂); plus 2 h retention without added heat, but with circulation (the "trickle phase" cooking concept) degassing for

1 h, and flushing out with washer filtrate. Spent liquor from this stage has a pH of 2 and 17-19% solids, if the Rauma pressure washer is used prior to 3-stage drum washing, or 15-17% if the drum washers are used alone.

The second (SO₂) stage hydrolyzes the lignin down to Kappa No. 15 and controls hemi- or alphacellulose content. Total cover to cover time for the 2-stage cook is 8 h. Previous digester time, and liquor color, are used for control. Normal pulp production is 500 t/d.

The digester is pumped out rather than blown under pressure. A stainless steel, ball type, motorized dump valve is located at the apex of the bottom cone of the digester. Flushing liquor (weak spent liquor) is added inside the vessel just above the cone, half way down the cone, and in the main dump line. A pump assists in transporting the pulp to either of two identical dump tanks.

The dump tanks are about 8.2 m in diameter by 17.4 m in height with 1.8 m at the bottom in the form of a truncated cone. The tanks sit on concrete shells through which piping to the bottom of the tank enters. The shell is of mild steel with a ceramic brick lining. Each has a capacity of 1000 m³.

The liquor drained from the first stage of cooking is stored in a spherical drawn liquor storage tank which is a typical "Horton Sphere" some 8.5 m in diameter with a total capacity of 337 m³. It is also lined with acid and carbon brick. This liquor is re-used in the next cook.

Unbleached pulp processing

Stock from the digester dump tanks is pumped to the Regulated Brown Stock Tank.

A magnetic flowmeter regulates the stock flow to two Hooper pressure Knotters followed by two typical Jonsson Knotter Screens. Approximately five tons of knots per day are burned in the steam and power department.

The chemical washed from the pulp is sufficiently acidic to prevent foaming. Brown stock is washed at 0.9% consistency countercurrently on three vacuum washers and one pressure washer with 12% consistency at discharge. The seal tanks are constructed of 319 stainless steel. Fresh (hot) water is added only on the last washer at a rate depending on washed pulp filtrate conductivity and solids level of spent liquor leaving the washing system.

The 12% stock from the last washer is diluted with water from the screened stock thickener's seal tank to 4%. This stock is stored in a 5.5 m diameter by 4.9 m all-tile tank. The tank is equipped with an agitator to avoid separation of the 4% stock.

After screening, the market pulp is bleached by a five-stage C_D-E-C_D/H-D sequence, screened and cleaned. It should be mentioned here that, in addition to the obvious need to choose materials of construction that will minimize corrosion, possible contamination of the product was a consideration of equal importance. Equipment coming into contact with the stock is either stainless steel or rubber lined, or ceramic brick. In some cases, carbon brick has also been used. The pulp is dried on a combination of regular Fourdrinier with a pre-dryer between the second and third press, and a Flakt dryer.

(b) High yield news furnish

The high yield sulfite plant (Fig. 278) is highly automated and is operated from a central control room by two operators per shift. It is located adjacent to the sulfite digester area and utilizes the existing modified chemical conversion department for the market pulp mill for preparation of cooking liquor and conversion of recovered chemicals from the recovery boiler. This simplifies the operation and supervision, as the area covered by the shift personnel includes the chemical conversion department and the market pulp digesters.

The process is supplied with chips from the same two silos serving the market pulp operation. The two $2c^\cap$ m³ digesters are equipped with heat exchangers and circulation pumps, as well as load cells for monitoring chip and liquor charges by weight, and programmed blow. They are charged with about 75 t wet chips by means of Swenson packers using 2268 kg of steam per hour.

The fresh liquor make-up system has been integrated with the acid preparation for the market pulp mill. Sodium bisulfite liquor at a pH of 5.8 with 80 g/L Na_2O is pumped through a pH sensor which adds SO_2 water (15 g SO_2/L) to control pH to about 5.6. The acid concentration is then reduced to about 18 g/L of Na_2O by use of a mixer and density chamber for addition of fresh water. This acid is pumped to the high yield plant for storage in a 5.5 m diameter, 280 m³ stainless steel tank.

Following presteaming for about 1 h, 98 m³ of fresh acid is added. The remaining space is filled with drawn liquor from the previous cooks. The resulting cooking acid has a pH of close to 4.8 and total SO_2 of 3%. Alkali charge is about 86 kg Na_2O per air dry ton of pulp. The cook is steamed to 900 kPa and 120°C for 30 min during the impregnation period after which 5.5 t of liquor is drawn off into a stainless steel clad tank of 23 m³ capacity. Temperature is raised to 160°C and held for about 1.5 h. The end point is estimated from experience of previous cooks and consumed SO_2 tests. As there is no top gas relief, the maximum pressure depends largely on the SO_2 formed during the cook, since there is no free SO_2 to start with. Before blowing, the pressure is reduced by flushing liquor, cooled to 45°C, through heat exchangers before storage; added at the collector ring and circulated to the top of the digester until the void space is filled. This quenches the cook and allows blowing at 276 kPa or less. The digester cooking and blow pit washing cycles are highly automated with timers and flow integrators.

There are two high yield blow pits, 8.5 m diameter, 13.4 m high with a volume of 453 m³ each. The pits are equipped with 74.6 kW bottom entering agitators and perforated drainer bottom having 2.4 mm diameter holes.

After the digester blow is complete the liquor

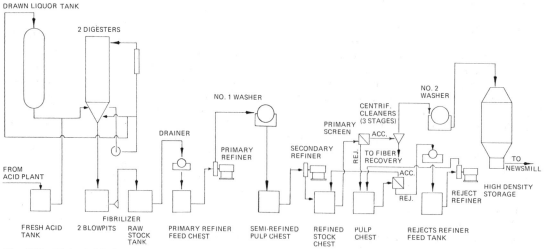

Fig. 278. High-yield plant.

in the blow pit is circulated through spray nozzles located at the top of the pit. This 15 min circulation results in some filtering of the liquor by building a mat of fines prior to drawing liquor off for further use in the high yield plant or recovery. As it is important to withdraw the spent cooking liquor without dilution, the principle of "displacement washing" is used. Wash liquor collected in a tank from the first washer filtrate is pumped through the nozzles at top of the blow pit and liquor from the bottom drained to the weak liquor tank at the same rate. When the displacement washing cycle is complete the cooked chips in the blow pit are diluted with filtrate. The complete displacement washing and dilution requires about 189.3 m³ of filtrate. The weak liquor, recovered from the blow pits is collected as make-up flushing liquor for quenching, cone flushing and initial blow pit dilution. The excess, about 300 L/min on a continuous basis with a dry content of about 7%, is used as dilution water in fresh liquor preparation for the market pulp digester system.

(c) Refining and processing

Cooked chips from the blow pits pass through a "Solvo" 112 kW fibrilizer, before storage in a stainless steel 310 m³ raw stock storage chest. From here on, the process is continuous. Thickening prior to refining is in a 2 m diameter by 2.5 m long Hedemora gravity-type thickener. First- and second-stage refining is performed in 106 cm Defibrator refiners equipped with 933 kW motors at 900 rpm. Two vacuum washers are used for interstage and final washing. Primary and secondary screening takes place after refining in two centrifugal pressurized screens with 1.5 mm perforations. Before final washing and thickening, the accepted stock from the primary screens are centrifugally cleaned. Secondary centrifugal cleaning takes place in sixteen cleaners and the tertiary in two more cleaners. Losses are estimated at about two tons/day (see Fig. 278).

Rejects from the primary and secondary screens are treated in a separate refining stage which consists of a Hedemora gravity type drainer 1.5 m diameter by 1.5 m long and pressurized Asplund Raffinator equipped with a 375 kW 900 rpm motor. The refined pulp is returned to a secondary screen for rescreening.

After the last stage of washing/thickening, the pulp is stored in a 448 m³ high density storage silo at 8% consistency. This is the only non-stainless steel chest in the entire high yield sulfite plant. After dilution with paper mill white water the high yield sulfite pulp is pumped to an intermediate storage chest in the paper mill.

Countercurrent washing

The liquor recovery has already been described in detail. Figure 279 shows the total countercurrent washing cycle and liquor recovery. The only point of addition is at the last washer when 1136 L/min of paper mill white water or fresh water is added to the shower. Filtrate from this washer is used for dilution in screens and centrifugal cleaners as well as for consistency control and showers on the first washer. Filtrate from the first washer is used basically for blow pit washing and dilution. The system results not only in limited water consumption but also contributes significantly to maintaining a high standard of pollution control by eliminating discharge of spent liquor to the receiving waters.

(d) Chemical recovery, conversion and balance (Fig. 280)

SO₂ recycle system

This unit operation is based on the Somer liquid SO₂ concept developed in the forties in Finland. The system ties together pulp digester relief, sulphur burning, liquid SO₂ production and storage for second pulping stage, pure (100%) SO₂ gas production for the Claus department, and bisulphite liquor production at two different pH levels for the two pulping processes.

The sulphur burners and spray guns are designed by Stebbins. The original molten sulphur feed system with a head tank and gravity flow control is used. This avoids pumps and gives always the correct pressure at the burners. Burner gas is cooled quickly to avoid SO_3 formation; it is first absorbed at 60% efficiency in cooled water (steam chiller) at maximum 10°C to make 2% SO_2 solution from which pure SO_2 gas for the Claus reaction is produced in a low pressure steam stripper at 121°C, 117 kPa. The overgas goes to bisulphite absorption towers to make pH 5.1-5.4 pH solution for the CO_2 liberation and sulfite making reaction at the bottom of the chemical conversion tower (see below). The SO_2 content of the vents is close to zero.

The digester relief gases are cooled, SO_2 is separated from condensate and absorbed to make 10% SO_2 solution which is stored under 83 kPa. The solution is stripped at high pressure and the SO_2 gas liquified by cooling to 30°C under 414 kPa. Compressors were eliminated at NSFI in 1972. The primary storage tank holds liquid SO_2 for 7 digesters. The quality of this SO_2 is

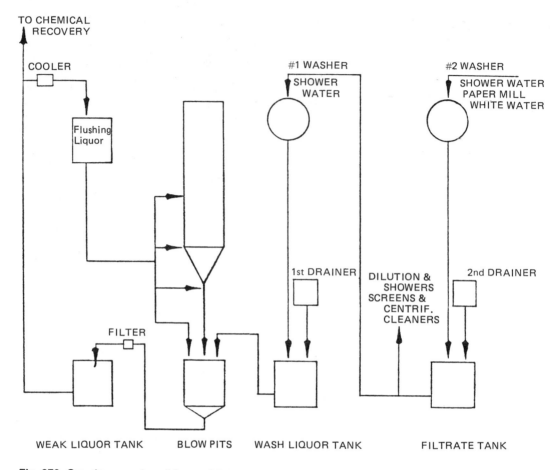

Fig. 279. Countercurrent washing and liquor recovery.

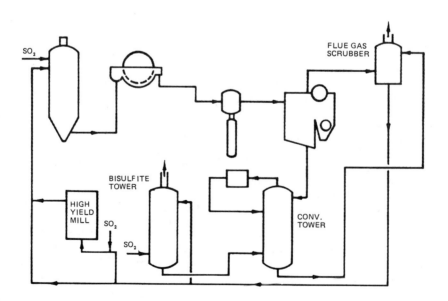

Fig. 280. Stora chemical recovery cycle at NSFI.

insufficient for Claus operation because of carry-over of organic material from the wood cooks. Organics are disallowed also in the bisulphite storage tank supplying the conversion tower.

Spent liquor evaporation

The spent liquor from the market pulpmill at 14-18% solids (and a bomb heat value of 17 MJ/kg DS) is preconcentrated in a Lockman multiple-stage flash evaporator to 20-21% solids. With the Rauma pressure washer operating, this stage is not essential. The Lockman can take liquor to 35% solids because of the small amount of liquor in the unit and very high flow rate. Its steam economy is 4.5 which is between quintuple and sextuple efficiency. A 26 cm vacuum is used and 45 360 L of water are removed per hour. The boiling point rise in evaporation depends on the inorganic/organic ratio. The Stora process has a low ratio, and only 2°C rise at 35% solids. Evaporation to 55% solids is completed in a 7-body Ramen vacuum set. The high temperature and high concentration units are washed daily with acid condensate which is sewered. There is an oxalate scale problem, common to all monovalent sulphite liquor evaporators. The pH during evaporation is held at 5.2 leaving the hot end to minimize formation of oxalate scale by increasing its solubility.

Burning of concentrated spent liquor

Sulfite spent liquor is usually burnt below 60% solids. At 17.6 MJ/kg of dry solids, it has a higher heat value than kraft liquor, due to low inorganic load. The heat value is also augmented by high yield spent liquor addition to the market pulp digesters. Based on a fuel oil price of $30 per barrel, the fuel value of the spent sulfite liquor is about twice its inorganic chemical value. The mole ratio S/Na_2O of the heavy sulphite liquor is 150-160%, which corresponds to about 2 parts $NaHSO_3$ and 1 part Na_2SO_3.

The Gotaverken (B&W) recovery furnace boiler burns 750 t dry solids per day and produces steam at 6200 kPa for electrical power generation. It supplies about 40% of the total process steam of the mill. The bottom and water tube wall sections of the furnace are protected by "Bailey" cast iron blocks laid in chromium cement. Occasional replacement of damaged blocks is required, but there is no serious corrosion of water pipes in this area. The only corrosion problem of the furnace wall is found about half way up at the level of the boiler screen tubes. It is felt that the bottom temperature is too high and the top temperature too low for corrosive effects to develop.

The recovery furnace is equipped with an electrostatic precipitator followed by an SF scrubber for SO_2 recovery. Typical of sulphite recovery technology, the flue gas scrubber closes the chemical cycle of the mill by returning SO_2 and particulates to liquor making. Thus minimizing SO_2 emission is an issue of economy rather than of air pollution control.

The resulting green liquor has about 80% sulfidity on the average. The smelt contains 14.5% carbonate, 3.0% sulfate, 2.7% thiosulfate, and 2.0% polysulfide, indicating a reduction efficiency of about 92%. The large smelt dissolving tank (9 h capacity) operates very quietly; there is no scrubber and emissions are negligible.

With the market pulp department alone the mill would be almost self-sufficient in electrial power and steam. The HY operation brings in the deficiency. News and GW use much power and steam. Practically all of the purchased power goes into groundwood production for newsprint.

Conversion of green liquor to sulphite liquor

The system is described and illustrated in the section on sodium base sulfite recovery in Chapter IX, D, 1a. The first important step is clarification of the green liquor since dregs carry-over can cause plugging of the heat exchangers at the bottom of the H_2S liberation and stripping tower.

With the present recovery boiler operating below capacity there is no carry-over of unburnt spent liquor into the reaction tower and consequently no foaming. It was found, however, that controlled addition of a small amount of spent liquor to the green liquor induces slight foaming in the top section; which causes more contact between gas and liquid phase, and facilitates green liquor carbonation with CO_2 gas and liberation of H_2S gas at 18% concentration. The reaction tower is cleaned routinely four times a year with green liquor or caustic, which removes sulfur deposits originating from sulfur vapor in the CO_2 gas stream returned from the Claus reactors. Sulfur is readily soluble in the green liquor or caustic which is returned to storage. The Stora system does not depend on CO_2 supply from furnace flue gas, as long as sufficient green liquor is in storage.

The Claus system uses two reactors in series [2], of which the second is critical for controlling the H_2S concentration in the recycled CO_2 gas. A second set of reactors is available as stand-by ready for use.

The temperature of the exothermic reaction is held at 216°C or higher by pre-heating only the

H₂S gas stream. If the temperature is too low, the pores in the alumina catalyst plug up gradually with sulfur and require frequent purges. No O_2 is allowed in the reactors. The sulfur vapor is condensed at 121-127°C with boiler feed water which is then used to make steam for the conversion reactor (added just below the CO_2-H_2S return line). The reactor system is under a small pressure of 55 kPa; the liquid sulfur discharges continuously and is fed to the burners.

As shown in Fig. 280, a large portion of the SO_2 gas produced in the burner is used directly in the second stage of the acid bisulfite cooks for market pulp, and the remainder for preparation of bisulfite cooking liquor with the effluent from the flue gas scrubber. Make-up sulfur is added to the sulfur burner, and make-up caustic soda to the flue gas scrubbing solution from the reaction tower, or at other suitable places in the cycle.

The power consumption of the conversion system is relatively small with a load of only 460 kW.

The chemical conversion plant including the Claus reactors is housed in a separate section of the main building and accessible from the main control room and a large service door at the ground floor. There is very little danger due to H_2S because the gas volume present in the equipment is less than 3 m³, with 82% CO_2 and 18% H_2S, or 450 L (680 g) of H_2S. Venting this to the outside would not be noticeable. In the case that the H_2S alarm should sound at 5 ppm ambient concentration, liquor feeding to the conversion tower is stopped and no further H_2S gas is produced so that the system clears of H_2S within 10 min. When a Claus reaction fails to operate, H_2S is not vented, but circulated for one or more hours, mainly to get all sulfur out of the system. The tail gas then contains some SO_2 with no, or only a trace of, H_2S. Repairs can then be made without protective equipment. The plant operates for months without interruption and is serviced by regular maintenance during shutdowns in spring, fall and holidays.

An important consideration in closed cycle systems is inactive load or "dead load" of inert chemicals, particularly sodium sulfate and sodium thiosulfate in the reactor tower, under inappropriate reaction conditions. Continuous monitoring and corrective adjustments of liquor streams is necessary to prevent build-up of undesirable compounds. For example, thiosulfate can form in the reaction tower during periods of abnormally poor sulfide removal. This is controlled by adjusting blower speed, use of weak liquor, or temporary reduction of GL feed rate. Sodium sulfate forms because of excessive

exposure of liquor to air. Improvements made between 1976 and 1978 are shown in Fig. 281.

At NSFI, thiosulfate content is quoted as % of active Na_2O, not as % in the cooking liquor.

Twice a shift, H_2S is tested in the gas streams from the conversion tower to the Claus reactor and back to the tower. Thio is tested every 2 h in the sulfite solution leaving the tower (5.1%) in the green liquor from the dissolving tank (2.7%), in the flue gas scrubber return liquor ($\cong$1%), in the spent HY liquor (some), and in the cooking liquors ($\cong$7%). The thio content in the cooking liquor is relatively small. A contributing factor is the vapor phase second-stage SO_2 cook for market pulp, which also helps to reduce the danger of liquor breakdown in the acid bisulfite stage.

(e) Mill make-up balance

The total cooking chemical make-up per adt bleaching pulp is 64 kg sulfur and 36 kg NaOH, not including lesser amounts for HY pulp.

The estimated washing efficiency is about 92%. About 9-14 kg organics/t (as Na_2O) go to the bleachery and about 18-23 kg are lost in the HY operation.

Total Na_2O losses in the recovery plant are 3 lb/t, of which 95% are in green liquor dregs.

Modifications for tightening the chemical cycle are pursued as part of the maintenance program. They include addition of a set of evaporators for correcting capacity problems. With the two pulping units and the recovery plant in operation the mill easily meets the current federal regulations.

Added flexibility

A sodium base sulfite mill with chemical recovery plant can consider alkaline sulfite operation without major modifications. At NSFI, the first full-scale AS-AQ cooks were made in 1982 and the resulting pulp and relevant operational aspects are being evaluated.

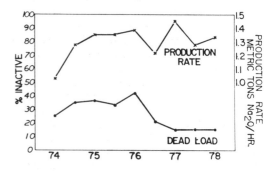

Fig. 281. Conversion system production rate and fresh liquor inactive load.

B. MAGNESIUM BASE BISULFITE PULP FOR COATED AND UNCOATED GROUNDWOOD PAPER, FINE PAPER, AND BLEACHED MARKET GRADE

In 1983, nine magnesium bisulfite mills were in operation in the United States, the same as in 1977. Canada counted three mills in this period, however two older mills were shut down and the two converted and updated mills described below started up. The geographical distribution of magnesium base mills here and abroad indicates clearly that the process thrives only in locations providing wood furnish of the spruce type (the spruce family, balsam fir, and hemlock). Some selected hardwood species, such as beech and the poplars, are compatible and used in minor quantities. In North America two mills add a small proportion of hardwood pulp to their production, and one mill uses hardwood only.

The two pulp mills chosen for this description of the most advanced state of the art of magnesium bisulfite pulping and recovery technology are both owned by Fraser Inc. of New Brunswick [4,5]. Extensive expansion, upgrading and modernization of the existing ammonium base acid bisulfite plants were completed for the older (Edmundston, 1917) and larger facility in 1980, and for the younger (Atholville, 1929) and smaller facility in 1983. Both plants use recovery systems of the same design, featuring magnesia ash collection by means of electrostatic precipitation and the new vertical four-stage multi-venturi absorption tower for liquor preparation (see Chap. IX, C, 8c). The most recent new magnesium bisulfite pulping plant using the classical combination of magnesia ash collection by electrostatic precipitation but liquor making in a horizontal five-stage single-venturi absorber set (B&W-type, see Chap. IX, C, 8a) was put on stream by Leykam, Inc. in Austria in 1978.

The history of the Edmundston mill in particular reflects the technological modifications during this century in the course of expanding and updating the manufacturing process with the objective of increasing productivity while improving pulp quality, and reducing cost by conserving energy while controlling liquid and gaseous emissions to meet environmental standards.

At the start the mill had four digesters producing 120 t/d calcium base acid bisulfite pulp; two digesters were added in 1920 to produce a total of 150 t/d of pulp. In 1925 production of fine papers started in Fraser's new paper mill across the St. John river in Madawaska, Maine and by 1928 the pulp mill supplied 200 t/d of chemical pulp from eight digesters and 150 t/d of groundwood pulp from a new plant to the growing paper mill via a one-mile pipeline under the river. To meet strong demand for paper in the forties, an accumulator was added to the acid liquor system and forced circulation installed on the digesters to bring the chemical pulp production to 330 t/d.

In 1953 the cooking base was switched from calcium to acid ammonium bisulfite with the economic benefit of boosting pulp production to 420 t/d, mainly due to better impregnation of the chips with cooking liquor. A new unbleached screening system and centricleaning were installed in 1968. In the early 1970s production was at 450 t/d and, as stated by the company, "at unreasonable production cost". It was time for a change, and as soon as possible.

Studies of four alternatives inclusive of conversion to kraft pulping extended to 1975 and resulted in the option for a change to the magnesium bisulfite process with chemical and heat recovery. Two options ruled unfeasible by the technical team were: closing the Edmundston pulp mill and purchasing all chemical pulp required for the Madawaska paper mill or installing ammonium base recovery facilities at Edmundston.

The kraft option was ruled unfeasible for a number of reasons [6], one major point being insufficient wood availability in the area for operating a kraft mill of economical size, that is, larger than 900 t/d pulp. The capital cost per ton pulp for the large kraft mill was twice that for the projected sulfite mill at 650 t/d, and much higher again for a kraft mill of the same size. Other reasons were: the Mg-bisulfite process provides a higher yield from wood than the kraft process, the difference amounting to 3% more pulp or better, which again lowers the cost per ton of pulp. High brightness of the Mg pulp completely eliminates the need for a separate line for semi-bleaching which would be required for part of the kraft pulp production. Since the associated paper mill at Madawaska uses a large amount of unbleached sulfite pulp, the high brightness of the new Mg pulp is a great advantage whereas with kraft there would be a major and costly disadvantage. There are also substantial energy savings because sulfite pulp requires less refining than kraft.

The decision to switch from ammonium acid bisulfite to magnesium bisulfite is well supported by the known improvement of unbleached pulp properties; brighter pulp at higher yield, 20% higher tear, and 25% higher tensile strength. Also, due to the higher brightness and better strength of the Mg pulp the amount of purchased kraft reinforcement pulp needed for lightweight grades is reduced considerably.

The three-year expansion and conversion project included modernization of the whole pulping plant, such as replacement of the existing unbleached washing, screening, and cleaning department, and renovation of the digester house. The wood room had been completely rebuilt in the early seventies and a new groundwood mill installed at the same time.

A similar project was undertaken to convert and update the ammonium base market pulp mill at Atholville [8,9,10] after it was shut down in 1982 because of high energy cost and other inefficiencies. In this case a four-stage bleach plant with a chlorine dioxide plant were also added. The kraft process option was duly considered and ruled as not suitable for reasons similar to those given in the previous case. In general, Mg-bisulfite cooking gives a better yield, lower operating costs, and a clean specialty pulp which makes a better furnish than a mixture of hardwood and softwood kraft.

Description of the new magnesium bisulfite plants

A schematic layout and a flow sheet of the new pulping and recovery facilities are shown in Figs. 282 and 283. From the data in Table 69 it is seen that there is a 40% difference in scale while the operations are rather similar.

1. Wood furnish

Most of the wood used in the Edmundston plant comes from company limits growing spruce and balsam fir at a ratio of 70:30. Wood handling is simplified by receiving a large portion of the requirement as chips by rail, mainly from large company-owned lumber mills. Hardwood chips will be purchased from local saw mills. There are also facilities for handling bark and sludge from the effluent treatment system.

At Atholville no roundwood is used. Facilities include receiving and storage of chips, both purchased and from company sawmills, and also receiving and handling of hogged hardwoods for the waste fuel boiler. A new mechanical chip conveying system replaces the previous pneumatic system, reducing digester filling time and energy usage.

2. The digester house

The same considerations and objectives were

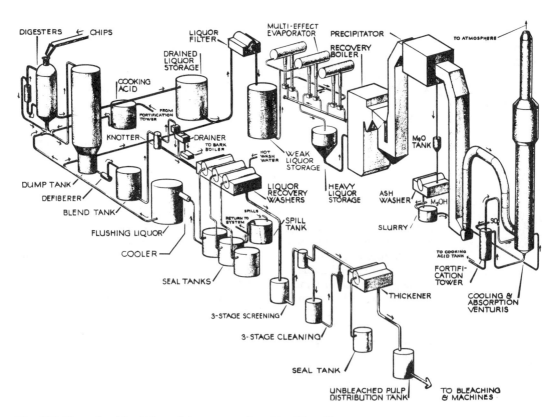

Fig. 282. Flow sheet for Edmundston magnesium bisulfite mill.

applied to the expansion and modernization of the cooking facilities of the two mills. Digesters were relined with carbon brick and fitted with bottom strainer plates and a much improved liquor circulation system. All are equipped with ball-type automatic discharge valves. Large heat exchangers (indirect liquor heaters) and circulation pumps permit greatly reduced time to reach maximum cooking temperature resulting in a

100 t/d increase in pulping capacity.

In order to meet the requirements of efficient chemical and heat recovery, the practice of blowing the digesters has been changed to cooling and draining the spent liquor and dumping the digester with the help of red stock washer effluent. The spent liquor is collected in a drain-down tank and sent to the evaporators. The pulp is flushed out of the digester into one of

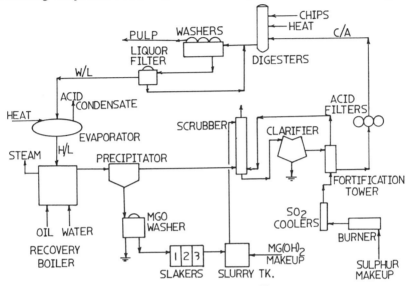

Fig. 283. Fraser Inc. liquor recovery cycle — Edmundston mill.

Table 69. Average operating data of two modern magnesium bisulfite plants

	Edmundston	Atholville
Wood supply	70/30 Spruce — Balsam	75/25 Spruce — Balsam
No. of digesters/volume, m³	8/239	5/245
Cooking liquor pH	3.8	3.8
Time to temperature, h	2	2
Cooking temperature, °C	165	160
Time at temperature, h	2.5-2.75	2.25
Cover to cover time, h	6-6.25	6.5
Digester yield, % o.d. wood	52	50
Kappa No.	32	28
Unbleached pulp, t/d	650	350
Weak spent liquor storage, m³	4666	3780
Heavy spent liquor storage, m³	708	756
Evaporator capacity, tH_2O/h (58% dry solids)	295	187
Bomb heat value, MJ/kg DS	15.1	15.2
Recovery boiler, tDS/h	42	23
Steam production, t/h	167	89
E. precipitator efficiency	95	95
MgO recovery, %	85	— (In start-up)
SO_2 recovery, %	82	— (In start-up)
Recovered liquor production, m³/h	197	100
MgO make-up, kg/adtp	23	23
SO_2 make-up, kg/adtp	30	30
Thermal self-sufficiency, %	65	85
El. power self-sufficiency, %	55	85
SO_2 emission, ppm	50-90	50-90
Effluent BOD, kg/adtp	10	—

two dump tanks with a capacity of three cooks.

The digester room is equipped with new instrumentation compatible with computer control.

3. Spent liquor collection and pulp processing

From the dump tanks the stock is pumped through equipment to a blending tank. Knots and uncooked chips are removed in pressure knotters and sent to the bark boiler. At Edmundston, the unbleached pulp is washed countercurrently in a set of three vacuum washers, screened, centricleaned, thickened, and stored. At Atholville, washing is done in two pressure washers under conditions affording efficient spent liquor recovery and followed by screening, thickening, and storage in a 55 t high consistency tank.

The bleaching sequence for the Atholville market pulp is C_DEHD with 35% ClO_2 replacement in the chlorination stage, including a Frotapulper treatment for cleanliness between the first and second stage, and followed by screening and thickening. Softwood pulp is stored in a 350 t high consistency tank. Good runnability of the hardwood pulp is obtained by means of a pulp blending system to mix 15% softwood with 85% hardwood ahead of the paper machines. A portion of the pulp made at Edmundston is bleached by the CEDH sequence for the production of chemical papers at Madawaska.

Unbleached sulfite and grondwood pulps from Edmundston furnish the on-site 100 t/d boxboard cylinder machine while the remainder is pumped to Madawaska, now expanded to a production level of 1250 t/d, for manufacture of coated and uncoated groundwood papers used in carbonless business forms, magazines, books, telephone directories, and catalogues.

4. Spent liquor processing

The spent red liquor drained from the digesters passes first a drum-type gravity filter made of 316 stainless steel to remove pulp fibers and any other solid material. It is then pumped to the horizontal spray film, six-body, five-effect vacuum evaporator to be concentrated from 10.5% to 58% dry solids. In this type of evaporator the spent liquor is sprayed onto the outside of the steam tubes, the reverse of common evaporator designs, which greatly facilitates cleaning for scale removal (see Chap. IX, B, 1, 2). It operates on 207 kPa steam at an evaporation ratio of 1:4. The evaporator at Atholville is a vapor recompression rather than a

vacuum unit; while using more power, it uses less steam and has therefore a higher energy efficiency (see Fig. 235). All steam condensate from the digesters and evaporators is collected and recycled to the steam plant where it is cleaned of contaminants and reused.

5. Liquor burning and recovery of chemicals (Fig. 284)

The recovery boiler at Edmundston burns concentrated magnesium bisulfite spent liquor at 58% dry solids and a bomb heat value of 15.1 MJ/kg DS with supplemental oil firing. With a dry solids capacity of 46 t/h and a steam capacity of 167 t/h at 8.62 MPa this is the largest MgO furnace-boiler built so far. Its design shown in Fig. 284 is similar to that of earlier models installed in various magnesium base sulfite mills by the same company. Typically, the furnace room is divided into two adjacent sections separated by a screen of boiler tubes. The burning chamber is fed from 18 spray burners, eight of which can handle red liquor or oil. Combustion is completed in the second section where the superheater and steam boiler are located. The remaining heat content of the combustion gases carrying all the MgO ash and SO_2 gas (2.3% per weight) from the pyrolysis of the lignosulfonate and residual magnesium bisulfite in the spent cooking liquor is utilized in the combustion air preheater and in the economizer before reaching the dust collector. The combustion air and the gas-dust phase after combustion are moved through the boiler and the subsequent MgO recovery stages by a forced draft fan of 5550 m^3/min capacity at 40°C driven by an electrical motor, and a twice as large induced draft fan at 11 300 m^3/min capacity at 218°C driven by a 3.75 MW steam turbine.

For the first time in North America this sytem uses an electrostatic precipitator for magnesia ash collection from the flue gases. The practice was developed in Sweden before 1977 and has been in use there in all 9 magnesium base mills. Advantages over the classical dry dust collection by multi-cyclones are higher efficiency, lower pressure drop, and no erosion problem of metal parts by the magnesia ash. The new single-chamber unit with two electrical fields in series has a design rating of 8127 m^3/min of flue gas at 204°C with an inlet dust loading of 10.7 g/m^3 and a guaranteed efficiency of 95%.

With most of the ash removed, the flue gas leaves the precipitator via the self cleaning draft fan and is cooled to 71°C in a venturi-type cooler with water spray. This last step removes the remaining ash as well as about 5 to 10% of the SO_2 content of the flue gas which is now ready for the preparation of fresh cooking liquor.

The magnesia ash is continuously removed from the bottom of the precipitator through a rotary valve. It contains traces of sodium, potassium, and calcium sulfates and chlorides originating from the wood and the $Mg(OH)_2$ make-up. To prevent build-up of these impurities in this closed chemical system, with resulting corrosion and boiler problems, the ash is slurried with water in a tank and washed on a belt filter

constructed of stainless steel. The filter cake at 15 - 20% solids is broken up and screw-fed to a set of three slaking tanks where it is treated with hot water and steam while being agitated for about 1 h in each tank. After adding milk of magnesia [$Mg(OH)_2$] make-up in another tank the slurry is ready for liquor making. It must be recirculated through an elevated tank to prevent settling in the slurry tank.

Again for the first time in North America, a

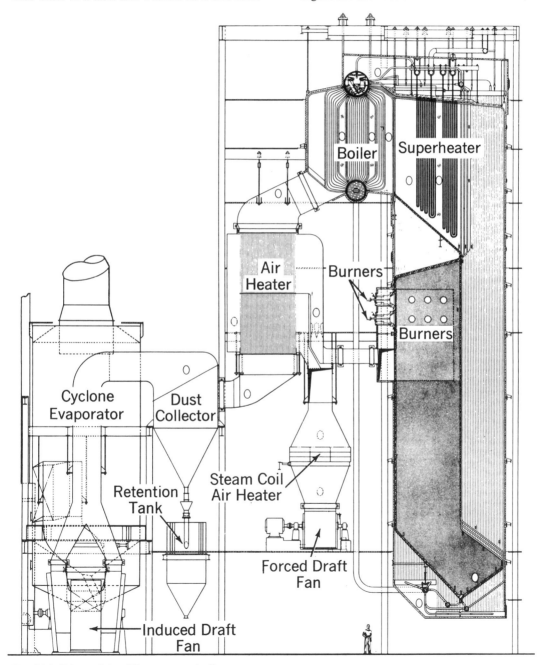

Fig. 284. Edmundston Mg recovery boiler.

vertical multi-venturi absorber was chosen for preparing fresh cooking liquor from hydrated magnesia slurry and SO_2 gas [7]. The companion set of electrostatic MgO precipitator and absorption tower shown in Fig. 285 is also the largest built so far. With one attached cooling stage and four superimposed countercurrent absorption stages the tower is designed to produce 197 m^3/h of raw magnesium bisulfite cooking liquor.

Details of the design of one complete absorption stage are seen in Fig. 286. Clean, cool flue gas containing 1.5% SO_2 gas by volume flows up the plenum against the bottom of the conical collection tray and up into the eight venturi tubes mounted in a circle through the tray. In the throat of each venturi (Fig. 287) a splash plate is supported horizontally against which a stream of $Mg(OH)_2$ slurry impinges. The upward spray of droplets formed is entrained by the gas passing rapidly up the throat. The upper half of the absorption stage is occupied by a bed of heavy duty plastic balls, about the size of ping pong balls. Nearly a million plastic balls are utilized in the four stages, supported and covered by horizontal wire mesh, and divided into eight wire mesh compartments positioned exactly above the eight venturis.

This bed serves to increase reaction surface area by collecting slurry entrained in the rising gas stream on the surface of the balls and also to fluidize the absorption process. Slurry reacting with SO_2 on the balls continually drips off and these larger drops fall back against the gas stream into the collection tray where 90% of the drip is circulated back to the splash plates in the venturis and 10% passes out of the tray through its overflow pipe to the stage below and is eventually withdrawn as "raw process acid" from the bottom of the tower. Reacted gas leaving the uppermost absorption stage passes through a de-mister spray unit into the stack and to atmosphere.

The key to successful operation of the scrubber-absorber unit is instrumentation, whereby pH, temperature, and flow rates of the stages are precisely controlled. Magnesium hydroxide is added to each absorption stage. If too little is added, SO_2 recovery drops off and SO_2 emission increases. If too much is added, insoluble magnesium monosulfite, $MgSO_3$, is produced which could plug the entire system. Control is critical because $MgSO_3$ deposition is favored by low temperature and high pH, the same factors which favor high absorption efficiency. Within the temperature range involved, the practical upper limit for scrubbing is about pH 6. Precautions against plugging include automatic cleaning of the pH electrodes with acid condensate, and continuous washing of sensitive areas with pH-controlled slurry. Generally, it is important to run the absorbers at full capacity which means that provision of ample

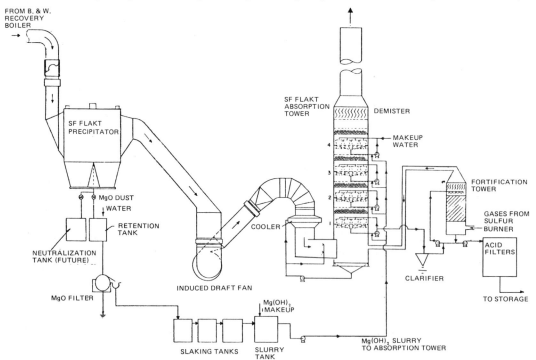

Fig. 285. Flow diagram for installation at Fraser's Edmundston mill.

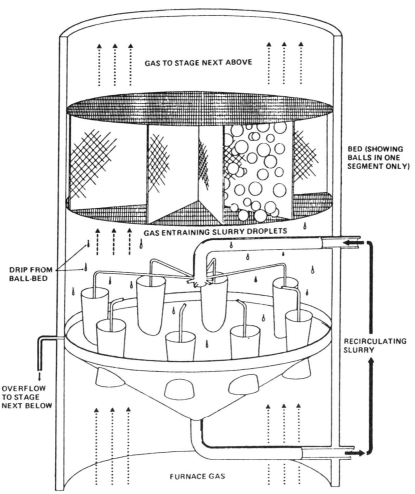

Fig. 286. One absorption stage complete — lower plenum, dished collection tray with eight venturis mounted on it, upper plenum, and ball-bed. Note that each of the eight sections of the bed is directly above one of the venturis.

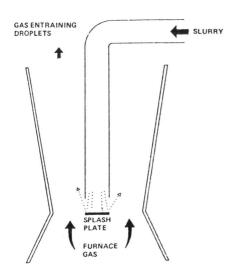

Fig. 287. One of the venturis, showing the splash plate in its throat.

red liquor storage capacity is a necessity.

The raw acid liquor from the bottom of the tower is fortified to cooking strength of 6.5% total SO_2, 3.1% combined SO_2, and pH 3.4 with sulfur burner gas at 14% SO_2 content in an absorption tower packed with ceramic saddles. Excess SO_2 from the top of the fortification tower is returned to the bottom of the flue gas scrubber. Both towers are constructed entirely of 316 stainless steel.

Solid material such as $MgSO_3$, elemental sulfur, and carbon particles are removed from the fresh cooking liquor in a lamella gravity settler having 230 m^2 of settling area, also made of stainless steel. This settler has successfully eliminated problems encountered by other mills in sand filtering operations. Final filtering out of monosulfite crystals and carbon specs is done in a deep-bed pressure sand filter where two units can handle the liquor flow while one unit is backwashed.

The smaller chemical recovery complex at Atholville (see Table 69) is essentially of the same design as the one at Edmundston. Only minor refinements have been made on the basis of the experience in two years of operation at Edmundston. In both plants the efficiency of chemical recovery is about 85% for $Mg(OH)_2$ and 82% for SO_2.

The Atholville plant is operated from two control centers, one for the recovery system and waste wood boiler, and the other for pulping and processing.

6. Energy economy

One of the major objectives guiding the modernization of the two pulping plants was to increase the degree of self-sufficiency in energy. In addition to the new recovery boiler, Edmundston has a bark boiler, installed in 1977, which burns 13.5 t/h of bark and sludge producing 68 t/h of steam at 4.14 MPa, and an older oil-fired boiler producing 113.5 t/h of steam at 8.62 MPa. With three steam turbines generating a total of 19.1 MW the mill is now capable of supplying 55% of its electrical power requirements. At the same time the new recovery boiler reduced the fuel oil demand by 48%.

At Atholville, the new recovery boiler burns 23 t/h DS to produce 89 t/h of steam. It is supported by a high pressure hog fuel boiler burning 40% of wood waste from local saw mills and 60% from one of the two saw mills owned by the company. The two-drum boiler is designed to burn 15.5 odt/h of hog fuel at 56% moisture producing 68.1 t/h of steam with wood alone or 90.8 t/h with oil or wood and oil combined. A radial steam turbogenerator having a nozzle pressure equal to that of the two boilers provides maximum power generation. The generator is tied into the provincial electric power system. With this arrangement, the mill is 85% self-sufficient in thermal and electrical energy.

7. Air and water quality protection

A general objective in the program for expanding, converting, and rebuilding the two plants was environmental compliance. The largest step in this direction was taken by installing properly sized recovery systems which reduce the amount of chemicals used by 80% and convert a large portion of the organic load of the spent liquor into process steam.

As for air quality, the vertical multi-cyclone scrubber with ball beds performs well above expectations. Designed for a maximum SO_2 emission of 250 ppm, which is one half the amount called for by government regulations, the actual value varies between 50 and 90 ppm

SO_2 from the stack.

As for water quality, the provincial authority announced in 1968 that release of spent sulfite liquor into the St. John river had to stop. Anticipating increased demand for pulp to meet the requirements of the Madawaska paper mill, a large primary and secondary effluent treatment system was built in 1971. The system consists of a settling tank, two aerated lagoons, two pump houses, and a neutralization building. Although large enough to lower the BOD at the present level of pulp production to the limits set by provincial regulation, it was found that carry-over of suspended bio-solids could be controlled below the allowable level only if retention time was in excess of 5.5 days. It was necessary, therefore, to reduce the volume of mill effluent from the increased pulp production. This was achieved by a program which cut water usage from 151 000 t/d to 95 000 t/d. At Atholville, SO_2 emission control is on the same level as at Edmundston.

Liquid effluent is treated in a clarifier of 46 m diameter which handles 38 m^3/min with solids removal efficiency of 90%. Clarified effluent is then neutralized in a system consisting of lime handling, storage, and two slaking tanks.

C. COMBINED KRAFT AND SULFITE RECOVERY WITH THE TAMPELLA SULFITE CONVERSION PROCESS.

This section describes an interesting development in joint sodium base recovery made possible by the introduction of the new Tampella sulfite recovery process (TRP) fifteen years ago.

It has been common practice for many kraft mills producing strong linerboard pulp to add a neutral sulfite high yield (NSSC) facility for adding the required amount of corrugated medium (CM) for the manufacture of the complete combined container wall board.

Since there is no independent sodium base sulfite recovery, the spent sulfite liquor is processed with the spent kraft liquor. The sulfite required for the NSSC cook is either prepared by absorbing SO_2 burner gas in carbonate solution (see Chap. V1, D, 2) or is purchased as technical grade. The sulfur and sodium values in the NSSC spent liquor then provide the necessary make-up for the kraft process (cross-recovery). In this way the size of the sulfite facility is tied to the recovery requirements and efficiency of the kraft mill cycle. Common size ratios of production capacity are 1:3 - 1:4, the problem being that with the greatly improving recovery efficiency of modern kraft mills, particularly for sulfur, the requirement of NSSC make-up liquor is shrink-ing and the size of the CM production capacity drops below the level of economic interest.

Some kraft container board mills in the U.S. have attempted to obviate this problem of unbalanced production capacity by producing CM in cooks with green liquor drawn from the kraft recovery system. Disadvantages of this process are losses in CM quality which are acceptable only in captive products; and reduced economic efficiency of the kraft recovery cycle which requires larger equipment capacity to meet the extra demand of the semichemical operation, but without producing an equivalent amount of prime kraft liner pulp.

A practical solution to the problem of balancing the recovery of chemicals in integrated kraft/sulfite operations was first demonstrated industrially in 1968 at the Kemi mill in Northern Finland near the Arctic Circle. In 1967 develop-ment work in a pilot plant attached to the Heinola NSSC mill in Finland had succeeded in modifying the original Sivola sodium base recovery process and to realize the objectives of safe operation, easy control, and relative simplic-ity to make the process competitive economical-ly [12]. Although the simple basic reactions of converting green liquor into sulfite liquor were known and used at that time, the novelty was in the combination and application of the stages for achieving the above objectives. The new version became known as the Tampella recovery process (TRP) (see Chap. IX, D, 1c).

The process can be applied to all types of sodium base pulping where sulfur is present, such as:

Conventional sulfite sodium-base
NSSC
Integrated kraft-sulfite and kraft-NSSC (cross-recovery)
Conventional kraft, where a lower sulfidity is desired
Polysulfide kraft

The Kemi mill was built in 1919 producing 30 000 t/y calcium base acid bisulfite pulp. By 1968 it produced 250 000 t/y of kraft pulp and 33 000 t/y of sodium bisulfite pulp. The spent liquor from the bisulfite mill was used as make-up for the whole system, the sulfidity of the kraft system being controlled by addition of caustic soda to the white liquor. At this point, recovery was modified by starting up the new sulfite conversion system and thereby integrat-ing kraft and sulfite operations by individual regeneration of chemicals for the processes

[13,14]. Figure 288 [15] shows the principle of the recovery process: separate stock washing and spent liquor evaporation, joint liquor combus-tion, separate conversion to the respective cooking liquors, and transfer of a portion of carbonate from sulfite recovery to the kraft cycle for maintaining correct sodium balance.

One practical reason for adding sulfite conver-sion at Kemi was that saltcake (Na_2SO_4) and sulfur could be used as the only make-up chemicals. There was a considerable cost advan-tage in covering total sodium and part of the sulfur losses by saltcake.

The second reason for this choice was the possibility of controlling the sulfidity of the white liquor independently of the bisulfite-kraft production ratio. The Kemi experience brought various improvements of the original design and other mill applications followed [16].

In 1975, a mature model of the Tampella system was installed in the kraft/sulfite mill of United Paper Mills Ltd. at Valkeakoski, Fin-land, again converting a cross-recovery opera-tion to an integrated kraft-sulfite operation. This mill produces 550 t/d kraft pulp from pine in 6 automated batch digesters and a continuous sawdust digester for the manufacture of sack and other industrial papers, both glazed and unglazed. Until 1981, 230 t/d of sodium bisulfite pulp at 55% yield were made in 4 batch digesters and sheeted wet for shipment as long fiber pulp to the UPM newsprint mill at Kaipola. Subse-quently, the sulfite mill was converted to alkaline sulfite-anthraquinone operation [17], adding one additional batch digester to the existing four. Only minor changes were required for adapting the bisulfite recovery operation to AS-AQ recovery. The size of the sulfite department of this mill corresponds with that of several Canadian sodium base bisulfite mills producing news furnish.

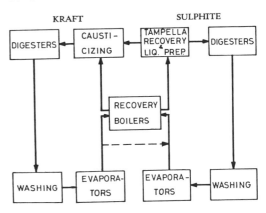

Fig. 288. Integrated kraft-sulphite recovery at Kemi.

On the flow sheet for unbleached pulp and cooking liquors in Fig. 289, plainly drawn components represent the original cross-recovery mill and shaded components the additions for individual sulfite recovery. The kraft liquor is collected by countercurrent washing on four drum washers. After evaporation and concentration to 60-62% solids the kraft liquor is burnt together with the spent sulfite liquor and the required portion of green liquor (about ⅔ of total) is passed through a conventional causticizing unit for regenerating kraft white liquor. The weak liquor from lime mud washing is used to dissolve the furnace smelt.

Sulfite spent liquor is recovered by in-digester displacement with filter water from the two pressurized sulfite pulp washers operating at 96-97% efficiency. Before 1981, the bisulfite spent liquor was neutralized with caustic soda and evaporated to 55-57% solids content at a bomb heat value of 15.6 MJ/kg dry solids. The present AS-AQ liquor is evaporated without neutralization to 68% solids at a bomb heat value of 12.8 MJ/kg dry solids. Due to the high degree of concentration, which is unattainable with common acid bisulfite or bisulfite liquor, the effective heat value of the AS-AQ liquor is higher than that of the bisulfite liquor.

Concentrated liquors from both processes are mixed in a tank and burnt in the Tampella recovery furnace boiler. When burning bisulfite or alkaline sulfite and kraft liquor together, the sulfidity in the furnace bed is 5-6% higher than in kraft systems, but no operational difficulties are apparent.

The bottom of the furnace is built of studded steel tubes with welded spacers, and filled with refractory material. The lower part of the sides exposed to smelt corrosion is built of prefabricated membrane wall units made of stainless steel clad (compound) tubes welded together to make large panels. Normal boiler steel is generally used for the steam tubes in the boiler banks above the hearth (see Chap. IX, C, 2). Compound tubes have been developed and introduced in recent years to meet the requirements of a major trend to raise the steam pressure of the boiler for optimum energy recovery. Older furnace boilers can be rebuilt with corrosion protection as routine maintenance for extending service life.

An existing kraft boiler could probably be used for sulfite operation, but it would have to be inspected to decide on any modifications. For instance, more studding at the bottom may be required in order to raise the temperature in the char bed to take care of the higher melting point of a smelt having higher sulfidity.

Bridging and corrosion of boiler banks may occur due to $NaHSO_4$ (sulfate) or $KHSO_4$ (bisulfate) formation in the upper part of the furnace. These bisulfates melt in the range of 200-300°C and can form sticky deposits on the boiler tubes which, together with ash components, can bridge tube sections and cause loss of boiler efficiency, metal failures and corrosion. It was found that there is a temperature peak for forming these deposits and that the problem disappears below 254°C, 4.14 MPa and above 301°C, 8.62 MPa. A recovery boiler operating at very high sulfidity can minimize SO_3 formation by using <3% O_2, just enough to prevent H_2S formation. Boiler tube banks are not normally rebuilt.

The balance of H_2S and SO_2 in liquor combustion depends entirely on the green liquor sulfidity. The transition from H_2S to SO_2 emission in the flue gas is a function of O_2 admission and only small changes of O_2 are required to produce it. Therefore effects on the heat economy of the boiler are negligible.

Flue gas scrubbing is mandatory for clean, economical continuous sulfite operation. Modern flue gas scrubbers feature low investment, operating and maintenance cost, reliability, and automatic control. They form an integrated part of the sulfite liquor making system, recovering ⅓ of the sulfur requirements in bisulfite as well as AS-AQ operations, while providing effective air quality control and freeing furnace operations from constraints. Emissions at Valkeakoski are small, for both SO_2 and H_2S, and for particulates. Note that the flue gas scrubber serving the sulfite department has corrected the odor emission of the kraft recovery boiler.

Thiosulfate formation in the scrubber can be controlled at low levels by maintaining high concentration, high temperature, low O_2 concentration, and pH in the range of 6.5 to 8.5. The scrubber can be equipped with a low heat recovery stage. The used scrubbing solution containing carbonate and some sulfite is returned to the absorption tower of the liquor-making plant where it is sulfited with SO_2 obtained from burning the H_2S and make-up sulfur.

Finally, the furnace smelt enters the dissolving tank and forms the common green liquor. Air quality from the smelt dissolver is controlled effectively by countercurrent spray washing of the exhaust with weak white liquor in a special chamber (not shown) located atop the dissolving tank. In new installations, a connection is made to the flue gas scrubber as an extra precaution. The green liquor has a concentration of 175 g Na_2O/L; the sulfidity was 38-40% in bisulfite

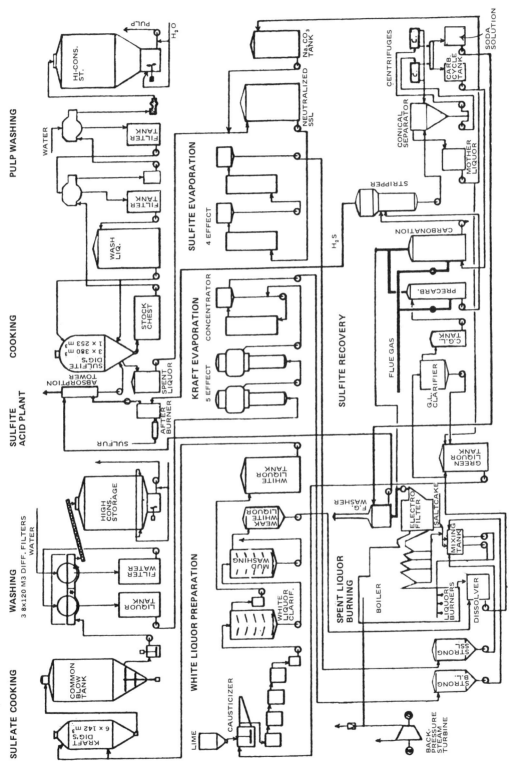

Fig. 289. Flow sheet for UPM-Valkeakoski integrated kraft-sulfite recovery.

operations and is now 40 to 45% in the AS-AQ operation.

The sulfite conversion unit operation begins with green liquor clarification and storage. Clarification is more efficient for sulfite than for kraft operations. The dregs are not washed but are added to the green liquor flow to kraft liquor preparation. The flow of green liquor to the sulfite unit is about ⅓ of total and is regulated according to analysis and pulp production rate. NaHS from the precarbonator (uses 2.5% of flue gas) and $NaHCO_3$ (uses about 10% of flue gas) enter the stripper where a suitable combination of steam heating, mixed phase reactions, vacuum application, and evaporation/concentration produces a mixture of 90% H_2O (steam) and 10% H_2S at the top, and a thin carbonate slurry at the bottom. Application of vacuum to the H_2S transfer line serves a triple purpose: it reduces the temperature in the stripper thus counteracting unwanted decomposition of $NaHCO_3$; removes H_2S from the reaction; and prevents any H_2S emission in case of line leaks. Also in the H_2S line is a surface condenser which removes the water and leaves the H_2S gas 90-95% pure for direct burning. The condensate is returned to the smelt dissolving tank.

The thin slurry in the stripper column has 2% solids content and is circulated by two large impeller type circulation pumps (Klaus, Germany). A more recent Tampella system (installed in Russia) uses only one circulation pump. Carbonate crystals from the slurry are allowed to settle in the separator and are pumped at 30% consistency (screw slurry pump, 200 RPM, Netsch, Germany) to the vertical centrifuges (Siebtechnik, Germany). The product is 99% pure, odorless, industrial grade carbonate. A major portion is recycled for sulfite liquor preparation via the flue gas scrubber and SO_2 absorption, a smaller portion is recycled to green liquor carbonation. The amount of mother liquor from a 500 t/d operation is less than 5 m^3/h which is returned to the green liquor flow to the kraft line. Thiosulfate and sulfate are concentrated in the mother liquor and are easily processed via evaporation and reduction in the furnace. Control of their build-up in the system is therefore an inherent feature of the sulfite recovery unit with carbonate crystallization. However, crystallization means additional investment and complication of the system, and it can be avoided in sulfite operations which do not require high purity of the cooking liquor and in operations which are not sensitive to thiosulfate in the liquor, viz. NSSC and alkaline sulfite.

Burning H_2S gas and make-up liquid sulfur is accomplished in equipment designed for this purpose, considering proper sizing, H_2S gas distribution, and both simultaneous and separate burning. The use of common sulfur burners as used in conventional sulfite operations is not recommended. Combustion temperature is 1400-1500°C with 5% O_2 in the flue gas. A boiler inserted after the burner produces 6 t/h superheated steam at 7 MPa. A quench cooler is used for washing out SO_3 and the effluent of 2 m^3/h is returned to the recovery process. The formation of SO_3 is on the same level as in common burners for elemental SO_2.

Chemical make-up requirements for AS-AQ production are 45 kg/adtp as salt cake, added in the form of caustic soda (25.4 kg/adt) and sulfur (10 kg/adt) to the sulfite recovery cycle. Make-up previously required for bisulfite production was somewhat lower in sodium (31 kg Na_2CO_3 equivalent to 23.4 kg caustic soda), and higher in sulfur (12 kg/adt). Recovery efficiency in the TRP system is 95% for sodium and 99% for sulfur with flue gas washing (70% without). For monitoring the TRP operation, analyses of active sodium and sulfur components are routinely performed on green liquor, precarbonated liquor, stripped liquor, bicarbonate solution, sulfite solution, and other samples.

There was practically no change in steam data when the mill was converted from bisulfite to AS-AQ:

Recovery boiler
Capacity with spent liquor, t/h	120
Capacity with oil addition, t/h	170
(Spent sulfite liquor burning, t/h)	(44)

Steam usage
Kraft evaporation	32
Sulfite evaporation	12
H_2S stripping	14
Remainder of the mill	62

In general, this sulfite recovery plant can serve as an example of high technological design and engineering standards. It is housed next to the new furnace boiler with the common control room between. The dimension of the plant was dicated by the size of the boiler feed water tank in the upper part of the structure; it could easily have been built in half of the space used. Components are arranged in a clear and functional manner with good accessibility for inspection, sampling, and maintenance. Stainless steel has been used throughout.

A comparison of the significant differences between the earlier bisulfite process and the new AS-AQ process and their products is afforded from the data in Table 70. At the higher pH level of AS-AQ the alkali charge is greatly increased, but not the sulfur charge. Due to the higher efficiency of chemical recovery achieved in this process, high chemical charge is of little consequence economically. The comparison is based on identical temperature of the cooks and higher yield of the AS-AQ pulp, in which case AS-AQ cooking time is more than twice and the Kappa No. is 20% lower than for bisulfite. The AS-AQ

Table 70. Comparison of sodium bisulfite and AS-AQ production data

	Bisulfite	AS-AQ
Na_2O charge on o.d. wood, %	5	16
Molar ratio of sulfur to Na_2O	2	0.85
Sulfur charge on o.d. wood, %	5.2	7.3
AQ charge on o.d. wood, %	—	0.1
Maximum temperature, °C	165	165
Cooking time at temperature, h	2	4.5
Screened yield, %	52	55
Kappa No.	50	40
Unbleached brightness, ISO	64-68	47-52
After single stage hypo, ISO	—	60-63
Tensile index, Nm/g	70	110
Breaking length, km	7.1	11.2
Tear index, mNm^2/g	8	9.5
Burst index, $kPam^2/g$	6	8.5
Evaporated spent liquor solids, %	55-57	68
Bomb heat value, MJ/kg	15.6	12.8
Effective heat value, MJ/kg	7.8	7.9

pulp is obtained at a much higher brightness level than kraft pulp and is used directly for news applications or, when bleached by a single hypostage to 75% brightness, for printing grades. The final product excels in all strength properties which greatly expands the range of application, including news reinforcement pulp similar to kraft pulp, semi-bleached pulp for printing grades, and market pulp.

The Valkeakoski experience confirms on the industrial level the flexibility of both the sulfite process and of the sodium base recovery system.

The data from production runs of supercalendered newsprint (Table 71) and machine glazed paper (Table 72) at the Kaipola and Valkeakoski mills show the excellent strength properties of AS-AQ [22].

D. AMMONIUM BASE BISULFITE PULP FOR HIGH QUALITY WOOD-FREE, BLEACHED PRINTING AND BUSINESS PAPERS

A description of the up-to-date ammonium base operation of an integrated fine paper mill serves to represent ten mills in the United States and one mill in Canada using this base. The decision for choosing ammonia as a base is made on premises of both convenience and economy for producing acid bisulfite and/or bisulfite

Table 71. Kaipola Trial, 1981, Super Calendar News, 45 g/m²

Chemical Pulp, %	11.5 SBSP + SBK[*]	10 SAP (AS-AQ)	9 SBK
Groundwood, %	39.5	36	32
TMP, %	49.0	54	59
Paper Strength			
Tensile, kN/m, MD	1.85	2.20	2.06
Tensile, kN/m, CD	0.68	0.72	0.75
Stretch, %, MD	1.0	1.0	1.1
Stretch, %, CD	2.3	2.0	1.9
Tear, mN, MD	167	164	165
Tear, mN, CD	230	243	241
Burst, $kPam^2/g$	1.48	1.68	1.45
Opacity, %	93.6	93.0	95.5
Light Scatt. Coeff.	52.6	55.2	53.7

[*] 21% Sodium bisulphite pulp + 75% semi-bleached kraft (65% brightness)

Table 72. Valkeakoski Trial, 1981, Machine Glazed Paper, PM 2
Pulps: Unbleached kraft 430 CSF
Unbleached SAP (AS-AQ) 520 CSF

	Kraft (before)	SAP	Kraft (after)
Grammage, g/m²	51.0	51.5	51.5
Density, kg/m³	650	710	680
Porosity, s/100 mL	40.1	48.0	56
Tensile Index, Nm/g, MD	83	106	81
Tensile Index, Nm/g, CD	47	55	46
Stretch, %, MD	1.9	1.6	2.0
Stretch, %, CD	3.9	2.7	3.2
Tear Index, mNm^2/g, MD	10.0	9.3	10.0
Tear Index, mNm^2/g, CD	11.3	12.3	11.0
Burst Index, $kPam^2/g$	3.8	5.3	3.8

pulp. Since the base is not inexpensive and is commonly not recovered as such, the proximity of the mill location to an ammonia synthesis plant is usually decisive.

The integrated sulfite mill of Finch, Pruyn & Co., Inc. is situated in Glens Falls, New York. About 400 t/d of ammonium bisulfite pulp from hardwood and/or softwood is produced in a continuous digester installed in 1969. The pulp is bleached in a CEH sequence and fine papers of different specifications are produced on four paper machines.

Spent cooking liquor is recovered, evaporated, and burnt to produce process steam. Sulfur dioxide is absorbed in dilute ammonia solution and returned to cooking. Ammonium compounds recovered in the effluent from the atmospheric mist eliminator supply the necessary amount of nutrient for the activated sludge unit for the secondary depollution treatment of mill effluent.

1. The wood furnish

The fiber source is predominately roundwood from local forests. Because bark contamination gives high dirt levels in the sulfite process, only low amounts of sawmill chips are purchased. Wood is received unbarked and is used fresh. It is debarked in drum barkers, by dry barking most

of the year but with a wet barking drum before the dry drum in winter when bark removal becomes difficult.

The clean wood is chipped in a conventional chipper to a length between 16 and 19 mm and screened through conventional chip screens to remove sawdust and fines below 8 mm, and oversize chips above 25 mm; oversize chips are reduced in size in a hammermill and returned to the chip screens. Bark and splinters from debarking plus sawdust and fines from the chip screens go to a wood burning boiler to produce steam. A small amount of oil (0 to 20% of total fuel) is burned in this boiler to give control of steam flow and pressure.

The pulp mill is capable of cooking and bleaching pure hardwood or pure softwood furnish or any mixture of these woods. Some SW species, i.e. red pine, may cause problems due to high pitch (extractables) content.

2. Cooking

Accepted chips are sent to a chip bin and hopper and then enter the continuous digester system through a low pressure feeder (Fig. 290). From the steaming vessel chips drop into a higher pressure feeder where cooking liquor from the recovery plant is added. Based on bone dry wood, 18 to 26% SO_2 (4.8 to 6.9% NH_3) is

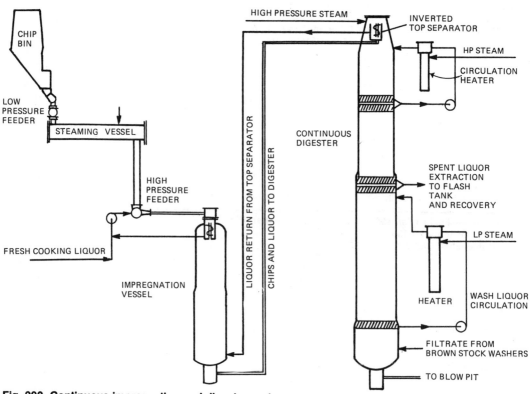

Fig. 290. Continuous impregnation and digester system.

added as ammonium bisulfite with a pH range between 3.0 and 5.0, typically 3.5. The total amount of liquid, including chip moisture, is adjusted to give a liquid to bone dry wood ratio in the range of 3.0 to 4.5.

The chip-liquor mix goes to the top of a down-flow impregnation vessel where a separator removes excess liquor to be re-cycled to the high pressure feeder. After 20-30 min residence time in this vessel the chips are diluted at the bottom and pumped to the top of the continuous digester (see Chap. V, B, 2).

A top separator in the digester removes excess liquor which is returned to the dilution zone of the impregnation vessel. The chips drop into the vapor space at the top of the digester where the temperature of the charge is increased to the cooking temperature of about 150°C with direct steam. The chips then pass into the liquid phase and through the cooking zone with about 2 h residence time. The actual cooking temperature is controlled by extracting liquor at the periphery of the digester, passing it through a heat exchanger and re-injecting it into the center of the chip mass (top circulation).

At the end of the cooking zone near the middle of the digester "red liquor" containing dissolved wood solids (primarily lignin) and spent cooking chemicals are extracted through screens and sent to recovery. The temperature of the chip mass is then quenched with preheated wash liquid and the remaining bottom section of the digester is devoted to diffusion washing. Wash liquor is extracted near the bottom, passed through the heat exchanger and re-injected near the top of the washing zone. Near the bottom of the digester, filtrate from brown stock washing is introduced to cool and dilute the chip mass before it is blown. An excess of filtrate is introduced to maintain a controlled upflow of wash liquor from the bottom of the digester to the spent liquor extraction level.

Pulp processing

Blown pulp is then washed on tandem vacuum washers, screened in a four-stage counter flow system and thickened before going to the bleach plant. Filtrate flows countercurrent to the pulp flow from the thickener, back through the two brown stock washers and the diffusion wash portion of the continuous digester, to the extraction level, giving, in effect, a four-stage countercurrent wash. The brown stock then enters the bleach plant where it is chlorinated, extracted, and bleached to 84-88% brightness with hypochlorite in a conventional system with three towers and three vacuum washers. After the final wash, the bleached pulp is stored at about 10% consistency in one of two 600 t storage tanks, ready for paper mill use.

3. Processing of spent cooking liquor

The spent liquor extracted from the digester has a temperature of about 150°C and contains between 12 and 18% solids depending on operating conditions. After leaving the digester, it is expanded in two flash tanks in series, the first at 103 kPa with the gases piped to the second effect of the five-effect evaporator, the second to atmospheric with the vapors passing a heat exchanger.

The evaporators are of the spray film type with the spent liquor recirculated over the outside of horizontal steam tubes (see Chap. IX, B, 2). The present arrangement is 6-body, 5-effect system. Scaling is controlled by filtration of extracted spent liquor and by washing the evaporator tube banks with acid condensate. About one third of the SO_2 applied leaves the evaporator system with non-condensibles. This is sent back to the second stage of the SO_2 absorption system.

Concentrated liquor at about 55% dry solids content and a wet heat value of 9350 MJ/kg is fired in three Combustion Engineering VP type boilers with Loddby muffle burners (see Chap. IX, C, 1) attached. Each recovery boiler is rated at 126 kg liquor solids per day. Gases from the boilers pass through air heaters, an ID fan, the primary SO_2 absorbers, a demister, and a final ID fan. Due to the low ash content of ammonium base spent liquor, the Loddby burners require cleaning only once a year.

4. Sulfur dioxide absorption and fresh liquor preparation

High SO_2 recovery efficiency is obtained with a countercurrent three-stage absorption system [19] as shown in Fig. 291. The absorbent ammonium hydroxide solution is prepared on site from liquid ammonia at a concentration of about 18%. There are three primary absorption towers, one for each recovery boiler. The two older recovery boilers feed tray-type absorption towers with three sections, each with an external heat exchanger, as shown. In the bottom section, water is used for cooling whereas in the two top sections ammonia solution is fed into the top and weak cooking liquor is withdrawn from the section below and pumped to the secondary absorber. The third and newer recovery boiler has an absorption system with a similar function but uses a separate tower for cooling the gases and a packed tower for absorption.

The primary absorber also receives the weak gas from the secondary absorber, which handles the SO_2 stream of highest strength (one third of

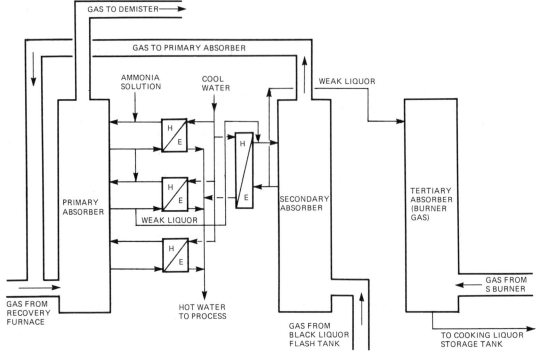

Fig. 291. Absorption and liquor making system.

SO_2 to be recovered) from the digester flash gases. The weak liquor is partially circulated through an external heat exchanger and then pumped to a tertiary absorption tower where SO_2 from the spray-type sulfur burner replaces losses, and brings the cooking liquor to final strength. If necessary, final adjustments of the cooking liquor pH are made in the storage tank following the absorption system. Basically this liquor making system uses control of SO_2 concentration for liquor strength, and adjustments with ammonia solution for final pH control.

The recovery efficiency in the absorption system is better than 99%. About 140 t/d of SO_2 are recovered and re-used while the total make-up from the burner is about 20 t SO_2 per day.

5. Depollution

In 1966 the company was producing NSSC pulp and had entered into an agreement with the Health Department of New York State to abate pollution at its site on the upper Hudson River. With the change to ammonium base bisulfite and implementation of continuous cooking, red liquor evaporation, recovery boilers, and SO_2 absorption, the problem of SO_2 emissions was soon brought under control, averaging less than 60 ppm which is well within the state standards.

Soon, however, an atmospheric haze problem peculiar to ammonium base mill emissions had to be faced. Under certain ambient air conditions a bluish haze develops which is found to be due to invisible sub-micron particles in a vaporous environment which reflect certain wavelengths in the violet light region of the spectrum. Since other companies with ammonium base facilities were also affected, an agreement was reached to underwrite a test program of available equipment at the Glens Falls plant. Of four systems found to be capable of eliminating the haze-creating particulates, a fiber-bed mist eliminator was found to be the simplest and most practical [18].

6. The demister (Fig. 292)

The heart of the fiber-bed mist eliminator is the element ("candle") which consists of two concentric cylindrical 316L SS screens with glass fibers filling the annular space. Each element measures about 3 m in length and 45 cm in diameter with a fiber bed thickness of 7.5 cm. This installation consists of two units of 28 vertical elements each in seven rows of four and housed in a stainless steel box 5.5 m wide, 10 m long and 6 m high. Pressure drop across the elements is about 20 cm water column. Ring-showers surround the top of each element to rinse away any insoluble particles, such as fuel oil residues from start-up firing of the recovery boilers. Gases from the three primary absorption towers enter a unit at its base and are pulled through the fiber bed elements by means of a variable-speed fan installed on a platform above the box.

Sub-micron mist particles such as ammonium sulfate and other substances are collected on the fiber surface, are irrigated down through the bed, and drain off continuously at the bottom into the sewer and on into the activated sludge treatment plant where they become nutrients. The efficiency of the system is 98% and the discharge load of dissolved solids 0.28-0.40 kg/h. The target of "no visible emission" had been reached. The gas from the demister at less than 60 ppm SO_2 is piped directly to the boiler house stack.

Regarding pollution of the small river by liquid mill effluents, the stipulation to the state

authority was met by spent liquor evaporation and burning of the solids. A further considerable improvement was the inclusion of waste water from the evaporator and absorption cooling systems, together with chlorination and extraction bleach room filtrates in the effluent stream to primary and secondary treatment.

There is no loss of ammonia. All ammonia applied to the process is utilized for SO_2 recovery, for the cooking reaction, for steam generation, and secondary effluent depollution treatment.

E. MILL OPERATIONS WITH HIGH YIELD PULPS [20-23]

A number of newsprint mills throughout the world and particularly in Eastern Canada were converted from low yield Ca-base acid sulfite to HY sodium or Mg based bisulfite pulping. The operation of a typical sodium and a typical magnesium base plant are described below.

1. Consolidated-Bathurst Inc.-Port Alfred mill [20a, 26] (HY sodium base bisulfite pulps)

This mill produces some 180 t/d of 70-75% yield sodium bisulfite pulp for use in standard and offset grade newsprint.

(a) Soda ash handling (Fig. 293)

Soda ash is stored in a tower and fed through a screw conveyor, to the dissolver. A level controller passes the 30% solution to a storage tank. From this tank, the solution is diluted to the desired operating concentration, and fed to the diluted soda ash tank. (The dissolver is operated about 8 h per day since the 91 000 L concentrated solution tank can supply about 32 h of production). Depending on market supply conditions, 50% caustic solution can be used instead of the soda ash handled at this mill.

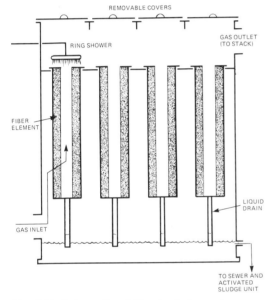

Fig. 292. Schematic of fiber bed element arrangement.

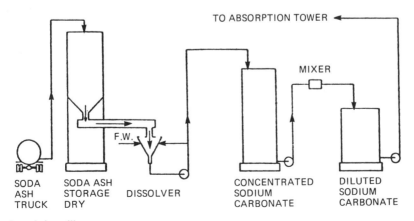

Fig. 293. Soda ash handling.

(b) Raw liquor preparation (Fig. 294)

After cooling, the SO_2 gas from the burner is sent to an absorption tower for reaction with the soda ash (sodium carbonate). A pH meter controls the amount of diluted soda ash solution fed to the tower. The tower is operated at atmospheric pressure, equipped with liquor recirculation and a vent at the top to exhaust gases. Under normal operating conditions, no SO_2 is released to the atmosphere.

(c) Cooking (Fig. 295)

The raw liquor is pumped first into an intermediary tank for the preparation of the cooking liquor of the desired concentration and pH prior to transfer into the digester. The side relief is taken from the digester circulating line and sent to this tank to a controlled level thus maintaining the desired amounts of liquor in the digester. The two vessels are kept under slight pressure but in parallel so that no pressure changes occur when transferring liquor. An automatic valve keeps the pressure at the set point.

Injection of large amounts of hot spent liquor was found necessary to clean blow the digester. This is accomplished by passing the liquor at a rate of 3000 gpm (227 L/s) during the blow through two top nozzles installed on the circulation system.

(d) Data on a typical mill HY bisulfite softwood cook [26]

Batch digester size = 8400 ft³ (238 m³) equipped with forced circulation, direct heating.

Weight of O.D. chip = 71 400 lb (32 t) — Spruce and Balsam blend

Amount of liquor introduced to fill the digester — 335 000 lb (152 t)

Amount of liquor introduced after the pressure impregnation = 365.000 lb (166 t)

Amount of liquor withdrawn via strainer = 180 000 lb (82 t)

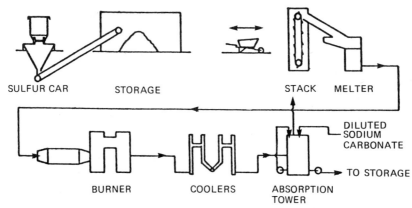

Fig. 294. Raw liquor preparation.

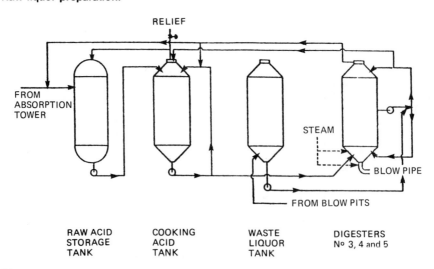

Fig. 295. Digesters.

Amount of liquor left in the digester for cooking
    = 185 000 lb (84 t)
Amount of chemicals applied as $NaHSO_3$ =
    15.9% based on O.D. wood
Amount of chemicals consumed as $NaHSO_3$ =
    11.1% based on O.D. wood
Liquor to wood ratio (including condensed
    steam) at maximum temperature = 4.62/1
Average digester pulp yield = 73%

(e) Refining (Fig. 296)

HY chip refining is carried out in two stages

with interstage pulp washing. At the beginning,
refining in the first stage was conducted at lower
consistencies of 8-10% by pumping the chips
through the refiners. Lately, a drainer was
installed to raise the chip consistency to approx.
20+% for improved refining.

(f) Pulp screening (Fig. 297)

This plant features double screening of the
refined stock, i.e. all the accepts from the
primary stage are fed to the secondary stage. In
addition to better efficiency, this system offers

Table 73. Liquor Analysis

| SAMPLE | pH | CONCENTRATION, SO_2 | | |
		TOTAL	FREE	COMBINED
Low pressure accumulator	4.15	2.98	1.52	1.46
After digester filling and pressure impregnation	3.85	2.48	1.26	1.22
At max. temp of 145 °C	3.25	1.54	0.80	0.74
After 2 h at 145 °C	3.30	0.84	0.52	0.32
After 4 h at 145 °C (before the blow)	3.25	0.64	0.42	0.22

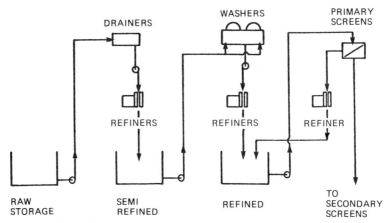

Fig. 296. High-yield sulphite refining.

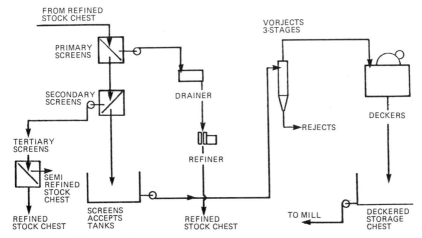

Fig. 297. High-yield sulphite screening.

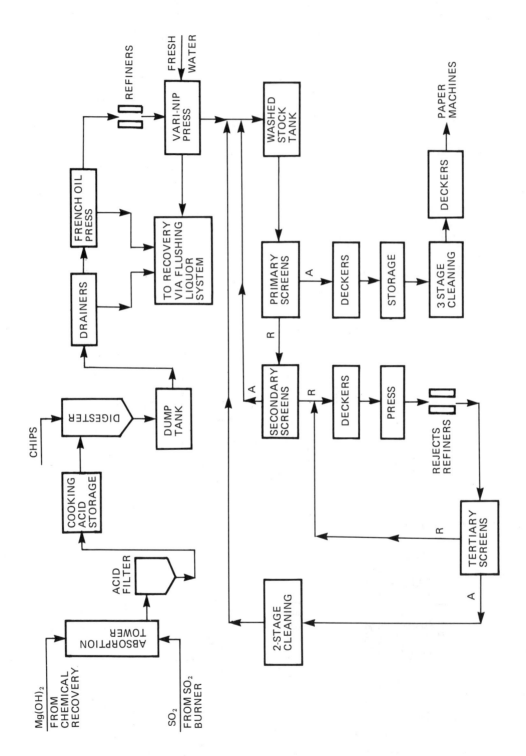

Fig. 298. Great Northern-Millinocket magnesium bisulphite process.

good protection against contamination of the pulp in case of breakdown. The rejects from the primary screens are sent to a separate "rejects" refiner and the rejects from the secondary screens are pumped to a "stripping" screen, where the accepts are fed to the refined stock chest and the rejects to the semi-refined stock chest for further refining. Conventional three-stage centrifugal cleaning and decker thickening operations are practiced.

2. Great Northern, Millinocket [23-25] (high-yield magnesium bisulfite pulping operation)

Introduction

One of the users of the magnefite process is Great Northern Paper Company at its Millinocket mill. The low yield Mg base sulfite pulps were initially supplying two mills: Millinocket and East Millinocket, 8 miles away.

Expansion of the mill in early 70's with a new paper machine created an extra load on the existing pulping and chemical recovery facilities. This led to the decision of having a combination of low yield and high yield (75%) magnesium bisulfite pulps to suit mill production. In this way the spent sulfite liquor from the HY operation can be economically recovered and burned after evaporation with the more concentrated liquor from the low yield operation.

The following is a description of the high yield pulping process.

Process description (Fig. 298)

a) *Acid making and recovery*

The acid making operation is basically similar to that for sodium base bisulfite i.e. an absorption tower is used to react recovered Mg (OH)₂ slurry with the gaseous SO_2 from the sulfur burner. In the absorption tower, the acid is fortified to 5.5% SO_2 and the pH controlled to 3.7 (this low pH allows production of one cooking liquor for both yield levels). Since the magnesium hydroxide also absorbs SO_2 from the recovery boiler flue gas, decantation and sand filtration are practiced to remove any traces of carbon.

b) *Digester*

The 198 m³ carbon brick lined digesters are filled with a mixture of 50/50 spruce/balsam fir,

using steam packing and presteaming. The volume prepared of cooking liquor at 5.5% concentration (corresponding to 12% SO_2 on wood) is insufficient for proper liquor circulation. Therefore flashing liquor is added as "make-up" before steaming.

The total cooking time for the production of 75% yield pulp is 2.5 h at maximum temperature of 155 °C (760 kPa). At the end of the cook, approximately 75 000 L of liquor are withdrawn and replaced by flushing liquor, which lowers the digester temperature to 115 °C.

c) *Pulp washing and refining*

The cooked chips are pumped out of a well-agitated dump tank to a rotary drainer, where the consistency is raised to 10-14%. This is followed by pressing in a French Oil press operating with an outlet consistency of 30%. The liquors extracted from the drainer and press are sent to the chemical recovery by the flushing liquor system. The solids content of the high yield spent liquors is approximately 10% as compared to 14% for the low yield spent liquor.

The three 1865 kW (2500 HP) refiners reduce pulp freeness to 700 CS in one stage with energy application of 1.2 GJ/t. The refined pulp, after dilution to 3% consistency, is pumped to an Impco twin-roll vari-nip press which serves as a washer by increasing pulp consistency to 38-40%. In this case, the amount of dissolved solids left with the pulp is approximately 33 kg/t.

d) *Screening and cleaning*

After dilution with mill white water, the pulp is pumped to the primary screens. The accept is centricleaned in three stages at the paper mill.

The secondary screens operate in cascade with the primary ones. Rejects from the secondary screens, after thickening to 20% consistency, are treated in a 1865 kW (2500 HP) reject refiner. The refined rejects are separately screened and centricleaned prior to recycling back to the primary screens.

Mill results showed that newsprint containing high yield magnesium sulfite pulp has higher bulk and tear but lower burst, tensile strength, and brightness than the corresponding newsprint containing low yield sulfite. In both cases, pressroom performance was competitive.

REFERENCES

1. T.M. Fitzsimmons, Nova Scotia Pulp Limited is Canada's newest market pulp producer. *Can. Pulp Paper Ind.* 16, 17-50 (Jan. 1963).
2. Johnston, D., Calder, K., Ho, L., Stora recovery process at NSFI. *Pulp Paper Can.* 80, T184-187 (June 1979).
3. Anon, N.S. Forest Industries new high yield sulphite pulp mill. *Paper Trade J.* 102, 22-26 (July 23, 1973).
4. Skory, L.D., Fraser Inc. reaps benefits of $91.5 million modernization. *Can. Pulp Paper Ind.*, 20-24 (Sept. 1979).
5. Evans, J.C.W., Fraser's Edmundston mill is world's largest magnesium base sulfite mill. *Pulp Paper,* 96-100 (Oct. 1979).
6. Evans, J.C.W., Sticking with sulfite: Fraser goes against trend toward kraft. *Pulp Paper,* 68-71 (Mar. 1982).
7. Stevens, F., Edmundston buys biggest SF-Fläkt magnesium system. *Pulp Paper Can.* 78, 15-17 (Feb. 1977).
8. Hohol, R., Atholville mill expansion nears completion. *Pulp Paper J.,* 18-19 (Sept. 1983).
9. Evans, J.C.W., Fraser completes major upgrade of its second sulfite pulp mill.
10. Williamson, P.N., Fraser's Atholville mill hits stride with sights on wider market share. *Pulp Paper Can.* 85, 24-29 (Apr. 1984).
11. Lange, H.B., Bick, M.P., McIlroy, R.A., and Foster, E.P., Chemical recovery and sulfur control using the Tampella recovery process. *Tappi* 58, 122-126 (Feb. 1975).
12. Karhola, A.K., Kejo, R.K., Heiki, A.K., and Saiha, E.V., Tampella reveals new recovery process for sodium base liquors. *Paper Trade J.* 151, 65-67 (Sept. 18, 1967).
13. Anon, Kemi starts multi-purpose Tampella recovery process. *Pulp Paper Intern.* 10, 32-33 (Aug. 1968).
14. Gullichsen, J., Saiha, E., and Westerberg, E.N., Recovery of sodium-base pulping chemicals by bicarbonation and crystallization. *Tappi* 51, 395-400 (Sept. 1968).
15. Ahava, T., Tutipaa (Kemi Oy), Romantschuk, H., Saiha, E. (Tampella Oy). Combined kraft and sulfite recovery with the Tampella process. *Paper Trade J.* 22-24 (Aug. 3, 1970).
16. Tanner, T., and Clement, J.L., Mill applications of the Tampella recovery process. *Tappi* 56, 63-66 (Sept. 1973).
17. Pusa, R., Neutral and alkaline anthraquinone sulfite pulping process. *Paperi ja Puu* 63, 663-666 (11, 1981).
18. Guerrier, J.J., Cooperative effort solves small mill's air problem. *Pulp & Paper* 38, 62-64 (Mar. 1974).
19. Schultz, W.C., SO₂ Recovery. Proceedings, 1982 International Sulfite Pulping Conference, pp. 251-52.
20. Refer to papers delivered at a Panel Presentation on "H.Y. Bisulfite Pulping" sponsored by the Sulfite Committee during the CPPA Annual Meeting.
    a) Meunier, G.A., re. Consolidated-Bathurst Inc. — Port Alfred Mill, Québec.
    b) Mosher, R.D. — re. Ontario-Minnesota — P. & P. Co. Ltd., Kenora Mill, Ont.
    c) Bieweg, Ward Allan, re. Domtar, Trois-Rivières Mill, Québec.
    d) Laberge, H.G., re. The Price Company, Alma, Québec.
    Convention issue: *Pulp & Paper Mag. Can.* (1971).
21. "Nova Scotia Forest Industries, New High Yield Sulfite Pulp Mill", reported in *Paper Trade J.,* pp.22-26, July 23/73.
22. Hassinen, Ilpo, "Continuous Digester for Bisulfite Pulp is Success at Finnish Mill", *Pulp & Paper,* Vol. 49, No. 12, pp. 92-94 (Nov. 1975).
23. Fowler, F., "More Production from Part Conversion of Low to High Yield Mg" *Pulp Paper Can.,* 79 (8): T260 (Aug. 78).
24. Keef, R.C., "Mg Bisulfite Recovery Start-up" *Tappi* 54 (4); 564 (April 71).
25. Binotta, A., "Effect of Chip Quality on High Yield Bisulfite Pulp". Data presented at 1980 CPPA Sulfite committee.
26. Ayroud, A.M., Internal reports covering some 20 years of unpublished studies. Research Centre, Consolidated-Bathurst Inc., Grand Mère, Québec, Canada.
27. Virkola, N.E., 1982 International Sulfite Pulping Conference, Toronto.
28. Sulfite Committee of the Technical Section of the Canadian Pulp & Paper Association.

APPENDICES

1. LIST OF SULFITE MILLS

U.S. Mills

Company	City	ST	Final Product	T/day	Wood Species	Type	Base	Recovery	Chem Product
Alaska Lumber & Pulp Co., Inc.	Sitka	AK	dissolving pulp	600	SW	acid sulfite	Mg	B&W	
Louisiana-Pacific Corp.	Ketchikan	AK	dissolving & paper pulp	600	SW	acid sulfite	Mg	B&W	
Mead Corporation	Stevenson	AL	corrugating medium	860	HW	NSSC	Na	SCA Billerud	
Louisiana-Pacific Corp.	Antioch	CA	corrugating medium	175	HW	NSSC	Na	B&W, C-E	Cross recovery
Container Corp of America	Fernandina Beach	FL	corrugating medium	350	HW	green liq	Na	B&W	Cross recovery
ITT Rayonier Inc.	Fernandina Beach	FL	dissolving pulp	450	SW	acid sulfite	NH3	B&W	
Great Southern Paper Co.	Cedar Springs	GA	corrugating medium	400	HW	NSSC	Na	B&W	Cross recovery
Union Camp Corp.	Savannah	GA	corrugating medium	300	HW	green liq	Na	C-E	Cross recovery
Celotex Corp.	Dubuque	IA	insulating board	250	HW	NSSC		City, Sewer	
Consolidated Packaged Co.	Ft. Madison	IA	corrugating medium	130	HW	NSSC	Na	Copeland	
Weston Paper & Mfg. Co.	Terre Haute	IN	corrugating medium	425	HW	non-sulfur	Na	Dorr Oliver	
Willamette Industries, Inc.	Hawesville	KY	corrugating medium	300	HW	green liq	Na	Copeland	
Continental Forest Industries	Hodge	LA	corrugating medium	250	HW	NSSC	Na	C-E	Cross recovery
Crown Zellerbach	Bogalusa	LA	corrugating medium	330	HW	green liq	Na	CE & B&W	Cross recovery
International Paper Co.	Bastrop	LA	corrugating medium	492	HW	NSSC	Na	Yes	Cross recovery
Manville Forest Products	West Monroe	LA	corrugating medium	320	HW	green liq	Na	Yes	Cross recovery
Great Northern Paper Co.	Millinocket	ME	printing paper	660	SW	bisulfite	Mg	B&W	
Champion International Corp.	Ontonagon	MI	corrugating medium	500	HW	non-sulfur	Na	Zimpro Wetox	
Menasha Corp.	Otsego	MI	corrugating medium	300	HW	non-sulfur	Na	Dorr Oliver	
Packaging Corp of America	Filer City	MI	corrugating medium	620	HW	non-sulfur	Na	Copeland	
Potlatch Corp.	Cloquet	MN	business paper	180	HW	acid sulfite	Ca	Sewer	
Brown Company	Berlin	NH	corrugating medium	210	HW	NSSC	Na	B&W	
Groveton Paper Co.	Groveton	NH	corrugating medium	120	HW	non-sulfur	Na	Kiln	
Finch, Pruyn & Company	Glen Falls	NY	business paper	380	HW	bisulfite	NH3	Loddby	
Georgia-Pacific Corp.	Lyons Falls	NY	business paper	120	HW	bisulfite	Na	Lagoon	
Georgia-Pacific Corp.	Plattsburg	NY	tissue	100	HW	bisulfite	Na	Municipal	

U.S. Mills

Company	City	ST	Final Product	T/day	Wood Species	Type	Base	Recovery	Chem Product
Container Corp. of America	Circleville	OH	corrugating medium	200	HW	NSSC	Na	Copeland	
Stone Container Corp.	Coshocton	OH	corrugating medium	650	HW	non-sulfur	Na	Copeland	
Weyerhaeuser Company	Valliant	OK	corrugating medium	500	HW	green liq	Na	C-E	Cross recovery
C Z Corporation (closed)	Lebanon	OR	card stock	105	SW	NSSC	Na	Yes	lignosulfonate
Georgia Pacific	Toledo	OR	corrugating medium	250	HW	green liq	Na	C-E	Cross recovery
Weyerhaeuser West Coast, Inc.	North Bend	OR	corrugating medium	300	HW	NSSC	Na	Dorr Oliver	
Publishers Paper Co.	Newberg	OR	newsprint	250	SW	bisulfite	Mg	B&W	Venturi scrubber closed 1984
Publishers Paper Company (closed)	Oregon City	OR	newsprint	100	SW	bisulfite	Mg	B&W	
Willamette Industries	Albany	OR	corrugating medium	250	HW	NSSC	Na	B&W	
Celotex Corp.	Sunbury	PA	insulation	90	HW	NSSC		City Sewer	
Proctor & Gamble Paper Products	Mehoopany	PA	tissue	400	HW	acid sulfite	NH3		lignosulfonate
International Paper Co.	Georgetown	SC	corrugating medium	100	HW	NSSC	Na	Yes	Cross recovery
Sonoco Products Co.	Hartsville	SC	cylinder board	280	HW	non-sulfur	Na	Kiln	
Harriman Paperboard Corp.	Harriman	TN	corrugating medium	185	HW	NSSC		City Sewer	
Inland Container Corp.	New Johnsonville	TN	corrugating medium	400	HW	NSSC	NH3	B&W	also conc liq.
Owens-Illinois	Big Island	VA	corrugating medium	575	HW	non-sulfur	Na	B&W	Cross recovery
Virginia Fibre	Riverville	VA	corrugating medium	600	HW	non-sulfur	Na	B&W	
Westvaco Corp.	Covington	VA	corrugating medium	300	HW	non-sulfur	Na	B&W	Cross recovery
Boise Cascade	Wallula	WA	corrugating medium	280	SW	NSSC	Na	B&W	Cross recovery
C Z Corporation	Camas	WA	tissue	440	SW	bisulfite	Mg	B&W	
Georgia-Pacific Corp.	Bellingham	WA	tissue	500	SW	acid sulfite	Ca	No	alcohol 15 000 USG/day; lignin
Georgia-Pacific Corp.	Bellingham	WA	tissue	60	HW	NSSC	Na	No	550T/day
ITT Rayonier Inc.	Port Angeles	WA	dissolving	510	SW	acid sulfite	NH3	B&W	
ITT Rayonier, Inc.	Hoquiam	WA	dissolving	500	SW	acid sulfite	Na	B&W	vanillin lignosulfonates
Longview Fibre Co.	Longview	WA	corrugating medium	220	HW	NSSC	Na		Cross recovery

U.S. Mills

Company	City	ST	Final Product	T/day	Wood Species	Type	Base	Recovery	Chem Product
Scott Paper	Everett	WA	tissue	525	SW	acid sulfite	NH3	B&W	
Weyerhaeuser Company	Cosmospolis	WA	dissolving and specialty pulps	450	SW/HW	acid sulfite	Mg	B&W	
Weyerhaeuser Company	Longview	WA	corrugating medium	220	SW/HW	green liq	Na	C-E	Cross recovery
James River Corp.	Green Bay	WI	tissue	200	HW	acid sulfite	Ca	No	lignosulfonates by Reed – Wisconsin
Badger Paper Mills, Inc.	Peshtigo	WI	business paper	105	SW/HW	acid sulfite	NH3	Marathon	conc. SSL
Proctor & Gamble Paper Products	Green Bay	WI	tissue	400	HW	acid sulfite	NH3	Incineration	conc. SSL
Flambeau Paper Company	Park Falls	WI	business paper	115	20-SW 80-HW	acid sulfite	Ca	No	lignosulfonates
Green Bay Packaging Inc.	Green Bay	WI	corrugating medium	220	HW	NSSC	Na	Dorr-Oliver	
Wausau Paper	Rhinelander	WI	glassine	70	SW	acid sulfite	Ca		torula yeast; lignosulfonates
Nekoosa Papers Inc.	Port Edwards	WI	business paper	230	HW	bisulfite	Mg	Copeland	
Owens Illinois	Tomahawk	WI	corrugating medium	840	HW	non-sulfur	Na	B&W	
Wausau Paper Mills Co.	Brokaw	WI	business paper	190	HW	bisulfite	Mg	Copeland	
Weyerhaeuser Company	Rothschild	WI	business paper	140	SW	acid sulfite	Ca	No	lignosulfonate by Reed – Wisconsin

+ short

LEGEND LyB – Low yield bisulfite, HyB – High yield bisulfite, LyA – Low yield acid sulfite, N – Neutral sulfite, SF – Sulfur free, GL – Green liquor, TS – 2 stage stora, HyN – High yield neutral sulfite

Canadian Mills

Company	City	PR	Final Product	T/day	Wood Species	Type	Base	Recovery	Species Product
Western Pulp Ltd. Partnership	Port Alice	BC	Dissolving pulp	450	SW	Ammonium	LyA	C-E	
Abitibi Price, Inc.	Pine Falls	NB	Newsprint	130	SW	Sodium	HyB	No	
Consolidated Bathurst	Bathurst	NB	Corrugating medium	315	HW	Sodium	NaOH	No	
Fraser, Inc.	Edmundston	NB	Market Pulp	592	SW	Magnesium	LyB	B&W/Flakt (1980)	
Fraser, Inc.	Atholville	NB	Market Pulp	360	SW/HW	Magnesium	LyB	B&W/Flakt (1983)	
Lake Utopia	St. George	NB	Corrugating medium	250	HW	Sodium	N	No	
New Brunswick Int'l Paper Co.	Dalhousie	NB	Newsprint		SW	Sodium	HyN	No (1982-HyN)	
Bowater Newfoundland Ltd.	Cornerbrook	NF	Newsprint	350	SW	Sodium	HyB	No	
Price (NFLD) Pulp & Paper Ltd.	Grand Falls	NF	Newsprint	230	SW	Sodium	HyB	No (1983-HyB)	
Bowater Mersey Paper Co. Ltd.	Liverpool	NS	Newsprint	160	SW	Sodium	LyB	No	
Nova Scotia Paper Industries	Point Tupper	NS	Newsprint	110	SW	Sodium	HyB	Stora	
Nova Scotia Paper Industries	Point Tupper	NS	Market Pulp	450	SW/HW	Sodium	TS	Stora	
Abitibi Price, Inc.	Thunder Bay	ON	Newsprint	105	SW	Sodium	HyB	Shutdown 6/82	
Abitibi Price, Inc., Mission	Thunder Bay	ON	Newsprint	140	SW	Sodium	HyN	No	
Abitibi Price, Inc.	Iroquois Falls	ON	Newsprint	300	SW	Sodium	HyB	No	
Boise Cascade Canada Ltd.	Kenora	ON	Newsprint	245	SW	Sodium	HyB	No	
CIP Inc. (closed)	Hawkesbury	ON	Dissolving Pulp	287	HW	Ammonium	LyA	Shutdown 11/82	
Domtar	Trenton	ON	Corrugating medium	140	HW	Sodium	SF	No	
Great Lakes Forest Products Co.	Thunder Bay	ON	Newsprint	315	SW	Sodium	HyB	No	Cross recovery
Macmillan Bloedel Ltd.	Sturgeon Falls	ON	Corrugating medium	275	HW	Sodium	N	No	
Spruce Falls Power & Paper Co.	Kapuskasing	ON	Newsprint & tissue	172	SW	Magnesium	LyB	B&W (CA-LyA Shutdown 1982)	
The Ontario Paper Co. Ltd.	Thorold	ON	Newsprint	200 200	SW HW	Sodium Sodium	LyA HyN	Copeland, Alcohol, Vanillin, Sodium Carbonate & Sulfate	
Abitibi Price, Inc.	Beaupre	PQ	Newsprint	120	SW	Sodium	HyB	No	Conv to sulfite treated TMP in 1983
CIP	Matane	PQ	Corrugating medium	230	HW	Sodium	N	No	
CIP Inc.	Trois-Rivières	PQ	Newsprint	260	SW	Sodium	LyB	No	
CIP, Inc.	Gatineau	PQ	Newsprint	470	SW	Sodium	HyN	No	

Canadian Mills

Company	City	PR	Final Product	T/day	Wood Species	Type	Base	Recovery	Species Product
Consolidated Bathurst	Shawinigan	PQ	Newsprint	270	SW	Sodium	HyB	No	
Consolidated Bathurst, Inc.	Grand'Mère	PQ	Newsprint	150	SW	Sodium	HyN	No	
Consolidated Bathurst, Inc.	Port Alfred	PQ	Newsprint	240	SW	Sodium	HyB	No	
Domtar Newsprint Ltd.	Dolbeau	PQ	Newsprint	180	SW	Calcium	LyB	No	
Domtar Newsprint Ltd.	Donnacona	PQ	Newsprint	200	SW	Calcium	LyA	No	
Gaspesia Pulp & Paper Co. Ltd.	Chandler	PQ	Newsprint	175	SW	Sodium	LyB	No (1983-HyB)	
Kruger Pulp & Paper Ltd.	Trois-Rivières	PQ	Newsprint	250	SW	Caustic Soda	LyB	Shutdown 9/82	
Papier Cascades	Cabano	PQ	Corrugating medium	240	HW	SF	Na	Copeland	incineration
Papeterie Reed Ltd.	Quebec	PQ	Newsprint/mkt pulp	500	SW	Calcium	LyA	No	Lignosulfonate
Quebec North Shore Paper Co.	Baie Comeau	PQ	Newsprint	250	SW	Sodium	HyB	No	Opco 275 t/day process
Rayonier Quebec, Inc.	Port Cartier	PQ	Dissolving pulp	750	SW	Ammonium	LyA	B&W	Shutdown since 1978
St. Raymond Paper Ltd.	Desbiens	PQ	Market pulp	150	SW	Magnesium	LyA	Shutdown 1982	
The Donohue Company, Ltd.	Clermont	PQ	Newsprint	150	SW	Sodium	LyB	Shutdown 4/83	
Tembec Inc.	Temiscaming	PQ	Diss/market pulp	550	SW/HW	Ammonium	LyA	1/2 of liquor burned to generate steam; lignosulfonate	
The James Maclaren Co. Ltd.	Buckingham	PQ	Newsprint	130	SW	Sodium	HyB	No	
Price Co. Ltd.	Alma	PQ	Newsprint	240	SW	Sodium	HyB	No	
The Price Company Ltd.	Kenogami	PQ	Newsprint	210	SW	Sodium	—	No (Sulfite Treated TMP)	

+ metric

European Mills

Company	Site	Final Product(s)	ADMT/Yr. Sulf.	Species	Base	Recovery	By-Product	Status
AUSTRIA								
PWA	Hallein	Coated wood free	95 000	Spruce	Ca	No	Alcohol	Mg conversion, 1985
PWA	Villach	Dissolving	45 000	Spruce	Ca	No	—	1985 closure
Chemiefraser	Lenzing	Dissolving	120 000	Spruce/Beech	Mg	B & W		
Leykam/Murztaler	Gratkorn	Wood containing	175 000	Spruce/Beech	Mg	Yes		1984 paper expansion
Neusiedler	Kematen	Wood free	25 000	Spruce	Ca	No		Mg conversion?
Steyrermuhl	Steyermuhl	Wood containing	51 000	Spruce	Mg	Copeland		
Brigl & Bergmeister	Niklasdorf	Market	32 000	Spruce	Ca			
BELGIUM								
SA Denaeyer	Willebrock	Market, paper	30 000		Ca			
BULGARIA								
CZECHOSLAVAKIA								
	Paskov	Market	200 000	Spruce	Mg	Yes	Yeast	
	Zilina	Dissolving, tissue, market	70 000	Spruce/Beech	Mg	Copeland		
	Sturovo		20 000		NSSC			
	Steti	Newsprint	20 000		Na	Cross recovery		
	Vetrni		90 000		Ca			
FRANCE								
GEC	Alizay	Writing paper	70 000					
GEC	Strasbourg	Tissue, paper	108 000					
Cellulose du Pin	Tartas	Market, fluff	108 000	Maritime pine, hwd	NH4			
GERMANY: FEDERAL REPUBLIC								
Baienfurt	Feldmuhle	Boxboard, wall paper	28 000	Spruce	Ca	Yes		
Hannoversche	Alsfeld	Wood free	70 000	Spruce	Ca	Yes	Lignosulfonates (Homens Bruk)	
Holtzmann	Karlsruhe	Newsprint, market	80 000	Spruce	Ca	Yes		
PWA	Stockstadt	Wood free	130 000	Beech	Mg	Yes		
PWA	Mannheim	Tissue, wood free	230 000	Spruce/Beech	Mg	Yes		
*PWA	Raubling	Paperboard	67 000					
*PWA	Aschaffenburg	Paperboard	67 000					
*Pfleiderer	Teisnach	Thin papers	13 000					

Company	Site	Final Product(s)	ADMT/Yr. Sulf.	Species	Base	Recovery	By-Product	Status
Bayerische	Kelheim	Dissolving	54 000	Spruce, pine	Mg	Yes		
Westfalische	Wildshausen	Dissolving	43 000	Beech	Ca	Yes		
Westfalische	Bonaforth	Dissolving	34 000	Beech	Ca	Yes		
Schwabische	Ehingen	Market	70 000	Beech/Spruce	Ca	Yes	Acetic acid, furfural, SSL	
*Not verified by Zellcheming								

GERMANY: DEMOCRATIC REPUBLIC

Company	Site	Final Product(s)	ADMT/Yr. Sulf.	Species	Base	Recovery	By-Product	Status
	Rosenthal	Market	200 000		Mg	Yes		
	Prina		50 000					

ITALY

Company	Site	Final Product(s)	ADMT/Yr. Sulf.	Species	Base	Recovery	By-Product	Status
Calabria	Crotone	Market	70 000		Tampella			
	Torviscosa	Market	150 000					

NETHERLANDS

Company	Site	Final Product(s)	ADMT/Yr. Sulf.	Species	Base	Recovery	By-Product	Status
Baihremann	Celtona	—	30 000		Ca			

POLAND

Company	Site	Final Product(s)	ADMT/Yr. Sulf.	Species	Base	Recovery	By-Product	Status
	Kwidzyn		60 000	Beech	NSSC			

PORTUGAL

Company	Site	Final Product(s)	ADMT/Yr. Sulf.	Species	Base	Recovery	By-Product	Status
Caima	Constancia	Market	18 000	Eucalyptus	Mg	Yes		
Caima	Abergaria	Market	15 000	Eucalyptus	Ca			

ROMANIA

SPAIN

Company	Site	Final Product(s)	ADMT/Yr. Sulf.	Species	Base	Recovery	By-Product	Status
SN IACE Nationales	Torrelavega	—	60 000	—	—	—	—	—

SWITZERLAND

Company	Site	Final Product(s)	ADMT/Yr. Sulf.	Species	Base	Recovery	By-Product	Status
Attisholz	Luterbach	—	120 000	Spruce/Beech	Ca	—		

USSR

Company	Site	Final Product(s)	ADMT/Yr. Sulf.	Species	Base	Recovery	By-Product	Status
	Kalingrad							
	Amur							
	Svetogorsk		140 000	Spruce/Beech	Na			
	Pilkaranta		60 000					To expand

Company	Site	Final Product(s)	ADMT/Yr. Sulf.	Species	Base	Recovery	By-Product	Status
			YUGOSLAVIA					
	Krsko		100 000	Soft/hwd	Mg			
	Bnja Luka	Dissolving	40 000					

Lignosulfonates (Homens Bruk)

Acetic acid, furfural, SSL

Sulfite Chemical and Semichemical Pulp Mills in Japan

Grades	Company (Mill)	Location (Prefecture)	Production Capacity A.D. Metric ton/day	End Products	Chemical Recovery Process
SP (yield 82%)	Oji Paper (Tomakomai)	Hokkaido	320	Newsprint	Tampella
SCP	Okayama Seishi	Okayama	90	Paperboard	RAS
	Koa Kogyo	Shizuoka	590	Paperboard	SCA
	Saga Paper Board Mfg.	Saga	250	Paperboard	Ebara
	Fukui Chemical	Fukui	430	Paperboard	SCA
	Hokuyo Seishi (Nayoro)	Hokkaido	580	Paperboard	Ebara
5 Mills		Total	1940		
CGP	Oji Paper (Tomakomai)	Hokkaido	270	Newsprint	Tampella
	Sanko Paper (Fuji)	Shizuoka	310	Paperboard	Ebara
	Jujo Paper (Kushiro)	Hokkaido	280	Newsprint	Ebara
	Settsu Itagami (Tokyo)	Saitama	470	Paperboard	Tampella
	Daishowa Paper Mfg. (Yoshinaga)	Shizuoka	650	Paperboard	SCA
	Hokuyo Seishi (Monbetsu)	Hokkaido	170	Paperboard	Green Liq. Cooking
	Howa Paper	Gurma	70	Paperboard	RAS
7 Mills		Total	2220		

SULFITE CHEMICAL AND SEMICHEMICAL PULP MILLS IN FINLAND AND SCANDINAVIA

Company/Mill	Production t/a	Market pulp	For own use	Dissolv. pulp	Bleach pulp	Unbl. pulp	NS-AQ pulp	Base	Raw material	Remarks
SWEDEN:										
Billerud Uddeholm Ab										
— Skoghall	80 000	x		x				Na	SW	Cross recovery
— Billeruds Bruk	25 000		x		x			Ca	SW	
— Saffle	7 000		x			x		Ca	SW	
Holmens Bruk Ab										
— Wargon	20 000				x			Ca	SW	lignosulfonate
	25 000	30 000	15 000			x				
Hyte Bruks Ab										
— Hyltebruk	110 000		x			x		Mg	SW	MgO recovery
Kopparfors Ab										
— Hammarby, Storvik	100 000	x						Mg	SW	Shut down 1982
Mo och Domsjo Ab										
— Domsjo, Ornskoldsvik	225 000	x			x	x		Na	SW	Stora-recovery
Norrlands Skogsagares Celluloosa Ab										
— Hornefors	90 000	70 000	20 000		x			Ca	SW	Shut down 1982
Nymolla Ab										
— Nymolla	250 000	180 000	70 000		x			Mg	HW	MgO recovery
Utansjo Bruk Ab										
— Utansjo	65 000	x				x		Mg	SW	MgO recovery
NORWAY:										
Borregaards Industries Ltd.										
— Sarpsborg	150 000	x			x			Ca	SW	lignosulfonate
A/S Follum Fabriker										
— Katfos	30 000	20 000	10 000			x		Mg	SW	Shut down
Hunsfos Fabrikker										
— Vennesla	60 000		x		x			Mg	SW	MgO recovery
	12 000		x			x		Mg		
Saugbrugsforeningen										
— Halden	70 000	x		x	x			Mg	SW	MgO recovery
A/S Union										
— Union Bruk	65 000	35 000	30 000			x		Mg	SW	MgO recovery

Company/Mill	Production t/a	Market pulp	For own use	Dissolv. pulp	Bleach pulp	Unbl. pulp	NS-AQ pulp	Base	Raw material	Remarks
DENMARK:										
A/S Fredericia Cellulosefabrik										
— Fredericia	32 000				x			Na	Straw	No recovery/alkaline sulfite
FINLAND:										
Kajaani Oy										
— Toppila/Oulu	80 000	x				x		Mg	SW	MgO recovery
Keskisuomen Selluloosa Oy										
— Lievestuore	55 000	x		x				Mg	SW	MgO recovery
Metsäliiton Teollisuus Oy										
— Aanekoski	60 000	x			x			Ca	SW/HW	
	5 000	x				x		Ca	SW	
Oy Nokia Ab										
— Nokia	80 000	60 000	60 000		x			Ca	SW	Will be shut down 1985
	40 000				x			Ca	SW	Will be shut down 1985
Rauma-Repola Oy										
— Rauma	155 000	x		x				Na	SW	Recovery
G.A. Serlachius Oy										
— Mantta	50 000	30 000	90 000			x		Ca	SW	Pekilo; lignosulfonate
— Lielahti	70 000	x		x	x			Ca	SW/HW	Will be shut down 1990
	100 000									
Yhtyneet Paperitehtaat Oy										
— Tervasaari	70 000	x	x				x	Na	SW	TRP/cross recovery
NSSC-MILLS:										
SWEDEN:										
Billerud Uddeholm Ab										
— Gruvon	220 000	x				x		Na	HW	Corr. medium Cross recovery
Fiskeby Ab										
— Skarblacka	70 000	x				x		Na	HW	Corr. medium Cross recovery
NORWAY:										
Sande Paper Mill A/S										
— Sande	55 000	x				x		Na	HW	Corr. medium No recovery
Treshow-Fritzoe										
— Larvik	30 000	x				x		Na	HW	No recovery

Company/Mill	Production t/a	Market pulp	For own use	Dissolv. pulp	Bleach pulp	Unbl. pulp	NS-AQ pulp	Base	Raw material	Remarks
DENMARK:										
F. Junckers Ind. A/S — Koge	75 000	x				x		Na	HW	No recovery
FINLAND:										
Oy Tampella Ab — Heinola Mills	150 000	x				x		Na	HW	Corr. medium recovery
Metsaliiton Teollisuus — Savon Sellu	165 000	x				x		NH_3	HW	Corr. medium, No recovery

2. CHARACTERISTICS OF SPENT SULFITE LIQUORS

Table 1. Comparison of spent sulfite liquors.

Mill	Base	Total Pulp yield, %	Vol.*, m³/odt	pH	TOC, kg/odt	BOD, kg/odt	COD, kg/odt	Spent liquor Diss. org., kg/odt	Diss. inorg., kg/odt	UV lignin, kg/odt	Total sugars, kg/odt	Reduc. sugars, kg/oat	Toxicity, TEF**
1	Mg	54	6.56	3.4	NT	169	807	651	173	469	94	32	316
2	Na	66	7.62	3.8	772	239	798	588	176	352	150	60	423
3	Ca	46	9.28	5.3	NT	357	1533	1043	250	800	264	238	422
4A	NH₃	45	9.11	3.3	NT	319	1553	1182	13	868	210	160	3719
4B	Mg	45	5.61	(3.3)	NT	275	1144	913	79	533	165	180	NT
5	Na	67	9.33	3.2	809	272	843	629	190	334	132	70	530
6A	Na	70	5.55	3.35	643	182	730	506	210	228	52	40	603
6B	Na	50	5.55	2.1	NT	371	1757	1178	222	700	267	218	1208
7A	NH₃	43	9.28	2.1	NT	461	1648	1167	7	822	329	239	3313
7B	NH₃	41	9.73	2.15	NT	410	1872	1283	10	868	296	257	3244
8	Na	50	10.67	4.2	1652	291	1674	779	299	853	192	NT	1067
9	Na	50	7.56	2.3	NT	343	1398	786	348	680	278	209	630
10	Na	51	6.00	3.1	NT	208	1058	598	323	497	127	86	797
11	Na	63	6.00	3.5	400	167	536	473	180	222	88	43	721
12	Na	80	5.78	4.8	NT	151	479	188	206	206	56	11	NT
13	Na	65	7.44	4.8	380	166	616	454	188	267	91	16	620
14	NH₃	41	9.73	1.5	NT	464	1840	1261	20	1009	317	192	4378
15	Na	58	8.76	3.2	600	276	889	667	280	379	142	47	730
16	Na	72	4.92	4.6	322	153	476	300	95	202	71	17	529

NT = not tested.

* estimated liquor volume at a few minutes before "blow"

** TEF = toxicity emission factor = $\dfrac{100\%}{96\text{ h LC50},\%} \times$ liquor volume in m³/odt pulp

Table 2. Elemental analysis of some SSL.

| Mill | Base | Total pulp yield, % | Est. SSL volume*, m³/odt pulp | SSL kg/odt pulp | | | Liquor Solids, % by wt | | | | | Ash | BHV**, BTU lb dry solids |
				Na	K	S	C	H	S	O 1)	O 2)		
2	Na	66	7.62	54.6	1.0	64.6	38.82	4.54	7.24	42.25	34.07	15.33	6497
5	Na	67	9.33	62.3	0.6	73.1	37.60	4.56	8.05	42.18	27.87	21.92	6146
6A	Na	70	5.55	66.6	3.9	79.3	35.83	4.23	7.74	42.90	24.50	27.70	NT
8	Na	50	10.67	97.1	2.9	114.4	41.55	4.46	7.40	37.58	27.15	19.44	6919
11	Na	63	6.00	54.6	0.7	61.3	34.98	4.19	7.75	44.72	27.84	25.24	5867
13	Na	65	7.44	56.2	1.6	72.5	34.36	4.17	9.57	43.15	23.74	28.16	5472
15	Na	58	8.76	81.9	1.4	109.9	33.68	4.01	9.95	43.71	24.34	28.02	5669
16	Na	72	4.92	30.7	0.6	33.4	38.99	4.54	7.35	41.35	27.70	21.42	6802
4B	Mg	45	5.61	<0.1	1.3	62.1	44.61	5.33	7.24	ND	35.99	6.83	7883

* estimated liquor volume at a few minutes before "blow"
** BHV = bomb heating value (corrected for sulfur to be in the form of SO2; for Na-base SSL, these BHV would be 4-7% less than BHV which have not been corrected for sulfur)

NT = not tested
ND = no data

1) calculated from TDS, Na, C, H and S data; assumed negligible amounts of inerts (e.g., K, Ca, etc)
2) calculated from C,H,S, and ash data; considerable 'oxygen' would be present in the ash as CO_3, SO_4, etc.

Table 3. Effect of SSL neutralization on evaporation condensate.

Mill	Base	SSL				Condensate (evaporation in the laboratory to 50% of original volume)							
		Total pulp yield, %	Est. volume, m³/odt pulp	"as is" pH	Total dissolved solids, kg/m³	Not neutralized				Neutralized to pH 7			
						pH	BOD	COD	SO_2	pH	BOD	COD	SO_2
							kg/odt pulp				kg/odt pulp		
8	Na	50	10.67	4.2	101.2	2.0	24.0	33.2	2.1	2.6	15.2	20.6	0.1
11	Na	63	6.00	3.5	108.8	1.7	7.6	11.4	4.8	2.8	4.7	6.7	0.1
13	Na	65	7.44	4.8	86.3	1.6	10.3	13.5	11.5	2.7	5.5	7.7	0.3
14	NH₃	41	9.73	1.5	131.7	1.7	52.5	86.3	0.4	3.8	25.8	45.1	0.1
15	Na	58	8.76	3.2	108.1	1.7	10.9	18.8	10.1	2.6	7.4	10.6	0.3

3. UNITS OF MEASUREMENT AND CONVERSION FACTORS

Quantity or Test	Trade or Customary Unit ×	Conversion Factor	=	SI Unit	(Symbol)
Area	Square Inch	6.45		Square Centimeter	(cm²)
	Square Foot	0.0929		Square Meter	(m²)
	Square Yard	0.836		Square Meter	(m²)
	Acre	0.405		Hectare	(ha)
Basis Weight* or Substance (500-sheet Ream)	17 × 22	3.760		Grams per Square Meter	(g/m²)
	24 × 36	1.627		Grams per Square Meter	(g/m²)
	25 × 38	1.480		Grams per Square Meter	(g/m²)
	25 × 40	1.406		Grams per Square Meter	(g/m²)
	Pounds per 1000 Square Feet (Paperboard)	4.831		Grams per Square Meter	(g/m²)
Breaking Length	Meters	0.001		Kilometers	(km)
Bulk	Grams per Cubic Centimeter	1.0		Grams per Cubic Centimeter	(g/cm³)
Burst Index	$\dfrac{g/cm^2}{g/m^2}$	0.0981		$\dfrac{(kPa)(m)}{g}$	(kPa)
Bursting Strength	Pounds per Square Inch	6.89		Kilopascal	(kPa)
Caliper	Mil	0.0254		Millimeter	(mm)
Capacity	Cubic Feet	0.0283		Cubic Meter	(m³)
	Gallon (U.S.)	3.785		Liter	(L)
Concora Crush	Pounds	4.45		Newtons	(N)
Density	Pounds per Cubic Foot	16.02		Kilograms per Cubic Meter	(kg/m³)
Edge Crush	Pounds per Inch	0.175		Kilonewtons per Meter	(kN/m)
Energy	Kilowatt-hour	3.6		Megajoule	(MJ)
	British Thermal Unit (BTU)	105		Joule	(J)
Energy (specific)	Horsepower-days per Short Ton	0.07		Megajoules per Kilogram	(MJ/kg)
Flat Crush	Pounds per Square Inch	6.89		Kilopascal	(kPa)
Flow	Gallons per Minute	0.00379		Cubic Meters per Minute	(m³/min)
Force	Kilogram	9.81		Newton	(N)
Length	Angstrom	0.1		Nanometer	(nm)
	Micron	1		Micrometer	(µm)
	Mil	0.0254		Millimeter	(mm)
	Foot	0.305		Meter	(m)
Mass	Ton (2000 lb)	0.907		Metric Ton	(t)
	Pound	0.454		Kilogram	(kg)
	Ounce	29.3		Gram	(g)
Mass per Unit Volume	Ounces per Gallon	7.49		Kilograms per Cubic Meter	(kg/m³)
	Pounds per Cubic Foot	1.60		Kilograms per Cubic Meter	(kg/m³)
NIP Pressure	Pounds per Lineal Inch Width	0.175		Kilonewtons per Meter	(kN/m)
Pressure	Pounds per Square Inch	6.89		Kilopasal	(kPa)
	Inches of Mercury	3.377		Kilopascal	(kPa)
Puncture Resistance	Foot Pounds	1.35		Joules	(J)
Power	Horsepower	0.746		Kilowatt	(kW)

Quantity or Test	Trade or Customary Unit ×	Conversion Factor =	SI Unit	(Symbol)
Ring Crush	Pounds	4.45	Newtons	(N)
Speed	Feet per Minute	0.3048	Meters per Minute	(m/min)
Stiffness (Taber)	Taber Units	2.03	Millinewtons	(mN)
Tear Strength	Grams	9.81	Millinewtons	(mN)
Tensile Breaking Load	Pounds per Inch	0.175	Kilonewtons per Meter	(kN/m)
	Kilograms per 15 mm	0.654	Kilonewtons per Meter	(kN/m)
Torque	Foot-Pounds	1.356	Newton-Meter	(N·m)
Volume	Ounce	29.6	Cubic Centimeter	(cm³)
	Gallon	0.00379	Cubic Meter	(m³)
	Liter	1000	Cubic Centimeter	(cm³)
	Milliliter	1	Cubic Centimeter	(cm³)
	Cubic Inch	16.4	Cubic Centimeter	(cm³)
	Cubic Foot	0.0283	Cubic Meter	(m³)
	Cubic Yard	0.765	Cubic Meter	(m³)
Viscosity	Poises	0.1	Pascal-seconds	(Pa·s)

INDEX

Production Editor: Catharine Findlay
Set in type by Southam Communications Limited
Printed by Harpell's Press Co-operative, Ste-Anne-de-Bellevue, Que.